Groups, Combinatorics & Geometry

DURHAM 2001

Groups, Combinatorics & Geometry

DURHAM 2001

A. A. Ivanov

Imperial College of Science, Technology and Medicine, UK

M. W. Liebeck

Imperial College of Science, Technology and Medicine, UK

J. Saxl

University of Cambridge, UK

World Scientific

New Jersey • London • Singapore • Hong Kong

Published by

World Scientific Publishing Co. Pte. Ltd.

5 Toh Tuck Link, Singapore 596224

USA office: Suite 202, 1060 Main Street, River Edge, NJ 07661

UK office: 57 Shelton Street, Covent Garden, London WC2H 9HE

British Library Cataloguing-in-Publication Data
A catalogue record for this book is available from the British Library.

GROUPS, COMBINATORICS AND GEOMETRY
Durham 2001

ISBN 981-238-312-3

Printed in Singapore by World Scientific Printers (S) Pte Ltd

Preface

This book contains the proceedings of the L.M.S. Durham Symposium on Groups, Geometry and Combinatorics, July 16-26, 2001, supported by the Engineering and Physical Sciences Research Council of Great Britain.

Over the past 20 years the theory of groups, in particular simple groups, finite and algebraic, has influenced a number of diverse areas of mathematics. Such areas include topics where groups have been traditionally applied, such as algebraic combinatorics, finite geometries, Galois theory and permutation groups, as well as several more recent developments. Among the latter are probabilistic and computational group theory, the theory of algebraic groups over number fields, and model theory, in each of which there has been a major recent impetus provided by simple group theory. In addition, there is still great interest in local analysis in finite groups, with substantial new input from methods of geometry and amalgams, and particular emphasis on the revision project for the classification of finite simple groups.

The symposium brought together about 70 leading experts in these areas, as well as 15 postdoctoral fellows and research students.

These proceedings contain 20 survey articles, most of which are expanded versions of lectures, or series of lectures, given at the symposium. Broadly speaking, the topics covered in the articles are:

Geometries, amalgams and recognition of simple groups: Bennett et al., Meierfrankenfeld et al., Timmesfeld

Groups of Lie type and representation theory: Brundan and Kleshchev, Liebeck and Seitz, Tiep

Probabilistic and asymptotic group theory: Diaconis, Pyber, Shalev

Algebraic combinatorics and permutation groups: Cameron, Fulman and Guralnick, Liebeck and Shalev, Praeger, Trofimov

Computational group theory and sporadic groups: Kantor and Seress, Norton, Wilson

Applications: Altinel et al., Müller, Segev

We wish to record our gratitude to the LMS and EPSRC for their financial support for the syposium, and to the staff at Durham University for their assistance with the organisation.

Sasha Ivanov, Martin Liebeck and Jan Saxl

CONTENTS

List of authors and addresses

Tuna Altınel, Institut Girard Desargues, Université Claude Bernard Lyon-1, . ment Braconnier, 21 Avenue Claude Bernard, 69622 Villeurbane Cedex, France; altinel@igd.univ-lyon1.fr

C.D. Bennett, Department of Mathematics and Statistics, Bowling Green State University, Bowling Green, OH 43403, USA; cbennet@bgnet.bgsu.edu

Alexandre V. Borovik, Department of Mathematics, UMIST, PO Box 88, Manchester M60 1QD; borovik@umist.ac.uk

Jonathan Brundan, Department of Mathematics, University of Oregon, Eugene, OR 97403, USA; brundan@darkwing.uoregon.edu,

Peter J. Cameron, School of Mathematical Sciences, Queen Mary College, London E1 4NS; p.j.cameron@qmul.ac.uk

G. Cherlin, Department of Mathematics, Rutgers University, Hill Center, Piscataway, NJ 08854, USA; cherlin@math.rutgers.edu

Persi Diaconis, Department of Statistics, Sequoia Hall, 390 Serra Mall, Stanford University, Stanford, CA 94305-4065, USA

Jason Fulman, University of Pittsburgh Mathematics Department, 301 Thackeray Hall, Pittsburgh, PA 15260, USA; fulman@math.pitt.edu

R. Gramlich, TU Darmstadt, Fachbereich Mathematik / AG 5, Schlossgartenstrasse 7, 64289 Darmstadt, Germany; gramlich@mathematik.tu-darmstadt.de

Robert Guralnick, Department of Mathematics, University of Southern California, Los Angeles, CA 90089-1113, USA; guralnic@math.usc.edu

C. Hoffman, Department of Mathematics and Statistics, Bowling Green State University, Bowling Green, OH 43403, USA; hoffman@bgnet.bgsu.edu

William M. Kantor, Department of Mathematics, University of Oregon, Eugene, OR 97403, USA; kantor@darkwing.uoregon.edu

Alexander Kleshchev,Department of Mathematics, University of Oregon, Eugene, OR 97403, USA; klesh@math.uoregon.edu

Martin W. Liebeck, Department of Mathematics, Imperial College, London SW7 2BZ; m.liebeck@ic.ac.uk

Ulrich Meierfrankenfeld, Deparment of Mathematics, Michigan State University, East Lansing, Michigan 48824, USA; meier@math.msu.edu

Thomas W. Müller, School of Mathematical Sciences, Queen Mary College, London E1 4NS; t.w.muller@qmul.ac.uk

Simon P. Norton, DPMMS, Centre for Mathematical Sciences, Cambridge University Wilberforce Road, Cambridge CB3 0WB; simon@dpmms.cam.ac.uk

Cheryl E. Praeger, Department of Mathematics and Statistics, The University of Western Australia, 35 Stirling Highway, Crawley, Western Australia 6009, Australia; praeger@maths.uwa.edu.au

László Pyber, A. Rényi Institute of Mathematics, Hungarian Academy of Sciences, P.O. Box 127, H–1364 Budapest, Hungary; pyber@renyi.hu

Gary M. Seitz, Department of Mathematics, University of Oregon, Eugene, OR 97403, USA; seitz@math.uoregon.edu

Yoav Segev, Department of Mathematics, Ben-Gurion University, Beer Sheva 84105, Israel; yoavs@cs.bgu.ac.il

Ákos Seress, The Ohio State University, Department of Mathematics, 231 W. 18th Avenue, Columbus, OH 43210, USA; akos@math.ohio-state.edu

Aner Shalev,Institute of Mathematics, Hebrew University, Jerusalem 91904, Israel; shalev@math.huji.ac.il

S. Shpectorov, Department of Mathematics and Statistics, Bowling Green State University, Bowling Green, OH 43403, USA; sergey@bgnet.bgsu.edu

Bernd Stellmacher, Mathematisches Seminar, Universität Kiel, Ludewig-Meyn-Str. 4, D-24098 Kiel, Germany; stellmacher@math.uni-kiel.de

Gernot Stroth, Fachbereich Mathematik, Universität Halle, 06099 Halle, Germany; stroth@coxeter.mathematik.uni-halle.de

Pham Huu Tiep, Department of Mathematics, University of Florida, Gainesville, FL 32611, USA; tiep@math.ufl.edu

F. G. Timmesfeld, Mathematisches Institut, Justus-Liebig-Universität Gießen, Arndtstraße 2, D-35392 Gießen, Germany; Franz.Timmesfeld@math.uni-giessen.de

V.I. Trofimov, Institute of Mathematics and Mechanics, Russian Academy of Sciences, S. Kovalevskoy, 16, 620219, Ekaterinburg, Russia; trofimov@imm.uran.ru

Robert A. Wilson, School of Mathematics and Statistics, University of Birmingham, Edgbaston, Birmingham B15 2TT; R.A.Wilson@bham.ac.uk

Classification of Simple K*-Groups of Finite Morley Rank and Even Type: Geometric Aspects

Tuna Altınel   Alexandre V. Borovik*   G. Cherlin†

Introduction

According to a long-standing conjecture in model theory, simple groups of finite Morley rank should be algebraic. The present paper outlines some of the last in a series of results aimed at proving the following:

Even Type Conjecture. *Let G be a simple group of finite Morley rank of even type, with no infinite definable simple section of degenerate type. Then G is a Chevalley group over an algebraically closed field of characteristic* 2.

See [13] for a brief informal introduction to the subject, [1] for the most recent survey of the classification programme, and [14] for further technical details on groups of finite Morley rank.

An infinite simple group G of finite Morley rank is said to be of *even type* if its Sylow 2-subgroups are infinite and of bounded exponent. It is of *degenerate type* if its Sylow 2-subgroups are finite. If the main conjecture is correct, then there should be no groups of degenerate type. So the flavour of the Even Type Conjecture is that the classification in the even type case reduces to an extended Feit-Thompson Theorem. Those who are skeptical about the main conjecture would expect degenerate type groups to exist. The Even Type Conjecture confirms that this is the heart of the matter.

In the present paper we outline some geometric arguments which play the crucial role at the final stages of analysis which has been undertaken in [3, 17, 4, 5, 7].

We work in the following context. Let G be a counterexample to the Even Type Conjecture of minimal Morley rank. This allows us to assume that every proper simple definable connected section of G is a Chevalley group over an algebraically closed field. We adopt the terminology of the classification of finite simple groups and say that G is a *K^*-group.* We take a 2-Sylow° subgroup S of G (that is, the connected component of a Sylow 2-subgroup), a Borel subgroup B containing S, and the set $\mathcal{M}$ of minimal 2-local° subgroups containing B as a proper subgroup. It is shown at some point of our analysis [5, 7] that if $P \in \mathcal{M}$ then $O^{2'}(P/O_2(P)) \simeq$

*Partially supported by the Royal Society Leverhulme Trust Senior Research Fellowship.
†Supported by NSF Grant DMS 0100794.

$\mathrm{SL}_2(K)$ for some algebraically closed field of characteristic 2 and $C_P(O_2(P)) \leqslant O_2(P)$.

We have the following natural case division:

Thin Groups: $|\mathcal{M}| \leqslant 1$. This case occurs in the nature only if $\mathcal{M} = \emptyset$ and $G \simeq \mathrm{SL}_2(K)$.

Quasithin Groups: $|\mathcal{M}| = 2$. In that case, we need to identify G with one of the Lie rank 2 Chevalley groups: $\mathrm{PSL}_3(K)$, $\mathrm{PSp}_4(K)$ or $\mathrm{G}_2(K)$ over an algebraically closed field of characteristic 2.

Generic Groups: $|\mathcal{M}| \geqslant 3$. In that case, G is a Chevalley group of Lie rank $|\mathcal{M}| \geqslant 3$.

Interestingly, each of these cases is resolved by an application of the amalgam method. In the case of *thin* groups, the crucial role is played by the Pushing-Up Theorem [5], proven, in our context, by essentially the same amalgam argument as its finite group prototype, due to Stellmacher [23].

Generic groups can be handled either by constructing a BN-pair in G of (Tits) rank at least 3 [11] and the subsequent application of the classification of BN-pairs of finite Morley rank (Kramer, van Maldeghem and Tent [18]), or by the analysis of the centralisers of p-elements for odd primes p [12] which eventually leads to the construction in G of a system of "root SL_2-subgroups" and application of the Curtis-Phan-Tits Theorem; see the paper by Bennett and Shpectorov [10] in this volume for the discussion of the underlying amalgams.

In this paper, we are dealing with *quasithin* groups and prove for them the following

Identification Theorem *Let G be a simple K^*-group of finite Morley rank and even type. Suppose that G is generated by two 2-local° subgroups P_1, P_2 each containing the connected component of the normaliser of a fixed Sylow° 2-subgroup of G. Assume that $O^{2'}(P_i)/O_2(P_i) \simeq \mathrm{SL}_2(F_i)$ with F_i an algebraically closed field of characteristic 2, for $i = 1, 2$, and that $C^\circ_{P_i}(O_2(P_i)) \leqslant O_2(P_i)$. Then G is a Lie rank 2 Chevalley group over an algebraically closed field of characteristic 2.*

The proof of the Even Type Conjecture itself from these ingredients will be the subject of one further paper.

The proof of the Identification Theorem relies very heavily on the amalgam method in the form used by Stellmacher in [22] and by Delgado and Stellmacher [15], particularly the former version. We have found that the type of arguments that are used in conjunction with the amalgam method can generally be adapted to the context of groups of finite Morley rank with comparatively little alteration, though some attention to detail is required, notably in conjunction with some basic facts of representation theory for which the analogs are obtained through some *ad hoc* arguments, and definability issues. Accordingly we will not devote much space here to the adaptation of those arguments, merely summarising the general flow, recording precisely the point to which they bring us, and pointing out a few issues that do require specific attention. A detailed account of the adapted argument will be found in the technical report [6]. A model for this sort of argument is also found in the appendix to [5], where an analog of the much shorter amalgam argument of [23] is presented.

The amalgam method delivers a great deal of information. We will show that once this information is in hand, the identification theorem can then be proved very efficiently on the basis of general principles, using two ingredients: a classification theorem for BN pairs of finite Morley rank and Tits rank 2 due to Kramer, Tent, and van Maldeghem [18] and a uniqueness result of Tits for parabolic amalgams for which Bennett and Shpectorov [9] have recently given a simple proof based on general principles.

1 Preliminaries

For general background on groups of finite Morley rank we refer to [13, 14]. A broader discussion of the problem to which the present paper is addressed is found in [1].

We will now present the main technical notions involved in the statement of the Identification Theorem.

Definition 1 Let G be a group of finite Morley rank.

1. A definable section of G is a quotient H/K with $K \triangleleft H$ and K, H definable in G. The section is proper if $K > 1$ or $H < G$.
2. G is a *K-group* if every infinite simple definable section of G is a Chevalley group over an algebraically closed field.
3. G is a *K^*-group* if every infinite simple definable proper section of G is a Chevalley group over an algebraically closed field.

Definition 2 Let G be a group of finite Morley rank and S a Sylow° 2-subgroup (the connected component of a Sylow 2-subgroup).

1. G is degenerate if $S = 1$.
2. G is of even type if G is nondegenerate and S is of bounded exponent and definable.
3. G is of odd type if G is nondegenerate and S is divisible abelian.

A simple K^*-group of finite Morley rank is of one of these three types: degenerate, odd type, or even type [16, 2].

Definition 3 Let G be a group of finite Morley rank and of even type.

1. A 2-local° subgroup of G is a group of the form $N_G^\circ(Q)$ where Q is a connected definable 2-subgroup of G.

2. A Borel subgroup of G is a maximal connected solvable subgroup of G.

3. A standard Borel subgroup of G is a Borel subgroup which contains a Sylow° 2-subgroup.

4. $O^{2'}(G)$ is the minimal definable normal subgroup of G such that $G/O^{2'}(G)$ contains no involutions.

4. $O_2(G)$ is the largest definable normal 2-subgroup of G.

A few remarks are in order. First, with regard to standard Borel subgroups, if G is a K^*-group of even type then the standard Borel subgroups are those of the form $N_G(S)$ with S a Sylow° 2-subgroup of G. Secondly, it is not immediately clear that O_2 exists, but when G is of even type this is the case. In practice we will take G to be a connected K^*-group of even type and in this case $O_2(G)$ is itself connected [5].

With these definitions, the Identification Theorem has a precise meaning. The underlying idea is to work with an appropriate notion of parabolic subgroup, and for our purposes "parabolic" is best taken to mean: 2-local°, and containing a standard Borel. Earlier papers have dealt with the existence of parabolics in this sense [5] and with their structure [7].

2 The amalgam method

2.1 The issues

The basis for the proof of the Identification Theorem is the *amalgam method* as applied in [22] and in greater generality in [15]. We will indicate how this method is used in our context, and what it produces. On the whole, this chapter of finite group theory goes over very smoothly to our context once the principles on which it relies are suitably translated. Accordingly we will not give the details of these arguments here; they may be found in [6]. On the whole we followed the line of [22] rather than the more general [15] as it is more efficient in our particular case.

The amalgam method has already been used in the context of groups of finite Morley rank in [5]; indeed, the original proofs of "pushing up" results in finite group theory [8, 19] do not seem to go over to our context, but the version given by Stellmacher in [23], based squarely on the amalgam method, goes over quite smoothly, as seen in the appendix to [5], where the argument is given in detail in the context of groups of finite Morley rank. We also refer to [5] for a verification that some of the key properties of SL_2 which are used in amalgam arguments hold in our context. We will indicate below what sorts of adaptations generally need to be made in our context.

In the context of the Identification Theorem, the amalgam method begins by introducing the graph Γ associated with the right cosets of P_1 and P_2, where two distinct cosets are linked by an edge if they meet. Thus this is a bipartite graph on which G acts naturally, and it should be thought of as a labeled graph in which every vertex and edge is labeled by its stabiliser under the action of G. The universal cover $\hat{\Gamma}$ of this graph in the topological sense is a tree which is associated to the free product $\hat{G}$ of P_1 and P_2 over their intersection, which is easily seen to be the standard Borel subgroup B. The objective is to show, after a lengthy analysis which can take place either in Γ or in $\hat{\Gamma}$, that $\hat{\Gamma}$ has a quotient Γ^* on which $\hat{G}$ acts (not faithfully) with the following properties:

(a) Γ^* is a generalized n-gon, and the image G^* of $\hat{G}$ in $\mathrm{Aut}(\Gamma^*)$ has a (B,N)-pair of rank 2.

(b) The triple (B, P_1, P_2) in G is isomorphic with the corresponding triple in G^* (one says that G is *parabolic isomorphic* with G^*).

The intent of course is to apply a classification theorem to G^* in order to determine the possible isomorphism types, at which point the isomorphism type of the triple (B, P_1, P_2) in G is known, and one can return to G and complete the identification of G without further use of the amalgam method.

An obvious and potentially serious drawback of this approach from the point of view of groups of finite Morley rank is that the group $\hat{G}$ will not be definable and will not be a group of finite Morley rank, and hence *a priori* the same problem arises in G^*. This is handled by showing that $\hat{G}$ is "locally" of finite Morley rank and that G^* is actually of finite Morley rank. We will deal with this more explicitly below.

The other issues that arise are merely technical, and are of two sorts. On the one hand certain chapters of finite group theory that are applied in this context have to be developed appropriately in our category, the most problematic one being the representation theory of the group SL_2, which is handled in a largely *ad hoc* way as the representations involved are taken over the field of 2 elements and are infinite dimensional; Morley rank has to replace dimension as the measure of size here, and the representation theory is inevitably in a rudimentary state, but sufficient for the limited needs of the amalgam method. The other point that bears watching is the role of connectedness in the analysis. This is absent in the finite case, but comes up naturally in the transposition to the context of groups of finite Morley rank, as can be seen quite clearly already in [5], where some care had to be taken around this point.

We will say no more about these technical points, but we will discuss the definability issue, and also state explicitly the result delivered by the amalgam method, which serves as the point of departure for the identification of the group G.

2.2 Definability

Definition 4 Let G be a group acting on a graph Γ.

1. For any vertex δ and any $k \geqslant 0$, $\Delta_k(\delta)$ is the set of vertices lying at distance at most k from δ in Γ, and $G_k(\delta)$ is the set of elements of G which can be expressed as a product of at most k elements of G, where each element stabilises some vertex in $\Delta_k(\delta)$.

2. The pair (G,Γ) is locally of finite Morley rank if for each k and δ the pair $(G_k(\delta),\Delta_k(\delta))$ has finite Morley rank, where the latter is a 2-sorted structure consisting of a partial group, a graph, and a partial action of the partial group on the graph. (A partial group is simply the restriction of a group, viewed as a relational system, to a subset.)

Because the amalgam method always works locally in the graph Γ, whatever can be done with groups of finite Morley rank can also be done with groups which are locally of finite Morley rank, in this context. As far as the universal cover is concerned, we have the following.

Lemma 1 *Let $\mathcal{P} = (P,Q,B)$ be a structure consisting of two groups P,Q. Let $G = P *_B Q$ be the free product with amalgamation and let Γ be the associated tree of cosets, on which G acts naturally. Then the structure $\mathcal{G} = (G,\Gamma)$ consisting of G acting on Γ is locally interpretable in $\mathcal{P}$ in the following sense: for any vertex $\delta \in V(\Gamma)$ and any $k \geqslant 0$ the graph $\Delta_k(\delta)$, the partial group $G_k(\delta)$, and the partial action of $G_k(\delta)$ on $\Delta_k(\delta)$ are all interpretable in $\mathcal{P}$.*

Proof. Let $X = P \cup Q$ and let $R_k(x_1,\ldots,x_k)$ be the relation on X defined by: "$x_1 \cdots x_k = 1$ in G". Everything comes down to the definability of this relation in $\mathcal{P}$, which is proved by induction based on the following property of free products with amalgamation: if $x_1,\ldots,x_k$ are alternately from $P \setminus B$ and $Q \setminus B$ then the product is nontrivial. In the remaining cases, either the product can be shortened, and induction applies, or else $k = 1$. Bearing in mind that the natural maps of P and Q into G are embeddings, the claim follows. □

Corollary 2 *Under the stated hypotheses, if $\mathcal{P}$ has finite Morley rank then (G,Γ) is locally of finite Morley rank.*

We must also look at the passage from the universal cover to a generalized n-gon. This is handled at the outset in [15] by two results, (3.6) (p. 77) and (3.7) (p. 79), most of which involve no finiteness hypothesis:

Fact 1 [15] *Let Γ be a tree and K a Cartan subgroup of G with apartment $T = T(K)$. Suppose that T fulfills the uniqueness and exchange conditions and that $s \geqslant 3$. Then there exists an equivalence relation $\approx$ on Γ which is compatible with the action of G so that:*

1. *$\tilde{\Gamma} = (\Gamma/\approx)$ is a generalised $(s-1)$-gon;*
2. *$G_{\tilde{\Gamma}} \cap G_\delta = 1$ for each $\delta \in V(\Gamma)$; here $G_{\tilde{\Gamma}}$ is the kernel of the action of G on $\tilde{\Gamma}$.*
3. *$G/G_{\tilde{\Gamma}}$ has a (B,N)-pair of rank 2.*

The precise meaning of the hypotheses is not really relevant here; for the most part they represent the conditions which must be verified in the course of a detailed analysis. Also, in quoting this statement verbatim, we have omitted the context, which is more general than that of the Identification Theorem, apart from a finiteness hypothesis that plays no role here. However it may be remarked that in the case which actually concerns us, the Borel subgroup B splits as $S \rtimes K$ with S a Sylow° 2-subgroup and K a complement which may be called a "maximal torus", and the apartment T may be defined as the fixed-point set of K, which will be a 2-way infinite path on which the normaliser of K acts, with two orbits. We will also enter somewhat more into the details below, in discussing the Moufang property.

What needs to be added to this fact is the following:

Lemma 3 *In the context of Fact* 1, *if* (G, Γ) *is locally of finite Morley rank then the "quotient"* $(G/G_{\tilde{\Gamma}}, \tilde{\Gamma})$ *has finite Morley rank.*

Proof. This requires an examination of the construction of $\tilde{\Gamma}$ as given in [15]. There are two points to be observed.

In the first place, the quotient $\tilde{\Gamma}$ is covered by $\Delta_{s-1}(\delta)$ for any vertex $\delta \in V(\Gamma)$. Secondly, with δ fixed, it needs to be seen that the equivalence relation $\approx$, which we factor out, is definable on $\Delta_{s-1}(\delta)$. In the proof of Fact 1 it is shown that equivalent pairs in $\tilde{\Gamma}$ lie at distance at least $2(s-1)$, and the argument shows that on $\Delta_{s-1}(\delta)$ the equivalence relation is given by:

$$\alpha \approx \beta \text{ iff } d(\alpha, \beta) = 2(s-1) \text{ and } \gamma(\alpha, \beta) \text{ is conjugate under } G \text{ to a subpath of } T$$

where $\gamma(\alpha, \beta)$ is the path from α to β. Here we may replace T by two fixed subpaths of T of length $2(s-1)$, and the problem of definability reduces to the relation: "(α, β) is conjugate to (α_0, β_0)" where the four vertices $\alpha, \beta, \alpha_0, \beta_0$ lie in a set of the form $\Delta_k(\delta)$.

The action of a vertex stabiliser G_δ is transitive on the set of neighbors $\Delta(\delta)$, so if α and α_0 are in fact conjugate and at distance $2d$, and δ is the midpoint of the path joining α and α_0, then there is an element of $G_d(\delta)$ carrying α to α_0. Thus the following serves as a definition of conjugacy, for such pairs: "α, α_0 lie at distance $2d$ for some (*bounded*) d, and with δ the midpoint of the path joining them, there is $g \in G_d(\delta)$ such that $\alpha^g = \alpha_0$ and β^g is conjugate to β_0 under G_{α_0}"; in the final clause we have a bound on $d(\beta^g, \beta_0)$, so this condition is also definable. □

This disposes of all definability issues: when the amalgam method succeeds, the group G^* (or $G/G_{\tilde{\Gamma}}$, in the current notation) has finite Morley rank, and for that matter is interpretable in the original group G, in the notation of the Identification Theorem.

2.3 Application of the amalgam method (the Moufang property)

We have already indicated the main thrust of the amalgam method in our context, namely:

Fact 2 [6] *Under the hypotheses of the Identification Theorem, there is a group* G^* *of finite Morley rank which is parabolic isomorphic to* G, *and which has a rank* 2 (B, N)*-pair.*

This leaves something to be desired however. We would like to apply the classification of Moufang (B, N) pairs of Tits rank at least 2 and of finite Morley rank, given in [18].

Definition 5 Let Γ be a generalised n-gon, $G = \text{Aut}\Gamma$.

1. For $\gamma = (\delta_0, \ldots, \delta_{n-1})$ a path of length $n-1$ in Γ, let $U(\gamma)$ be the intersection of $G_{\Delta_1(\delta)}$ for $\delta = \delta_1, \ldots, \delta_{n-1}$.
2. Γ is Moufang if for every path $\gamma = (\delta_0, \ldots, \delta_{n-1})$ in Γ, the group $U(\gamma)$ operates transitively on $\Delta(\delta_0) \setminus \{\delta_1\}$.

As it happens the Moufang property follows from general principles for the generalised $(s-1)$-gons delivered by Fact 1 in the context of the Identification theorem.

At this point one should actually invoke the definition of s:

Definition 6 In the context of the amalgam method (e.g., the Identification Theorem):

1. A path $\gamma = (\delta_0, \ldots, \delta_n)$ in Γ is regular if G_γ (the pointwise stabiliser) operates transitively on $\Delta(\delta_0) \setminus \{\delta_1\}$ and on $\Delta(\delta_n) \setminus \{\delta_{n-1}\}$.
2. s is the minimum length of a non-regular path.

It follows easily from the definition of s, and induction, that any two paths of equal length l, with $l \leqslant s$, are conjugate under the action of G.

Now the following is contained in the analog of [15, (14.1)]:

Fact 3 *Let $\gamma = (\delta_0, \ldots, \delta_{s-1})$ be a path of length $s - 1 \geqslant 2$. Then $O_2(G_\gamma)$ acts transitively on $\Delta(\delta_0) \setminus \{\delta_0\}$ and $\Delta(\delta_{s-1}) \setminus \{\delta_{s-1}\}$.*

Here the notation $O_2(G_\gamma)$ simply represents a Sylow 2-subgroup of G_γ since G_γ is contained in a Borel subgroup, of the form $S \rtimes K$ with S a Sylow 2-subgroup and K a torus; furthermore some conjugate of K is a complement to $O_2(G_\gamma)$ (any path of length at most $s-1$ is conjugate to a path contained in $T = T_K$).

Now to verify the Moufang property for the generalised $(s-1)$-gon furnished by Fact 1 in the context of the Identification Theorem, let $\gamma = (\delta_0, \ldots, \delta_{s-2})$ be a path of length $s-2$ in $\hat{\Gamma}$, and extend it to a path $\tilde{\gamma} = (\delta_0, \ldots, \delta_{s-1})$ of length $s-1$. Let $Q = O_2(G_\gamma)$. It suffices to show that Q fixes the neighbors of each δ_i for $1 \leqslant i \leqslant s-2$. Or, more simply:

Lemma 4 *If $\delta \in V(\Gamma)$ and α, β are distinct neighbors of δ, then $O_2(G_{\alpha\beta})$ fixes $\Delta(\delta)$.*

Now $G_{\alpha\beta} = G_{\alpha\delta} \cap G_{\beta\delta}$ is the intersection of two Borel subgroups of G_δ, and $O_2(G_{\alpha\beta})/O_2(G_\delta)$ is the intersection of two distinct Sylow subgroups of SL_2, hence trivial, that is: $O_2(G_{\alpha\beta}) \leqslant O_2(G_\delta)$, and as G_δ acts transitively on $\Delta(\delta)$, it follows that $O_2(G_\delta)$ fixes all neighbors of δ. Thus the lemma is immediate.

3 Identification

We apply a very general result from [18, Theorem 3.14]:

Fact 4 *Let G^* be an infinite simple group of finite Morley rank with a spherical Moufang BN-pair of Tits rank 2. Then $G^* \simeq \mathrm{PSL}_3(F)$, $\mathrm{PSp}_4(F)$, or $G_2(F)$ for some field F.*

This field must of course be algebraically closed as it will also have finite Morley rank.

Thus we now have as a corollary of the amalgam analysis sketched in the previous section:

Lemma 5 *Under the hypotheses of the identification theorem, the triple (B, P_1, P_2) is isomorphic to a triple consisting of a Borel subgroup and the two minimal parabolic subgroups containing it, in one of the groups $G^* \simeq \mathrm{PSL}_3(F)$, $\mathrm{PSp}_4(F)$, or $\mathrm{G}_2(F)$ for some algebraically closed field F (of characteristic 2, as we work with even type).*

Evidently we now want to identify G itself with the appropriate one of these three groups. We use a theorem of Tits found in [20, Chapter II, Theorem 8]; for an alternative proof, based on Tits' Lemma [24], see Bennett and Shpectorov [9].

Fact 5 *Let G^* be a Chevalley group of Lie rank 2 and let P_1, P_2 be minimal parabolics containing a common Borel subgroup B. Let N be the normaliser of a Cartan subgroup of B. Then G^* is the universal closure of the amalgam of P_1, P_2, and N.*

(The idea of the proof given by Bennett and Shpectorov is to adjoin to the natural point/line geometry associated with G^* a third kind of object, the set of apartments, where an apartment is incident with its elements. This has the effect of making the geometry simply connected, and a very general result of Tits [24] on groups acting flag-transitively on simply connected geometries then applies.)

To complete the proof of the Identification Theorem we may therefore proceed as follows. Let G^* be the target group $\mathrm{PSL}_3(F)$, $\mathrm{PSp}_4(F)$, or $\mathrm{G}_2(F)$. Working in the original group G, fix a maximal torus K in B and let $N = N_G(K)$.

Lemma 6 *If $C_G(K) = K$ then $G \simeq G^*$.*

Proof. Let K be a maximal torus in B. $P_i = O_2(P_i) \rtimes (L_i \times K_i)$ with $L_i \simeq \mathrm{SL}_2(F)$ and $K = (K \cap L_i)K_i$. Let $w_i \in L_i$ be an involution inverting $K \cap L_i$ and let $W = \langle w_1, w_2 \rangle$, $a = w_1 w_2$. Evidently the structure of P_1 and P_2 determine the map $W \to \mathrm{Aut}(K)$, so as G and G^* are parabolic isomorphic, W acts on K like D_{s-1}. In particular $a^{s-1} \in C_G(K) = K$, and a is inverted by both w_1 and w_2. It follows that $a^{s-1} = 1$.

Thus $KW \simeq N_{G^*}(K)$. By Fact 5 the subgroup of G generated by P_1, P_2, KW is isomorphic with G^* and as P_1, P_2 already generate G, we have $G \simeq G^*$. □

In the proof of the next lemma we make use of information on centralisers of semisimple elements in semisimple algebraic groups found in [21].

Fact 6 [21, Corollary 4.6] *Let G^* be a semisimple algebraic group and x a semisimple element of G^* of prime order p. Let $\pi : \tilde{G} \to G^*$ be the canonical map from the simply connected cover. If p does not divide $|\ker \pi|$ then $C_{G^*}(x)$ is connected.*

Fact 7 [21, 3.19] *Let G^* be a semisimple algebraic group and and y any semisimple element. Then $C_{G^*}(y)$ is reductive.*

Combining these two:

Corollary 7 *With the hypotheses and notation of Fact 6, $C_{G^*}(x)$ is connected and reductive. In particular if G^* is one of the groups SL_3, Sp_4, or G_2 over an algebraically closed field of characteristic 2 and x is a semisimple element of prime order $p > 3$, then $C_{G^*}(x)$ is a torus or the product of a torus with SL_2.*

Proof. $C_{G^*}(x)$ is reductive of Lie rank 2, and contains a central element of order greater than 3. The claim follows. □

Lemma 8 $C_G(K) = K$.

Proof. We will make free use of the parabolic isomorphism of G and G^*.

With K_i, L_i as in the preceding proof, $C_{P_i}(K_i) = L_i \times K_i$ with $L_i \simeq \mathrm{SL}_2(F)$, with F the base field of G^*. More exactly, $L_i \simeq \mathrm{SL}_2(F_i)$ with F_1 and F_2 definably isomorphic, but this amounts to the same thing.

Take $a \in K$ of order greater than 3. As observed above $C_{G^*}(a)$ is reductive and is either a torus, or the product of a torus with $\mathrm{SL}_2(F)$.

In particular the rank of $C_S(a)$ is at most $f = \mathrm{rk}(F)$ for any such element a. Accordingly the same applies to $C_Q(a)$ for any Sylow° 2-subgroup Q of G, and any a normalising Q of order greater than 3. Let U be a Sylow° 2-subgroup of $C_G(K_i)$ ($i = 1$ or 2). It follows that $\mathrm{rk}(U) \leqslant f$. As $\mathrm{rk}(S \cap L_i) = f$, we conclude that $S \cap L_i$ is a Sylow° 2-subgroup of $C_G^\circ(K_i)$.

Let $U_i = S \cap L_i$. Then we have

$$(*) \qquad U_i \leqslant L_i \leqslant C_G^\circ(K_i)$$

and $C_G^\circ(K_i)$ is a connected K-group, with U_i as a Sylow 2-subgroup.

By $(*)$ we have $O_2(C_G^\circ(K_i)) = 1$, and by an elementary result on K-groups [4, 2.33] it follows that $C_G^\circ(K_i) = E(C_G^\circ(K_i)) * O(C_G^\circ(K_i))$. Here $E = E(C_G^\circ(K_i))$ is a central product of quasisimple algebraic groups, U_i is a Sylow 2-subgroup of E, and $U_i \leqslant L_i \leqslant E$. It is then easy to see that $L_i = E$. As a result, L_i is normalised by $C_G(K_i)$ for $i = 1, 2$ and hence:

Both L_1 and L_2 are normalised by $C_G(K)$.

The groups $L_i \simeq \mathrm{SL}_2(F)$, $i = 1, 2$, do not allow definable groups of outer automorphisms [14, Theorem 8.4]. Hence $C_G(K)$ must act on L_i via inner automorphisms commuting with $K \cap L_i$ and hence $C_G(K) = (K \cap L_i) \times C_G(KL_i)$. Let $H_i = C_G(KL_i)$. Since $(K \cap L_1)(K \cap L_2) \leqslant K$, it follows that $C_G(K) = K(H_1 \cap H_2)$.

Now $H = H_1 \cap H_2$ centralises $\langle U_1, U_2 \rangle = S$ and H centralises each L_i, hence also each P_i, hence G. As G is simple, $H = 1$ and $C_G(K) = K$. □

This completes the identification of G.

References

[1] T. Altınel, *Simple groups of finite Morley rank of even type*, in "Model Theory of Groups and Tits Buildings" (K. Tent, ed.), London Math. Soc. Lect. Notes Ser. vol. 291, 2002.

[2] T. Altınel, A. Borovik and G. Cherlin, *Groups of mixed type*, J. Algebra 192 (1997), 524–571.

[3] T. Altınel, A. Borovik and G. Cherlin, *On groups of finite Morley rank with weakly embedded subgroups*, J. Algebra 211 (1999), 409–456.

[4] T. Altınel, A. Borovik and G. Cherlin, *Groups of finite Morley rank and even type with strongly closed abelian subgroups*, J. Algebra 232 (2000), 420–461.

[5] T. Altınel, A. Borovik and G. Cherlin, *Pushing up and $C(G,T)$ in groups of finite Morley rank of even type*, J. Algebra 247 (2002), 541–576.

[6] T. Altınel, A. Borovik and G. Cherlin, *An analog of a theorem of Stellmacher for groups of finite Morley rank*, Dimacs Technical Report, in preparation.

[7] T. Altınel, A. Borovik, G. Cherlin and L.-J. Corredor, *Parabolic 2-local subgroups in groups of finite Morley rank of even type*, submitted.

[8] B. Baumann, *Über endliche Gruppen mit einer zu $L_2(2^n)$ isomorphen Factorgruppe*, Proc. Amer. Math. Soc. 74 (1979), 215–222.

[9] C. D. Bennett and S. Shpectorov, *A remark on a theorem of J. Tits*, Proc. Amer. Math. Soc. 129 (2001), 2571–2579.

[10] C. D. Bennett, R. Gramlich, C. Hoffman and S. Shpectorov, *Curtis-Phan-Tits theory*, these proceedings.

[11] A. Berkman and A. Borovik, *An identification theorem for groups of finite Morley rank and even type*, to appear in J. Algebra.

[12] A. Berkman and A. Borovik, *A generic identification theorem for groups of finite Morley*, submitted.

[13] A. Borovik, *Tame groups of odd and even type*, in "Algebraic groups and their representations, Cambridge 1997" (R. Carter, ed.), NATO Adv. Sci. Inst. Ser. C, vol. 517, Kluwer, Dordrecht, 1998, pp. 341–366.

[14] A. Borovik and A. Nesin, "Groups of Finite Morley Rank", Oxford University Press, 1994.

[15] A. Delgado and B. Stellmacher, *Weak (B,N)-pairs of rank 2*, in "Groups and Graphs: New Results and Methods", by A. Delgado, D. Goldschmidt, and B. Stellmacher, Birkhaüser, DMV Seminar 6, Basel, 1985.

[16] E. Jaligot, *Groupes de type mixte*, J. Algebra 212 (1999), 753–768.

[17] E. Jaligot. *Groupes de type pair avec un sous-groupe faiblement inclus*, J. Algebra 240 (2001), 413–444.

[18] L. Kramer, H. van Maldeghem and K. Tent, *Simple groups of finite Morley rank and Tits buildings*, Israel J. Math. 109 (1999), 189–224.

[19] R. Niles, *Pushing-up in finite groups*, J. Algebra 57 (1979), 26–63.

[20] J.-P. Serre, "Trees", Springer-Verlag, New York, 1980.

[21] T. A. Springer and R. Steinberg, *Conjugacy classes*, in "Seminar on Algebraic Groups and related finite groups", (A. Borel et al., eds.), Lecture Notes Math. 131, Springer Verlag, New York, 1970, pp. 167–266.

[22] B. Stellmacher, *On graphs with edge-transitive automorphism groups*, Illinois J. Math. 28 (1984), 211–266.

[23] B. Stellmacher, *Pushing Up*, Arch. Math. 46 (1986), 8–17.

[24] J. Tits, *Ensembles ordonnés, immeubles et sommes amalgameés*, Bull. Soc. Math. Belg. A38 (1986), 367–387.

Curtis-Phan-Tits Theory

C.D. Bennett, R. Gramlich, C. Hoffman, S. Shpectorov

Abstract

We demonstrate that there is a relation between the Curtis-Tits theorem and Phan's theorems that goes beyond the similarity in appearance. We present a geometric construction connecting those theorems and suggesting that Phan's theorems can be thought of as "twisted versions" of the Curtis-Tits theorem. The construction itself further suggests that Phan's theorems are only some of many possible such theorems. We make this explicit by presenting a new Phan-type theorem for the symplectic groups.

1 Introduction

An important step of the classification of finite simple groups announced in 1981 and of the ongoing Gorenstein-Lyons-Solomon revision of the classification is the identification of the "minimal counterexample" with one of the known simple groups. This step follows the local analysis step, when inside the minimal counterexample G one reconstructs one or more of the proper subgroups using the inductive assumptions and available techniques. Thus the input of the identification step is a set of subgroups of G that resemble certain subgroups of some known simple group $\hat{G}$ referred to as the *target group*. The output of the identification step is the statement that G is isomorphic to $\hat{G}$. Two of the most widely used identification tools are the Curtis-Tits theorem (see [GLS], Theorem 2.9.3) and Phan's theorem [Ph1].

The Curtis-Tits theorem allows the identification of G with a simple Chevalley group $\hat{G}$ provided that G contains a system of subgroups identical to the system of appropiately chosen rank two Levi factors from $\hat{G}$. In the particular case where $\hat{G}$ is of type A_n, the system in question consists of all the groups $SL(3,q)$ and $SL(2,q) \times SL(2,q)$ lying in $\hat{G} \cong (P)SL(n+1,q)$ block-diagonally.

Phan's theorem deals with the case $\hat{G} = (P)SU(n+1,q^2)$ and the system of block-diagonal subgroups $SU(3,q^2)$ and $SU(2,q^2) \times SU(2,q^2)$ of $\hat{G}$. Thus, Phan's theorem appears to be similar to the A_n case of the Curtis-Tits theorem. However, unlike the case of A_n, the block-diagonal $SU(3,q^2)$ and $SU(2,q^2) \times SU(2,q^2)$ are not Levi factors in $SU(n+1,q^2)$. Consequently, Phan's theorem is not a special case of the Curtis-Tits theorem.

One of the purposes of this paper is to demonstrate that the relation between the Curtis-Tits theorem for the type A_n and Phan's theorem goes beyond a similarity in appearance. To this end, we present a geometric construction revealing a deeper connection between these theorems and suggesting that Phan's theorem is simply a "twisted" version of the Curtis-Tits theorem for A_n. Furthermore, from this point of view, there appears to be a much broader variety of "Phan-type" theorems that includes Phan's theorem and his further results from [Ph2] as special cases corresponding to particular diagrams (such as A_n) and particular "twists". We stress this point in the by presenting a new Phan-type theorem for the case of $\hat{G} = (P)Sp(2n,q)$ and a system of semisimple subgroups of rank two which again do not come from Levi factors of $\hat{G}$ (cf. [GHSh]).

It should also be noted that the construction generalizes well to the case of infinite fields and/or nonspherical diagrams. In fact, there already exists a version of the Curtis-Tits theorem for a broad class of Kac-Moody groups (cf. [M]). We believe it to be a very interesting problem to develop a parallel Phan-type theory for arbitrary diagrams.

The structure of the paper is as follows: In Section 2 we introduce some notions from the areas of diagram geometry, chamber systems and amalgams of groups. In Section 3 we discuss the proof of Phan's theorem from [BSh]. In Section 4 we introduce the language of buildings and twin buildings and present an overview of Mühlherr's geometric proof of the Curtis-Tits theorem. Finally in Section 5 we present our construction and discuss the new Phan-type theorem for $Sp(2n,q)$ from [GHSh] and further examples. Along the way we pose a number of open problems.

2 Geometries and amalgams

2.1 Geometries

A *pregeometry* over I is a set of elements Γ together with a type function t and a reflexive and symmetric incidence relation $\sim$. The type function maps Γ onto the type set I, and for any two elements $x, y \in \Gamma$ with $x \sim y$ and $t(x) = t(y)$ we have $x = y$. A *flag* in Γ is a set of pairwise-incident elements. Notice that the type function injects any flag into the type set. A *geometry* is a pregeometry for which t induces a bijection between any maximal flag of Γ and I.

The *residue* $\mathrm{res}_\Gamma(F)$ of a flag F in a geometry Γ is the set of elements from $\Gamma \setminus F$ that are incident to all elements of F. It follows that the residue $\mathrm{res}_\Gamma(F)$ is a geometry with type set $I \setminus t(F)$. The *rank* of the geometry Γ is the cardinality of its type set I. We will only consider the case where I is finite. The rank of the residue of a flag F is called the *corank* of F. The geometry Γ is *connected* if the graph with vertex set Γ and edges given by $\sim$ is connected. The geometry Γ is *residually connected* if the residue in Γ of every flag of corank at least 2 is connected.

An automorphism of a geometry Γ is a permutation of its elements that preserves type and incidence, and we denote the group of all automorphisms of Γ by $\mathrm{Aut}\,\Gamma$. A subgroup $G \leq \mathrm{Aut}\,\Gamma$ acts *flag-transitively* on Γ if G is transitive on the set of maximal flags. A geometry that possesses a flag-transitive automorphism group is also called flag-transitive. Finally, a *parabolic subgroup* (or simply a *parabolic*) H of G is the stabilizer in G of a non-empty flag F of Γ. The *rank* of the parabolic H is the corank of F.

2.2 Simplicial complexes

A *simplicial complex* $\mathcal{S}$ is a pair (X, Δ) where X is a set and Δ is a collection of subsets of X such that if $A \in \Delta$ and $B \subset A$ then $B \in \Delta$. The subsets from Δ are called *simplices.* A *morphism* from a complex $\mathcal{S} = (X, \Delta)$ to a complex $\mathcal{S}' = (X', \Delta')$ is a map between X and X' that takes simplices to simplices. The *star* of a simplex $A \in \Delta$ is the set of all subsets $B \in \Delta$ such that $A \subseteq B$, and we define a *covering* to be a surjective morphism ϕ from $\mathcal{S}$ to $\mathcal{S}'$ such that for every $A \in \Delta$ the function ϕ maps the star of A bijectively onto the star of $\phi(A)$.

A *path* on a complex $\mathcal{S}$ is a sequence $x_0, x_1, \ldots, x_n$ of elements of X such that x_{i-1} and x_i are contained in a simplex for all $i = 1, \ldots, n$. We do not allow repetitions, so $x_{i-1} \neq x_i$ for all i. The complex $\mathcal{S}$ is *connected* if every two elements of X can be connected by a path. The following two operations on paths are called *elementary homotopies*: (a) substituting a subsequence x, y, x (a return) by just x, and (b) substituting a subsequence x, y, z, x (a triangle) by x, provided that x, y and z lie in a common simplex. Two paths are *homotopically equivalent* if they can be obtained from one another by a finite sequence of elementary homotopies. A *loop* is a closed path, that is, a path with $x_0 = x_n$. We say that the loop is *based* at the point $x_0 = x_n$. A loop is called *null-homotopic* if it is homotopically equivalent to the trivial path x_0. The *fundamental group* $\pi_1(\mathcal{S}, x)$, where $x \in X$, is the set of equivalence classes of loops based at x with respect to the homotopical equivalence. The product is defined by the concatenation of loops. Notice that the fundamental group is independent up to isomorphism of the choice of the base vertex x inside a fixed connected component. The coverings of $\mathcal{S}$, taken up to a certain natural equivalence, correspond bijectively to the subgroups of $\pi_1(\mathcal{S}, x)$. A connected complex $\mathcal{S}$ is called *simply connected* if it has no proper coverings, or, equivalently, if $\pi_1(\mathcal{S}, x) = 1$.

To every geometry Γ one can associate its *flag complex* $\mathcal{F}(\Gamma)$. This is the simplicial complex defined on the set Γ, whose simplices are the flags of Γ. We will say that Γ is *simply connected* if $\mathcal{F}(\Gamma)$ is simply connected.

2.3 Chamber systems

A *chamber system* over a type set I is a set $\mathcal{C}$, called the set of chambers, together with equivalence relations $\sim_i$, $i \in I$, on $\mathcal{C}$. For $i \in I$ and chambers $c, d \in \mathcal{C}$, we say that c and d are *i-adjacent* if $c \sim_i d$. More generally, we say that c and d are *adjacent* if they are i-adjacent for some $i \in I$. A chamber system $\mathcal{C}$ is called *thick* if for every $i \in I$ and every chamber $c \in \mathcal{C}$, there are at least three chambers (c and two further chambers) i-adjacent to c. A chamber system is called *thin* if c is i-adjacent to exactly two chambers (itself and one further chamber) for all $i \in I$ and $c \in \mathcal{C}$.

If Γ is a geometry with type set I then one can construct a chamber system $\mathcal{C} = \mathcal{C}(\Gamma)$ over I as follows: The chambers are the maximal flags of Γ, and two maximal flags are i-adjacent if and only if they contain the same element of type j for all $j \in I \setminus \{i\}$. A chamber system is called *geometric* if it can be obtained in this way from some geometry.

If Γ is residually connected, it can be recovered from the associated chamber system $\mathcal{C}(\Gamma)$ as follows: For $J \subseteq I$, a *J-cell* is an equivalence class of the minimal equivalence relation containing the relations $\sim_i$ for all $i \in J$. The poset of all cells ordered by reverse inclusion is naturally isomorphic to the poset of the flags of Γ ordered by inclusion. Under this isomorphism the cell corresponding to a flag F consists of all chambers (maximal flags) containing F. In particular, the elements of type i of Γ correspond to the $(I \setminus \{i\})$-cells.

2.4 Amalgams of groups

An *amalgam of groups* is a set $\mathcal{A} = \bigcup_{i \in I} G_i$ with a partial operation of multiplication such that

(A1) the restriction of the multiplication to every G_i makes G_i a group;

(A2) the product ab is defined if and only if $a, b \in G_i$ for some $i \in I$; and

(A3) $G_i \cap G_j$ is a subgroup of G_i and G_j for all $i, j \in I$.

A *completion* of the amalgam $\mathcal{A}$ is a group G together with a mapping ϕ from $\mathcal{A}$ to G such that (i) the restriction of ϕ to every G_i is a homomorphism and (ii) $\phi(\mathcal{A})$ generates G. The *universal completion* of $\mathcal{A}$ is the group $U(\mathcal{A})$ with generators $\{t_s \mid s \in \mathcal{A}\}$ and relations $t_x t_y = t_{xy}$ for all pairs of elements $x, y \in \mathcal{A}$ such that $x, y \in G_i$ for some i. The corresponding mapping is given by $x \mapsto t_x$. By abuse of notation we identify the completion (G, ϕ) with just the group G, and in this sense we can think of every completion as a quotient of the universal completion $U(\mathcal{A})$.

In terms of amalgams, the identification problem (see the introduction) amounts to finding the universal completions of certain amalgams arising in Chevalley groups. An important observation due to Jacques Tits connects completions of amalgams with geometries, and we finish this section with a discussion of this result.

2.5 Tits' lemma

Given a geometry Γ and a flag-transitive group $G \leq \operatorname{Aut}\Gamma$, we associate an amalgam $\mathcal{A}$ with them as follows. Let F be a maximal flag of Γ. Then $\mathcal{A} = \bigcup_{i \in I} G_i$, where G_i is the stabilizer in G of the element of type i from F. This amalgam $\mathcal{A}$ is called the *amalgam of maximal parabolics*, and notice that $\mathcal{A}$ is independent of the choice of F if we consider it up to isomorphism. Furthermore, if Γ is connected then $\mathcal{A}$ generates G so that G is a completion of $\mathcal{A}$.

The following proposition (*Tits' lemma*) is a restatement for the case of geometries of *Corollaire* 1 from [T1]

Proposition 2.1. *Let Γ be a connected geometry and let $G \leq \operatorname{Aut}\Gamma$ be a flag-transitive group of automorphisms. Moreover, let F be a maximal flag of Γ. Then G is the universal completion of the amalgam $\mathcal{A}$ of maximal parabolics with respect to F if and only if the geometry Γ is simply connected.* □

This result reduces the problem of identifying the universal completion of certain amalgams to proving that the corresponding geometries are simply connected. As we have mentioned above, simple connectedness can be verified by proving that the fundamental group of the corresponding flag complex is trivial, that is, by proving that every loop on the flag complex is null-homotopic.

3 Phan's theorem

3.1 History

In 1975, Kok-Wee Phan gave a method for identifying an unknown group G with a quotient of the unitary group $SU(n+1,q^2)$, by finding in G a generating configuration of subgroups $SU(3,q^2)$ and $SU(2,q^2) \times SU(2,q^2)$. We begin by looking at a configuration of such subgroups in $SU(n+1,q^2)$ to motivate our later definition.

Suppose $n \geq 2$ and q is a prime power. Let $G = SU(n+1,q^2)$, and let $U_i \cong SU(2,q^2)$, $i = 1, 2, \ldots, n$, be the subgroups of G corresponding to the 2×2 blocks along the main diagonal. Define D_i to be the diagonal subgroup of U_i and notice that D_i is a maximal torus of U_i of size $q+1$. When $q \neq 2$, the group G is generated by the subgroups U_i, and the following hold for $1 \leq i,j \leq n$:

(P1) if $|i-j| > 1$ then $[x,y] = 1$ for all $x \in U_i$ and $y \in U_j$;

(P2) if $|i-j| = 1$ then $\langle U_i, U_j \rangle$ is isomorphic to $SU(3,q^2)$; and

(P3) $[x,y] = 1$ for all $x \in D_i$ and $y \in D_j$.

Suppose now that G is an arbitrary group containing a system of subgroups $U_i \cong SU(2,q^2)$, and suppose a maximal torus D_i of size $q+1$ is chosen in each U_i. If the conditions (P1)–(P3) above hold true for G, we will say that G contains a *Phan system of rank n*. In [Ph1] Kok-Wee Phan proved the following result:

Theorem 3.1. *If G contains a Phan system of rank $n \geq 3$ with $q > 4$, then G is isomorphic to a factor group of $SU(n+1,q^2)$.*

Phan's proof of this result, however, is somewhat incomplete. Much of the proof is calculation-based, and many of these calculations are left to the reader. Moreover, while Phan apparently deals with the question of what the Phan system generates if the amalgam $\mathcal{A}$ formed by the subgroups $U_{ij} = \langle U_i, U_j \rangle$ is exactly as in $SU(n+1,q^2)$, he never addresses the question of the uniqueness of $\mathcal{A}$. Unfortunately, this is crucial. Indeed, nothing in the conditions (P1)–(P3) tells us right away that $\mathcal{A}$ must be as in $SU(n+1,q^2)$. Potentially, there may be many such amalgams, in which case G could be a quotient of the universal completion of any one of those amalgams. Thus, the proof of the uniqueness of $\mathcal{A}$ must be an important part of the proof of Phan's theorem.

3.2 Strategy

Let us assume for now that the uniqueness of $\mathcal{A}$ is known so that $\mathcal{A}$ can be identified with the amalgam formed by block-diagonal subgroups $SU(3,q^2)$ and $SU(2,q^2) \times SU(2,q^2)$ of $\hat{G} = SU(n+1,q^2)$. Under this assumption, what remains to be shown is that the universal completion of $\mathcal{A}$ coincides with $\hat{G}$. A natural way to show this is via Tits' lemma.

In order to apply Tits' lemma we need a geometry on which G acts flag-transitively, so that $\mathcal{A}$ is (or at least, is related to) the corresponding amalgam of maximal parabolics. Such a geometry has, in fact, already appeared in the literature (e.g. see [A]). This geometry, $\mathcal{N} = \mathcal{N}(n, q^2)$, is defined as follows. Let V be the $(n+1)$-dimensional unitary space over $GF(q^2)$. The elements of $\mathcal{N}$ are the proper non-singular subspaces U of V. The type of U is given by its dimension and the incidence is defined by containment. Fixing an orthonormal basis $\{e_1, \ldots, e_{n+1}\}$ in V, we make $\hat{G}$ act on $\mathcal{N}$, and it is easy to see that this action is flag-transitive. The next key fact is that $\mathcal{N}$ is almost always simply connected. Deferring the exact statement and a discussion of the proof until the next subsection, we just mention that the case where $q > 3$ is odd was first done in [D].

Once $\mathcal{N}$ is known to be simply connected, Tits' lemma implies that $\hat{G}$ is the universal completion of the amalgam $\hat{\mathcal{A}}$ of maximal parabolics associated with $\mathcal{N}$. Choosing the maximal flag consisting of all the subspaces $U_i = \langle e_1, \ldots, e_i \rangle$, the amalgam $\hat{\mathcal{A}}$ is the union of the block-diagonal subgroups

$$(GU(i, q^2) \times GU(n+1-i, q^2))^+,$$

where the plus indicates that within this direct product we only take matrices with determinant equal to one. In particular, $\mathcal{A}$ is completely contained in $\hat{\mathcal{A}}$. Unfortunately, $\mathcal{A}$ is not equal to $\hat{\mathcal{A}}$, which means that we have to do more work.

Let G be the universal completion of $\mathcal{A}$. Notice that $\hat{G} = SU(n+1, q^2)$ is generated by $\mathcal{A}$ and hence $\hat{G}$ is a completion of $\mathcal{A}$. This means that $\hat{G}$ is a quotient of G. Thus, it suffices to show that G cannot be larger than $\hat{G}$. We accomplish this by finding a copy of $\hat{\mathcal{A}}$ inside G, that extends $\mathcal{A}$. This implies that G is in turn a quotient of $\hat{G}$ and hence G cannot be larger than $\hat{G}$.

Let $\hat{\mathcal{A}}_s$ be the subamalgam of $\hat{\mathcal{A}}$ formed by all parabolics of rank at most s. Recall that in each U_i we have a torus D_i of order $q+1$. Viewing $\mathcal{A}$ as embedded in G, define $D = \prod D_i$. We show that D is in fact the direct product of the D_i's and that $U_{ij}D$ is isomorphic to the full rank 2 parabolic from $\hat{\mathcal{A}}$. Furthermore, the union of the subgroups $U_{ij}D$ in G produces an amalgam isomorphic to the subamalgam $\hat{\mathcal{A}}_2$ of $\hat{\mathcal{A}}$. The remaining part is easy, as we inductively extend every $\hat{\mathcal{A}}_s$ to $\hat{\mathcal{A}}_{s+1}$ using the case $s = 2$ as a base of induction. Notice that the simple connectedness of $\mathcal{N} = \mathcal{N}(s+1, q)$ is used in extending $\hat{\mathcal{A}}_s$.

At this point we turn to the question of how the simple connectedness is proven.

3.3 Simple Connectedness

Recall that simple connectedness can be shown by proving that every loop of the flag complex of $\mathcal{N}$ is null-homotopic. Fixing a base element x to be a point (an element of type 1), a standard technique is to reduce every loop of $\mathcal{N}$ based at x to a loop in the point-line incidence graph (lines are elements of type 2). This technique requires that the geometry in question contains sufficiently many connected residues, which is the case for the geometry $\mathcal{N}$. In fact, for $q \neq 2$, $\mathcal{N}$ is residually connected.

Thus, we only need to consider loops fully contained in the point-line incidence graph. Every such loop can be understood as a loop in the collinearity graph Σ of $\mathcal{N}$. The vertices of Σ are the points of $\mathcal{N}$ and two points are adjacent if and only if they are collinear (*i.e.*, incident to a common line).

A loop in Σ that is contained entirely within the residue of an element of $\mathcal{N}$ (such a loop is called *geometric*) is null-homotopic. Thus, proving that $\mathcal{N}$ is simply connected requires showing that every loop in Σ can be decomposed into a product of geometric loops. In fact, we only use geometric triangles for this.

The key fact that allows us to proceed is that, with few exceptions, Σ has diameter two. By induction every loop in Σ is a product of loops of length up to five. Hence it suffices to show that every loop γ of length 3, 4, and 5 is null-homotopic. For large N, one can always find a point that is perpendicular to all the points on γ. This produces a decomposition of γ into geometric triangles. Hence the claim is essentially obvious for large n. All the difficulty of the proof lies in the case of small n, where we resort to a case-by-case analysis and the proof at times becomes rather intricate.

We end this section with the exact statement from [BSh].

Proposition 3.2. *The geometry* $\mathcal{N} = \mathcal{N}(n+1, q^2)$ *is simply connected if* (n, q) *is not one of* $(3, 2)$ *and* $(3, 3)$.

Our proof of this proposition is computer-free with the exception of the case $n = 5$ and $q = 2$, which was handled by Jon Dunlap using a Todd-Coxeter coset enumeration in GAP ([GAP]). Notice that neither one of the exceptions above is simply connected, so that the result is (in a sense) best possible.

3.4 Uniqueness of $\mathcal{A}$

Notice that Phan does not address the cases $q \leq 4$ at all. Furthermore his definitions do not even make sense for $q = 2$. To include all possible cases in our theorem, we need to modify Phan's setup.

We say that a group G possesses a *weak Phan system* if G contains subgroups $U_i \cong SU(2, q^2)$, $i = 1, 2, \ldots, n$, and $U_{i,j}$, $1 \leq i < j \leq n$, so that the following hold:

(wP1) If $|i - j| > 1$ then $U_{i,j}$ is a central product of U_i and U_j;

(wP2) For $i = 1, 2, \ldots, n-1$, the groups U_i and U_{i+1} are contained in $U_{i,i+1}$, which is isomorphic to $SU(3, q^2)$ or $PSU(3, q^2)$; moreover, U_i and U_{i+1} form a standard pair in $U_{i,i+1}$; and

(wP3) The subgroups $U_{i,j}$, $1 \leq i < j \leq n$, generate G.

Here a *standard pair* in $SU(3, q^2)$ denote a pair of subgroups $SU(2, q^2)$ conjugate as a pair to the pair of block-diagonal $SU(2, q^2)$'s. Standard pairs in $PSU(3, q^2)$ are defined as the images under the natural homomorphism of the standard pairs from $SU(3, q^2)$.

This definition leaves a lot of possibilities for the members of the amalgam $\mathcal{A} = \bigcup U_{ij}$, producing a variety of amalgams so that we are unable to make any claims of uniqueness in the general case. We call an amalgam $\mathcal{A}$ *unambiguous* if every

U_{ij} is isomorphic to just $SU(3,q^2)$ or $SU(2,q^2) \times SU(2,q^2)$ (rather than a quotient of these groups). Using some "scissors-and-glue" methods, one can associate to every amalgam $\mathcal{A}$ of weak Phan type an unambiguous amalgam whose universal completion has $U(\mathcal{A})$ as a quotient, reducing the analysis of $\mathcal{A}$ to the case where $\mathcal{A}$ is unambiguous. However even in this case we cannot claim uniqueness, and we must impose another restriction. A *non-collapsing* amalgam is an amalgam such that $U(\mathcal{A}) \neq 1$ (this simple definition works in all cases except for $q = 2$; the latter case requires the stronger condition that every U_i embeds into $U(\mathcal{A})$). Clearly, from the point of view of Phan's theorem, we are only interested in the non-collapsing amalgams. It is interesting that although many unambiguous amalgams exist, only one of them is non-collapsing.

Proposition 3.3. *If $\mathcal{A} = \bigcup U_{ij}$ is unambiguous and non-collapsing, then it is isomorphic to the canonical amalgam of block-diagonal subgroups of the group $SU(n+1,q^2)$.*

We use the non-collapsing condition as follows. For $\epsilon = \pm 1$, define $D_i^\epsilon = N_{U_i}(U_{i+\epsilon})$. Note that this normalizer makes sense in $U_{i,i+\epsilon}$. Assuming that $\mathcal{A}$ is non-collapsing, we have a completion H in which every member of $\mathcal{A}$ embeds. Working in H we show that $D_i^{+1} = D_i^{-1}$ for all $i = 2, \ldots, n-1$. This extra condition makes $\mathcal{A}$ unique. It also enables us to introduce the tori $D_i = D_i^{+1} = D_i^{-1}$ as in Phan's original setup.

The main part of the uniqueness proof splits into the cases $n = 3$ and $n > 3$. In the first case we use Goldschmidt's Lemma (cf. 2.7 of [G]) to prove that the amalgam of U_{12} and U_{23} with joint subgroup U_2 is unique up to isomorphism. To identify $\mathcal{A}$ we need to decide which subgroups of U_{12} and U_{23} can serve as U_1 and U_3. Once these subgroups are found, the remaining member U_{13} is added to $U_{12} \cup U_{23}$ as $U_1 \times U_3$.

The condition on U_1 and U_3 is that each must form a standard pair with U_2. It can be seen that U_2 acts transitively by conjugation on the candidates for U_1 and on candidates for U_3. Since conjugation by an element of U_2 is an automorphism of the amalgam $U_{12} \cup U_{23}$, we can assume that U_1 is a fixed subgroup. On the other hand, for U_3 we have many possibilities that lead to many amalgams. Fortunately we have the extra condition arising from the assumption that $\mathcal{A}$ is non-collapsing. This condition leaves only two candidates for U_3 and we complete the proof by finding an automorphism of $U_{12} \cup U_{23}$ that stabilizes U_1 and permutes the two candidates for U_3.

For the $n > 3$ case, we now appeal to induction using the case $n = 3$ as the base. In the end, combining all the above we obtain the following two theorems.

Theorem 3.4. *If G contains a weak Phan system of rank n at least three with $q > 3$ then G is isomorphic to a factor group of $SU(n+1,q^2)$.*

Theorem 3.5. *Suppose G contains a weak Phan system of rank n specified below with $q = 2$ or 3.*

(1) *Suppose $q = 3$, $n \geq 4$, and additionally, for $i = 1, 2, \ldots, n-2$, the subgroup generated by $U_{i,i+1}$ and $U_{i+1,i+2}$ is isomorphic to a factor group of $SU(4,9)$. Then G is isomorphic to a factor group of $SU(n+1,9)$.*

(2) *Suppose* $q = 2$, $n \geq 5$ *and, for* $i = 1, 2, \ldots, n-3$, *the subgroup generated by* $U_{i,i+1}$, $U_{i+1,i+2}$ *and* $U_{i+2,i+3}$ *is isomorphic to a factor group of* $SU(5,4)$. *Then* G *is isomorphic to a factor group of* $SU(n+1,4)$.

Notice that the extra conditions are required, for $q \leq 3$ and small n, as the geometry $\mathcal{N}$ is not simply connected and in the case, $n = 2$ and $q = 2$, it is not even connected.

4 The Curtis-Tits theorem

The following formulation of the Curtis-Tits theorem is taken from [GLS].

Theorem 4.1. *Let* G *be the universal version of a finite Chevalley group of (twisted) rank at least* 3 *with root system* Σ, *fundamental system* Π, *and root groups* X_α, $\alpha \in \Sigma$. *For each* $J \subseteq \Pi$ *let* G_J *be the subgroup of* G *generated by all root subgroups* X_α, $\pm\alpha \in J$. *Let* D *be the set of all subsets of* Π *with at most* 2 *elements. Then* G *is the universal completion of the amalgam* $\bigcup_{J \in D} G_J$.

We first discuss the similarities and differences between Phan's theorem and the Curtis-Tits theorem. Let us consider the case of the Chevalley group of type A_n, which is $G = SL(n+1,q)$. With the usual choice of the root subgroups in G, the subgroups $G_J = G_{ij}$ are the block-diagonal subgroups $SL(3,q)$ and $SL(2,q) \times SL(2,q)$, which we note are similar to the subgroups in the amalgam in Phan's theorem. The main difference between the two theorems is that the Curtis-Tits theorem merely claims that the universal completion of the known amalgam (the one found in $SL(n+1,q)$, *i.e.*, $\bigcup_{J \in D} G_J$) is $SL(n+1,q)$, while Phan's theorem makes a claim about the completion of an arbitrary Phan amalgam.

Clearly, as we are again trying to find the universal completion of an amalgam, Tits' lemma appears to be a natural tool for this task. To use it, one needs to find a suitable geometry on which G acts flag-transitively with the correct amalgam of maximal parabolics, and then prove that the geometry is simply connected. We begin by modifying the amalgam so as to replace the rank 2 subgroups, G_J, with the maximal ones. Consider the amalgam $\mathcal{A} = \bigcup_{\alpha \in \Pi} G_{\Pi \setminus \{\alpha\}}$. By induction on the rank, the Curtis-Tits theorem is equivalent to the following.

Theorem 4.2. *Under the assumptions of Theorem 4.1, the group* G *is the universal completion of the amalgam* $\mathcal{A}$.

In the rest of this section we will discuss a geometric proof of this theorem given by Mühlherr in [M].

Recall that a finite Chevalley group G acts on its natural finite geometry called a building. Let I be a set and M be a Coxeter matrix over I. Let (W, S) be the Coxeter system of type M, where $S = \{s_i \mid i \in I\}$. A *building of type* M is a pair $\mathcal{B} = (\mathcal{C}, \delta)$ where $\mathcal{C}$ is a set and $\delta : \mathcal{C} \times \mathcal{C} \longrightarrow W$ is a distance function satisfying the following axioms. Let $x, y \in \mathcal{C}$ and $w = \delta(x, y)$. Then

(B1) $w = 1$ if and only if $x = y$;

(B2) if $z \in \mathcal{C}$ is such that $\delta(y,z) = s \in S$, then $\delta(x,z) = w$ or ws; furthermore if $l(ws) = l(w) + 1$, then $\delta(x,z) = ws$; and

(B3) if $s \in S$, there exists $z \in \mathcal{C}$ such that $\delta(y,z) = s$ and $\delta(x,z) = ws$.

In this survey we will concentrate (unlike Mühlherr) on finite buildings, in which case the diagram is spherical, although a number of results that we state also apply to the non-finite case.

Given a building $\mathcal{B} = (\mathcal{C}, \delta)$ we can define a chamber system on the set of chambers $\mathcal{C}$ (we denote the chamber system by $\mathcal{C}$ as well) where two chambers c and d are i-adjacent if and only if $\delta(c,d) = s_i$. Conversely, the building $\mathcal{B}$ can be recovered from its chamber system $\mathcal{C}$. We will only consider those buildings $\mathcal{B}$ for which $\mathcal{C}$ is thick. If $\mathcal{B}$ is a building, its chamber system contains a class of thin subsystems called *apartments*. In an apartment Σ, for any $c \in \Sigma$ and $w \in W$, there is a unique chamber $d \in \Sigma$ such that $\delta(c,d) = w$. Every pair of chambers of $\mathcal{C}$ is contained in an apartment. Notice that the chamber system $\mathcal{C}$ defined by a building is always geometric. Let $\Gamma = \Gamma(\mathcal{B})$ be the corresponding geometry. It is well known that Γ is simply connected. Unfortunately, we cannot use this to prove the Curtis-Tits theorem because it corresponds to the wrong amalgam. So we need to find a different geometry.

Given two buildings $\mathcal{B}_+ = (\mathcal{C}_+, \delta_+)$, $\mathcal{B}_- = (\mathcal{C}_-, \delta_-)$ of the same type M, a *codistance* (*twinning*) is a map $\delta_* : (\mathcal{C}_+ \times \mathcal{C}_-) \cup (\mathcal{C}_- \times \mathcal{C}_+) \longrightarrow W$ such that the following axioms hold where $\epsilon = \pm$, $x \in \mathcal{C}_\epsilon, y \in \mathcal{C}_{-\epsilon}$ and $w = \delta_*(x,y)$:

(T1) $\delta_*(y,x) = w^{-1}$;

(T2) if $z \in \mathcal{C}_{-\epsilon}$ such that $\delta_{-\epsilon}(y,z) = s \in S$ and $l(ws) = l(w) - 1$, then $\delta_*(x,z) = ws$; and

(T3) if $s \in S$, there exists $z \in \mathcal{C}_{-\epsilon}$ such that $\delta_{-\epsilon}(y,z) = s \in S$ and $\delta_*(x,z) = ws$.

A *twin building* of type M is a triple $(\mathcal{B}_+, \mathcal{B}_-, \delta_*)$, where $\mathcal{B}_+$ and $\mathcal{B}_-$ are buildings of type M and δ_* is twinning between $\mathcal{B}_+$ and $\mathcal{B}_-$.

Tits showed (cf. Proposition 1 of [T2]) that every spherical twin building can be obtained as follows from some building $\mathcal{B} = (\mathcal{C}, \delta)$ of the same type M. Let $\mathcal{B}_+ = (\mathcal{C}_+, \delta_+)$ be a copy of $\mathcal{B}$, define $\mathcal{B}_- = (\mathcal{C}_-, \delta_-)$ as $(\mathcal{C}, w_0 \delta w_0)$, and let δ_* be defined as $w_0 \delta$ and δw_0 on $\mathcal{C}_+ \times \mathcal{C}_-$ and $\mathcal{C}_- \times \mathcal{C}_+$ respectively. Here w_0 is the longest element of the Weyl group W.

Given a twin building $\mathcal{T} = (\mathcal{B}_+, \mathcal{B}_-, \delta_*)$, one can define a chamber system $\mathrm{Opp}(\mathcal{T}) = \{(c_+, c_-) \in \mathcal{C}_+ \times \mathcal{C}_- \mid \delta_*(c_+, c_-) = 1_W\}$. Chambers $x \in \mathcal{C}_+$ and $y \in \mathcal{C}_-$ with $\delta_*(x,y) = 1_W$ are called *opposite*, hence the notation. Note that $\mathrm{Opp}(\mathcal{T})$ is a geometric chamber system. Its corresponding geometry is denoted by Γ_{op} and is called the *opposites geometry*. It can be described as follows. Let Γ_+ and Γ_- be the building geometries that correspond to $\mathcal{B}_+$ and $\mathcal{B}_-$. Elements $x_+ \in \Gamma_+$ and $x_- \in \Gamma_-$ of the same type $i \in I$ are called *opposite* if they are contained in opposite maximal flags (*i.e.*, chambers). The elements of Γ_{op} of type i are pairs (x_+, x_-) of opposite elements of type i. Two pairs (x_+, x_-) and (x'_+, x'_-) are incident in Γ_{op} if both x_+ and x'_+ are incident in Γ_+ and x_- and x'_- are incident in Γ_-. Clearly, a pair $(c_+, c_-) \in \mathrm{Opp}(\mathcal{T})$ produces a maximal flag in Γ_{op}, and it can be shown that every maximal flag is obtained in this way.

We now give some examples.

Example 1a. Let $G \cong PSL(n+1, q)$, *i.e.*, M is of type A_n. Then the building geometry Γ is the projective space, whose elements of type i, $1 \leq i \leq n$, are all the i-dimensional subspaces in the corresponding $(n+1)$-dimesional vector space V. The geometries Γ_+ and Γ_- are isomorphic respectively to Γ and the dual geometry of Γ (same as Γ except that the type of the i-dimensional subspace is $n+1-i$). Elements (subspaces) $x_+ \in \Gamma_+$ and $x_- \in \Gamma_-$ of type i are opposite if they intersect trivially and thus form a direct sum decomposition $V = x_+ \oplus x_-$. It follows that these decompositions are the elements of Γ_{op}.

Example 2a. Let $G \cong PSp(2n, q)$, which corresponds to the diagram C_n. Then Γ is the geometry of all totally isotropic subspaces of a nondegenerate $2n$-dimensional symplectic space V. In this case, both Γ_+ and Γ_- are isomorphic to Γ. Two i-dimensional totally isotropic subspaces x_+ and x_- are opposite if x_- intersects trivially with the orthogonal complement of x_+. Such pairs (x_+, x_-) are the elements of Γ_{op}.

In general, if the twin building consists of two isomorphic parts $\mathcal{B}_+ \cong \mathcal{B} \cong \mathcal{B}_-$, which is the case for a spherical diagram, the automorphism group $\mathrm{Aut}\,(\mathcal{B})$ of the building acts on the twin building $\mathcal{T}$ by automorphisms, in particular, it preserves the opposition relation, and hence it also acts on Γ_{op}. It can be shown that the action of $\mathrm{Aut}\,(\mathcal{B})$ on the set of pairs of opposite chambers is transitive, thus it is flag-transitive on Γ_{op}. The stabilizers of the elements of a maximal flag of Γ_{op} are Levi factors in the maximal parabolic subgroups (in the sense of Chevalley groups) of G. The Levi factors differ from the members of the amalgam of Theorem 4.2 only by the Cartan subgroup. To be precise, the full Levi factors are the products of the subgroups $G_{\Pi\setminus\{\alpha\}}$ with the Cartan subgroup H. This is not a major impediment as the Cartan subgroup can be recovered piecewise from the initial amalgam $\mathcal{A}$. Therefore the Curtis-Tits theorem is equivalent to the following:

Theorem 4.3. *If $\mathcal{T} = (\mathcal{B}_+, \mathcal{B}_-, \delta_*)$ is a spherical twin building of rank at least three, then the geometry Γ_{op} is simply connected.*

This was proved by Mühlherr in [M] for twin buildings with arbitrary (that is, not only spherical) Coxeter matrix M. His proof is case-independent, short and elegant. The claim is derived directly from the axioms of twin buildings, properties of apartments in buildings, and certain connectivity properties of buildings. However his proof does not cover a number of exceptional (small field) cases where the connectivity fails. In particular, in the spherical case, the groups $G \cong Sp(2n, 2)$ and $F_4(2)$ are not covered by his proof. In the nonspherical case Mühlherr has to exclude tree residues and rank 2 residues related to the buildings of type $B_2(2)$, ${}^2F_4(2)$, $G_2(2)$, and $G_2(3)$. Mühlherr remarks that in the nonspherical situation there appear to be counterexamples. Hence a general proof for all M may not be possible. In the spherical case we know by the original Curtis-Tits proof that there are no counterexamples. Thus the following seems to be an interesting problem.

Problem 1. *Generalize Mühlherr's proof to cover all spherical matrices M.*

As we have already noticed, the Curtis-Tits theorem is not concerned with the question of the uniqueness of the amalgam $\mathcal{A} = \bigcup_{\alpha\in\Pi} G_{\Pi\setminus\{\alpha\}}$. In our opinion this makes applying the Curtis-Tits theorem more complicated. Indeed, in order to apply it one has to show that inside the group G under consideration there is an exact copy of the amalgam $\mathcal{A}$. Thus it would be advantageous to strengthen the Curtis-Tits theorem by solving the following problem.

Problem 2. *Prove that any non-collapsing amalgam of groups isomorphic to $G_{\Pi\setminus\{\alpha\}}$ with given isomorphism types of their intersections is in fact isomorphic to $\mathcal{A}$.*

5 Flipflop geometries

We will start with an example.

Example 1b. Consider the situation of Example 1a, but change the field of definition to $GF(q^2)$, so that $G \cong PSL(n+1, q^2)$. Consider a unitary polarity σ, that is, an involutory isomorphism from Γ onto the dual of Γ which is defined by a nondegenerate Hermitian form Φ on V. More precisely, σ sends every subspace of V to its orthogonal complement with respect to Φ. This σ produces an involutory automorphism of the twin building $\mathcal{T}$ that switches $\mathcal{C}_+$ and $\mathcal{C}_-$ (or else, Γ_+ and Γ_-). It is an automorphism in the sense that it transforms δ_+ into δ_- and *vice versa*, and preserves δ_*. Note that σ induces an automorphism of G, which, by abuse of notation, will also be denoted by σ. Consider $G_\sigma = C_G(\sigma)$ and $\Gamma_\sigma = \{(x_+, x_-) \in \Gamma_{op} \mid x_+^\sigma = x_-\}$. Then $G_\sigma \cong PSU(n+1, q^2)$ acts on Γ_σ. Notice that the elements of Γ_σ are of the form (x_+, x_-) where $x_- = x_+^\sigma = x_+^\perp$ and $V = x_+ \oplus x_- = x_+ \oplus x_+^\perp$. Thus, the mapping $(x_+, x_-) \mapsto x_+$ establishes an isomorphism between Γ_σ and the geometry of all proper nondegenerate subspaces of the unitary space V, as defined by Φ. This is exactly the geometry from Section 3 that was used for a new proof of Phan's first theorem.

This suggests the following general construction. Let $\mathcal{T} = (\mathcal{B}_+, \mathcal{B}_-, \delta_*)$ be a twin buiding. Consider an involutory automorphism σ of $\mathcal{T}$ with the following properties:

(F1) $\mathcal{C}_+^\sigma = \mathcal{C}_-$;

(F2) σ flips the distances, *i.e.*, $\delta_\epsilon(x, y) = \delta_{-\epsilon}(x^\sigma, y^\sigma)$ for $\epsilon = \pm$; and

(F3) σ preserves the codistance, *i.e.*, $\delta_*(x, y) = \delta_*(x^\sigma, y^\sigma)$.

We additionally require that there be at least one chamber $c \in \mathcal{C}_\pm$ such that $\delta_*(c, c^\sigma) = 1_W$. Such σ's will be called *flips*.

Let $\mathcal{C}_\sigma$ be the chamber system whose chambers are pairs (c, c^σ) that belong to $Opp(\mathcal{T})$. Note that by our assumption $\mathcal{C}_\sigma$ is non-empty. We do not know if $\mathcal{C}_\sigma$ is geometric in general, however this is the case in each of our examples with the possible exception of exception of Example 5. If $\mathcal{C}_\sigma$ is geometric, let Γ_σ denote the corresponding geometry. It will be referred to as the *flipflop geometry*.

In case of a spherical twin building, we can compute the action of σ on the Coxeter diagram of the building, as has been done in Section 3.3 of [Gr]. Indeed, using Tits' characterization of spherical twin buildings (Proposition 1 of [T2]), we have $\delta(c,d) = \delta_+(c,d) = \delta_-(c^\sigma, d^\sigma) = w_0\delta(c^\sigma, d^\sigma)w_0$. Therefore, the flip σ acts on the Coxeter diagram via conjugation with the longest word w_0 of the Weyl group. This gives the following characterization of a flip of a spherical twin building.

Proposition 5.1. *Let $\mathcal{T} = (\mathcal{B}_+, \mathcal{B}_-, \delta_*)$ be a spherical twin building. An adjacency-preserving involution σ that interchanges $\mathcal{B}_+$ and $\mathcal{B}_-$ and maps some chamber onto an opposite chamber is a flip if and only if the induced map $\hat{\sigma}$ on the building $\mathcal{B} = (\mathcal{C}, \delta)$ satisfies $\delta(c,d) = w_0\delta(c^{\hat{\sigma}}, d^{\hat{\sigma}})w_0$ for all chambers $c, d \in \mathcal{C}$ where w_0 is the longest word in the Weyl group W.*

Note that, in the case of a building of type A_n, a flip gives a polarity σ. The condition that a chamber is mapped to the opposite chamber implies in particular that there is a 1-space not incident to it polar. This excludes symplectic polarities and orthogonal polarities in characteristic two. Conversely given any unitary or an orthogonal polarity in odd characteristic, we can find an orthogonal basis for the corresponding form. This in turn will give an apartment in which each chamber is mapped to its opposite.

Here are some additional examples.

Example 2b. Consider the situation of Example 2a, but with the field of definition of order q^2. Let $\{e_1, \ldots, e_n, f_1, \ldots, f_n\}$ be a hyperbolic basis of the symplectic space V. (So that $(e_i, f_j) = \delta_{ij}$.) Consider the semilinear transformation σ of V which is the composition of the linear transformation given by the Gram matrix of the form and the involutory field automorphism applied to the coordinates with respect to the above basis. It can be shown that σ produces a flip of $\mathcal{T}$. Furthermore, $\mathcal{C}_\sigma$ is geometric and $G_\sigma \cong PSp(2n,q)$ acts flag-transitively on the corresponding flipflop geometry Γ_σ. The geometry Γ_σ can be described as follows. For $u, v \in V$ let $((u,v)) = (u, v^\sigma)$, where $(\cdot,\cdot)$ is the symplectic form on V. Then $((\cdot,\cdot))$ is a nondegenerate Hermitian form. The flipflop geometry Γ_σ can be identified (via $(x_+, x_-) \mapsto x_+$) with the geometry of all subspaces of V which are totally isotropic with respect to $(\cdot,\cdot)$ and, at the same time, nondegenerate with respect to $((\cdot,\cdot))$.

The configuration of Example 2b was looked at in [GHSh]. It is proved there that Γ_σ is almost always simply connected. Here is the main theorem from that paper.

Theorem 5.2. *The flipflop geometry Γ_σ described in Example 2b is simply connected if $n \geq 5$ or $n = 4$, $q \geq 3$ or $n = 3$, $q \geq 8$.*

We expect that some of the larger q's on this list of exceptions are there only because of the shortcomings of our particular proof, so that the final list of exceptions will be shorter.

The above theorem leads to a new "Phan-type" result on groups generated by subgroups $U_i \cong SU(2,q^2)$. Here we have that $\langle U_i, U_{i+1}\rangle \cong SU(3,q^2)$ for all $1 \leq i < n-1$, while $\langle U_{n-1}, U_n\rangle \cong Sp(4,q)$. As in Phan's original situation U_i and U_j with $|i-j| > 1$ commute elementwise. An amalgam of subgroups as indicated here is called a *Phan system of type C_n*. For the exact statements and other applications, see [GHSh]. We have to point out that the uniqueness of amalgams is not addressed in [GHSh] leaving the following an open problem.

Problem 3. *If q is sufficiently large prove that any non-collapsing Phan system of type C_n is in fact isomorphic to the canonical Phan system inside the group $Sp(2n, q)$.*

We expect that this problem can be solved by using the same methods as given in [BSh]. Consequently, for small q one first has to introduce the notion of a weak Phan system of type C_n as in Section 3 and then study unambiguous, non-collapsing weak Phan systems.

Example 3. For $G = PSO(2n, q^2, +)$ and $PSO(2n+1, q^2)$ (diagrams D_n and B_n, respectively) flips can be constructed by the same algorithm as in Example 2b, that is, σ can be defined as the composition of the linear transformation given by the Gram matrix, say, taken with respect to a hyperbolic basis (the actual requirement is that all entries of the Gram matrix must be in the subfield $GF(q)$) and the involutory field automorphism with respect to the same basis. In both cases we checked that this σ produces a flipflop geometry on which G_σ acts flag-transitively. While we have not obtained an exact result on the simple connectivity of Γ_σ, it is clear that Γ_σ is simply connected for all sufficiently large n and q, leading to new "Phan-type" theorems, cf. [BGHSh]. Notice that the D_n case here is likely to lead to Theorem 1.9 from Phan's second paper [Ph2]. This conjecture is underscored by our above observation (before Proposition 5.1) that a flip acts via conjugation with the longest word of the Weyl group on the diagram D_n. Indeed, for n even, Phan's target group is $Spin^+(q)$ (the universal Chevalley group of type $D_n(q)$) and conjugation with the longest word leaves the diagram invariant, while for n odd, Phan's target group is $Spin^-(q)$ (the universal Chevalley group of type ${}^2D_n(q^2)$) and conjugation with the longest word interchanges the two nodes representing the two classes of maximal totally singular subspaces. Another flip is induced by the linear transformation given by the Gram matrix with respect to a hyperbolic basis alone, without applying the involutory field automorphism.

Example 4. Now consider the group $G = PSO(2n, q, -)$ acting on the flag complex $\mathcal{C}$ of totally singular subspaces of a nondegenerate orthogonal form of $-$ type on the vector space V of dimension $2n$ over $GF(q)$. Choose two opposite chambers c and d of that flag complex and let U be the subspace of V that is perpendicular to the $n-1$ dimenional subspaces that appear in c, d. Fix a hyperbolic basis

$$\{e_1, \ldots, e_{n-1}, f_1, \ldots, f_{n-1}\}$$

of the vector space $c \oplus d$ such that $c = (\langle e_1 \rangle, \ldots, \langle e_1, \ldots, e_{n-1} \rangle)$ and $d = (\langle f_1 \rangle, \ldots, \langle f_1, \ldots, f_{n-1} \rangle)$ and, moreover, fix some orthogonal basis of U. Then there exists a linear map on V that preserves the form, maps e_i onto f_i and vice versa, and acts by scalar multiplication on each of the vectors of the orthogonal basis of U, e.g., the Gram matrix of the form with respect to the given basis. This linear map induces a flip σ of the twin building belonging to the flag complex $\mathcal{C}$. Notice, unlike Example 3, that we cannot compose this flip σ with an involutory field automorphism that acts entrywise on the vectors with respect to the given basis in order to obtain another flip, because this field automorphism would not commute with σ.

Example 5. Let G be the universal Chevalley group of type $E_6(q^2)$ and consider its 27-dimensional module V, a vector space over $GF(q^2)$. For sake of simplicity let us assume that q is not divisible by two. A vector $x \in V$ is represented by the triple $(x^{(1)}, x^{(2)}, x^{(3)})$ where $x^{(i)}$, $1 \le i \le 3$, is a (3×3)-matrix over $GF(q^2)$. The shadow space $E_{6,1}(q^2)$ can be described as the geometry on certain subspaces of V, cf. Section 5.2 of Cohen's Chapter 12 of [Bu]. There exists a nondegenerate bilinear form $(\cdot, \cdot)$ on V defined by

$$(x, y) = \text{trace}\,(x^{(1)}y^{(1)} + x^{(2)}y^{(3)} + x^{(3)}y^{(2)}).$$

Define $g^\sharp \in \mathrm{GL}\,(V)$ o be the adjoint of g^{-1} with respect to the form $(\cdot, \cdot)$. More precisely $g^\sharp$ is characterized by $(gx, g^\sharp y) = (x, y)$ for all $x, y \in V$. The map $\sharp : \mathrm{GL}\,(V) \to \mathrm{GL}\,(V) : g \mapsto g^\sharp$ induces an involutory automorphism α of the group G. This automorphism α in turn induces a correlation β of the geometry $E_{6,1}(q^2)$, *i.e.*, an incidence-preserving permutation of $E_{6,1}(q^2)$ that does not necessarily preserve types. In fact, β induces the involutory graph automorphism on the Coxeter diagram E_6. The composition of β and the involutory field automorphism acting entrywise on the representation $(x^{(1)}, x^{(2)}, x^{(3)})$ of any vector $x \in V$ induces a map σ on the corresponding twin building $\mathcal{T}$ that satisfies the axioms of a flip except that we did not check whether there exists a chamber that is mapped to an opposite chamber. We do, however, strongly believe that such a chamber exists. This observation is underscored by the fact that the centralizer in G of the composition of α and the involutory field automorphism equals ${}^2E_6(q^2)$ and, thus, the present setting is likely to lead to an alternative proof of Phan's Theorem 2.6 of [Ph2]. The correlation β can be expected to induce a flip as well.

We do not have a concrete example of a flip for an F_4 twin building, but we will discuss a general method for finding flips in the case where conjugation with the longest word of the Weyl group acts trivially on the diagram, which, for example, applies in the F_4 case. As a concrete example, one would hope to find a flip that centralizes the group $F_4(q)$ inside the group $F_4(q^2)$; the resulting flipflop geometry should admit the flipflop geometry of type B_3 from [BGHSh] and the flipflop geometry of type C_3 from [GHSh] as residues.

Let $\mathcal{T} = (\mathcal{B}_+, \mathcal{B}_-, \delta_*)$ be a twin building. Define the *automorphism group* $\mathrm{Aut}\,(\mathcal{T})$ to be the set of all permutations α of $\mathcal{T}$ with

- $\delta_\epsilon(c, d) = \delta_\epsilon(c^\alpha, d^\alpha)$ for all $c, d \in \mathcal{C}_\epsilon$ if α preserves $\mathcal{C}_+$ and $\mathcal{C}_-$,
- $\delta_\epsilon(c, d) = \delta_{-\epsilon}(c^\alpha, d^\alpha)$ for all $c, d \in \mathcal{C}_\epsilon$ if α interchanges $\mathcal{C}_+$ and $\mathcal{C}_-$, and
- $\delta_*(c, d) = \delta_*(c^\alpha, d^\alpha)$ for all $c \in \mathcal{C}_\epsilon$, $d \in \mathcal{C}_{-\epsilon}$,

where $\epsilon = \pm$. Clearly, if $\alpha, \beta \in \mathrm{Aut}\,(\mathcal{T})$ both interchange $\mathcal{C}_+$ and $\mathcal{C}_-$ then their product $\alpha\beta$ preserves $\mathcal{C}_+$ and $\mathcal{C}_-$. So, $\mathrm{Aut}\,(\mathcal{T})$ is of the form $\mathrm{Aut}\,(\mathcal{B}).2$. If there exists a flip or any other distance-switching and codistance-preserving involution of $\mathcal{T}$, then $\mathrm{Aut}\,(\mathcal{T})$ even is a semidirect product.

Now suppose we have a spherical twin building with a Coxeter diagram such that conjugation with the longest word w_0 acts as the trivial automorphism on the diagram. Then the map τ assigning to each chamber c of $\mathcal{C}_\pm$ the unique chamber d of $\mathcal{C}_\mp$ with $\delta_*(c, d) = w_0$ (called the *closest* chamber to c) is contained in $\mathrm{Aut}\,(\mathcal{T})$. Moreover, τ commutes with any automorphism of $\mathcal{T}$ that preserves $\mathcal{C}_+$ and $\mathcal{C}_-$, so $\mathrm{Aut}\,(\mathcal{T})$ is even a direct product. This implies the following.

Proposition 5.3. *Let $\mathcal{T} = (\mathcal{B}_+, \mathcal{B}_-, \delta_*)$ be a spherical twin building such that conjugation with the longest word w_0 of the Weyl group acts trivially on its Coxeter diagram. Then* $\mathrm{Aut}\,(\mathcal{T}) = \mathrm{Aut}\,(\mathcal{B}) \times \langle \tau \rangle$, *where τ is the automorphism assigning to each chamber $c \in \mathcal{C}_\pm$ the unique closest chamber $d \in \mathcal{C}_\mp$. Moreover, any flip of $\mathcal{T}$ is the product $\alpha\tau$ for an involutory $\alpha \in$* $\mathrm{Aut}\,(\mathcal{B})$ *such that there exists a chamber $c \in \mathcal{C}$ with $\delta(c, c^\alpha) = w_0$. Conversely, every such $\alpha\tau$ is a flip.*

This partial result motivates the following problem.

Problem 4. *Classify all flips for all spherical twin buildings. For each flip investigate Γ_σ and its simple connectivity.*

Of course, it would be much nicer to have general building-theoretic arguments (Mühlherr's type) in place of a case-by-case analysis. In particular, this concerns showing that $\mathcal{C}_\sigma$ is always geometric.

Besides the spherical case the investigation of flips might be interesting for the nonspherical case as well.

Problem 5. *Find an interesting flip for a nonspherical twin building.*

A flip might be considered interesting if it either centralizes or flips an interesting geometry or if it has an interesting centralizer. Also, Mühlherr's proof of the Curtis-Tits theorem has established a Curtis-Tits-type theorem for certain Kac-Moody groups. It might be worth the effort to investigate whether interesting Phan-type theorems can be proved for Kac-Moody groups as well. A starting point for the search of flips of nonspherial twin buildings might be [B] on diagram automorphisms induced by certain root reflections.

References

[A] M. Aschbacher, Simple connectivity of p-group complexes, *Israel J. Math.* **82** (1993), 1–43.

[B] C. D. Bennett, Imaginary roots of a Kac-Moody Lie algebra whose reflections preserve root multiplicities, *J. Algebra* **158** (1993), 244–267.

[BSh] C. D. Bennett and S. Shpectorov, A new proof of Phan's theorem, preprint.

[BGHSh] C. D. Bennett, R. Gramlich, C. Hoffman and S. Shpectorov, A Phan-type theorem for $SO(2n+1, q)$, in preparation.

[Bu] F. Buekenhout (editor), *Handbook of Incidence Geometry*, Elsevier, Amsterdam 1995.

[D] K. M. Das, Simple connectivity of the Quillen complex of $\mathrm{GL}_n(q)$, *J. Algebra* **178** (1995), 239–263.

[GAP] The GAP Group, GAP – Groups, Algorithms, and Programming, Version 4.2; 2000, (`http://www.gap-system.org`).

[G] D. M. Goldschmidt, Automorphisms of trivalent graphs, *Annals of Math.* **111** (1980), 377–406.

[GLS] D. Gorenstein, R. Lyons and R. Solomon, *The classification of the finite simple groups. Number 3. Part I. Chapter A. Almost simple K-groups*, Mathematical Surveys and Monographs 40.3. American Mathematical Society, Providence 1998.

[Gr] R. Gramlich, *On Graphs, Geometries, and Groups of Lie Type*, PhD thesis, Technische Universiteit Eindhoven 2002, (`http://cage.rug.ac.be/~nick/Theses/theses.html`).

[GHSh] R. Gramlich, C. Hoffman and S. Shpectorov, A Phan-type theorem for $Sp(2n, q)$, *J. Algebra*, to appear.

[M] B. Mühlherr, On the simple connectedness of a chamber system associated to a twin buiding, preprint.

[Ph1] K. W. Phan, On groups generated by three-dimensional special unitary groups, I, *J. Austral. Math. Soc. Ser. A* **23** (1977), 67–77.

[Ph2] K. W. Phan, On groups generated by three-dimensional special unitary groups, II, *J. Austral. Math. Soc. Ser. A* **23** (1977), 129–146.

[S] J.-P. Serre, *Arbres, amalgames,* SL_2, Astérisque 46, Soc. Math. France, Paris 1977

[T1] J. Tits, Ensembles Ordonnés, immeubles et sommes amalgamées, *Bull. Soc. Math. Belg. Sér. A* **38** (1986), 367–387.

[T2] J. Tits, Twin buildings and groups of Kac-Moody type. In: *Groups, Combinatorics and Geometry*, LMS Lecture Note Series 165, 249–286, Cambridge University Press, Cambridge 1992.

REPRESENTATION THEORY OF SYMMETRIC GROUPS AND THEIR DOUBLE COVERS

JONATHAN BRUNDAN AND ALEXANDER KLESHCHEV

1. Introduction

In this article we will give an overview of the new Lie theoretic approach to the p-modular representation theory of the symmetric groups and their double covers that has emerged in the last few years. There are in fact two parallel theories here: one for the symmetric groups S_n involving the affine Kac-Moody algebra of type $A_{p-1}^{(1)}$, and one for their double covers $\widehat{S}_n$ involving the twisted algebra of type $A_{p-1}^{(2)}$. In the case of S_n itself, the theory has been developed especially by Kleshchev [19], Lascoux-Leclerc-Thibon [21], Ariki [1] and Grojnowski [9], while the double covers are treated for the first time in [4] along the lines of [9], after the important progress made over $\mathbb{C}$ by Sergeev [35, 36] and Nazarov [30, 31].

One of the most striking results at the heart of both of the theories is the explicit description of the modular branching graphs in terms of Kashiwara's crystal graph for the basic module of the corresponding affine Lie algebra. Note that the results described are just a part of a larger picture: there are analogous results for the cyclotomic and affine Hecke algebras, and their twisted analogues, the cyclotomic and affine Hecke-Clifford superalgebras. However we will try here to bring out only those parts of the theory that apply to the symmetric group, since that is the most applicable to finite group theory.

2. The symmetric group

In this section, we describe the representation theory of the symmetric group S_n over a field F of arbitrary characteristic p.

2.1. Formal characters.

For $k = 1, \ldots, n$, we define the *Jucys-Murphy element*

$$x_k := \sum_{i=1}^{k-1} (i\ k) \in FS_n, \tag{1}$$

see [15, 28]. It is straightforward to show that the elements $x_1, x_2, \ldots, x_n$ commute with one another. Moreover, we have by [15] or [29, 1.9]:

Theorem 2.1. *The center of the group algebra FS_n is precisely the set of all symmetric polynomials in the elements $x_1, x_2, \ldots, x_n$.*

Now let M be an FS_n-module. Let $I = \mathbb{Z}/p\mathbb{Z}$ identified with the prime subfield of F. For $\underline{i} = (i_1, \ldots, i_n) \in I^n$, define

$$M[\underline{i}] := \{v \in M \mid (x_r - i_r)^N v = 0 \text{ for } N \gg 0 \text{ and each } r = 1, \ldots, n\}.$$

Thus, $M[\underline{i}]$ is the simultaneous generalized eigenspace for the commuting operators $x_1, \dots, x_n$ corresponding to the eigenvalues $i_1, \dots, i_n$ respectively.

Lemma 2.2. *Any FS_n-module M decomposes as $M = \bigoplus_{\underline{i} \in I^n} M[\underline{i}]$.*

Proof. It suffices to show that all eigenvalues of x_r on M lie in I, for each $r = 1, \dots, n$. This is obvious if $r = 1$ (as $x_1 = 0$). Now assume that all eigenvalues of x_r on M lie in I, and consider x_{r+1}. Let $v \in M$ be a simultaneous eigenvector for the commuting operators x_r and x_{r+1}. Consider the subspace N spanned by v and $s_r v$.

Suppose that N is two dimensional. Then the matrix for the action of x_r on N with respect to the basis $\{v, s_r v\}$ is $\begin{pmatrix} i & c \\ 0 & j \end{pmatrix}$ for some $i, j \in I$ and $c \in F$ (by assumption on the eigenvalues of x_r). Hence, the matrix for the action of $x_{r+1} = s_r x_r s_r + s_r$ on N is $\begin{pmatrix} j & 1 \\ c+1 & i \end{pmatrix}$. Since v was an eigenvector for x_{r+1}, we see that $c = -1$, hence v has eigenvalue j for x_{r+1} as required.

Finally suppose that N is one dimensional. Then, $s_r v = \pm v$. Hence, if $x_r v = iv$ for $i \in I$, then $x_{r+1} v = (s_r x_r s_r + s_r) v = (i \pm 1) v$. Since $i \pm 1 \in I$, we are done. □

We define the *formal character* ch M of a finite dimensional FS_n-module M to be

$$\operatorname{ch} M := \sum_{\underline{i} \in I^n} \dim(M[\underline{i}]) e^{\underline{i}}, \tag{2}$$

an element of the free $\mathbb{Z}$-module on basis $\{e^{\underline{i}} \mid \underline{i} \in I^n\}$. This is a useful notion, since ch is clearly additive on short exact sequences and we have the following important result proved in [38, §5.5]:

Theorem 2.3. *The formal characters of the inequivalent irreducible FS_n-modules are linearly independent.*

Given $\underline{i} = (i_1, \dots, i_n) \in I^n$, define its *weight* $\mathrm{wt}(\underline{i})$ to be the tuple $\gamma = (\gamma_i)_{i \in I}$ where γ_j counts the number of i_r $(r = 1, \dots, n)$ that equal j. Thus, γ is an element of the set Γ_n of I-tuples of non-negative integers summing to n. Clearly $\underline{i}, \underline{j} \in I^n$ lie in the same S_n-orbit (under the obvious action by place permutation) if and only if $\mathrm{wt}(\underline{i}) = \mathrm{wt}(\underline{j})$, hence Γ_n parametrizes the S_n-orbits on I^n.

For $\gamma \in \Gamma_n$ and an FS_n-module M, we let

$$M[\gamma] := \sum_{\underline{i} \in I^n \textit{ with } \mathrm{wt}(\underline{i}) = \gamma} M[\underline{i}]. \tag{3}$$

Unlike the $M[\underline{i}]$, the subspaces $M[\gamma]$ are actually FS_n-submodules of M. Indeed, as an elementary consequence of Theorem 2.1 and Lemma 2.2, we have:

Lemma 2.4. *The decomposition $M = \bigoplus_{\gamma \in \Gamma_n} M[\gamma]$ is precisely the decomposition of M into blocks as an FS_n-module.*

We will say that an FS_n-module M *belongs to the block* γ if $M = M[\gamma]$.

2.2. **Induction and restriction operators.** Now that we have the notion of formal character, we can introduce the *i-restriction* and *i-induction operators* e_i and f_i. Suppose that $\gamma \in \Gamma_n$. Let $\gamma + i \in \Gamma_{n+1}$ be the tuple $(\delta_i)_{i \in I}$ with $\delta_j = \gamma_j$ for

$j \neq i$ and $\delta_i = \gamma_i + 1$. Similarly, assuming this time that $\gamma_i > 0$, let $\gamma - i \in \Gamma_{n-1}$ be the tuple $(\delta_i)_{i \in I}$ with $\delta_j = \gamma_j$ for $j \neq i$ and $\delta_i = \gamma_i - 1$.

If M is an FS_n-module belonging to the block $\gamma \in \Gamma_n$, define

$$e_i M := (\mathrm{res}^{S_n}_{S_{n-1}} M)[\gamma - i] \quad \text{(interpreted as 0 in case } \gamma_i = 0), \tag{4}$$

$$f_i M := (\mathrm{ind}^{S_{n+1}}_{S_n} M)[\gamma + i]. \tag{5}$$

Extending additively to arbitrary FS_n-modules M using Lemma 2.4 and making the obvious definition on morphisms, we obtain exact functors

$$e_i : FS_n\text{-mod} \to FS_{n-1}\text{-mod} \qquad \text{and} \qquad f_i : FS_n\text{-mod} \to FS_{n+1}\text{-mod}.$$

The definition implies:

Lemma 2.5. *For an FS_n-module M we have*

$$\mathrm{res}^{S_n}_{S_{n-1}} M \cong \bigoplus_{i \in I} e_i M, \qquad \mathrm{ind}^{S_{n+1}}_{S_n} M \cong \bigoplus_{i \in I} f_i M.$$

Note that $e_i M$ can be described alternatively as the generalized eigenspace of x_n acting on M corresponding to the eigenvalue i. This means that the effect of e_i on characters is easy to describe:

$$\text{if} \quad \mathrm{ch}\, M = \sum_{\underline{i} \in I^n} a_{\underline{i}} e^{\underline{i}} \qquad \text{then} \qquad \mathrm{ch}\,(e_i M) = \sum_{\underline{i} \in I^{n-1}} a_{(i_1, \dots, i_{n-1}, i)} e^{\underline{i}}. \tag{6}$$

Let us also mention that there are higher *divided power functors* $e_i^{(r)}, f_i^{(r)}$ for each $r \geq 1$. To define them, start with an FS_n-module M belonging to the block γ. Let $\gamma + i^r = \gamma + i + i + \cdots + i$ (r times), and define $\gamma - i^r$ similarly (assuming $\gamma_i \geq r$). View M instead as an $F(S_n \times S_r)$-module by letting S_r act trivially. Embedding $S_n \times S_r$ into S_{n+r} in the obvious way, we then define

$$f_i^{(r)} M := (\mathrm{ind}^{S_{n+r}}_{S_n \times S_r} M)[\gamma + i^r]. \tag{7}$$

Extending additively, we obtain the functor $f_i^{(r)} : FS_n\text{-mod} \to FS_{n+r}\text{-mod}$. This exact functor has a two-sided adjoint $e_i^{(r)} : FS_{n+r}\text{-mod} \to FS_n\text{-mod}$. It is defined on a module M belonging to block γ by

$$e_i^{(r)} M := (M^{S_r})[\gamma - i^r] \quad \text{(interpreted as zero if } \gamma_i < r), \tag{8}$$

where M^{S_r} denotes the space of fixed points for the subgroup $S_r < S_{n+r}$ that permutes $n+1, \dots, n+r$, viewed as a module over the subgroup $S_n < S_{n+r}$ that permutes $1, \dots, n$. The following lemma relates the divided power functors $e_i^{(r)}$ and $f_i^{(r)}$ to the original functors e_i, f_i:

Lemma 2.6. *For an FS_n-module M we have*

$$e_i^r M \cong (e_i^{(r)} M)^{\oplus r!}, \qquad f_i^r M \cong (f_i^{(r)} M)^{\oplus r!}.$$

The functors $e_i^{(r)}$ and $f_i^{(r)}$ have been defined in an entirely different way by Grojnowski [9, §8.1], which is the key to proving their properties including Lemma 2.6.

2.3. **The affine Kac-Moody algebra.** Let R_n denote the character ring of FS_n, i.e. the free $\mathbb{Z}$-module spanned by the formal characters of the irreducible FS_n-modules. In view of Theorem 2.3, the map ch induces an isomorphism between R_n and the Grothendieck group of the category of all finite dimensional FS_n-modules. Similarly, let R_n^* denote the $\mathbb{Z}$-submodule of R_n spanned by the formal characters of the projective indecomposable FS_n-modules. This time, the map ch induces an isomorphism between R_n^* and the Grothendieck group of the category of all finite dimensional projective FS_n-modules.

Let

$$R = \bigoplus_{n\geq 0} R_n, \qquad R^* = \bigoplus_{n\geq 0} R_n^* \subseteq R. \tag{9}$$

The exact functors e_i and f_i induce $\mathbb{Z}$-linear operators on R. Since induction and restriction send projective modules to projective modules, Lemma 2.5 implies that e_i and f_i do too. Hence, $R^* \subseteq R$ is invariant under the action of e_i and f_i.

Extending scalars we get $\mathbb{C}$-linear operators e_i and f_i on $R_{\mathbb{C}} := \mathbb{C}\otimes_{\mathbb{Z}} R = \mathbb{C}\otimes_{\mathbb{Z}} R^*$. There is also a non-degenerate symmetric bilinear form on $R_{\mathbb{C}}$, the usual *Cartan pairing*, with respect to which the characters of the projective indecomposables and the irreducibles form a pair of dual bases.

Theorem 2.7. *The operators e_i and f_i ($i \in I$) on $R_{\mathbb{C}}$ satisfy the defining relations of the Chevalley generators of the affine Kac-Moody Lie algebra $\mathfrak{g}$ of type $A_{p-1}^{(1)}$ (resp. A_∞ in case $p = 0$), see [16]. Moreover, viewing $R_{\mathbb{C}}$ as a $\mathfrak{g}$-module in this way,*

(i) *$R_{\mathbb{C}}$ is isomorphic to the basic representation $V(\Lambda_0)$ of $\mathfrak{g}$, generated by the highest weight vector e^0 (the character of the irreducible FS_0-module);*
(ii) *the decomposition of $R_{\mathbb{C}}$ into blocks coincides with its weight space decomposition with respect to the standard Cartan subalgebra of $\mathfrak{g}$;*
(iii) *the Cartan pairing on $R_{\mathbb{C}}$ coincides with the Shapovalov form satisfying $(e^0, e^0) = 1$;*
(iv) *the lattice $R^* \subset R_{\mathbb{C}}$ is the $\mathbb{Z}$-submodule of $R_{\mathbb{C}}$ generated by e^0 under the action of the operators $f_i^{(r)} = f_i^r/r!$ $(i \in I, r \geq 0)$;*
(v) *the lattice $R \subset R_{\mathbb{C}}$ is the dual lattice to R^* under the Shapovalov form.*

This was essentially proved by Lascoux-Leclerc-Thibon [21] and Ariki [1] (for a somewhat different situation), and another approach has been given more recently by Grojnowski [9, 14.2],[10].

2.4. **The crystal graph.** In view of Theorem 2.7, we can identify $R_{\mathbb{C}}$ with the basic representation of the affine Kac-Moody algebra $\mathfrak{g} = A_{p-1}^{(1)}$. Associated to this highest weight module, Kashiwara has defined a purely combinatorial object known as a *crystal*, see e.g. [18] for a survey of this amazing theory. We now review the explicit description of this particular crystal, due originally to Misra and Miwa [26]. This contains all the combinatorial notions we need to complete our exposition of the representation theory.

Let $\lambda = (\lambda_1 \geq \lambda_2 \geq)$ be a partition. We identify λ with its *Young diagram*

$$\lambda = \{(r, s) \in \mathbb{Z}_{>0} \times \mathbb{Z}_{>0} \mid s \leq \lambda_r\}.$$

Elements $(r,s) \in \mathbb{Z}_{>0} \times \mathbb{Z}_{>0}$ are called *nodes*. We label each node $A = (r,s)$ of λ with its *residue* $\operatorname{res} A \in I$ defined so that $\operatorname{res} A \equiv (s-r) \pmod{p}$, see Example 2.8 below.

Let $i \in I$ be some fixed residue. A node $A \in \lambda$ is called *i-removable* (for λ) if

(R0) $\operatorname{res} A = i$ and $\lambda - \{A\}$ is the diagram of a partition.

Similarly, a node $B \notin \lambda$ is called *i-addable* (for λ) if

(A0) $\operatorname{res} B = i$ and $\lambda \cup \{B\}$ is the diagram of a partition.

Now label all i-addable nodes of the diagram λ by $+$ and all i-removable nodes by $-$. The *i-signature* of λ is the sequence of pluses and minuses obtained by going along the rim of the Young diagram from bottom left to top right and reading off all the signs. The *reduced i-signature* of λ is obtained from the i-signature by successively erasing all neighbouring pairs of the form $-+$.

Example 2.8. Let $p = 3$ and $\lambda = (11, 10, 9, 9, 5, 1)$. The residues are as follows:

0	1	2	0	1	2	0	1	2	0	1
2	0	1	2	0	1	2	0	1	2	
1	2	0	1	2	0	1	2	0		
0	1	2	0	1	2	0	1	2		
2	0	1	2	0						
1										

The 2-addable and 2-removable nodes are as labelled in the diagram:

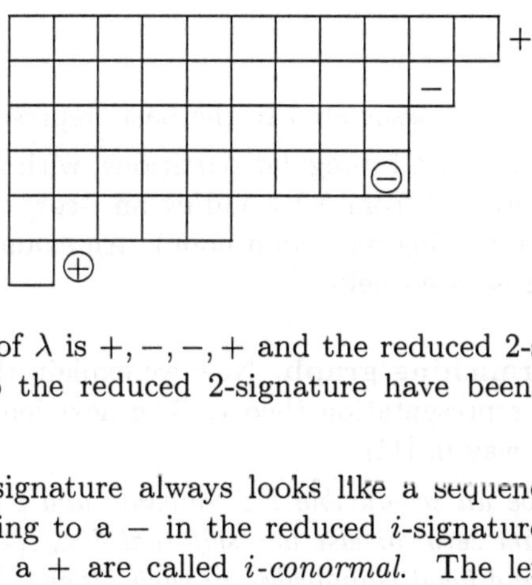

Hence, the 2-signature of λ is $+, -, -, +$ and the reduced 2-signature is $+, -$ (the nodes corresponding to the reduced 2-signature have been circled in the above diagram).

Note the reduced i-signature always looks like a sequence of $+$'s followed by $-$'s. Nodes corresponding to a $-$ in the reduced i-signature are called *i-normal*, nodes corresponding to a $+$ are called *i-conormal*. The leftmost i-normal node (corresponding to the leftmost $-$ in the reduced i-signature) is called *i-good*, and the rightmost i-conormal node (corresponding to the rightmost $+$ in the reduced i-signature) is called *i-cogood*.

We recall finally that a partition λ is called *p-regular* if it does not have p non-zero equal parts. It is important to note that if λ is p-regular and A is an i-good node, then $\lambda - \{A\}$ is also p-regular. Similarly if B is an i-cogood node, then $\lambda \cup \{B\}$ is p-regular.

The basic crystal graph of type $A_1^{(1)}$

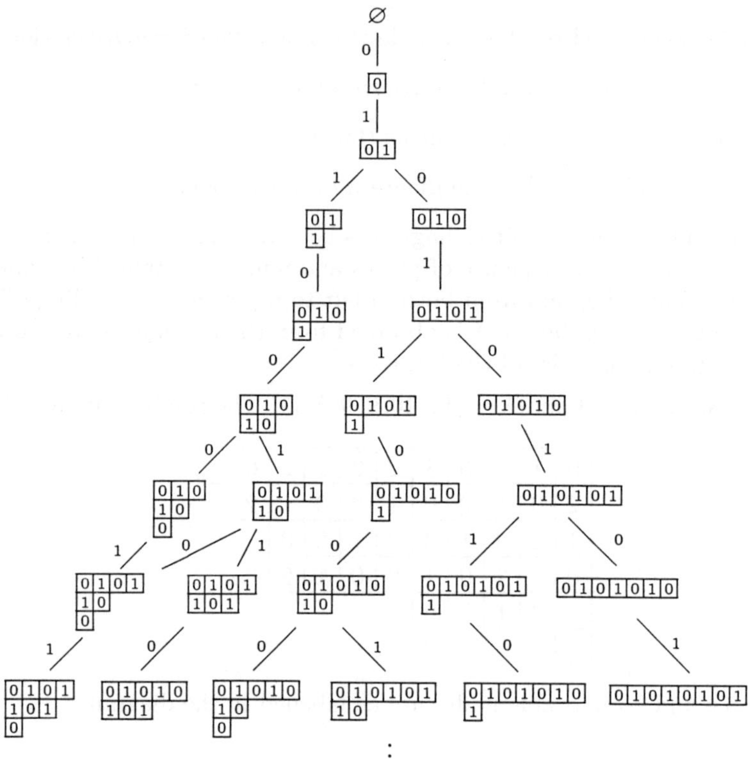

By [26], the *crystal graph* associated to the basic representation $V(\Lambda_0)$ of $\mathfrak{g}$ can now be realized as the set of all p-regular partitions, with a directed edge $\lambda \xrightarrow{i} \mu$ of color $i \in I$ if μ is obtained from λ by adding an i-cogood node (equivalently, λ is obtained from μ by removing an i-good node). An example showing part of the crystal graph for $p = 2$ is listed below.

2.5. **The modular branching graph.** Now we explain the relationship between the crystal graph and representation theory. The next lemma was first proved in [20], and in a different way in [11].

Lemma 2.9. *Let D be an irreducible FS_n-module and $i \in I$. Then, the module $e_i D$ (resp. $f_i D$) is either zero, or else is a self-dual FS_{n-1}- (resp. FS_{n+1}-) module with irreducible socle and head isomorphic to each other.*

Introduce the *crystal operators* $\tilde{e}_i, \tilde{f}_i$: for an irreducible FS_n-module D, let

$$\tilde{e}_i D := \mathrm{socle}(e_i D), \qquad \tilde{f}_i D := \mathrm{socle}(f_i D). \tag{10}$$

In view of Lemma 2.9, $\tilde{e}_i D$ and $\tilde{f}_i D$ are either zero or irreducible. Now define the *modular branching graph*: the vertices are the isomorphism classes of irreducible FS_n-modules for all $n \geq 0$, and there is a directed edge $[D] \xrightarrow{i} [E]$ of color i if

$E \cong \tilde{f}_i D$ (equivalently by Frobenius reciprocity, $D \cong \tilde{e}_i E$). The fundamental result is the following:

Theorem 2.10. *The modular branching graph is uniquely isomorphic (as an I-colored, directed graph) to the crystal graph of §2.4.*

This theorem was first stated in this way by Lascoux, Leclerc and Thibon [21]: they noticed that the combinatorics of Kashiwara's crystal graph as described by Misra and Miwa [26] is exactly the same as the modular branching graph first determined in [19]. A quite different and independent proof of Theorem 2.10 follows from the more general results of [9].

Theorem 2.10 has some important consequences. To start with, it implies that the isomorphism classes of irreducible FS_n-modules are parametrized by the vertices in the crystal graph, i.e. by p-regular partitions. For a p-regular partition λ of n, we let D^λ denote the corresponding irreducible FS_n-module. To be quite explicit about this labelling, choose a path

$$\varnothing \xrightarrow{i_1} \square \xrightarrow{i_2} \cdots \xrightarrow{i_n} \lambda$$

in the crystal graph from the empty partition to λ, for $i_1, \dots, i_n \in I$. Then,

$$D^\lambda := \tilde{f}_{i_n} \dots \tilde{f}_{i_1} D^\varnothing, \tag{11}$$

where $D^\varnothing$ denotes the irreducible FS_0-module. Note the labelling of the irreducible module D^λ defined here is known to agree with the standard labelling of James [13], although James' construction is quite different.

Let us state one more result about the structure of the modules $e_i D^\lambda$ and $f_i D^\lambda$, see [2, Theorems E, E′] for this and some other more detailed results.

Theorem 2.11. *Let λ be a p-regular partition of n.*

(i) *Suppose that A is an i-removable node such that $\mu := \lambda - \{A\}$ is p-regular. Then, $[e_i D^\lambda : D^\mu]$ is the number of i-normal nodes to the right of A (counting A itself), or 0 if A is not i-normal.*

(ii) *Suppose that B is an i-addable node such that $\nu := \lambda \cup \{B\}$ is p-regular. Then, $[f_i D^\lambda : D^\nu]$ is the number of i-conormal nodes to the left of B (counting B itself), or 0 if B is not i-conormal.*

2.6. **More on characters.** Let M be an FS_n-module. Define

$$\varepsilon_i(M) = \max\{r \geq 0 \mid e_i^r M \neq 0\} \qquad \varphi_i(M) = \max\{r \geq 0 \mid f_i^r M \neq 0\}. \tag{12}$$

Note $\varepsilon_i(M)$ can be computed just from knowledge of the character of M: it is the maximal r such that $e^{(\dots,i^r)}$ appears with non-zero coefficient in $\operatorname{ch} M$. Less obviously, $\varphi_i(M)$ can also be read off from the character of M. By additivity of f_i, we may assume that M belongs to the block $\gamma \in \Gamma_n$. Then

$$\varphi_i(M) = \varepsilon_i(M) + \delta_{i,0} - 2\gamma_i + \gamma_{i-1} + \gamma_{i+1}, \tag{13}$$

see [9, 12.6]. We note the following extremely useful lemma from [11], see also [9, §9]:

Lemma 2.12. *Let D be an irreducible FS_n-module, $\varepsilon = \varepsilon_i(D), \varphi = \varphi_i(D)$. Then, $e_i^{(\varepsilon)} D \cong \tilde{e}_i^\varepsilon D$, $f_i^{(\varphi)} D \cong \tilde{f}_i^\varphi D$.*

The lemma implies that

$$\varepsilon_i(D) = \max\{r \geq 0 \mid \tilde{e}_i^r D \neq 0\}, \qquad \varphi_i(D) = \max\{r \geq 0 \mid \tilde{f}_i^r D \neq 0\}.$$

Thus, $\varepsilon_i(D)$ can also be read off directly from the combinatorics: if $D \cong D^\lambda$, then $\varepsilon_i(D)$ is the number of '$-$'s in the reduced i-signature of λ. Similarly, $\varphi_i(D)$ is the number of '$+$'s in the reduced i-signature of λ.

Now we can describe an inductive algorithm to determine the label of an irreducible FS_n-module D purely from knowledge of its character $\mathrm{ch}\, D$. Pick $i \in I$ such that $\varepsilon := \varepsilon_i(D)$ is non-zero. Let $E = e_i^{(\varepsilon)} D$, an irreducible $FS_{n-\varepsilon}$-module with explicitly known character thanks to Lemmas 2.12, 2.6 and (6). By induction, the label of E can be computed purely from knowledge of its character, say $E \cong D^\lambda$. Then, $D \cong \tilde{f}_i^\varepsilon E \cong D^\mu$ where μ is obtained from λ by adding the rightmost ε of the i-conormal nodes.

We would of course like to be able to reverse this process: given a p-regular partition λ of n, we would like to be able to compute the character of the irreducible FS_n-module D^λ. One can compute a quite effective *lower bound* for this character inductively using the branching rules of Theorem 2.11. But only over $\mathbb{C}$ is this lower bound always correct: indeed if $p = 0$ then D^λ is equal to the Specht module S^λ and

$$\mathrm{ch}\, S^\lambda = \sum_{(i_1,\dots,i_n)} e^{(i_1,\dots,i_n)} \tag{14}$$

summing over all paths $\varnothing \xrightarrow{i_1} \square \xrightarrow{i_2} \cdots \xrightarrow{i_n} \lambda$ in the characteristic zero crystal graph (a.k.a. Young's partition lattice) from $\varnothing$ to λ. (Reducing the residues in (14) modulo p in the obvious way gives the formal characters of the Specht module in characteristic p.) We refer to [32] for a concise self-contained approach to the complex representation theory of S_n along the lines described here.

Now we explain how Lemma 2.12 can be used to describe some composition factors of Specht modules—this provides new non-trivial information on decomposition numbers which is difficult to obtain by other methods. The following result follows easily from Lemma 2.12.

Lemma 2.13. *Let M be an FS_n-module and set $\varepsilon = \varepsilon_i(M)$. If $[e_i^{(\varepsilon)} M : D^\mu] = m > 0$ then $\tilde{f}_i^\varepsilon D^\mu \neq 0$ and $[M : \tilde{f}_i^\varepsilon D^\mu] = m$.*

Example 2.14. Let $p = 3$. By [13, Tables], the composition factors of the Specht module $S^{(6,4,2,1)}$ are $D^{(12,1)}$, $D^{(9,4)}$, $D^{(9,2^2)}$, $D^{(7,4,2)}$, $D^{(6,5,2)}$, $D^{(6,4,3)}$, and $D^{(6,4,2,1)}$, all appearing with multiplicity 1. As $\varepsilon_1(S^{(6,4,2^2)}) = 1$ (by (14) reduced modulo 3) and $e_1 S^{(6,4,2^2)} = S^{(6,4,2,1)}$, application of Lemma 2.13 implies that the following composition factors appear in $S^{(6,4,2^2)}$ with multiplicity 1: $D^{(12,1^2)}$, $D^{(9,4,1)}$, $D^{(9,3,2)}$, $D^{(8,4,2)}$, $D^{(6^2,2)}$, $D^{(6,4^2)}$, and $D^{(6,4,2^2)}$.

Given $\underline{i} = (i_1, \dots, i_n) \in I^n$ we can gather consecutive equal terms to write it in the form

$$\underline{i} = (j_1^{m_1} \dots j_r^{m_r}) \tag{15}$$

where $j_s \neq j_{s+1}$ for all $1 \leq s < r$. For example $(2,2,2,1,1) = (2^3 1^2)$. Now, for an FS_n-module M, the tuple (15) is called *extremal* if

$$m_s = \varepsilon_{j_s}(e_{j_{s+1}}^{m_{s+1}} \dots e_{j_r}^{m_r} M)$$

for all $s = r, r-1, \ldots, 1$. Informally speaking this means that among all the n-tuples $\underline{i}$ such that $M[\underline{i}] \neq 0$ we first choose those with the longest j_r-string in the end, then among these we choose the ones with the longest j_{r-1}-string preceding the j_r-string in the end, etc. By definition $M[\underline{i}] \neq 0$ if $\underline{i}$ is extremal for M.

Example 2.15. The formal character of the Specht module $S^{(5,2)}$ in characteristic 3 is

$$e^{(0210201)} + 2e^{(0120201)} + 2e^{(02120^21)} + 4e^{(012^20^21)}$$
$$+e^{(0212010)} + 2e^{(012^2010)} + e^{(0120210)} + e^{(0120120)}.$$

The extremal tuples are (012^20^21), (012^2010), (0120210), and (0120120).

Our main result about extremal tuples is

Theorem 2.16. *Let $\underline{i} = (i_1, \ldots, i_n) = (j_1^{m_1} \ldots j_r^{m_r})$ be an extremal tuple for an irreducible FS_n-module D^λ. Then $D^\lambda = \tilde{f}_{i_n} \ldots \tilde{f}_{i_1} D^{\varnothing}$, and $\dim D^\lambda[\underline{i}] = m_1! \ldots m_r!$. In particular, the tuple $\underline{i}$ is not extremal for any irreducible $D^\mu \not\cong D^\lambda$.*

Proof. We apply induction on r. If $r = 1$, then by considering possible n-tuples appearing in the Specht module S^λ, of which D^λ is a quotient, we conclude that $n = 1$ and $D = D^{(1)}$. So for $r = 1$ the result is obvious. Let $r > 1$. By definition of an extremal tuple, $m_r = \varepsilon_{j_r}(D^\lambda)$. So, in view of Lemmas 2.6 and 2.12, we have

$$e_{j_r}^{m_r} D^\lambda = m_r! \tilde{e}_{j_r}^{m_r} D^\lambda.$$

Moreover, $(j_1^{m_1} \ldots j_{r-1}^{m_{r-1}})$ is clearly an extremal tuple for the irreducible module $\tilde{e}_{j_r}^{m_r} D^\lambda$. So the inductive step follows. □

Corollary 2.17. *If M is an FS_n-module and $\underline{i} = (i_1, \ldots, i_n) = (j_1^{m_1} \ldots j_r^{m_r})$ is an extremal tuple for M then the multiplicity of $D^\lambda := \tilde{f}_{i_n} \ldots \tilde{f}_{i_1} D^{\varnothing}$ as a composition factor of M is* $\dim M[\underline{i}]/(m_1! \ldots m_r!)$.

We note that for any tuple $\underline{i}$ represented in the form (15) and any FS_n-module M we have that $\dim M[\underline{i}]$ is divisible by $m_1! \ldots m_r!$. This follows from the properties of the principal series modules ('Kato modules') for degenerate affine Hecke algebras, see [11] for more details.

Example 2.18. In view of Corollary 2.17 extremal tuple (012^20^21) in Example 2.15 yields the composition factor $D^{(5,2)}$ of $S^{(5,2)}$, while the extremal tuple (0120120) yields the composition factor $D^{(7)}$. It turns out that these are exactly the composition factors of $S^{(5,2)}$, see e.g. [13, Tables].

For more non-trivial examples let us consider a couple of Specht modules for $n = 11$ in characteristic 3. For $S^{(6,3,1^2)}$, Corollary 2.17 yields composition factors $D^{(6,3,1^2)}$, $D^{(7,3,1)}$, and $D^{(8,2,1)}$ but 'misses' $D^{(11)}$, and for $S^{(4,3,2^2)}$ we get hold of $D^{(4,3,2^2)}$, $D^{(5,3,2,1)}$, $D^{(8,2,1)}$, and $D^{(8,3)}$, but 'miss' $2D^{(11)}$ and $D^{(5,4,1^2)}$, cf. [13, Tables].

We record here one other useful general fact about formal characters which follows from the Serre relations satisfied by the operators e_i:

Lemma 2.19. *Let M be an FS_n-module. Assume $i, j, i_1, \ldots, i_{n-2} \in I$ and $i \neq j$.*

(i) *Assume that* $|i-j|>1$. *Then for any* $1 \le r \le n-2$ *we have*

$$\begin{aligned} &\dim M[(i_1,\dots,i_r,i,j,i_{r+1},\dots,i_{n-2})] \\ =\ &\dim M[(i_1,\dots,i_r,j,i,i_{r+1},\dots,i_{n-2})]. \end{aligned}$$

(ii) *Assume that* $|i-j|=1$ *and* $p>2$. *Then for any* $1 \le r \le n-3$ *we have*

$$\begin{aligned} &2\dim M[(i_1,\dots,i_r,i,j,i,i_{r+1},\dots,i_{n-3})] \\ =\ &\dim M[(i_1,\dots,i_r,i,i,j,i_{r+1},\dots,i_{n-3})] \\ &+\dim M[(i_1,\dots,i_r,j,i,i,i_{r+1},\dots,i_{n-3})]. \end{aligned}$$

(iii) *Assume that* $|i-j|=1$ *and* $p=2$. *Then for any* $1 \le r \le n-4$ *we have*

$$\begin{aligned} &\dim M[(i_1,\dots,i_r,i,i,i,j,i_{r+1},\dots,i_{n-4})] \\ &+3\dim M[(i_1,\dots,i_r,i,j,i,i,i_{r+1},\dots,i_{n-4})] \\ =\ &\dim M[(i_1,\dots,i_r,j,i,i,i,i_{r+1},\dots,i_{n-4})] \\ &+3\dim M[(i_1,\dots,i_r,i,i,j,i,i_{r+1},\dots,i_{n-4})]. \end{aligned}$$

2.7. **Blocks.** Finally we discuss some properties of blocks, assuming now that $p \neq 0$. In view of Theorem 2.7(ii), the blocks of the FS_n for all n are in 1–1 correspondence with the non-zero weight spaces of the basic module $V(\Lambda_0)$ of $\mathfrak{g} = A^{(1)}_{p-1}$. So let us begin by describing these following [16, ch.12].

Let $P = \bigoplus_{i\in I} \mathbb{Z}\Lambda_i \oplus \mathbb{Z}\delta$ denote the weight lattice associated to $\mathfrak{g}$. Let α_i $(i \in I)$ be the simple roots of $\mathfrak{g}$, defined from

$$\alpha_0 = 2\Lambda_0 - \Lambda_1 - \Lambda_{p-1} + \delta, \qquad \alpha_i = 2\Lambda_i - \Lambda_{i+1} - \Lambda_{i-1} \quad (i \neq 0). \tag{16}$$

There is a positive definite symmetric bilinear form $(.|.)$ on $\mathbb{R} \otimes_{\mathbb{Z}} P$ with respect to which $\alpha_0,\dots,\alpha_{p-1},\Lambda_0$ and $\Lambda_0,\dots,\Lambda_{p-1},\delta$ form a pair of dual bases. Let W denote the Weyl group of $\mathfrak{g}$, the subgroup of $GL(\mathbb{R} \otimes_{\mathbb{Z}} P)$ generated by s_i $(i \in I)$, where s_i is the reflection in the hyperplane orthogonal to α_i. Then, by [16, (12.6.1)], the weight spaces of $V(\Lambda_0)$ are the weights

$$\{w\Lambda_0 - d\delta \mid w \in W,\ d \in \mathbb{Z}_{\ge 0}\}.$$

For a weight of the form $w\Lambda_0 - d\delta$, we refer to $w\Lambda_0$ as the corresponding *maximal weight*, and d as the corresponding *depth.*

There is a more combinatorial way of thinking of the weights. Following [24, I.1, ex.8] and [14, §2.7], to a p-regular partition λ one associates the corresponding p-core $\tilde{\lambda}$ and p-weight d: $\tilde{\lambda}$ is the partition obtained from λ by successively removing as many hooks of length p from the rim of λ as possible, in such a way that at each step the diagram of a partition remains. The number of p-hooks removed is the p-weight d of λ. The p-cores are in 1–1 correspondence with the maximal weights, i.e. the weights belonging to the W-orbit $W\Lambda_0$, and the p-weight corresponds to the notion of depth introduced in the previous paragraph, see [21, §5.3] and [22, §2] for the details.

Now Theorem 2.7(ii) gives yet another proof of the *Nakayama conjecture*: the FS_n-modules D^λ and D^μ belong to the same block if and only if λ and μ have the same p-core. We will also talk about the *p-weight* of a block B, namely, the p-weight of any λ such that D^λ belongs to B.

The Weyl group W acts on the $\mathfrak{g}$-module $R_{\mathbb{C}}$ from §2.3, the generator s_i ($i \in I$) of W acting by the familiar formula

$$s_i = \exp(-e_i)\exp(f_i)\exp(-e_i).$$

The resulting action preserves the Shapovalov form, and leaves the lattices R and R^* invariant. Moreover, W permutes the weight spaces of $R_{\mathbb{C}}$ in the same way as its defining action on the weight lattice P. Since W leaves δ invariant, it follows that the action is transitive on all weight spaces of the same depth. So using Theorem 2.7(iii) we see:

Theorem 2.20. *Let B and B' be blocks of symmetric groups with the same p-weight. Then, B and B' are isometric, in the sense that there is an isomorphism between their Grothendieck groups that is an isometry with respect to the Cartan form.*

The existence of such isometries was first noticed by Enguehard [8]. Implicit in Enguehard's paper is the following conjecture, made formally by Rickard: *blocks B and B' of symmetric groups with the same p-weight should be derived equivalent.* This has been proved by Rickard for blocks of p-weight ≤ 5. Moreover, it is now known by work of Marcus [25] and Chuang-Kessar [6] that the famous Abelian Defect Group Conjecture of Broué for symmetric groups follows from the Rickard's conjecture above.

There is one situation that is particularly straightforward, when there is actually a *Morita* equivalence between blocks of the same p-weight. This is a theorem of Scopes [34], though we are stating the result in a more Lie theoretic way following [22, §8]:

Theorem 2.21. *Let $\Lambda, \Lambda + \alpha_i, \dots, \Lambda + r\alpha_i$ be an α_i-string of weights of $V(\Lambda_0)$ (so $\Lambda - \alpha_i$ and $\Lambda + (r+1)\alpha_i$ are not weights of $V(\Lambda_0)$). Then the functors $f_i^{(r)}$ and $e_i^{(r)}$ define mutually inverse Morita equivalences between the blocks parametrized by Λ and by $\Lambda + r\alpha_i$.*

Proof. Since $e_i^{(r)}$ and $f_i^{(r)}$ are both left and right adjoint to one another, it suffices to check that $e_i^{(r)}$ and $f_i^{(r)}$ induce mutually inverse bijections between the isomorphism classes of irreducible modules belonging to the respective blocks. This follows by Lemma 2.12. □

Let us end the discussion with one new result here: we can in fact explicitly compute the determinant of the Cartan matrix of a block. The details of the proof will appear in [5]. Note in view of Theorem 2.20, the determinant of the Cartan matrix only depends on the p-weight of the block. Moreover, by Theorem 2.7(iii), we can work instead in terms of the Shapovalov form on $V(\Lambda_0)$. Using the explicit construction of the latter module over $\mathbb{Z}$ given in [7], we show:

Theorem 2.22. *Let B be a block of p-weight d of FS_n. Then the determinant of the Cartan matrix of B is p^N where*

$$N = \sum_{\lambda=(1^{r_1}2^{r_2}\dots)\vdash d} \frac{r_1 + r_2 + \dots}{p-1} \binom{p-2+r_1}{r_1}\binom{p-2+r_2}{r_2}\cdots.$$

3. The double covers

We turn now to the representation theory of the group algebra $F\widehat{S}_n$, where $\widehat{S}_n$ denotes one of the double covers of the symmetric group and F is a field of characteristic $p \neq 2$. We will assume that F contains square roots of (the images of) all integers, since that ensures that F is a splitting field for $\widehat{S}_n$ for all n (see [3, Remark 10.5]).

3.1. Analogues of the Jucys-Murphy elements. For definiteness, we work with the double cover $\widehat{S}_n$ defined by generators $\zeta, \hat{s}_1, \dots, \hat{s}_{n-1}$ subject to the relations

$$\zeta^2 = 1, \quad \zeta\hat{s}_i = \hat{s}_i\zeta, \quad \hat{s}_i^2 = 1, \quad \hat{s}_i\hat{s}_j = \zeta\hat{s}_j\hat{s}_i, \quad \hat{s}_i\hat{s}_{i+1}\hat{s}_i = \hat{s}_{i+1}\hat{s}_i\hat{s}_{i+1},$$

for all admissible i, j with $|i-j| > 1$. Note right away that $1 = \zeta_+ + \zeta_-$ is a decomposition of the identity as a sum of mutually orthogonal central idempotents, where $\zeta_\pm = (1 \mp \zeta)/2$. So we can decompose

$$F\widehat{S}_n = \zeta_+ F\widehat{S}_n \oplus \zeta_- F\widehat{S}_n.$$

The algebra $\zeta_+ F\widehat{S}_n$ is isomorphic to the group algebra FS_n itself, so we focus our attention instead on the summand $S(n) := \zeta_- F\widehat{S}_n$.

The algebra $S(n)$ is the *twisted group algebra* of S_n over F. It can be realized directly as the algebra generated by the elements $t_i := \zeta_-\hat{s}_i$ subject only to the relations

$$t_i^2 = 1, \quad t_it_j = -t_jt_i, \quad t_it_{i+1}t_i = t_{i+1}t_it_{i+1},$$

for admissible i, j with $|i-j| > 1$. For $1 \leq i < j \leq n$, let

$$[i\, j] = -[j\, i] = (-1)^{j-i-1}t_{j-1}\dots t_{i+1}t_it_{i+1}\dots t_{j-1}. \tag{17}$$

These 'transpositions' satisfy the relations

$$[i\, j]^2 = 1, \quad [i\, j][k\, l] = -[k\, l][i\, j] \text{ if } \{i, j\} \cap \{k, l\} = \varnothing,$$
$$[i\, j][j\, k][i\, j] = [j\, k][i\, j][j\, k] = [k\, i] \text{ for distinct } i, j, k$$

(cf. [36, (1.1)]). Finally, for distinct $1 \leq i_1, \dots, i_r \leq n$, let

$$[i_1\, i_2\, \dots\, i_r] = (-1)^{r-1}[i_2\, \dots\, i_r\, i_1] = [i_{r-1}\, i_r][i_{r-2}\, i_r]\dots[i_1\, i_r],$$

giving the 'r-cycles'.

For $1 \leq k \leq n$, the analogue of the Jucys-Murphy element is

$$y_k := \sum_{i=1}^{k-1} [i\, k], \tag{18}$$

in particular, $y_1 = 0$. This definition appears in Sergeev [36]; Nazarov originally used a different approach [31]. One checks using the relations that:

$$t_iy_k = \begin{cases} -y_kt_i & \text{if } i \neq k-1, k, \\ -y_{k-1}t_i + 1 & \text{if } i = k-1, \\ -y_{k+1}t_i + 1 & \text{if } i = k. \end{cases} \tag{19}$$

It follows that $y_ky_l = -y_ly_k$ if $k \neq l$. Now using these facts, it is easy to show:

(a) for $1 \leq k, l \leq n$, y_k^2 and y_l^2 commute;
(b) t_i commutes with y_k^2 for $k \neq i, i+1$;
(c) t_i commutes with $y_i^2 + y_{i+1}^2$ and $y_i^2y_{i+1}^2$.

This implies:

Lemma 3.1. *The symmetric polynomials in the elements $y_1^2, y_2^2, \dots, y_n^2$ belong to the center of $S(n)$.*

However it is not in general true that center of $S(n)$ *equals* the set of symmetric polynomials in the $y_1^2, \dots, y_n^2$. We need to view $S(n)$ instead as a $\mathbb{Z}_2$-graded algebra – a *superalgebra* $S(n) = S(n)_{\bar{0}} \oplus S(n)_{\bar{1}}$ – by declaring that the generators t_i are odd. Then:

Theorem 3.2. *The even center of $S(n)$ (i.e. the space $Z(S(n))_{\bar{0}}$ of all central elements of degree $\bar{0}$) is the set of all symmetric polynomials in the $y_1^2, \dots, y_n^2$.*

Proof. For each $w \in S_n$, make a fixed choice for a representation of w as a product of disjoint cycles (all of length > 1). If $w = (i_1 \ \dots \ i_a)(j_1 \ \dots \ j_b) \dots$ is this choice, define $[w] := [i_1 \ \dots \ i_a][j_1 \ \dots \ j_b] \cdots \in S(n)$. The $\{[w] \mid w \in S_n\}$ then form a basis for $S(n)$. We will say that $w \in S_n$ appears in $x \in S(n)$ if the coefficient of $[w]$ is non-zero when x is expanded in terms of this basis.

Let $\lambda = (\lambda_1 \geq \cdots \geq \lambda_h > 0)$ be an *odd* partition of n, i.e. all its non-zero parts are odd. Define

$$p_\lambda := \sum_{w \in S_n / S_\lambda} y_{w1}^{\lambda_1 - 1} y_{w2}^{\lambda_2 - 1} \dots y_{wh}^{\lambda_h - 1} \in S(n),$$

where S_λ denotes the stabilizer of the n-tuple $(\lambda_1 - 1, \lambda_2 - 1, \dots, \lambda_h - 1, 0, \dots, 0)$ under the natural action of S_n on n-tuples by place permutation. Also let

$$u_\lambda := (n - \hat{\lambda}_1 + 1 \ \dots \ n)(n - \hat{\lambda}_2 + 1 \ \dots \ n - \hat{\lambda}_1) \dots \\ (n - \hat{\lambda}_h + 1 \ \dots \ n - \hat{\lambda}_{h-1}) \in S_n,$$

where $\hat{\lambda}_i = \lambda_1 + \cdots + \lambda_i$. Fix a total order $>$ on the odd partitions of n so that $\lambda > \mu$ if either λ has more non-zero parts than μ, or if λ, μ have the same number of non-zero parts but $\lambda \rhd \mu$ in the usual dominance ordering. By exactly the same argument as in the proof of [29, 1.9], $[u_\lambda]$ appears in p_λ but not in any p_μ with $\mu > \lambda$. This implies that the p_λ are linearly independent, as λ runs over all odd partitions of n.

Finally, the p_λ are symmetric polynomials in the $y_1^2, \dots, y_n^2$ by definition. So we have shown that the dimension of the subspace of $S(n)$ spanned by the symmetric polynomials in the $y_1^2, \dots, y_n^2$ is at least the number of odd partitions of n. On the other hand, by consideration of the conjugacy classes of even elements in $\widehat{S}_n$ (see [37, 2.1] or [33, p.172]), $\dim Z(S(n))_{\bar{0}}$ is equal to the number of odd partitions of n. So an application of Lemma 3.1 completes the proof. $\square$

3.2. **Formal characters.** Motivated in part by Theorem 3.2, we will be interested from now on in the $S(n)$-*supermodules*, i.e. the $\mathbb{Z}_2$-graded $S(n)$-modules where $S(n)$ is viewed as a $\mathbb{Z}_2$-graded algebra as before. We refer to [3, §2] for basic notions here. Let us just recall here that there are two sorts of irreducible $S(n)$-supermodule D: type $\mathtt{M}$ or type $\mathtt{Q}$ according to whether the endomorphism algebra $\mathrm{End}_{S(n)}(D)$ is one or two dimensional. In case D has type $\mathtt{M}$, it is irreducible when viewed as an ordinary $S(n)$-module. But if D is of type $\mathtt{Q}$ it decomposes as $D = D_+ \oplus D_-$ where $D_\pm$ are non-isomorphic irreducible $S(n)$-submodules – but not subsupermodules – of D. Providing we keep track at all times of the *type* of an irreducible supermodule, we can easily recover results about ordinary representation theory. Incidentally, if D has type $\mathtt{M}$ then $D \cong D \otimes \mathrm{sgn}$, and if D has type $\mathtt{Q}$ then $D_+ \cong D_- \otimes \mathrm{sgn}$.

Now we proceed along the lines of §2.1. Let $\ell = (p-1)/2$ (resp. $\ell = \infty$ if $p = 0$). Let $I = \{0, 1, \dots, \ell\}$. Given a tuple $\underline{i} = (i_1, \dots, i_n) \in I^n$ and an $S(n)$-supermodule M, we let

$$M[\underline{i}] = \left\{ v \in M \;\middle|\; \left(y_r^2 - \frac{i_r(i_r+1)}{2} \right)^N v = 0 \text{ for } N \gg 0 \text{ and } r = 1, \dots, n \right\}.$$

Lemma 3.3. *Any $S(n)$-supermodule M decomposes as $M = \bigoplus_{\underline{i} \in I^n} M[\underline{i}]$.*

Proof. This follows from [4, 4.9,9.9] on noting that the image of our element y_k^2 under the map $\varphi : S(n) \to W(n)$ from [4, 9.8] is equal to one half of the image of the element denoted x_k^2 in [4]. □

We let Γ_n denote the set of all I-tuples of non-negative integers summing to n, and define the weight of $\underline{i} \in I^n$ in the same way as in §2.1. Given $\gamma \in \Gamma_n$ and an $S(n)$-supermodule M, we set

$$M[\gamma] := \sum_{\underline{i} \in I^n \text{ with } \mathrm{wt}(\underline{i}) = \gamma} M[\underline{i}]$$

as before. Theorem 3.2 and Lemma 3.3 imply:

Lemma 3.4. *The decomposition $M = \bigoplus_{\gamma \in \Gamma_n} M[\gamma]$ is the precisely the decomposition of M into superblocks as an $S(n)$-supermodule.*

Now fix $\underline{i} \in I^n$ of weight γ. Consider the *Clifford superalgebra* with odd generators $c_1, \dots, c_n$ subject to the relations

$$c_r c_s = -c_s c_r \quad (r \neq s), \qquad c_r^2 = \frac{i_r(i_r+1)}{2}. \tag{20}$$

By [3, 2.7,2.9,2.10], it has a unique irreducible supermodule $U(\underline{i})$, of type $\mathtt{M}$ if $(n - \gamma_0)$ is even, type $\mathtt{Q}$ if $(n - \gamma_0)$ is odd. Moreover,

$$\dim U(\underline{i}) = 2^{\lfloor \frac{n-\gamma_0+1}{2} \rfloor}. \tag{21}$$

Now suppose that M is an $S(n)$-supermodule. The subspace $M[\underline{i}]$ is obviously invariant under the action of the subalgebra of $S(n)$ generated by the y_k. Moreover, these y_k satisfy the above relations (20) on every irreducible constituent of $M[\underline{i}]$. This shows that $\dim M[\underline{i}]$ is divisible by $\dim U(\underline{i})$. Now define the *formal character* of M by

$$\mathrm{ch}\, M := \sum_{\underline{i} \in I^n} \frac{\dim M[\underline{i}]}{\dim U(\underline{i})} e^{\underline{i}}, \tag{22}$$

an element of the free $\mathbb{Z}$-module on basis $\{e^{\underline{i}} \mid \underline{i} \in I^n\}$. By [4, 5.12,9.10], we have:

Theorem 3.5. *The characters of the pairwise inequivalent irreducible $S(n)$-supermodules are linearly independent. Moreover, the type of an irreducible $S(n)$-supermodule D can be read off from its character: if D belongs to the block γ then D is of type $\mathtt{M}$ if $(n - \gamma_0)$ is even, type $\mathtt{Q}$ if $(n - \gamma_0)$ is odd.*

3.3. **Induction and restriction operators.** Next we introduce the analogues of the i-induction and i-restriction functors. Note $S(n-1)$ is naturally embedded in $S(n)$ as the subalgebra generated by $t_1, \dots, t_{n-2}$. So we have natural restriction and induction functors $\mathrm{res}^{S(n)}_{S(n-1)}$ and $\mathrm{ind}^{S(n+1)}_{S(n)} := S(n+1) \otimes_{S(n)} ?$.

Let M be an $S(n)$-supermodule belonging to the block $\gamma \in \Gamma_n$. Given $i \in I$, define

$$\mathrm{res}_i M := (\mathrm{res}^{S(n)}_{S(n-1)} M)[\gamma - i] \quad \text{(interpreted as zero in case } \gamma_i = 0), \tag{23}$$

$$\mathrm{ind}_i M := (\mathrm{ind}^{S(n+1)}_{S(n)} M)[\gamma + i], \tag{24}$$

where the notation $\gamma \pm i$ is as in §2.2. These definitions extend in an obvious way to give exact functors res_i and ind_i. We note in particular that $\mathrm{res}_i M$ is the generalized eigenspace of eigenvalue $i(i+1)/2$ for the action of y_n^2. By the definition and Lemma 3.3, we have:

Lemma 3.6. *For an $S(n)$-supermodule M,*

$$\mathrm{res}^{S(n)}_{S(n-1)} M \cong \bigoplus_{i \in I} \mathrm{res}_i M, \qquad \mathrm{ind}^{S(n+1)}_{S(n)} M \cong \bigoplus_{i \in I} \mathrm{ind}_i M.$$

The next elementary lemma, proved rather indirectly in [4, 9.13,9.14], shows how the functors res_i and ind_i can be refined to obtain the correct definition of the operators e_i and f_i in this setting.

Lemma 3.7. *Let D be an irreducible $S(n)$-supermodule, and $i \in I$.*

(i) *There is an $S(n-1)$-supermodule $e_i D$, unique up to isomorphism, such that*

$$\mathrm{res}_i D \cong \begin{cases} e_i D \oplus e_i D & \text{if } i \neq 0 \text{ and } D \text{ is of type } \mathtt{Q}, \\ e_i D & \text{if } i = 0 \text{ or } D \text{ is of type } \mathtt{M}. \end{cases}$$

(ii) *There is an $S(n+1)$-supermodule $f_i D$, unique up to isomorphism, such that*

$$\mathrm{ind}_i D \cong \begin{cases} f_i D \oplus f_i D & \text{if } i \neq 0 \text{ and } D \text{ is of type } \mathtt{Q}, \\ f_i D & \text{if } i = 0 \text{ or } D \text{ is of type } \mathtt{M}. \end{cases}$$

We have now defined the operators $e_i, f_i (i \in I)$ on irreducible $S(n)$-supermodules (but note they are *not* functors defined on arbitrary supermodules, unlike before). Extending linearly, they induce operators also denoted e_i, f_i at the level of characters. The effect of e_i on characters is exactly the same as before:

$$\text{if} \quad \mathrm{ch}\, M = \sum_{\underline{i} \in I^n} a_{\underline{i}} e^{\underline{i}} \qquad \text{then} \qquad \mathrm{ch}\,(e_i M) = \sum_{\underline{i} \in I^{n-1}} a_{(i_1, \dots, i_{n-1}, i)} e^{\underline{i}}. \tag{25}$$

This is one reason we have chosen to normalize characters the way we did in (22).

There are also divided power operators $e_i^{(r)}$ and $f_i^{(r)}$. Again we just state a lemma characterizing them uniquely, rather than giving their explicit definition:

Lemma 3.8. *Let D be an irreducible $S(n)$-supermodule, and $i \in I$.*

(i) *There is an $S(n-r)$-supermodule $e_i^{(r)} D$, unique up to isomorphism, such that*

$$(\mathrm{res}_i)^r D \cong \begin{cases} (e_i^{(r)} D)^{\oplus r!} & \text{if } i = 0, \\ (e_i^{(r)} D)^{\oplus 2^{\lfloor r/2 \rfloor} r!} & \text{if } i \neq 0 \text{ and } D \text{ is of type } \mathtt{M}, \\ (e_i^{(r)} D)^{\oplus 2^{\lfloor (r+1)/2 \rfloor} r!} & \text{if } i \neq 0 \text{ and } D \text{ is of type } \mathtt{Q}. \end{cases}$$

(ii) *There is an $S(n+r)$-supermodule $f_i^{(r)}D$, unique up to isomorphism, such that*

$$(\mathrm{ind}_i)^r D \cong \begin{cases} (f_i^{(r)}D)^{\oplus r!} & \text{if } i = 0, \\ (f_i^{(r)}D)^{\oplus 2^{\lfloor r/2 \rfloor} r!} & \text{if } i \neq 0 \text{ and } D \text{ is of type } \mathtt{M}, \\ (f_i^{(r)}D)^{\oplus 2^{\lfloor (r+1)/2 \rfloor} r!} & \text{if } i \neq 0 \text{ and } D \text{ is of type } \mathtt{Q}. \end{cases}$$

Note comparing Lemmas 3.6 and 3.8, we see that

$$e_i^r = r! e_i^{(r)}, \qquad f_i^r = r! f_i^{(r)} \tag{26}$$

at the level of characters.

3.4. **The affine Kac-Moody algebra.** Now things go in almost exactly the same way as §2.3. Let R_n denote the character ring of $S(n)$, i.e. the free $\mathbb{Z}$-module spanned by the formal characters of the irreducible FS_n-supermodules, and let R_n^* denote the $\mathbb{Z}$-submodule of R_n spanned by the formal characters of the projective indecomposable $S(n)$-supermodules. Let

$$R = \bigoplus_{n \geq 0} R_n, \qquad R^* = \bigoplus_{n \geq 0} R_n^* \subseteq R. \tag{27}$$

The e_i and f_i induce $\mathbb{Z}$-linear operators on R, stabilizing R^*. Extending scalars we get $\mathbb{C}$-linear operators e_i and f_i on $R_{\mathbb{C}} := \mathbb{C} \otimes_{\mathbb{Z}} R = \mathbb{C} \otimes_{\mathbb{Z}} R^*$. Finally, we have the symmetric Cartan form on $R_{\mathbb{C}}$, with respect to which the characters of the projective indecomposable supermodules and the irreducible supermodules form a pair of dual bases.

Theorem 3.9. *The operators e_i and f_i $(i \in I)$ on $R_{\mathbb{C}}$ satisfy the defining relations of the Chevalley generators of the affine Kac-Moody Lie algebra $\mathfrak{g}$ of type $A_{p-1}^{(2)}$ (resp. B_∞ in case $p = 0$), see [16]. Moreover, viewing $R_{\mathbb{C}}$ as a $\mathfrak{g}$-module in this way,*

(i) *$R_{\mathbb{C}}$ is isomorphic to the basic representation $V(\Lambda_0)$ of $\mathfrak{g}$, generated by the highest weight vector e^0 (the character of the irreducible $FS(0)$-module);*
(ii) *the decomposition of $R_{\mathbb{C}}$ into superblocks coincides with its weight space decomposition with respect to the standard Cartan subalgebra of $\mathfrak{g}$;*
(iii) *the Cartan form on $R_{\mathbb{C}}$ coincides with the Shapovalov form satisfying $(e^0, e^0) = 1$;*
(iv) *the lattice $R^* \subset R_{\mathbb{C}}$ is the $\mathbb{Z}$-submodule of $R_{\mathbb{C}}$ generated by e^0 under the action of the operators $f_i^{(r)} = f_i^r / r!$ $(i \in I, r \geq 0)$;*
(v) *the lattice $R \subset R_{\mathbb{C}}$ is the dual lattice to R^* under the Shapovalov form.*

This was proved in [4, 7.16].

3.5. **The crystal graph.** Next we describe the crystal underlying the basic representation $V(\Lambda_0)$ of the affine Kac-Moody algebra $\mathfrak{g} = A_{p-1}^{(2)}$. This explicit combinatorics is due to Kang [17]. We will work now with the set of all *p-strict partitions*, i.e. the partitions $\lambda = (\lambda_1, \lambda_2, \dots)$ with the property that p divides λ_r whenever $\lambda_r = \lambda_{r+1}$. For example, the 0-strict partitions are the partitions with no repeated non-zero parts.

Given a p-strict partition λ, we label its nodes with residues taken from the set $I = \{0, 1, \dots, \ell\}$ (recall $\ell = (p-1)/2$ or ∞ if $p = 0$). The labelling depends only on the column and follows the repeating pattern

$$0, 1, \dots, \ell - 1, \ell, \ell - 1, \dots, 1, 0,$$

starting fom the first column and going to the right, see Example 3.10 below. The residue of the node A is denoted $\operatorname{res} A$.

Let $i \in I$ be some fixed residue. A node $A = (r, s) \in \lambda$ is called *i-removable* (for λ) if one of the following holds:

(R1) $\operatorname{res} A = i$ and $\lambda - \{A\}$ is again a p-strict partition;
(R2) the node $B = (r, s + 1)$ immediately to the right of A belongs to λ, $\operatorname{res} A = \operatorname{res} B = i$, and both $\lambda - \{B\}$ and $\lambda - \{A, B\}$ are p-strict partitions.

Similarly, a node $B = (r, s) \notin \lambda$ is called *i-addable* (for λ) if one of the following holds:

(A1) $\operatorname{res} B = i$ and $\lambda \cup \{B\}$ is again an p-strict partition;
(A2) the node $A = (r, s - 1)$ immediately to the left of B does not belong to λ, $\operatorname{res} A = \operatorname{res} B = i$, and both $\lambda \cup \{A\}$ and $\lambda \cup \{A, B\}$ are p-strict partitions.

We note that (R2) and (A2) above are only possible in case $i = 0$.

Now label all i-addable nodes of the diagram λ by $+$ and all i-removable nodes by $-$. The *i-signature* of λ is the sequence of pluses and minuses obtained by going along the rim of the Young diagram from bottom left to top right and reading off all the signs. The *reduced i-signature* of λ is obtained from the i-signature by successively erasing all neighbouring pairs of the form $+-$. *Warning:* for historical reasons, the rule for obtaining the reduced i-signature here is different from in §2.4: there one deleted pairs of the form $-+$.

Note the reduced i-signature always looks like a sequence of $-$'s followed by $+$'s. Nodes corresponding to a $-$ in the reduced i-signature are called *i-normal*, nodes corresponding to a $+$ are called *i-conormal*. The rightmost i-normal node (corresponding to the rightmost $-$ in the reduced i-signature) is called *i-good*, and the leftmost i-conormal node (corresponding to the leftmost $+$ in the reduced i-signature) is called *i-cogood*.

Example 3.10. Let $p = 5$, so $\ell = 2$. The partition $\lambda = (16, 11, 10, 10, 9, 5, 1)$ is p-strict, and its residues are as follows:

0	1	2	1	0	0	1	2	1	0	0	1	2	1	0	0
0	1	2	1	0	0	1	2	1	0	0					
0	1	2	1	0	0	1	2	1	0						
0	1	2	1	0	0	1	2	1	0						
0	1	2	1	0	0	1	2	1							
0	1	2	1	0											
0															

The 0-addable and 0-removable nodes are as labelled in the diagram:

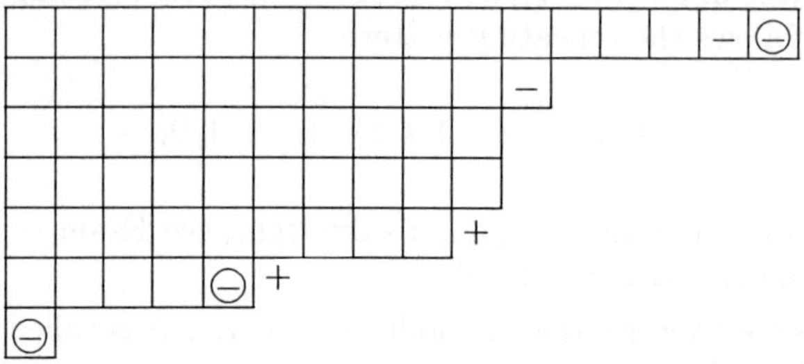

Hence, the 0-signature of λ is $-,-,+,+,-,-,-$ and the reduced 0-signature is $-,-,-$. Note the nodes corresponding to the $-$'s in the reduced 0-signature have been circled in the above diagram. So, there are three 0-normal nodes, the rightmost of which is 0-good; there are no 0-conormal or 0-cogood nodes.

Finally we call a p-strict partition λ *restricted* if either $p = 0$ or

$$\begin{aligned} \lambda_i - \lambda_{i+1} &< p \qquad \text{if } p \mid \lambda_i, \\ \lambda_i - \lambda_{i+1} &\le p \qquad \text{if } p \nmid \lambda_i, \end{aligned}$$

for each $i = 1, 2, \ldots$. The *crystal graph* associated to the basic representation $V(\Lambda_0)$ of $\mathfrak{g}$ is now the set of all restricted p-strict partitions of n, for all $n \ge 0$, with a directed edge $\lambda \stackrel{i}{\longrightarrow} \mu$ of color $i \in I$ if μ is obtained from λ by adding an i-cogood node (equivalently, λ is obtained from μ by removing an i-good node). For an example in case $p = 3$, see below.

3.6. **The modular branching graph.** The connection between the crystal graph and the representation theory of $S(n)$ now proceeds in exactly the same way as in §2.5. The starting point is the following lemma proved in [4, 6.6,9.13,9.14]:

Lemma 3.11. *Let D be an irreducible $S(n)$-supermodule and $i \in I$. Then, the supermodule e_iD (resp. f_iD) is either zero, or else is a self-dual $S(n-1)$- (resp. $S(n+1)$-) supermodule with irreducible socle and head isomorphic to each other.*

The *crystal operators* $\tilde{e}_i, \tilde{f}_i$ are defined on an irreducible $S(n)$-supermodule D by

$$\tilde{e}_iD := \text{socle}(e_iD), \qquad \tilde{f}_iD := \text{socle}(f_iD). \tag{28}$$

In view of Lemma 3.11, $\tilde{e}_iD$ and $\tilde{f}_iD$ are either zero or irreducible. The *modular branching graph* has vertices equal to the isomorphism classes of irreducible $S(n)$-supermodules for all $n \ge 0$, and there is a directed edge $[D] \stackrel{i}{\longrightarrow} [E]$ of color i if $E \cong \tilde{f}_iD$ (equivalently by Frobenius reciprocity, $D \cong \tilde{e}_iE$). The fundamental result is the following:

Theorem 3.12. *The modular branching graph is uniquely isomorphic (as an I-colored, directed graph) to the crystal graph of §3.5.*

As before, Theorem 3.12 yields a parametrization of the isomorphism classes of irreducible $S(n)$-supermodules by the vertices of the crystal graph. Precisely, if λ is a restricted p-strict partition of n, choose a path

$$\varnothing \stackrel{i_1}{\longrightarrow} \square \stackrel{i_2}{\longrightarrow} \cdots \stackrel{i_n}{\longrightarrow} \lambda$$

The basic crystal graph of type $A_2^{(2)}$

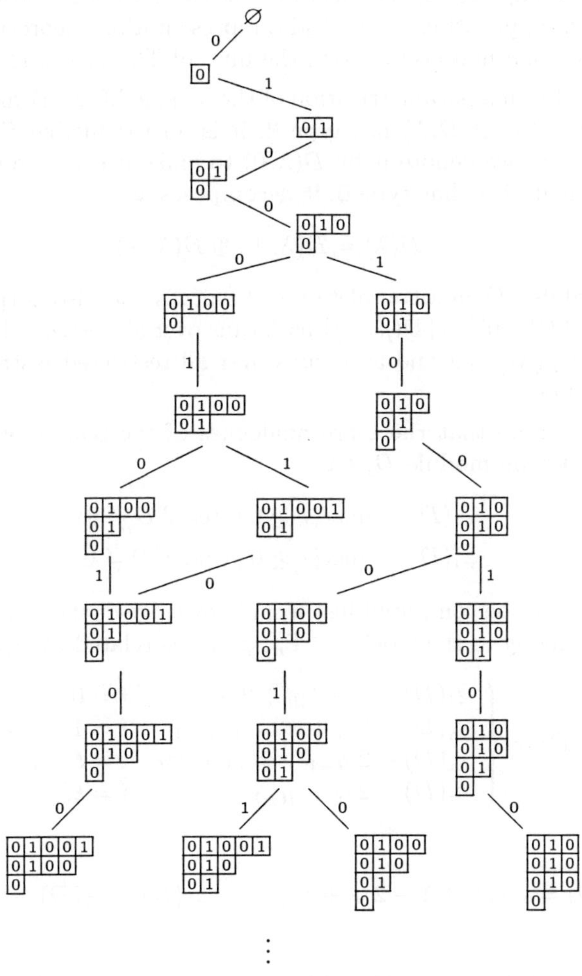

in the crystal graph from the empty partition to λ, for $i_1, \dots, i_n \in I$. Define

$$D(\lambda) := \tilde{f}_{i_n} \dots \tilde{f}_{i_1} D(\varnothing), \tag{29}$$

where $D(\varnothing)$ denotes the irreducible $S(0)$-supermodule. Using Theorem 3.5 for the second statement, we have:

Corollary 3.13. *The supermodules*

$$\{D(\lambda) \mid \lambda \text{ a restricted } p\text{-strict partition of } n\}$$

form a complete set of inequivalent irreducible $S(n)$-supermodules. Moreover, letting $h_{p'}(\lambda)$ denote the number of parts of λ not divisible by p, $D(\lambda)$ has type M *or* Q *according to whether $(n - h_{p'}(\lambda))$ is even or odd respectively.*

This corollary solves the labelling problem for irreducible representations of the double covers of the symmetric group. That the restricted p-strict partitions should be suitable for this was suggested first by Leclerc and Thibon [23]. Note we gave an entirely different construction of the irreducible $S(n)$-supermodules, also labelled by restricted p-strict partitions, in [3]. At present we cannot prove that the two labellings agree, though we expect this to be the case. Another problem, which would be very useful in applications, is to find a representation theoretic interpretation of the normal and conormal nodes along the lines of Theorem 2.11.

It is easy to obtain a parametrization of the irreducible $S(n)$-modules (not super) from Corollary 3.13. If $D(\lambda)$ has type $\mathtt{M}$, it is an irreducible $S(n)$-module in the ordinary sense, but we denote it by $D(\lambda, 0)$ to make it clear we have forgotten the $\mathbb{Z}_2$-grading. But if $D(\lambda)$ has type $\mathtt{Q}$, it decomposes as

$$D(\lambda) = D(\lambda, +) \oplus D(\lambda, -)$$

as an $S(n)$-module. Then a complete set of pairwise non-isomorphic $S(n)$-modules is given by the $\{D(\lambda, 0)\} \cup \{D(\mu, \pm)\}$ as λ runs over all restricted p-strict partitions of n with $n - h_{p'}(\lambda)$ even and as μ runs over all restricted p-strict partitions of n with $n - h_{p'}(\mu)$ odd.

Let us finally note that there are analogues of the results of §2.6 too. For an irreducible $S(n)$-supermodule D, set

$$\varepsilon_i(D) = \max\{r \geq 0 \mid (\mathrm{res}_i)^r D \neq 0\}, \tag{30}$$

$$\varphi_i(D) = \max\{r \geq 0 \mid (\mathrm{ind}_i)^r D \neq 0\}. \tag{31}$$

As before $\varepsilon_i(D)$ can be computed just from knowledge of the character of D. Moreover, for D belonging to the block $\gamma \in \Gamma_n$, $\varphi_i(D)$ is related to $\varepsilon_i(D)$ by the formula

$$\varphi_i(D) = \begin{cases} \varepsilon_i(D) + 1 - 2\gamma_0 + 2\gamma_1 & i = 0, \\ \varepsilon_i(D) - 2\gamma_i + \gamma_{i-1} + \gamma_{i+1} & i = 1, \dots, \ell - 2, \\ \varepsilon_i(D) - 2\gamma_{\ell-1} + \gamma_{\ell-2} + 2\gamma_\ell & i = \ell - 1, \\ \varepsilon_i(D) - 2\gamma_\ell + \gamma_{\ell-1} & i = \ell, \end{cases} \tag{32}$$

for $\ell \neq 1$, or

$$\varphi_0(D) = \varepsilon_0(D) + 1 - 2\gamma_0 + 4\gamma_1, \qquad \varphi_1(D) = \varepsilon_1(D) - 2\gamma_1 + \gamma_0. \tag{33}$$

if $\ell = 1$.

The analogue of Lemma 2.12 holds exactly as stated before, so $\varepsilon_i(D(\lambda))$ and $\varphi_i(D(\lambda))$ can also be read off directly from the crystal graph as the number of '$-$'s resp. '$+$'s in the reduced i-signature of λ. So one obtains an inductive algorithm to determine the label of an irreducible $S(n)$-supermodule D purely from knowledge of its character $\mathrm{ch}\, D$, in exactly the same way as before.

3.7. **Blocks.** To state some results about blocks, let $\mathfrak{g} = A_{p-1}^{(2)}$ and let $P = \bigoplus_{i \in I} \mathbb{Z}\Lambda_i \oplus \mathbb{Z}\delta$ be the associated weight lattice. The simple roots α_i $(i \in I)$ can be defined by

$$\begin{aligned} \alpha_0 &= 2\Lambda_0 - \Lambda_1, \\ \alpha_1 &= -2\Lambda_0 + 2\Lambda_1 - \Lambda_2, \\ \alpha_i &= -\Lambda_{i-1} + 2\Lambda_i - \Lambda_{i+1} \qquad (i = 2, \dots, \ell - 1), \\ \alpha_\ell &= -2\Lambda_{\ell-1} + 2\Lambda_\ell + \delta \end{aligned}$$

if $\ell \neq 1$, and

$$\alpha_0 = 2\Lambda_0 - \Lambda_1, \qquad \alpha_1 = -4\Lambda_0 + 2\Lambda_1 + \delta$$

if $\ell = 1$. Let $(.|.)$ be the positive definite symmetric bilinear form on $\mathbb{R} \otimes_{\mathbb{Z}} P$ with respect to which $2\alpha_0, \alpha_1, \ldots, \alpha_{\ell-1}, \frac{1}{2}\alpha_\ell, \Lambda_0$ and $\Lambda_0, \ldots, \Lambda_\ell, \delta$ form a pair of dual bases. The Weyl group W is the subgroup of $GL(\mathbb{R} \otimes_{\mathbb{Z}} P)$ generated by the reflections $s_i\,(i \in I)$ in the hyperplanes orthogonal to the α_i. Then, by [16, (12.6.1)], the weights of the $\mathfrak{g}$-module $V(\Lambda_0)$ are the

$$\{w\Lambda_0 - d\delta \mid w \in W, d \in \mathbb{Z}_{\geq 0}\}.$$

There are combinatorial notions paralleling this description of weights, namely, Morris' notions of p-bar core and p-bar weight [27]. Let λ be a p-strict partition. By a *p-bar* of λ, we mean one of the following:

(B1) the rightmost p nodes of row i of λ if $\lambda_i \geq p$ and either $p|\lambda_i$ or λ has no row of length $(\lambda_i - p)$;
(B2) the set of nodes in rows i and j of λ if $\lambda_i + \lambda_j = p$.

If λ has no p-bars, it is called a *p-bar core.* In general, the p-bar core $\tilde{\lambda}$ of λ is obtained by successively removing p-bars, reordering the rows each time so that the result is still a p-strict partition, until it is reduced to a core. The *p-bar weight* d of λ is then the total number of p-bars that get removed.

Now we get the classification of superblocks from Theorem 3.9(ii): irreducible $S(n)$-supermodules $D(\lambda)$ and $D(\mu)$ belong to the same superblock if and only if λ and μ have the same p-bar core. Moreover, exactly as for Theorem 2.20, superblocks B and B' of the same p-bar weight are isometric, in the sense that there is an isomorphism between their Grothendieck groups (induced by the action of W) which is an isometry with respect to the Cartan pairing.

It is more natural from the point of view of finite group theory to ask for a description of the ordinary (not super) blocks of $S(n)$. This does not seem to follow easily from the present theory, unless we invoke the work of Humphreys [12] (in fact, all we need from [12] is to know the *number* of ordinary blocks). There are two sorts of superblocks, of type `M` or `Q` according to whether all the irreducible supermodules belonging to the superblock are of type `M` or `Q`. Corresponding to superblocks of type `M`, there are ordinary blocks all of whose irreducible modules are of the form $D(\lambda, 0)$, and $D(\lambda, 0)$ and $D(\mu, 0)$ belong to the same block if and only if λ and μ have the same p-bar core. Corresponding to superblocks of type `Q`, there are ordinary blocks consisting of irreducible modules of the form $D(\lambda, \pm)$. Again, $D(\lambda, \varepsilon)$ and $D(\mu, \delta)$ belong to the same block if and only if λ and μ have the same p-bar core, with one exception: if λ is itself a p-bar core, then $D(\lambda, +)$ and $D(\lambda, -)$ are in different blocks.

Finally let us state the analogue of Theorem 2.22 giving the Cartan determinant of a superblock, see [5]:

Theorem 3.14. *Let B be a superblock of p-bar weight d of $S(n)$. Then the determinant of the Cartan matrix of B is p^N where N equals*

$$\sum \frac{2r_1 + 2r_3 + 2r_5 + \ldots}{p-1} \binom{\frac{p-3}{2} + r_1}{r_1} \binom{\frac{p-3}{2} + r_2}{r_2} \binom{\frac{p-3}{2} + r_3}{r_3} \cdots,$$

the sum being over all partitions $\lambda = (1^{r_1}2^{r_2}\dots) \vdash d$.

Note for superblocks of type M, this same formula gives the Cartan determinant of the corresponding ordinary block of $S(n)$. It appears that the same is true for superblocks of type Q, but we do not see how to deduce this from the theorem: the problem is that the ordinary block has twice as many irreducibles as in the corresponding superblock.

References

[1] S. Ariki, On the decomposition numbers of the Hecke algebra of type $G(m,1,n)$, *J. Math. Kyoto Univ.* **36** (1996), 789–808.

[2] J. Brundan and A. Kleshchev, Translation functors for general linear and symmetric groups, *Proc. London Math. Soc.* **80** (2000), 75–106.

[3] J. Brundan and A. Kleshchev, Projective representations of symmetric groups via Sergeev duality, *Math. Z.* **239** (2002), 27–68.

[4] J. Brundan and A. Kleshchev, Hecke-Clifford superalgebras, crystals of type $A_{2\ell}^{(2)}$ and modular branching rules for $\widehat{S}_n$, *Represent. Theory* **5** (2001), 317–403.

[5] J. Brundan and A. Kleshchev, Cartan determinants and Shapovalov forms, to appear in *Math. Ann.*, 2002.

[6] J. Chuang and R. Kessar, Symmetric groups, wreath products, Morita equivalences, and Broué's abelian defect group conjecture, Bull. London Math. Soc. **34** (2002), 174–184.

[7] C. De Concini, V. Kac and D. Kazhdan, "Boson-Fermion correspondence over $\mathbb{Z}$", in: Infinite-dimensional Lie algebras and groups (Luminy-Marseille, 1988), *Adv. Ser. Math. Phys.* **7** (1989), 124–137.

[8] M. Enguehard, Isométries parfaites entre blocs de groupes symétriques, *Astérisque* **181-182** (1990), 157–171.

[9] I. Grojnowski, Affine $\widehat{\mathfrak{sl}}_p$ controls the modular representation theory of the symmetric group and related Hecke algebras, preprint, 1999.

[10] I. Grojnowski, Blocks of the cyclotomic Hecke algebra, preprint, 1999.

[11] I. Grojnowski and M. Vazirani, Strong multiplicity one theorem for affine Hecke algebras of type A, *Transf. Groups.* **6** (2001), 143–155.

[12] J. F. Humphreys, Blocks of projective representations of the symmetric groups, *J. London Math. Soc.* **33** (1986), 441–452.

[13] G. D. James, *The representation theory of the symmetric groups*, Lecture Notes in Math., vol. 682, Springer-Verlag, 1978.

[14] G. James and A. Kerber, *The Representation Theory of the Symmetric Groups*. Addison-Wesley, London, 1980.

[15] A. Jucys, Symmetric polynomials and the center of the symmetric group ring, *Report Math. Phys.* **5** (1974), 107–112.

[16] V. Kac, *Infinite dimensional Lie algebras*, Cambridge University Press, third edition, 1995.

[17] S.-J. Kang, Crystal bases for quantum affine algebras and combinatorics of Young walls, preprint, Seoul National University, 2000.

[18] M. Kashiwara, "On crystal bases", in: Representations of groups (Banff 1994), *CMS Conf. Proc.* **16** (1995), 155–197.

[19] A. Kleshchev, Branching rules for modular representations of symmetric groups II, *J. reine angew. Math.* **459** (1995), 163–212.

[20] A. Kleshchev. Branching rules for modular representations of symmetric groups. III. Some corollaries and a problem of Mullineux. *J. London Math. Soc. (2)* **54** (1996), 25–38.

[21] A. Lascoux, B. Leclerc and J.-Y. Thibon, Hecke algebras at roots of unity and crystal bases of quantum affine algebras, *Comm. Math. Phys.* **181** (1996), 205–263.

[22] B. Leclerc and H. Miyachi, Some closed formulas for canonical bases of Fock spaces, *preprint*, 2001.

[23] B. Leclerc and J.-Y. Thibon, q-Deformed Fock spaces and modular representations of spin symmetric groups, *J. Phys. A* **30** (1997), 6163–6176.

[24] I. G. Macdonald, *Symmetric functions and Hall polynomials*, Oxford Mathematical Monographs, second edition, OUP, 1995.

[25] A. Marcus, On equivalences between blocks of group algebras: reduction to the simple components, *J. Algebra*, **184** (1996), 372-396.

[26] K. Misra and T. Miwa, Crystal base for the basic representation of $U_q(\mathfrak{sl}(n))$, *Comm. Math. Phys.* **134** (1990), 79–88.

[27] A.O. Morris, The spin representations of the symmetric group, *Canad. J. Math.* **17** (1965), 543-549.

[28] G. Murphy, A new construction of Young's seminormal representations of the symmetric groups, *J. Algebra* **69** (1981), 287–297.

[29] G. Murphy, The idempotents of the symmetric group and Nakayama's conjecture, *J. Algebra* **81** (1983), 258–265.

[30] M. Nazarov, Young's orthogonal form of irreducible projective representations of the symmetric group, *J. London Math. Soc. (2)* **42** (1990), 437–451.

[31] M. Nazarov, Young's symmetrizers for projective representations of the symmetric group, *Advances Math.* **127** (1997), 190–257.

[32] A. Okounkov and A. Vershik, A new approach to representation theory of symmetric groups. *Selecta Math. (N.S.)* **2** (1996), 581–605.

[33] I. Schur, Über die Darstellung der symmetrischen und der alternierenden Gruppe durch gebrochene lineare Substitutionen, *J. Reine Angew. Math.* **139** (1911), 155–250.

[34] J. Scopes, Cartan matrices and Morita equivalence for blocks of the symmetric groups. *J. Algebra* **142** (1991), 441–455.

[35] A. N. Sergeev, Tensor algebra of the identity representation as a module over the Lie superalgebras $GL(n, m)$ and $Q(n)$, *Math. USSR Sbornik* **51** (1985), 419–427.

[36] A. N. Sergeev, The Howe duality and the projective representations of symmetric groups, *Represent. Theory* **3** (1999), 416–434.

[37] J. Stembridge, Shifted tableaux and the projective representations of symmetric groups, *Advances in Math.* **74** (1989), 87–134.

[38] M. Vazirani, *Irreducible modules over the affine Hecke algebra: a strong multiplicity one result*, Ph.D. thesis, UC Berkeley, 1999.

Coherent configurations, association schemes and permutation groups

Peter J. Cameron

Abstract

Coherent configurations are combinatorial objects invented for the purpose of studying finite permutation groups; every permutation group which is not doubly transitive preserves a non-trivial coherent configuration. However, symmetric coherent configurations have a much longer history, having been used in statistics under the name of *association schemes.*

The relationship between permutation groups and association schemes is quite subtle; there are groups which preserve no non-trivial association scheme, and other groups for which there is not a unique minimal association scheme.

This paper gives a brief outline of the theory of coherent configurations and association schemes, and reports on some recent work on the connection between association schemes and permutation groups.

1 Coherent configurations

This section contains the definitions of coherent configurations and of various specialisations (including association schemes), and their connection with finite permutation groups. It is by no means a complete survey of this topic, but it includes some historical remarks.

1.1 Definitions

Let Ω be a finite set. A *coherent configuration* on Ω is a set $\mathcal{P} = \{R_1, \ldots, R_s\}$ of binary relations on Ω (subsets of Ω^2) satisfying the following four conditions:

(a) $\mathcal{P}$ is a partition of Ω^2;

(b) there is a subset $\mathcal{P}_0$ of $\mathcal{P}$ which is a partition of the diagonal $\Delta = \{(\alpha, \alpha) : \alpha \in \Omega\}$;

(c) for every relation $R_i \in \mathcal{P}$, its *converse* $R_i^\top = \{(\beta, \alpha) : (\alpha, \beta) \in R_i\}$ is in $\mathcal{P}$; say $R_i^\top = R_{i^*}$.

(d) there exist integers p_{ij}^k, for $1 \le i, j, k \le s$, such that,for any $(\alpha, \beta) \in R_k$, the number of points $\gamma \in \Omega$ such that $(\alpha, \gamma) \in R_i$ and $(\gamma, \beta) \in R_j$ is equal to p_{ij}^k (and, in particular, is independent of the choice of $(\alpha, \beta) \in R_k$).

The numbers p_{ij}^k are called the *intersection numbers* of the coherent configuration $\mathcal{P}$. (They are so-called because p_{ij}^k is the cardinality of the intersection $R_i(\alpha) \cap R_j^\top(\beta)$ for $(\alpha, \beta) \in R_k$, where $R(\alpha) = \{\beta \in \Omega : (\alpha, \beta) \in R\}$.)

We can represent a binary relation R on Ω by its *basis matrix* $A(R)$, whose rows and columns are indexed by Ω, and whose (α, β) entry is 1 if $(\alpha, \beta) \in R$, 0 otherwise. Using these matrices, and letting I and J be the identity and all-1 matrices, the axioms take the following form:

(a) $\sum_{i=1}^{s} A(R_i) = J$.

(b) $\sum_{i=1}^{t} A(R_i) = I$, where $\{R_1, \ldots, R_t\}$ is the subset referred to in (b) above.

(c) For each i, there exists i^* such that $A(R_i)^\top = A(R_{i^*})$.

(d) For each pair i, j, we have

$$A(R_i)A(R_j) = \sum_{k=1}^{s} p_{ij}^k A(R_k). \tag{1}$$

It follows from (b) and (d) that the span of $\{A(R_1), \ldots, A(R_s)\}$ (over the complex numbers) is an algebra, and from (c) that this algebra is semisimple (and so is isomorphic to a direct sum of matrix algebras over $\mathbb{C}$). This algebra is called the *basis algebra* of the configuration. We denote the basis algebra of $\mathcal{P}$ by $\mathrm{BA}(\mathcal{P})$. Note that $\mathrm{BA}(\mathcal{P})$ consists of all the functions from Ω^2 to $\mathbb{C}$ which are constant on the parts of $\mathcal{P}$.

Moreover, if P_j is the $s \times s$ matrix with (i, k) entry p_{ij}^k, then the map $A(R_j) \mapsto P_j$ for $j = 1, \ldots, s$ extends linearly to an algebra isomorphism. (Indeed, Equation (1) shows that this map is the regular representation of $\mathrm{BA}(\mathcal{P})$, written with respect to the basis matrices.) Thus the matrices $P_1, \ldots, P_s$ also span an algebra, called the *intersection algebra* of $\mathcal{P}$.

The irreducible modules for the intersection algebra, and their multiplicities in the module $\mathbb{C}\Omega$, can be calculated from the intersection numbers. The multiplicities must of course be non-negative integers. This is one of the most powerful methods for showing nonexistence of coherent configurations with given intersection numbers. See Higman [16] for an early application to permutation groups.

Many familiar algebraic and combinatorial objects are coherent configurations. These include strongly regular and distance-regular graphs, symmetric and quasi-symmetric designs, partial geometries, generalised polygons, difference sets, and Schur rings. Moreover, as Delsarte [9] showed, the study of special subsets of coherent configurations provides a common context for much of design theory and coding theory. Delsarte also introduced new methods (such as linear programming) in this general context and applied them to both codes and designs.

Two extreme examples of coherent configurations will be important to us:

- The *trivial configuration* on Ω consists of the two relations E and $\Omega^2 \setminus E$, where $E = \{(\alpha, \alpha) : \alpha \in \Omega\}$ is the diagonal (the relation of equality).

- The *discrete configuration* on Ω is the partition of Ω^2 into singleton sets.

1.2 Permutation groups

If G is any permutation group on Ω, then the partition of Ω^2 into orbits of G is a coherent configuration, which we denote by $\mathcal{K}(G)$. We refer to this as the *group case*; a coherent configuration of the form $\mathcal{K}(G)$ is called *Schurian.* The trivial and discrete c.c.s are Schurian, corresponding to the symmetric group and the identity group respectively. Indeed, D. G. Higman [16, 17] introduced coherent configurations in order to study permutation groups, as the title of his early lecture notes [18] suggests.

An *automorphism* of a partition $\mathcal{P}$ of Ω^2 is a permutation of Ω which fixes every set in $\mathcal{P}$; a *weak automorphism* is a permutation which maps each member of $\mathcal{P}$ to a member of $\mathcal{P}$. The automorphisms of $\mathcal{P}$ form a group, the *automorphism group* of $\mathcal{P}$, denoted by $\mathrm{Aut}(\mathcal{P})$. We will be concerned almost exclusively with automorphisms (which are sometimes called *strict automorphisms*); but we note the following fact. (The order relation on partitions is the usual one, which will be discussed further in the next section.)

Proposition 1.1 *Let G be a group of weak automorphisms of the coherent configuration $\mathcal{P}$ on Ω. Let $\mathcal{P}^G$ be the partition of Ω whose parts are the unions of the G-orbits on the parts of $\mathcal{P}$. Then $\mathcal{P}^G$ is a coherent configuration; it is the unique finest coherent configuration coarser than $\mathcal{P}$ which admits G as a group of automorphisms.* ■

We note also that G is a group of automorphisms of the coherent configuration $\mathcal{K}(G)$; that is, $G \leq \mathrm{Aut}(\mathcal{K}(G))$. The group G is said to be 2-*closed* if $\mathrm{Aut}(\mathcal{K}(G)) = G$: that is, any permutation of Ω which fixes every G-orbit on pairs belongs to G. There is thus a bijection between Schurian coherent configurations and 2-closed permutation groups on Ω.

1.3 Some special coherent configurations

Let $\mathcal{P}$ be a coherent configuration on Ω. The sets F such that $\{(\alpha, \alpha) : \alpha \in F\}$ belong to $\mathcal{P}$ are called the *fibres* of $\mathcal{P}$; they form a partition of Ω. We say that $\mathcal{P}$ is *homogeneous* if there is only one fibre. If $\mathcal{P} = \mathcal{K}(G)$, the fibres of $\mathcal{P}$ are the orbits of G on Ω; so $\mathcal{K}(G)$ is homogeneous if and only if G is transitive.

Table 1 gives the numbers of homogeneous coherent configurations on small numbers n of points ($n \leq 30$, $n \neq 29$). These configurations have been computed by A. Hanaki and I. Miyamoto [15], and are available from Hanaki's Web page (which gives the configurations explicitly). The numbers up to 23 are cited by Bannai [5, p 48]. By checking which of the configurations are Schurian, we obtain the numbers in the third column of the table. These numbers can also be calculated in another way. Alexander Hulpke [8, 20, 21] has computed the transitive permutation groups of degree at most 29. Those of degree at most 23 are included in GAP [13]; I am grateful to Alexander for providing me with data on larger degrees. We can check which of these are 2-closed, either by using a built-in GAP function, or more efficiently using `nauty` [28], which is interfaced with GAP using the share package GRAPE [32]. Reassuringly, all methods give the same answer! The values for $n = 29$ follow from another result of Hanaki and Miyamoto (personal communication) that any non-Schurian homogeneous coherent configuration on 29 points is a complementary pair of strongly regular graphs, together with the result of Spence [33] that there are 41 strongly regular graphs on 29 points, of which only one is self-complementary. The last two columns of the table will be described later in this section.

The table shows that, on small numbers of points, most homogeneous coherent configurations arise from groups. (Indeed, the smallest non-Schurian example has 15 points; it is a "strongly regular tournament", whose automorphism group has order 21 and has three orbits on points.) This pattern is unlikely to hold in general; it is plausible that the proportion which are Schurian tends to zero as the number of points increases. Pyber [30] gives some results on the number of subgroups of S_n; it is likely that only a small proportion of these are 2-closed. However, only in special cases such as particular strongly regular graphs do good estimates for the numbers of coherent configurations exist.

The combinatorial explosion is not revealed by the data in Table 1, although it probably begins shortly after this point. For example, for $n = 36$, McKay and Spence [29] have shown that there are 32 548 strongly regular graphs with parameters $(36, 15, 6, 6)$ (these are particular homogeneous coherent configurations), of which only one is Schurian!

A homogeneous c.c. is called *thin* if all basis matrices have row and column sums 1. A thin homogeneous c.c. is Schurian, and arises from a regular permutation group. For this reason, homogeneous c.c.s are sometimes called *generalized groups* [36].

A coherent configuration is called *symmetric* if all the relations are symmetric. A symmetric c.c. is homogeneous. (For, given any relation R in a c.c. with fibres $F_1, \ldots, F_t$, there are indices i, j such that $R \subseteq F_i \times F_j$.) If $\mathcal{P} = \mathcal{K}(G)$, then $\mathcal{P}$ is symmetric if and only if G is *generously transitive*, that is, any two points of Ω are interchanged by some element of G.

Let $\mathcal{P}$ be a c.c. on Ω. The *symmetrisation* $\mathcal{P}^{\mathrm{sym}}$ of $\mathcal{P}$ is the partition of Ω^2 whose parts are all unions of the parts of $\mathcal{P}$ and their converses. It may or may not be a c.c.; if it is, we say that $\mathcal{P}$ is *stratifiable*. The name, arising in statistics [2], will be explained later. It can be shown that, if $\mathcal{P} = \mathcal{K}(G)$, then $\mathcal{P}$ is stratifiable if and only if the permutation representation of G is *real-multiplicity-free*, that is, if it is decomposed into irreducibles over $\mathbb{R}$, they are pairwise non-isomorphic. (Equivalently, the complex irreducibles have multiplicity at most one except for those of quaternionic type, that is, Frobenius–Schur index -1, which may have multiplicity 2.)

Finally, a coherent configuration is called *commutative* if its basis matrices commute with one another. It can be shown that, if $\mathcal{P} = \mathcal{K}(G)$, then $\mathcal{P}$ is commutative if and only if the permutation representation is *(complex)-multiplicity-free*.

Thus, the following implications hold:

Proposition 1.2 *A symmetric c.c. is commutative; a commutative c.c. is stratifiable; and a stratifiable c.c. is homogeneous.* ∎

Number of points	Homogeneous c.c.s	2-closed trans. groups	Association schemes	2-closed, gen. trans. groups
1	1	1	1	1
2	1	1	1	1
3	2	2	1	1
4	4	4	3	3
5	3	3	2	2
6	8	8	4	4
7	4	4	2	2
8	21	21	10	10
9	12	12	6	6
10	13	13	8	8
11	4	4	2	2
12	59	59	21	21
13	6	6	4	4
14	16	16	8	8
15	25	24	10	10
16	222	206	63	56
17	5	5	4	4
18	95	93	32	32
19	7	6	3	3
20	95	95	41	41
21	32	32	11	11
22	16	16	8	8
23	22	4	2	2
24	750	669	157	136
25	45	32	33	20
26	34	24	24	14
27	502	122	39	38
28	185	124	106	47
29	26	6	24	4
30	243	228	79	73

Table 1: Homogeneous c.c.s and 2-closed groups

None of these implications reverses.

We note also that, if $\mathcal{P} = \mathcal{K}(G)$, then $\mathcal{P}$ is trivial if and only if G is doubly transitive.

1.4 History and terminology

A symmetric coherent configuration is usually known as an *association scheme.* Association schemes were first used in the context of experimental design in statistics, by R. C. Bose and his school, as "carriers" of partially balanced designs [7]. The basis algebra of an association scheme, and the isomorphism to the intersection algebra, were constructed by Bose and Mesner [6], and for this reason the basis algebra is often called the *Bose–Mesner algebra* of the association scheme.

There are several reasons why only symmetric c.c.s are used. First, the relations which arise in practice in treatment and plot structures and the covariance matrices that arise are almost always symmetric. Indeed, these relations are often defined by concurrences in blocks. Secondly, statistical data consists of real numbers, and a large part of the work consists in computing orthogonal decompositions of real vector spaces; association schemes are a valuable tool for this purpose. In an association scheme, the basis matrices are commuting symmetric real matrices, and so the vector space $\mathbb{R}^{\Omega}$ has an orthogonal decomposition into common eigenspaces of these matrices. These eigenspaces are called *strata.* (This is the origin of the term "stratifiable".) Further, the Moore–Penrose inverse of any matrix in $\mathrm{BA}(\mathcal{P})$ is also in $\mathrm{BA}(\mathcal{P})$.

General coherent configurations were defined at about the same time by Higman [17] and by Weisfeiler and Leman [34]; the latter used the term *cellular algebra* for the algebra generated by the basis matrices. Subsequently, Delsarte in his thesis [9] showed the importance of association schemes in coding theory. Though his discussion applies to any commutative c.c. (and he extended the usage of the term "association scheme" to this class), his important examples are all symmetric. Bannai [5], in a recent survey, uses the term "non-commutative association scheme" for a homogeneous coherent configuration. More recently, there have been signs that the term "association scheme" or "scheme" is being applied to any coherent configuration (see [11], for example).

I propose that this term should be restricted to its original meaning of "symmetric coherent configuration". This proposal is motivated in part by the large numbers of papers on association schemes in the statistical literature (see the references in [4]). I have adopted my proposal in this paper.

Symmetric matrices are so pervasive in statistics that, in the study of estimation of variance components, the fourth condition in the definition of a coherent configuration (closure under multiplication) is sacrificed; instead, closure under the *Jordan product* $A \circ B = \frac{1}{2}(AB + BA)$ is required. (This product is commutative and preserves symmetry of matrices.) Then Wedderburn's theorem on simple associative algebras must be replaced by some form of the Jordan–von Neumann–Wigner theorem on simple Jordan algebras. See Jacobson [23] for the theory of Jordan algebras, and Malley [27] for the statistical applications.

It is worth recording here a question to which I don't know the answer. Define a *Jordan scheme* to be a partition $\mathcal{P}$ of Ω^2 with the properties that the diagonal is a single part, every part is symmetric, and there are numbers q_{ij}^k such that

$$A(R_i)A(R_j) + A(R_j)A(R_i) = \sum_{k=1}^{s} q_{ij}^k A(R_k). \tag{2}$$

The span of the symmetric matrices $A(R_i)$ over the real numbers is thus a Jordan algebra, and contains the Moore–Penrose inverse of each of its elements.

Clearly, the partition $\mathcal{Q}^{\text{sym}}$ obtained by symmetrising a homogeneous coherent configuration is a Jordan scheme. *Are there any others?*

The fourth column in Table 1 gives the number of association schemes on n points for $n \le 30$. This number is obtained from the information provided by Hanaki and Miyamoto [15], simply by checking which of the configurations are symmetric. (As noted earlier, in the case $n = 29$, the classification of strongly regular graphs by Spence [33] is also used.) The fifth column lists the number of generously transitive groups of degree n which are 2-closed, or equivalently the number of Schurian association schemes. The smallest number of points in a non-Schurian association scheme is 16; an example of such a scheme is the Shrikhande strongly regular graph [31].

Note also that the term "cellular algebra" has been used with an entirely different meaning by Graham and Lehrer [14] and in a number of subsequent papers.

2 The partial order

The set of partitions of $\Omega \times \Omega$ forms a lattice, under the ordering given by $\mathcal{P} \preceq \mathcal{Q}$ if $\mathcal{P}$ refines $\mathcal{Q}$ (that is, every $\mathcal{P}$-class is contained in a $\mathcal{Q}$-class). We use the symbol $\vee$ to denote the join in this lattice. Thus, (α_1, α_2) and (β_1, β_2) lie in the same $\mathcal{P} \vee \mathcal{Q}$-class if it is possible to move from (α_1, α_2) to (β_1, β_2) by a sequence of moves each within either a $\mathcal{P}$-class or a $\mathcal{Q}$-class.

The following result is due to Higman [19]; the short proof here is due to Bailey [3].

Theorem 2.1 *The join (in the partition lattice) of two c.c.s is a c.c. The same holds for homogeneous, stratifiable, commutative, or symmetric c.c.s.*

Proof A function is constant on the classes of $\mathcal{P} \vee \mathcal{Q}$ if and only if it is constant on both the $\mathcal{P}$-classes and the $\mathcal{Q}$-classes. So $\mathrm{BA}(\mathcal{P} \vee \mathcal{Q}) = \mathrm{BA}(\mathcal{P}) \cap \mathrm{BA}(\mathcal{Q})$. ■

From this, we deduce the following result.

Theorem 2.2 *The coherent configurations on Ω form a lattice.*

Proof We must show that any two coherent configurations have a greatest lower bound. But the g.l.b. of $\mathcal{P}$ and $\mathcal{Q}$ is the join of all the coherent configurations lying below both $\mathcal{P}$ and $\mathcal{Q}$; and this set is non-empty, since it contains at least the discrete c.c. (Note that the meet of two coherent configurations is not the same as their meet as partitions!) ■

This argument fails for any of the other classes of coherent configurations, since there may be no c.c. in the appropriate class below any given $\mathcal{P}$ and $\mathcal{Q}$. We will see examples for association schemes later.

Theorem 2.1 also has the following consequence.

Corollary 2.3 *Let $\mathcal{P}$ be any partition of Ω^2. Then there is a unique coherent configuration $\mathcal{P}^*$ on Ω which is maximal with respect to being a refinement of $\mathcal{P}$. Moreover,* $\mathrm{Aut}(\mathcal{P}) = \mathrm{Aut}(\mathcal{P}^*)$.

Proof The set of coherent configurations below $\mathcal{P}$ is non-empty (since it contains the discrete c.c.) and so has a supremum. The construction shows that any permutation preserving $\mathcal{P}$ will preserve $\mathcal{P}^*$. The converse holds since $\mathcal{P}^*$ refines $\mathcal{P}$. ∎

The coherent configuration $\mathcal{P}^*$ can be computed efficiently from $\mathcal{P}$. First, replacing $\mathcal{P}$ by $\mathcal{P} \wedge \mathcal{P}^\top$ if necessary (where $\mathcal{P}^\top$ is the partition whose parts are the converses of the parts of $\mathcal{P}$), we can assume that $\mathcal{P} = \mathcal{P}^\top$. Let $\mathcal{P}'$ be the partition in which two ordered pairs (α_1, β_1) and (α_2, β_2) are in the same part whenever, for any two parts X, Y of $\mathcal{P}$, the numbers of points $\gamma \in \Omega$ for which $(\alpha_i, \gamma) \in X$ and $(\gamma, \beta_i) \in Y$ are equal for $i = 1, 2$. Clearly, if $\mathcal{P} = \mathcal{P}^\top$, then $\mathcal{P}' = (\mathcal{P}')^\top$. Now set $\mathcal{P}_0 = \mathcal{P}$ and $\mathcal{P}_{n+1} = \mathcal{P}'_n$ for $n \geq 0$. There exists n such that $\mathcal{P}_n = \mathcal{P}_{n+1}$; set $\mathcal{P}^* = \mathcal{P}_n$.This is the required coherent configuration.

This provides a useful reduction for the problem of finding the automorphism group of a partition of Ω^2: we may assume that the partition is a coherent configuration. It also explains why coherent configurations provide difficult test cases for this problem, since this reduction gives no extra information in this case. This problem includes the problem of finding the automorphism group of a graph (consider the partition of Ω^2 into the diagonal, the set of edges, and the set of non-edges of the graph). This is the original context in which coherent configurations arose in the work of Weisfeiler and Leman [34].

Coherent configurations were also used by the Soviet school for establishing the maximality of certain subgroups of symmetric and alternating groups, using the following principle:

Proposition 2.4 *Let G be a 2-closed permutation group on Ω. Suppose that no coherent configuration lies strictly between $\mathcal{K}(G)$ and the trivial configuration on Ω. Then any proper supergroup of G in* $\mathrm{Sym}(\Omega)$ *is doubly transitive.* ∎

For example, Kaluznin and Klin [25] verified the maximality in this sense of $\mathcal{K}(G)$, where G is the symmetric group of degree n in its action on m-sets, for n sufficiently large in terms of m, and thus showed the maximality of G in the symmetric or alternating group of degree $\binom{n}{m}$.

Much more is known now about maximality, but the proofs typically use the classification and subgroup structure of the finite almost-simple groups (see [26]). As a result, many interesting questions on the maximality of $\mathcal{K}(G)$ for various groups G remain unanswered! (See Faradžev *et al.* [12] for an account of some of the results that have been obtained.)

At about the same time, Kageyama [24] was considering the same question for association schemes, motivated by the question of when the number of associate classes in a partially balanced incomplete block design could be reduced.

We conclude this section with a curiosity. Let Λ_n denote the lattice of c.c.s on a fixed n-set, say $\{0, 1, \ldots, n-1\}$.

Proposition 2.5 *For any n, there is an embedding of Λ_n as a down-set in Λ_{n+1} preserving zero, join and meet.*

Proof Let $\mathcal{P}$ be a c.c. on $\{0, 1, \ldots, n-1\}$, with fibres $F_1, \ldots, F_t$. We define $F(\mathcal{P})$ to be the partition of $\{0, \ldots, n\}^2$ consisting of the following sets of pairs: all the parts of $\mathcal{P}$; the sets $\{(x, n) : x \in F_i\}$ and their converses for $i = 1, \ldots, t$; and the set $\{(n, n)\}$. It is not difficult to show that $F(\mathcal{P})$ is a c.c., and that every c.c. that has $\{(n, n)\}$ as one of its parts arises in this way. Moreover, F maps the discrete c.c. to the discrete c.c. and preserves join and meet. ∎

Note that the direct limit of these lattices and maps is the "lattice of all finite coherent configurations". (The direct limit contains no top element; this has to be adjoined.)

3 AS-friendly and AS-free groups

In the remainder of this paper, we will be concerned with association schemes. Although there is a coherent configuration $\mathcal{K}(G)$ associated with every permutation group G (and this configuration is trivial if and only if the group is doubly transitive), this is far from being the case for association schemes. Since any association scheme is a homogeneous coherent configuration, I will consider only transitive permutation groups in this section, which reports on work of Alejandro *et al.* [1]. I have sketched some of the proofs, but refer to [1] for full details.

We begin with some definitions. Let G be a transitive permutation group on the finite set Ω.

(a) We say that G is *AS-free* if the only G-invariant association scheme on Ω is the trivial scheme.

(b) We say that G is *AS-friendly* if there is a unique minimal G-invariant association scheme on Ω.

Of course, if we replaced "AS" by "CC" in the above definitions, then every group would be CC-friendly, and the CC-free groups would be precisely the doubly transitive groups.

Note that a 2-homogeneous group G (one which is transitive on the 2-element subsets of Ω) is AS-free, since the symmetrisation of $\mathcal{K}(G)$ is the trivial configuration.

It is difficult, both theoretically and computationally, to decide whether a transitive group G of large rank is AS-friendly or AS-free: we have to merge the symmetrised orbitals in all possible ways and check whether an association scheme is obtained. The following idea can be used to reduce the search for G-invariant association schemes. Suppose that the permutation character of G has the form $\pi = \sum e_i\chi_i$, where the χ_i are distinct irreducible characters. Then the rank of G (the number of orbits on Ω^2) is equal to $\sum e_i^2$, while the dimension of the largest commutative semisimple subalgebra of $\mathrm{BA}(\mathcal{K}(G))$ is $\sum e_i$. So, if some of the multiplicities e_i are large, then a lot of merging is required.

3.1 Basic theory

We begin with an example of a group which is not AS-friendly. Let G be the symmetric group S_n (for $n \geq 5$), acting on the set Ω of ordered pairs of distinct elements from the set $\{1,\ldots,n\}$: we write the pair (i,j) as ij for brevity. The coherent configuration $\mathcal{K}(G)$ consists of the following parts:

$$\begin{aligned} R_1 &= \{(ij,ij) : i \neq j\}, \\ R_2 &= \{(ij,ji) : i \neq j\}, \\ R_3 &= \{(ij,ik) : i,j,k \text{ distinct}\}, \\ R_4 &= \{(ij,kj) : i,j,k \text{ distinct}\}, \\ R_5 &= \{(ij,ki) : i,j,k \text{ distinct}\}, \\ R_6 &= \{(ij,jk) : i,j,k \text{ distinct}\}, \\ R_7 &= \{(ij,kl) : i,j,k,l \text{ distinct}\}. \end{aligned}$$

We have $R_5^\top = R_6$; all other relations are symmetric. The symmetrised partition is not an association scheme, but we find three minimal association schemes as follows:

- the *pair* scheme: $\{R_1, R_2, R_3 \cup R_4, R_5 \cup R_6, R_7\}$ (see [24, pp. 576–578]);
- two "divisible" schemes $\{R_1, R_3, R_2 \cup R_4 \cup R_5 \cup R_6 \cup R_7\}$ and $\{R_1, R_4, R_2 \cup R_3 \cup R_5 \cup R_6 \cup R_7\}$.

These are all incomparable, so there is not a unique minimal association scheme.

The next result shows how the concepts just defined are related to more familiar concepts of permutation group theory.

Theorem 3.1 *The following implications hold between properties of a permutation group G:*

$$\begin{array}{ccccccc} \textit{2-transitive} & \Rightarrow & \textit{2-homogeneous} & \Rightarrow & \textit{AS-free} & \Rightarrow & \textit{primitive} \\ \Downarrow & & \Downarrow & & \Downarrow & & \Downarrow \\ \textit{gen. trans.} & \Rightarrow & \textit{stratifiable} & \Rightarrow & \textit{AS-friendly} & \Rightarrow & \textit{transitive} \end{array}$$

None of these implications reverses, and no further implications hold. ∎

Since the properties of a permutation group G listed in the theorem depend only on $\mathcal{K}(G)$, it suffices to consider 2-closed groups. Table 2 gives the numbers of 2-closed groups of small degree which are respectively 2-transitive, 2-homogeneous, AS-free, primitive, generously transitive, stratifiable, AS-friendly, and transitive.

n	2T	2H	AΦ	Pr	GT	St	AF	Tr
1	1	1	1	1	1	1	1	1
2	1	1	1	1	1	1	1	1
3	1	2	2	2	1	2	2	2
4	1	1	1	1	3	4	4	4
5	1	1	1	3	2	3	3	3
6	1	1	1	1	4	7	7	8
7	1	2	2	4	2	4	4	4
8	1	1	1	1	10	20	20	21
9	1	1	1	2	6	12	12	12
10	1	1	1	2	8	11	11	13
11	1	2	2	4	2	4	4	4
12	1	1	1	1	21	47	47	59
13	1	1	1	6	4	6	6	6
14	1	1	1	1	8	14	14	16
15	1	1	1	2	10	23	23	24
16	1	1	1	4	56	171	171	206
17	1	1	1	5	4	5	5	5
18	1	1	1	1	32	71	71	93
19	1	2	2	6	3	6	6	6
20	1	1	1	1	41	73	73	95
21	1	1	1	3	11	29	29	32
22	1	1	1	1	8	14	14	16
23	1	2	2	4	2	4	4	4
24	1	1	1	1	136	454	454	669
25	1	1	1	9	20	32	32	32
26	1	1	1	1	14	20	20	24
27	1	2	2	5	38	112	112	122
28	1	1	1	4	47	103	103	124
29	1	1	1	6	4	6	6	6
30	1	1	1	1	73	166	166	228

Table 2: Small 2-closed permutation groups

Note that up to degree 30, every AS-free group is 2-homogeneous and every AS-friendly group is stratifiable.

The smallest 2-closed primitive group which is not AS-friendly is PSL(2, 11), with degree 55. The smallest 2-closed primitive groups which are AS-friendly but not stratifiable are PSL(2, 13), with degrees 78 and 91. These groups are numbers (55, 1), (78, 1), (91, 1) and (91, 3) in the list of primitive groups available in GAP. The smallest examples of AS-free groups which are not stratifiable have degree 234, and are isomorphic to PSL(3, 3) and PSL(3, 3) : 2, numbers (234, 1) and (234, 2) in the list. (Further examples of such groups will be given later.) 2-homogeneous groups which are not generously transitive are well known (such groups must have prime power degree congruent to 3 mod 4). I hope to make the GAP code used for these tests available shortly.

The class of AS-friendly groups is also closed under taking supergroups, wreath products, and primitive components:

Theorem 3.2 *(a) If a group has an AS-friendly subgroup, then it is AS-friendly.*

(b) The class of AS-friendly permutation groups is closed under wreath product (with the imprimitive action).

(c) Let G be imprimitive; let Γ be a system of imprimitivity and Δ a block in Γ, and let H be the permutation group induced on Δ by its setwise stabiliser and K the group induced on Γ by G, so that $G \leq H \wr K$. If G is AS-friendly, then so are H and K.

(d) The same assertions hold with "stratifiable" or "generously transitive" in place of "AS-friendly". ■

3.2 Regular groups

Groups whose regular representation is AS-friendly have been determined. In particular, the properties "AS-friendly" and "stratifiable" coincide for regular groups.

Following Bailey [3], a partition $P = \{P_1, \ldots, P_s\}$ of a group G is called a *blueprint* if the partition $\mathcal{P} = \{R_1, \ldots, R_s\}$ of $G \times G$ given by

$$R_i = \{(x, y) : xy^{-1} \in P_i\}$$

is a coherent configuration on G. Note that this coherent configuration is invariant under right translation by G. This condition is equivalent to the assertion that the sums (in the group ring $\mathbb{Z}G$) of the classes $P_1, \ldots, P_s$ span a *Schur ring*: see Wielandt [35].

The *inverse partition* of a group G is the partition whose parts are the sets $\{g, g^{-1}\}$ for $g \in G$.

Theorem 3.3 *For a finite group G, the following four conditions are equivalent:*

(a) the regular action of G is AS-friendly;

(b) the regular action of G is stratifiable;

(c) the inverse partition of G is a blueprint;

(d) either G is abelian, or $G \cong Q \times A$ where Q is the quaternion group of order 8 and A is an elementary abelian 2-group.

Proof I will outline briefly the two non-trivial parts of the proof.

Clearly (c) implies (b) implies (a). To show that (a) implies (c), we proceed as follows. Let P be the minimal blueprint on G; we must show that P is the inverse partition. So take an element $g \in G$; we must show that $(1, h)$ lies in the same class as $(1, g)$ if and only if $h = g^{\pm 1}$. Now the right cosets of $H = \langle g \rangle$ form a system of imprimitivity, and so $G \leq H \wr S_m$, where $m = |G : H|$. This group preserves the association scheme obtained by "nesting" the cyclic scheme on H in the trivial scheme on m points, so the unique minimal G-invariant scheme is contained in this one. But in this scheme, we see that the result holds.

Also, the proof that (d) implies (b) uses relatively straightforward representation theory. Suppose that a group satisfies (c). It is easy to see that, given any two elements $g, h \in G$, either g and h commute, or each inverts the other. Thus every subgroup of G is normal, and the structure is determined by the theorem of Dedekind (see [22], Satz 7.12 on p. 308) and a little more work. ■

3.3 Primitive groups

We have seen that 2-homogeneous groups are AS-free. Are there any other transitive AS-free groups?

A permutation group is called *non-basic* if there is a bijection between Ω and Γ^Δ (the set of functions from Δ to Γ) for some finite sets Γ and Δ, which induces an isomorphism from G to a subgroup of $\mathrm{Sym}(\Gamma) \wr \mathrm{Sym}(\Delta)$ with the product action. This concept arises in the O'Nan–Scott classification of primitive permutation groups, see [10], p. 106. Of course, a group is *basic* if it is not non-basic.

Theorem 3.4 *Let G be a transitive AS-free group. Then G is primitive and basic, and is 2-homogeneous, diagonal or almost simple.*

Proof An imprimitive permutation group G preserves the "divisible" association scheme whose parts are the diagonal, the G-congruence with the diagonal removed, and the rest of $\Omega \times \Omega$, while a non-basic group preserves a *Hamming scheme* (see Delsarte [9]). By the O'Nan–Scott theorem, basic primitive groups are affine, diagonal, or almost simple. An affine group has an abelian regular normal subgroup and thus is stratifiable; so if such a group is AS-free, then it is 2-homogeneous. ■

Almost simple AS-free groups which are not 2-homogeneous do exist. This can be seen from the paper of Faradžev *et al.* [12]. These authors consider the following problem. *Let G be a simple primitive permutation group of order at most 10^6 but not* $\mathrm{PSL}(2, q)$. *Describe the coherent configurations above $\mathcal{K}(G)$.* Table 3.5.1 on p. 115 gives their results. In several cases, no non-trivial configuration consists entirely of symmetric matrices: such groups are of course AS-free. The smallest example is the group $\mathrm{PSL}(3, 3)$, acting on the right cosets of $\mathrm{PO}(3, 3)$ (a subgroup isomorphic to S_4), with degree 234; as we have seen, this is the smallest AS-free group which is not 2-homogeneous. Other examples of AS-free groups in this list are M_{12}, degree 1320; J_1, degree 1463, 1540 or 1596; and J_2, degree 1800. The situation is not well understood!

The table also gives another example of a primitive group which is not AS-friendly; this is M_{12}, with degree 495.

No AS-free primitive diagonal group is known at present. It is known that the socle of such a group must have at least four simple factors:

- A primitive diagonal group whose socle has two factors is a group of weak automorphisms of the *conjugacy class configuration* of a simple group (corresponding to the blueprint formed by the conjugacy classes), and indeed $\mathcal{K}(G)$ is commutative for such groups G (so they are stratifiable).

- A primitive diagonal group whose socle has three factors preserves a *Latin square scheme* based on the Cayley table of the simple group. However, it is not known whether such groups are AS-friendly.

Thus, the smallest possible degree of an AS-free diagonal group is 216 000.

In the next subsection I report on a more general investigation of diagonal groups (not necessarily primitive).

3.4 Diagonal groups

The *diagonal group* $D(T,n)$, where T is a group and n a positive integer, is defined as the permutation group on the set

$$\Omega = T^n = \{[x_1,\ldots,x_n] : x_1,\ldots,x_n \in T\}$$

generated by the following permutations:

(a) the group T^n acting by right translation, that is, the permutations

$$[x_1,\ldots,x_n] \mapsto [x_1t_1,\ldots,x_nt_n]$$

for $t_1,\ldots,t_n \in T$;

(b) the automorphism group of T, acting coordinatewise, that is,

$$[x_1,\ldots,x_n] \mapsto [x_1^\alpha,\ldots,x_n^\alpha]$$

for $\alpha \in \mathrm{Aut}(T)$;

(c) the symmetric group S_n, acting by permuting the coordinates, that is,

$$\pi : [x_1,x_2,\ldots,x_n] \mapsto [x_{1\pi},x_{2\pi},\ldots,x_{n\pi}]$$

for $\pi \in S_n$;

(d) the permutation

$$\tau : [x_1,x_2,\ldots,x_n] \mapsto [x_1^{-1},x_1^{-1}x_2,\ldots,x_1^{-1}x_n].$$

The group $D(T, n)$ is a "maximal diagonal group"; by Theorem 3.2, if any diagonal group having a normal subgroup T^{n+1} acting on the cosets of the diagonal is generously transitive, stratifiable, or AS-friendly, then $D(T, n)$ will have this property. If T is abelian, then $D(T, n)$ has an abelian regular normal subgroup consisting of the permutations of type (a), and so is AS-friendly (and even stratifiable). For non-abelian groups T, it is not known when $D(T, n)$ can be AS-friendly. However, the following result is known about generous transitivity and stratifiability.

Theorem 3.5 *Let T be a non-abelian finite group.*

(a) If $D(T, n)$ is stratifiable, then $n \leq 8$; and if $D(T, n)$ is generously transitive, then $n \leq 7$.

(b) The group $D(T, 7)$ is generously transitive if and only if T is the quaternion group of order 8.

Proof The group $D(T, n)$ is generously transitive if and only if every n-tuple of elements of T can be inverted by a combination of transformations of types (b)–(d). If T is non-abelian, it is shown that this is impossible for suitably chosen 8-tuples, while for 7-tuples we find that any two non-commuting elements generate Q_8. The argument for stratifiability is similar but a bit more complicated. ∎

The group $D(T, n)$ is primitive if and only if T is characteristically simple, and is AS-free only if T is simple. It seems likely that the bounds in the theorem can be improved for non-abelian simple groups T.

Acknowledgment I am grateful to P. P. Alejandro, R. A. Bailey, A. Hanaki, A. Hulpke, M. H. Klin, L. H. Soicher and E. Spence for helpful comments.

References

[1] P. P. Alejandro, R. A. Bailey and P. J. Cameron, Association schemes and permutation groups, *Discrete Math.*, to appear.

[2] R. A. Bailey, Strata for randomized experiments, *J. Royal Statist. Soc.* (B) **53** (1991), 27–78.

[3] R. A. Bailey, Suprema and infima of association schemes, *Discrete Math.* **248** (2002), 1–16.

[4] R. A. Bailey, *Association Schemes: Designed Experiments, Algebra and Combinatorics*, Cambridge University Press, Cambridge, to appear.

[5] E. Bannai, An introduction to association schemes, *Methods of Discrete Mathematics* (S. Löwe, F. Mazzocca, N. Melone and U. Ott, eds.), Quaderni di Mathematica **5**, Seconda Università di Napoli, Napoli, 1999, pp. 1–70.

[6] R. C. Bose and D. M. Mesner, On linear associative algebras corresponding to association schemes of partially balanced designs, *Ann. Math. Statist.* **30** (1959), 21–38.

[7] R. C. Bose and K. R. Nair, Partially balanced incomplete block designs, *Sankhyā* **4** (1939), 337–372.

[8] J. H. Conway, A. Hulpke and J. McKay, On transitive permutation groups, *London Math. Soc. J. Comput. Math.* **1** (1998), 1–8, `http://www.lms.ac.uk`

[9] Ph. Delsarte, An algebraic approach to the association schemes of coding theory, *Philips Research Reports Suppl.* **10** (1973).

[10] J. D. Dixon and B. Mortimer, *Permutation Groups*, Graduate Texts in Mathematics **163**, Springer-Verlag, New York, 1996.

[11] S. Evdokimov and I. Ponomarenko, Separability number and Schurity number of coherent configurations, *Electronic J. Combinatorics* **7**(1) (2000), #R31, `http://www.combinatorics.org`

[12] I. A. Faradžev, M. H. Klin and M. E. Muzichuk, Cellular rings and groups of automorphisms of graphs, in *Investigations in Algebraic Theory of Combinatorial Objects* (I. A. Faradžev, A. A. Ivanov, M. H. Klin and A. J. Woldar, eds.), Kluwer, Dordrecht, 1994, pp. 1–152.

[13] The GAP Group, GAP — Groups, Algorithms, and Programming, Version 4.2; Aachen, St Andrews, 1999, `http://www-gap.dcs.st-and.ac.uk/~gap`

[14] J. J. Graham and G. I. Lehrer, Cellular algebras, *Invent. Math,* **123** (1996), 1–34.

[15] A. Hanaki and I. Miyamoto, Classification of association schemes with small vertices, `http://kissme.shinshu-u.ac.jp/as/`

[16] D. G. Higman, Finite permutation groups of rank 3, *Math. Z.* **86** (1964), 145–156.

[17] D. G. Higman, Intersection matrices for finite permutation groups, *J. Algebra* **6** (1967), 22–42.

[18] D. G. Higman, *Combinatorial Considerations about Permutation Groups*, Mathematical Institute, Oxford, 1971.

[19] D. G. Higman, Coherent algebras, *Linear Algebra Appl.* **93** (1987), 209–239.

[20] A. Hulpke, Constructing transitive permutation groups, in preparation.

[21] A. Hulpke, Transitive groups of small degree, `http://www.math.colostate.edu/~hulpke/smalldeg.html`

[22] B. Huppert, *Endliche Gruppen I*, Grundlehren der mathematischen Wissenschaften **134**, Springer-Verlag, Berlin, 1967.

[23] N. Jacobson, *Structure and Representation of Jordan Algebras*, American Mathematical Society, Providence, RI, 1968.

[24] S. Kageyama, Reduction of associate classes for block designs and related combinatorial arrangements, *Hiroshima Math. J.* **4** (1974), 527–618.

[25] L. A. Kaluznin and M. H. Klin, On certain maximal subgroups of symmetric and alternating groups, *Math. USSR Sbornik* **16** (1972), 95–123.

[26] M. W. Liebeck, C. E. Praeger and J. Saxl, The classification of the maximal subgroups of the finite symmetric and alternating groups, *J. Algebra* **111** (1987), 365–383.

[27] J. D. Malley, *Optimal Unbiased Estimation of Variance Components*, Lecture Notes in Statistics **39**, Springer–Verlag, Berlin, 1986.

[28] B. D. McKay, `nauty` user's guide (version 1.5), Technical report TR-CS-90-02, Computer Science Department, Australian National University, 1990.

[29] B. D. McKay and E. Spence, in preparation; see `http://gauss.maths.gla.ac.uk/~ted/srgraphs.html`

[30] L. Pyber, Asymptotic results for permutation groups, *Groups and Computation* (L. Finkelstein and W M. Kantor, eds.), DIMACS Series in Discrete Mathematics and Theoretical Computer Science **11**, American Mathematical Society, Providence, RI, 1993, pp. 197–219.

[31] S. S. Shrikhande, The uniqueness of the L_2 association scheme, *Ann. Math. Statistics* **30** (1959), 781–798.

[32] L. H. Soicher, GRAPE: A system for computing with graphs and groups, in *Groups and Computation* (L. Finkelstein and W. M. Kantor, eds.), DIMACS Series in Discrete Mathematics and Theoretical Computer Science **11**, American Mathematical Society, Providence, RI, 1993, pp. 287–291.

[33] E. Spence, Strongly regular graphs on at most 64 vertices, `http://gauss.maths.gla.ac.uk/~ted/srgraphs.html`

[34] B. Yu. Weisfeiler and A. A. Leman, Reduction of a graph to a canonical form and an algebra which appears in the process, *Scientific-Technological Investigations* (2) **9** (1968), 12–16.

[35] H. Wielandt, *Finite Permutation Groups*, Academic Press, New York, 1964.

[36] P.-H. Zieschang, Homogeneous coherent configurations as generalized groups and their relationship to buildings, *J. Algebra* **178** (1995), 677–709.

MATHEMATICAL DEVELOPMENTS FROM THE ANALYSIS OF RIFFLE SHUFFLING

Persi Diaconis

1. *Introduction* The most common method of mixing cards is the ordinary riffle shuffle, in which a deck of n cards (often $n = 52$) is cut into two parts and the parts are riffled together. A sharp mathematical analysis for a natural model of riffle shuffling was carried out by Bayer and Diaconis (1992). This gives closed form expressions for the chance of any permutation and allows analytic approximation and exact numerical evaluation to show things like "seven shuffles are necessary and suffice to approximately randomize 52 cards". These results are carefully stated in Section 2A.

The shuffling work builds on earlier studies of Jordan (magic tricks), Borel (bridge), Gilbert, Shannon, Reeds (basic model) and D. Aldous (coupling). This background is described in Section 2B. The "seven shuffles" result is mildly dependent on the choice of metric and a number of alternative measures of randomness are discussed in Section 2C.

There is a mathematical reason that allows riffle shuffles to be analyzed so completely. The basic shuffling model falls squarely into Solomon's descent algebra (and indeed gives an independent development). This allows shuffling theorems to be translated into permutation enumeration results (e.g. how many permutations have a given number of descents and a given cycle structure). The eigenvalues of the Markov chain underlying shuffling were actually first determined in an investigation of Hochschild homology(Hanlon). There is an intimate connection with free Lie algebras and the Poincare-Birkoff-Witt Theorem (Bergeron-Bergeron-Garsia). The chance of a given cycle structure after riffle shuffling *equals* the chance that a random degree n polynomial has a given number of irreducible factors. This in turn is explained by considering the connection between shuffling and the action of the associated Lie type group $SL_n(F_q)$ on its Lie algebra (Fulman). Finally, shuffling gives a fairly direct interpretation of Schur symmetric functions (Stanley-Fulman). These results are described in section three.

The analyses above seem so rich and natural that they call out for generalization. A sweeping generalization of the theory was discovered by Bidigare, Hanlon and Rockmore. This involves random walk on the chambers of a hyperplane arrangement. The classical braid arrangement gives riffle shuffles but there are many other hyperplane arrangements where the chambers can be labeled by natural combinatorial objects and much (but not all) of the theory goes through. In an amazing synthesis, Ken Brown has shown that almost everything can be pushed through to random walks on idempotent semi-groups. This allows analysis of natural random walk on the chambers of spherical buildings. These results are described in Section 4.

The final section describes ten open problems.

Throughout I have tried to show the links with algebra. To be fair, many of the authors cited have no interest in the card shuffling implication of their work. This paper has improved from detailed comments of Ken Brown, Nantel Bergeron, Jason Fulman, Adriano Garsia and J.C. Uyemura-Reyes.

2A. Basic Shuffling.

The basic riffle shuffling model was introduced by Gilbert and Shannon (See Gilbert (1955)) and independently by Reeds (1981). It can be described as a probability distribution $Q(w)$ on the symmetric group S_n – the GSR Distribution. One description of Q is as follows: cut the deck into two piles according to the binomial distribution so the chance that pile one has j cards is $\binom{n}{j}/2^n$. Then, sequentially drop cards from the bottoms of the two piles according to the following rule: if at some stage pile one has "A" cards and pile two has "B" cards, drop the next card from pile one with probability $A/(A+B)$. This is continued until the two piles are exhausted and then the piles are pushed together. An equivalent description in terms of inverse riffle shuffles is due to Gilbert, Shannon and Reeds. An inverse shuffle begins by labeling each of n cards zero or one by a flip of a fair coin. Then, all the cards labeled zero are removed and placed on top keeping the cards in the same relative order. It is a simple exercise to show that the forward and backward descriptions are the same. From either description, given the cut, all ways of interleaving are equally likely, so the GSR shuffle is a maximum entropy model. The identity has probability $\frac{n+1}{2^n}$ while all other possible permutations have probability $1/2^n$.

Repeated shuffles are modeled by convolution:

$$Q * Q(w) = \sum_u Q(wu^{-1})Q(u)$$

Thus the chance of w after two shuffles is calculated as the chance of first choosing u and then choosing the permutation resulting in w. Similarly, $Q^{*k}(w) = \Sigma_u Q^{*(k-1)}(wu^{-1})Q(u)$. These ingredients complete the description of the GSR measure $Q^{*k}(w)$. Of course, shuffling is an example of random walk on a group and of a finite state-space Markov chain. See Saloff-Coste (2001, 2002), for an extensive overview with relevance to shuffling.

Repeated shuffling converges to the uniform distribution $U(w) = 1/n!$. The earliest works on Markov chains, due to Markov (1906) and Poincare (1912), used shuffling cards as an example. They gave results which allow us to conclude that

$$Q^{*k}(w) \to U(w) \quad \text{as} \quad k \to \infty.$$

It is natural to try to quantify this statement. The usual distance to stationarity is the total variation distance

$$\|Q^{*k} - U\| = \max_{A \subset s_n} |Q^{*k}(A) - U(A)| = \frac{1}{2}\sum_w |Q^{*k}(w) - U(w)|.$$

Consider the middle term. Its interpretation is this: let A be any subset of S_n (e.g. the set of all permutations where the ace of spades is in the top half). Calculate the chance of A after k shuffles (that is $Q^{*k}(A)$). Calculate the uniform measure of A (namely $|A|/n!$). Take the difference between these numbers and then take the A which makes this difference as large as possible. This is a very non-forgiving distance. If $\|Q^{*k} - U\| \le \epsilon$ then shuffling is close to uniform for any set A.

These definitions translate the basic question "how many times should a deck of cards be shuffled to adequately mix it?" into a well posed math problem: given $\epsilon > 0$ how large should k be to have $\|Q^{*k} - U\| < \epsilon$? A historical review of progress on this problem is contained in the following section. Here we state the basic result.

Theorem 1. [Bayer-Diaconis]. When $n = 52$, the distance to uniformity is

k:	1	2	3	4	5	6	7	8	9	10
$\|Q^{*k} - U\|$	1.000	1.000	1.000	1.000	.924	.614	.334	.167	.085	.043

For general n, and $k = \frac{3}{2}\log_2 n + c$

$$\|Q^{*k} - U\| = 1 - 2\Phi\left(\frac{-2^{-c}}{4\sqrt{3}}\right) + 0\left(\frac{1}{n^{1/4}}\right) \quad \text{with} \quad \Phi(x)\frac{1}{\sqrt{2\pi}}\int_{\infty}^{x} e^{-t^2/2}dt.$$

Remarks The total variation distance is a number between zero and one. A graph of Q^{*k} vs k shows it stays close to its maximum until a bit before $\frac{3}{2}\log_2 n$ and then falls exponentially fast to zero. The analysis shows that for $k = \frac{3}{2}\log_2 n + c$ the distance goes to one doubly exponentially fast as $c \to -\infty$ and to zero exponentially fast as $c \to \infty$. These asymptotic results are borne out by the data for $n = 52$. The cutoff occurs at about seven shuffles. From six shuffles on, the variation distance falls by a factor of 2 for each extra shuffle. It is worth noting that the table is derived from exact results from Theorem 2 below not from an asymptotic approximation.

It is natural to wonder if this mathematics has much to do with real shuffles. People used to think that cards were suitably well mixed after three, four or five shuffles. Like many things people believe, this is simply not true. In Section 2B a classical card trick and some extensive analysis of bridge hands are used to prove this point.

Theorem 1 is a consequence of a more central result. To explain it, it is useful to have a geometric description of riffle shuffles. Picture n points dropped uniformly and independently into the unit interval. Label the ordered points, left to right, as $x_1, x_2, \ldots, x_n$. Now perform the baker's transform of $[0,1]$ to itself. This takes $x \to 2x (\text{mod } 1)$. The points x_i are permuted inducing a permutation w. Note that there are a binomial number of the x_i in $[0, 1/2]$. The baker's map stretches these out to $[0,1]$. The same holds for the points in $[\frac{1}{2}, 1]$. These two sets of points are interrleaved. It is not hard to see that the induced permutation has exactly the GSR distribution $Q(w)$.

This geometric description suggests a variant which will prove useful. For positive integer a, consider the a-shuffle which results from n random points under the map $x \to ax$ (mod 1). In shuffling language one may cut the deck into a-packets according to the multinomial distribution $\binom{n}{n_1 \ldots n_a}/a^n$ with $0 \leq n_i \leq n$, $\Sigma_{k=1}^{a} n_i = n$. The a packets are sequentially mixed, dropping a card from the ith packet with probability proportional to packet size. These equivalent definitions result in a probability $Q_a(w)$. In present notation $Q_2(w) = Q(w)$. From the geometric description, it is easy to see that an a-shuffle followed by a b-shuffle is the same as an ab shuffle thus $Q_a * Q_b = Q_{ab}$ and $Q_2^{*k} = Q_2k$. It is enough to study only a-shuffles. The main result of Bayer-Diaconis can now be stated.

Theorem 2 For all positive integers n and a, $Q_a(w) = \dfrac{\binom{n+a-r}{n}}{a^n}$ with $r = r(w)$ the number of rising sequences in w.

To explain, consider a permutation w as an arrangement of a deck of cards, with w_i the label of the card at position i. Decompose w into disjoint rising sequences by finding card labeled 1, and then card labeled 2 if label 2 is below label 1. Continue until label k stopping if label $k+1$ is above one of $\{1, 2, \ldots, k\}$. Remove cards labeled $\{1, 2, \ldots, k\}$. This is the first rising sequence. Continue with the reduced deck, finding $\{k+1, \ldots, k+\ell\}$ a second rising sequence and so on. Thus, for $n = 9$, the permutation 716248359 has rising sequences 123, 45, 6, 789. Let $r = r(w)$ be the number of rising sequences obtained. Thus $1 \leq r \leq n$. Another description: $r(w) = d(w^{-1}) + 1$ with $d(w^{-1})$ the number of descents in w^{-1}. Descents will make a major appearance in Section 2C.

We will not give a proof of Theorem 2 here (see Bayer-Diaconis (1992) or the clear elementary treatment of Mann (1995)). It is straightforward from the geometric description using the "stars and bars" argument of elementary combinatorics. The hard part was discovering the result. We did this by looking at exact computer calculations for small decks (size 3, 4, 5) and noticing a pattern.

Theorem 2 gives yet another description of the GSR measure

$$Q(w) = \begin{cases} (n+1)/2^n & \text{If } w = id \\ 1/2^n & \text{If } w \text{ has 2 rising sequences} \\ 0 & \text{otherwise.} \end{cases}$$

Theorem 2 reduces the calculation of total variation to evaluating

$$\|Q^{*k} - u\| = \frac{1}{2}\sum_{j=1}^{n} k_n(j)\left|\frac{\binom{n+2^k-j}{n}}{2^{kn}} - \frac{1}{n!}\right|$$

with $k_n(j)$ the number of permutations with j rising sequences. At this point another surprise occurred. The $k_n(j)$ are very well studied as the coefficients of Eulerian polynomials (Stanley, 1997). This allowed careful asymptotic analysis and led to Theorem 1.

This concludes our overview of the basic shuffling story. We turn to a bit of history and then some more mathematical consequences.

2B. History and Practical Consequences.

The earliest treatments of Markov chains treat card shuffling as a leading example (Markov (1906), Poincare (1912), Doob (1954)). These treatments show that shuffling cards *eventually* results in a well mixed deck. It is very hard to guess how many shuffles are needed. When $n = 52$, $n! \doteq 8 \times 10^{67}$. For the other popular method of shuffling (overhand) Pemantle (1984) shows order $n^2 \log n$ shuffles suffice. This is more than 2500 when $n = 52$.

Emil Borel in Borel and Cheron (1940) began a quantitative investigation by studying how long individual cards and pairs of cards take to randomize. This allowed him to conclude that at least six or seven shuffles are needed. Similar conclusions are drawn by Keller (1995).

Independently, magicians had discovered that rising sequences allowed good card tricks to be performed if cards are not well shuffled. Details and references appear in Bayer-Diaconis (1992). Bridge players went from hand shuffling to computer generated shuffling in their tournaments. A comparison of before and after suit distribution shows that the standard four or five riffle shuffles followed by a cut are grossly inadequate Berger (1973). Thorpe (1972) is an early survey detailing ways of taking advantage of poor shuffling in casino games.

Modern work on the mathematics of riffle shuffling begins with work of Gilbert (1955). He reports joint work with Shannon on the GSR model. They proved some combinatorial properties of GSR shuffles and suggested $\log_2 n$ would be enough. The model was independently discovered by Reeds (1981) who made extensive computer studies. Aldous (1983) gave a coupling argument which proves that $\frac{3}{2} \log_2 n$ shuffles are sufficient for n large. Aldous and Diaconis (1986) carefully prove that

$$\|Q^{*k} - U\| \leq 1 - \prod_{i=1}^{n-1} \left(1 - \frac{i}{2^k}\right).$$

This bound becomes less than $\frac{1}{2}$ for $k = 11$ when $n = 52$.

An empirical study of the GSR model compared to actual shuffles appears in Diaconis (1988). This concludes that the model is a good fit. Of course, much depends on the shuffler – casino dealers (along with the present author) can shuffle close to perfectly and eight perfect shuffles recycle the deck! See Diaconis-Graham-Kantor (1983) or Morris (1990) for more of this. There is much further work to do in developing tractable models with a few parameters which allow individual tuning. Because of its maximum entropy property the GSR model offers a provable lower bound to any less uniform distribution.

As a final practical note, Diaconis-Holmes (2000) analyze a class of mechanical 'shelf-shufflers' used in casino games. In these, a deck of n cards is distributed randomly onto a shelves. At each stage, cards are placed at random above or below previously placed cards on a shelf. At the end, the packets are output in random order (it turns out not to matter). The shuffle is *not* repeated. It turned out that the theory developed for type B (hyperoctahedral group) gave a complete analysis.

2C. Other Measures of Randomness

The results of Theorem 2 allow computation in various alternatives to the total variation metric. Aldous and Diaconis (1983) derive results for separation distance

$$s(k) = \max_w 1 - \frac{Q^{*k}(w)}{U(w)} = 1 - \frac{n!\binom{2^k}{n}}{2^{nk}} = 1 - \prod_{i=1}^{n-1} \left(1 - \frac{i}{2^k}\right).$$

As discussed above this needs $k = 11$ to make it small when $n = 52$. Su (1995), Trefethan-Trefethan (2002) and Stark et al. (2000) derive results for entropy distance that suggest $k = 5$ or 6 shuffles suffice when $n = 52$. The theorem of Stark et al. (2000) shows that the entropy distance decreases by a constant factor up to $\log_2 n$ shuffles when it goes to zero exponentially. A graph of the distance versus entropy for small values of n seems to show a discontinuous derivative at $\log_2(n)$. If true, this would be a new kind of phase transition. Lovasz and Winkler (1995) use Theorem 2 to show that a very different distance, the expectation of the fastest strong stationary time will be small after $k = 11$.

All of the above are global measures of uniformity. In explaining the convergence results to a popular audience, the following notion seemed useful. Consider playing the following game. A deck of cards is on the table. Guess at the top card. This card is then shown and discarded. Then guess at the next card (which is then shown and discarded) and so on. If the deck is perfectly mixed, the chance that the first guess is correct is $1/n$, the chance the second guess is correct is $1/(n-1)$, etc. Thus $1/n + 1/(n-1) + \ldots + 1$ correct guesses are expected. When $n = 52$ this is about 4.5. Suppose instead that k riffle shuffles have been carried out. A conjectured optimal strategy for guessing was derived by McGrath (see Bayer-Diaconis (1992)). Using the strategy yields about 5.01 correct guesses after seven shuffles with 4.97 correct following seven shuffles and a cut. In related work, Ciucu (1998) studies the optimal guessing strategy following k-riffle shuffles when *no* feedback is given. He proves that for $2n$ cards if $k \geq 2\log_2(2n) + 1$, the best strategy is to guess card 1 for the first half of the deck and card $2n$ for the second half. For $k < 2\log(2n)$, there are better strategies. In particular, after one shuffle he shows that guessing $1, 2, 2, 3, 3, 4, 4, \ldots$ in order gives $\sqrt{8n/\pi}$ correct guesses asymptotically. His analysis rests on an explicit diagonalization of the Markov chain which tracks the position of the card labeled 1. This is closely related to work in Section 4B below.

The above study suggested looking at classical permutation enumeration questions (e.g. number of fixed points or cycles) after an a-shuffle. This turned out to be surprisingly neat. For example, the expected number of fixed points is

$$E_a(Fp) = 1 + \frac{1}{a} + \frac{1}{a^2} + \ldots + \frac{1}{a^{n-1}}.$$

For cycles, the full story was derived by Diaconis-McGrath-Pitman (1995). Let $Q_a(n_1, n_2, \ldots n_n)$ be the chance that an a-shuffle results in a permutation with n_i i-cycles. They proved

$$(2.1) \qquad Q_a(n_1 \ldots n_n) = \frac{1}{a^n} \prod_{i=1}^{n} \binom{n_i + f_i(a)}{n_i} \quad \text{with} \quad f_i(a) = \frac{1}{i} \sum_{d|i} \mu(d) a^{i/d}$$

The proof uses a remarkable bijection of Gessel and indeed gives a self-contained proof of Gessel's results – see Gessel-Reutenauer (1993) for Gessel's version with extensive application to enumerating permutations by descents and cycle structure. The formula 2.1 and some analysis show that features of a permutation that only depend on cycle structure become random before $\frac{3}{2}\log_2 n$-shuffles; the length of the longest cycle is close to its uniform distribution after one shuffle.

In a different direction, discussed further in Section 3B, Fulman (2002) has shown that the length of the longest increasing subsequence has its correct limiting distribution after $\frac{5}{6}\log_2 n$ shuffles. These results also imply that the patience sorting solitaire game described by Aldous-Diaconis (1995) will then behave as if the deck was random.

Uyemura-Reyes (2002) has studied the number of riffle shuffles required to randomize just a few cards e.g. the original top card. He derives bounds using coupling and remarkable formulas for how the eigen-values of the GSR shuffles split by representations. His results generalize earlier profound work of Bergeron, Bergeron, Garsia (1989), and Hanlon (1990). They are discussed further in 4B below.

All of this shows that "seven shuffles suffice" is just a rough guide. From Theorem 1, it is where the cutoff happens.

To finish off this part of the shuffling story we note that the analysis has been broadened to show that the age old custom of following shuffling by a random cut does not help appreciably in convergence. This is illustrated in Bayer-Diaconis (1992) and much more sharply in Fulman (2000B). This last paper connects shuffling with cuts to cyclic descent theory.

3. Some Mathematical Connections

A. Descent Theory

A permutation w has a descent at i if $w_{i+1} < w_i$. The set of all such i makes up the descent set $D(w) \subseteq \{1, 2, \ldots, n-1\} = [n-1]$. Descents record the up down pattern in permutations and are a natural object of combinatorial study. Stanley [1972, 1986] lays out the classical theory and Buhler et al. (1994) make a fascinating connection to the mathematics of juggling. Stadler (1997) develops links between descents, shuffling and juggling for permutations of multisets.

Let $S \subseteq [n-1]$ and let $a_S = \sum_{w:D(w)=S} w$. Louis Solomon (1976) observed that as elements of the group algebra $Q[S_n]$, the a_S are the basis for a subalgebra now called Solomon's descent algebra. In particular $a_S a_T = \Sigma_u c^u_{ST} a_u$ for $c^u_{ST} \in$ **Z**. Solomon's motivation was to give a group theoretic interpretation of Mackey's induction theorem. He did this in a unified way for classical Weyl groups. The development he started now has a life of its own.

The connection to shuffling cards comes through the following observations. The set of permutations with a single descent at position i (along with the identity) are exactly the permutations realized by removing an i element subset of $1, 2, \ldots, n$ and placing them to the left (keeping all else in its same relative order). This is exactly the inverse riffle shuffles consonant with i cards cut. Summing in i, let $A_1 = \Sigma_{i=1}^{n-1} a_i$ this is the sum of all permutations with a single descent. Excepting the identity, it is also the result of an arbitrary inverse riffle. If Q is the Gilbert-Shannon-Reeds measure then, as an element in $Q[S_n]$,

$$\sum_w Q(w^{-1})w = \frac{n+1}{2^n}\, id + \frac{1}{2^n} A_1.$$

Thus the neat convolution properties of the GSR measure show that if A_i is the sum of permutations with exactly i descents (so $A_0 = id$), then $A_0, A_1, \ldots, A_{n-1}$

are a basis for a commutative subalgebra of the descent algebra. In particular, $A_iA_j = A_jA_i = \Sigma c_{ij}^k A_k$. This commutative subalgebra of the descent algebra appears in Bayer-Diaconis (1992). As explained there, close relatives had been discovered by Gerstenhaber-Schack [1987] in their development of Hochschild Homology and by Loday [1988], Hanlon [1990] in their development of cyclic homology. The idempotents of this algebra act naturally on a complex constructed from the usual bar resolution and, for commutative algebras, commute with the boundary maps. Hence their kernel and image offer a natural Hodge-type splitting of the associated homology.

It would take us too far afield to explain the connections between the descent algebra, the free-Lie algebra, and Philip Hall's commutator calculus. Fortunately, this has been splendidly carried out by Garsia (1990) and Garsia-Reutenauer (1989) as summarized by Reutenauer [1993]. This book contains a central chapter on shuffle algebras. It omits most of the topics discussed in the present review! A number of other appearances of shuffling are in the series of papers by Nantel Bergeron (with several sets of coauthors) listed in the bibliography. These extend previous results to more general Coxeter groups, include applications to Vassiliev invariants and much else.

B. Connections with Symmetric Function Theory

The theory of symmetric functions as developed by Stanley (1972, 1999) and Macdonald (1985) has had a great unifying effect on combinatorics. Many seemingly isolated facts about balls in boxes, permutations and partitions are nowadays seen as formulae for change of basis. Schurs symmetric functions are at the heart of this theory. A charming discovery of Stanley (2001) developed by Fulman (2002) shows how Schur functions arise in a natural way from riffle shuffling. Let $\theta_1, \theta_2, \ldots$ be non-negative numbers that sum to one. Drop n balls into a set of boxes with θ_i the chance of a ball dropping into box i. Suppose the box counts are $N_1, N_2, \ldots$ with $N_1 + N_2 + \ldots = n$. Take a deck of n cards; cut off the top N_1, cards, then the next N_2 cards (forming a separate pile), etc. of course, many of the piles may be empty. Riffle shuffle these piles together as in Section 2a. This results in a final permutation w. Apply the Schensted map to w to get a pair of standard Young-tableaux of the same shape λ.

Proposition The probability that the above procedure results in the partition λ is the Schur function s_λ times the dimension f_λ of the associated representation of the symmetric group:

$$s_\lambda(\theta_1, \theta_2, \ldots) f_\lambda.$$

Stanley's proof of this proposition uses quasi-symmetric functions, an emerging tool in algebraic combinatorics. Fulman's proof of the proposition uses only classical facts from symmetric function theory. Both authors develop corollaries and variations. One striking application to shuffling due to Fulman shows that the distribution of features of a permutation dependent on the shape of the associated Young-tableaux- e.g. the longest increasing subsequence – have the correct limiting distribution after $\frac{5}{6}\log_2 n$ shuffles. Stanley (1999) (2002) is a good place to start reading about quasi-symmetric functions. Aguiar and Sottile (2002), Billera, Hsiao and Van Willigenburg (2001) and Garsia, Wallach (2002) are relevant, significant studies. All have shuffles as part of their combinatorial essence.

3C Work of Fulman Some profound connections between shuffling and the enumerative theory of finite groups of Lie type have been developed by Jason Fulman. Some of this has already made an appearance above in Sections 2B and 3B. This section describes some further developments. Yet others appear in the rich collection of papers listed in the bibliography.

One striking result of Fulman explains a mystery. A main result in Diaconis-McGrath-Pitman (1995) is a closed formula for the cycle structure of a permutation after an a-shuffle (see (2.1) in Section 2C). It was also observed that this formula answers a second question: pick a random monic degree n polynomial $x^n + a_{n-1}x^{n-1} + \ldots + a_0$ with coefficients in $\mathbf{F}_q$ by choosing $a_0, a_1, \ldots, a_{n-1}$ from the uniform distribution. Factor this polynomial into irreducible factors and suppose there are n_i irreducibles of degree i. The chance of a given $n_1, n_2, \ldots, n_n$ occurring is given by (2.1) with $a = q$. This was proved by observing that two formulae agreed – that is, without understanding. Fulman [1998] found a conceptual explanation and an extension to other groups and shuffling schemes.

Fulman's explanation begins with a simply connected, semi-simple group G defined over $\mathbf{F}_q$. Let $\mathcal{G}$ be the Lie algebra. Consider the orbits of semi-simple elements of $\mathcal{G}$ under the adjoint action of $\mathcal{G}$. For example, for groups of type A, $G = SL_n(\mathbf{F}_q), \mathcal{G} = s\ell(n, q)$ and semi-simple elements correspond to monic degree n polynomials with coefficient of x^{n-1} vanishing. For types A and B, Fulman shows that there is a natural map Φ from the semi-simple orbits to the conjugacy classes of the Weyl group W such that a uniformly chosen orbit maps to the measure induced by a-shuffling with $a = q$. Thus a randomly chosen polynomial maps to an a-shuffle and the factors map to cycles. For shuffles of type B, the correspondence is with symmetric polynomials $f(z) = f(-z)$

In algebraic group theory there is an analog of the map Φ which carries semi-simple conjugacy classes of the group G to conjugacy classes of the Weyl group. Picking a semi-simple class uniformly induces a probability distribution on conjugacy classes. Fulman [1997] managed to find a card shuffling interpretation of this map as well and give an enumerative theory that works for all split semi-simple groups. His work uses results of Cellini and Carter's work on the Brauer complex. Indeed, Carter (2002) has recently extended Fulman's work to more general groups.

We give the card shuffling version of Fulman's work for type A. Define an F-shuffle of a deck of $2n$ cards as follows: choose an even number j, between 1 and $2n$ with probability $\binom{2n}{2j}/2^{2n-1}$. Remove the top j cards of the deck. Remove the bottom j cards of the deck and place them on top of the original top j cards to form a packet of size $2j$. Shuffle this packet with the remaining $2n - 2j$ cards. Fulman derives remarkable closed form generating functions for the cycles of a permutation after an F-shuffle. He also shows that F-shuffles convolve nicely and, for special deck sizes, gives an alternate description in terms of a riffle followed by a cut.

The analogous developments for type B yield closed formulae for the cycles of randomly chosen unimodal permutations. These arise in dynamical systems and in social choice theory.

One further aspect of Fulman's work deserves special mention (and follow-up!). The shuffling work in Diaconis-McGrath-Pitman (1995) leans on a remarkable bijection of Gessel between multisets of primitive necklaces and permutations with cycle structure equal to that of the necklace. Fulman shows that by refining the correspondence Φ described above to a map to the Weyl group (instead of just to conjugacy classes) one recovers Gessel's bijection in a group theoretically natural way.

3C. Work of Lalley

Steve Lalley has written a series of papers studying extensions of the basic Gilbert-Shannon-Reeds model to less uniform methods of riffle shuffling. Even changing the method of cutting the deck in two from a fair binomial distribution to a skewed binomial distribution with parameter $p < \frac{1}{2}$ destroys a basic symmetry. For this case, Lalley [2000] conjectures that there is a sharp threshold for the mixing time at $C_p \log n$ for $C_p = (3 + \theta_p)/\log(1/p^2 + q^2)$ with θ_p the unique solution of $p^\theta + q^\theta = (p^2 + q^2)^2$. Observe that $C_{\frac{1}{2}} = \frac{3}{2} \log_2 n$ in agreement with Theorem 1. Lalley [2000] and Fulman (1998) give upper and lower bounds of this form for the mixing time but sharp results are conjectural.

Lalley [1996], [1999] expands the basic interlacing mechanism underlying the GSR shuffle. To explain, recall the dynamical systems description of GSR shuffles as the permutation induced by n uniform points in $[0, 1]$ under the baker's transformation $x \longmapsto 2x \bmod(1)$. This results in all interleaving being equally likely. It is natural to consider more general maps $f : [0, 1] \to [0, 1]$ which preserve Lebesgue measure. Lalley works with piecewise C^2 maps which are piecewise monotone increasing. He shows that several interpretable shuffles can be so described. For example, the biased cut shuffles described above or shuffles where the left card is dropped with probability $uA/(uA + wB)$ when packets are of size A, B, here u, w are fixed parameters. When $u = w = \frac{1}{2}$ this becomes the original GSR shuffle.

The main result of Lalley [1996] shows that when n is large, for fixed i, the number N_i of cycles of length i after an f-shuffle are approximately independent geometric random variables with $P(N_i = k) = (1 - w)w^k$ the parameter w depends on i and on the map f in a simple way. Further, the N_i are approximately independent. The main result of Lalley [1999] gives a lower bound for the number of f-shuffles required to mix N cards; at least $h^{-1} \log N$ shuffles are needed where h is the 'fiber entropy' associated to f. The proofs are a marvelous mix of ergodic-theoretic symbolic dynamics and combinatorics.

One interesting aspect of these f-shuffles is that, aside from a-shuffles, the successive permutations chosen for repeated convolution are not independent. They form a stationary sequence. This is not necessarily bad; perhaps real shufflers remember a few steps back – if a particularly lumpy shuffle was just made the next shuffle might be neater. See also Dubrow-Fill (1995). There is much to follow up from Lalley's work. Perhaps the leading problem is to prove any kind of upper bound for f-shuffles or better, to determine where cutoffs appear.

3D. Early Shuffling

The basic combinatorial shuffling of two sequences, one with m letters $x_1, \ldots, x_m$

and one with n letters $y_1, \ldots, y_n$, into the formal sum of sequences of $n+m$ letters in all orders that preserve the order of the x's and the order of the y's (thus $\binom{n+m}{m}$ terms) appears in other areas of algebra.

Perhaps earliest is the classical wedge product of two alternating forms. If V is a vector space and $f : V^m \to \mathcal{R}$, $g : V^n \to \mathcal{R}$ are alternating multilinear functions, then $f \wedge g : V^{n+m} \to \mathcal{R}$ may be constructed as the function

$$f \wedge g(x_1, \ldots, x_{n+m}) = \sum_{\sigma} sgn(\sigma) f(x_{\sigma_1}, \ldots, x_{\sigma_m}) g(x_{\sigma_{m+1}}, \ldots, x_{\sigma_{n+m}})$$

where the sum is over all shuffles. A splendid account of this classical subject appears in Cartan (1967, pg. 179-188). The shuffling construction guarantees that $f \wedge g$ is alternating, that $f \wedge g = (-1)^{mn} g \wedge f$ and that the wedge product is associative. Cartan's proof of this last statement results from the following fact: with three packets of cards of sizes ℓ, m, n, shuffling ℓ into m and then the n into this joint packet results in the same distribution as shuffling in any of the other orders, or indeed shuffling the 3 packets together simultaneously as in the 3-shuffles described in Section 2d. More general shuffles appear when studying flag manifolds. A flag is an increasing sequence of subspaces. If the successive dimensions are $n_1, n_1 + n_2, \ldots$ then shuffles based on cutting off packets of size $n_1, n_2, \ldots$ appear. In particular, such shuffles index a basis for the homology of the associated flag variety. See Fulton (1997) or Shahshahani (2002) for textbook descriptions.

Eilenberg-MacLane (see MacLane 1950) used the shuffle construction as a basic building block for constructing a chain complex giving an appropriate cohomology theory for Abelian groups. They get $H^2(\pi, G)$ as the group of Abelian extensions of G by π.

Shuffles appear frequently in other basic constructions in algebraic topology. For example, if X is a space with an associative, commutative product, Milgram (1967) defined a product on the classifying space $B(X)$ using shuffles. This work was systematized by Steenrod (1967) and further by MacLane (1970). Shuffles appear in the Eilenberg-Zilber Theorem and in explicit proofs of the Künneth formula giving a chain equivalence between a chain complex for the product of two spaces and the tensor product of the two chain complexes. See Hatcher (2002, pg. 278) for details and Dupont (2001, pg. 29) for a charming appearance in the world of scissors congruences! The essence of much of this is that the shuffling map gives a natural triangulation of the product of two simplices.

From a modern view, many of these appearances of shuffling occur because of the many natural Hopf algebras in mathematics. See Schneider-Sternberg (1993) for references and pointers to Rees' shuffle algebras and Chen's iterated integrals. Perhaps even more basic, the permutatedron is the convex hull of all permutations of the vector $(1, 2, 3, \ldots, n)$ in $\mathbf{R}^n$. It is a convex polytope with vertices indexed by permutations. It may be seen that the edges and faces of various dimensions are indexed by shuffles. See Billera and Sarangarajan (1996) for a clear statement and proof. It would be marvelous if some of what we know about shuffling illuminates these applications or vice versa.

4. Some Generalizations

There are a bewildering variety of extensions of riffle shuffling where much of the successful analysis goes through. It is easiest to lead into this by considering inverse riffle shuffles where a subset is selected at random and moved to the top. A natural generalization is to partition $[n]$ into ordered blocks $(B_1, B_2, \ldots, B_k)$. Then remove all cards with labels in block one and move these to the top (keeping the cards within a block in their original relative order). Next cards with labels in B_2 are removed and put directly below those in B_1, and so on with cards having labels in block k finishing at the bottom. Let $\mathcal{B}$ be the space of all ordered blocks of any shape if a weight $w(B), B \in \mathcal{B}$ is specified with $w(B) \geq 0$ $\Sigma w(B) = 1$, then a random walk can proceed.

Inverse riffle shuffles and the GSR model proceed from the uniform distribution on the set of 2^n partitions into two blocks. A widely studied special case puts weights $w_1, w_2, \ldots, w_n$ on each card and then removes card i to top. This arises as a method of rearranging files so that frequently called for items are near the top. See Fill [1996] for an extensive survey. Curiously, the special case with $w_i = 1/n$ for all i is central in Wallach (1986) and Garsia-Wallach (2002).

As will emerge, there is a relatively complete theory for this class of walks – a description of stationary distribution, reasonable rates of convergence and a complete description of the associated eigenvalues. This will follow from the following sweeping generalization.

A. Hyperplane Walks

Bidigare-Hanlon-Rockmore (1999) introduced a class of walks on chambers of a hyperplane arrangement which includes the walks above as a special case. Their works was completed in various ways by Bidigare (1997), Brown [2000, 2001], Brown-Diaconis [1998]. Billera-Brown-Diaconis (1999) offer an introduction.

The story begins with a set $\mathcal{A}$ of affine hyperplanes in $\mathbf{R}^d$. This cuts $\mathbf{R}^d$ into regions called chambers. These chambers are polyhedra with sides called the faces of the arrangement. For example, three lines in the plane in general position yield 7 chambers (2-dimension), 9 one dimensional faces and three zero-dimensional faces (the three points of intersection). Given a face F and a chamber C, the projection of F on C, written FC, is the unique chamber with F as a face and closest to C. Here closeness if measured by the number of hyperplanes in $\mathcal{A}$ one must cross in moving from C to FC.

Let $w_F \geq 0$ $\Sigma w_F = 1$ be weights on the faces of the arrangement $\mathcal{A}$. Define a random walk on the set of chambers by moving from C to FC when F is chosen with probability w_F. The theory depends on the lattice $\mathcal{L}$ of all possible intersections of elements in $\mathcal{A}$. Here are the main theorems of Bidigare-Hanlon-Rockmore [1999], Brown-Diaconis [1998].

Theorem 1 Let $\mathcal{A}$ be a hyperplane arrangement in $\mathbf{R}^d$. Let $\mathcal{L}$ be the intersection lattice of $\mathcal{A}$ and w_F a probability measure on the faces. Then, the transition matrix of the Markov chain is diagonalizable. For each $W \in \mathcal{L}$ there is an eigenvalue

$$\lambda_W = \sum_{F \leq W} w_F$$

with multiplicity

$$m_W = |\mu(W,V)| = (-1)^{\dim(W,V)}\mu(W,V)$$

where μ is the Möbius function of $\mathcal{L}$.

Theorem 2

(a) The Markov chain of Theorem 1 has a unique stationary distribution π if and only if for each $H \in \mathcal{A}$ there is a face F not in H with $w_F > 0$.

(b) The stationary distribution in (a) can be described by sampling faces *without replacement* from w_F to get an ordering $F_1, F_2, \ldots$. Then, for any chamber C_0, the product $F_1 F_2 F_3 \ldots C_0$ is a chamber distributed from π.

(c) For π as in (a), (b), and starting chamber C_0

$$\|K^{\ell}_{C_0} - \pi\| \leq \sum_{H\in\mathcal{A}} \lambda^{\ell}_H$$

To complete this section, let us show how these hyperplane walks extend riffle shuffles. The *braid arrangement* $\mathbf{A}_d$ consists of hyperplanes $H_{ij} = \{x \in \mathbf{R}^d : x_i = x_j\}$. All points within the same chamber have the same relative order so the chambers may be labeled with permutations. The faces are points in $\mathbf{R}^d$ which lie on some of the H_{ij} and on various sides of the rest. These may be labeled by block ordered partitions $(B_1, B_2, \ldots, B_k)$ of $[n]$. Finally, the action FC of a block ordered partition on the permutation corresponding to C results from removing cards from the first block and moving to the top, etc., as described in the introduction to this section.

The present description does not do justice to the wealth of examples of hyperplane arrangements where the chambers have natural names and the walk has a natural interpretation. We can only hope that the reader will consult the references above.

B. Some Representation Theory

I want to describe work of Bergeron-Bergeron-Garsia (1989), Hanlon [1990] and Uyemura-Reyes (2002) which shows a deep interplay between the shuffling schemes of Section A and the representation theory of the symmetric group. To keep things manageable, consider random walks on the braid arrangement driven by invariant face weights: $w(F) = w(\sigma F)$. This includes (uniform) random to top and inverse riffle shuffles as special cases. Let $Q(\sigma) = \Sigma_{Fid=\sigma} w(F)$. These walks may be described via repeated convolution by the probability measure Q.

It is natural to ask how the eigenvalues of the walk split up by representation. Recall that the irreducible representations of S_n are indexed by partitions ν of n. If $\rho_\nu(\sigma)$ is the associated matrix representation, we are asking about the eigenvalues of the matrix $\widehat{Q}(\nu) = \Sigma_\sigma Q(\sigma)\rho_\nu(\sigma)$. By general theory (Diaconis (1988, Chapter 3E)) these are a subset of the eigenvalues from Theorem 1 in Section 4A Above. For the braid arrangement, the eigenvalues are indexed by block ordered partitions. However, because of the symmetry $w(F) = w(\sigma F)$, the eigenvalues only depend on the underlying number partition. Thus for each pair of partitions (μ, ν) we may ask how many times the eigenvalue λ_μ occurs in the matrix $\widehat{Q}(\rho_\nu)$. To describe the answer we need both the usual irreducible characters χ_ν of S_n and the Lie characters ψ_μ (Reutenauer (1993, Chapter 8)). These Lie characters may be described by taking a permutation of cycle type μ in S_n. Its centralizer is a product of Wreath products $S_k wr C_j$. Take a ξ primitive jth root of 1, consider the one dimensional character of C_j^k which takes $x_j, \ldots, x_n$ to $\xi^{x_1 + \cdots + x_k}$. This induces a one dimensional character of the Wreath Product. Taking a product of these 1-dimensional characters over all factors in the centralizer and then inducing up from the centralizer to S_n gives ψ_μ. The main theorem below was proved by Hanlon [1990] for the case of GSR shuffles. Richard Stanley (personal communication) conjectured the general result which was proved by Uyemura-Reyes (2002).

Theorem 3

For an S_n invariant hyperplane walk on the braid arrangement the multiplicity of the eigenvalue λ_μ of Theorem 1 in the νth irreducible representation of S_n equals

$$\langle \chi_\nu, \psi_\mu \rangle.$$

Remarks

(a) Lie characters have been extensively investigated when $\mu = (n)$, see Stembridge (1989), where an explicit decomposition formula is given. For general partitions μ, much less is known. Theorem three shows that the S_n invariant shuffles are equivalent objects in the group algebra. Any such shuffle is a linear combination of what may be called μ shuffles as described in the introduction to this section. As shown in Diaconis-Fill-Pitman [1992, Sec. 5], these μ shuffles form a basis for the descent algebra.

(b) Uyemura-Reyes (2002) shows how the numbers described above allow bounds on how many shuffles of a given type are required to randomize a subset of cards, e.g. the original top card or top 13 cards. Here is one example. If $k = \log_2(n/c)$, after k inverse GSR shuffles, let Q_k be the probability distribution of the position of the original top card. Then Q_k is close to uniform if c is small:

$$\|Q_k - u\| \leq 1 - (1 - 2^{-k})^n$$

(c) These connections to representation theory are crucially used in Fulman (2000B) to get nice formulae for the cycle structure of shuffles followed by a cut.

C. Brown's Semigroup Walks

Ken Brown [2000, 2001] has given a marvelous extension of the hyperplane walks which leads to interesting special cases *and* a conceptual explanation of why the eigenvalues of these non-symmetric Markov chains are non-negative real numbers. The brief treatment given here is a shuffling together of two of Brown's papers and the reader is strongly encouraged to read the originals.

Let S be a finite semigroup satisfying $x^2 = x$ for all $x \in S$. A random walk is driven by a probability distribution $w(x), x \in S$. At each stage, one picks x from $w(x)$ and multiplies on the left. Thus the transition matrix is

$$K(s,t) = \sum_{x \cdot s = t} w(x)$$

In all the examples, the state space of the walk is restricted to a left ideal I in S.

Example 1. Hyperplane Walks. Let S be the set of faces of a hyperplane arrangement with I the set of chambers under the product of Section 4A. This product is idempotent and the results of Section 4A will be seen as special cases of the main theorem below.

Example 2. q analogs Let $MAT(n, \ell, q)$ be the set of $n \times \ell$ matrices of rank ℓ with coefficients in $\mathbf{F}_q$. Let $S = \cup_{\ell=1}^{n} MAT(n, \ell, q)$ and $I = GL_n(q) = MAT(n, n, q)$. Define a product on S as follows: If s has columns $(s_1, \ldots, s_\ell)$ and t has columns $(t_1, \ldots, t_m)$ form $s \cdot t$ by appending the columns of t to the columns of s in order $t_1, t_2, \ldots$ deleting a t_i if it is linearly dependent on the columns already there. This is an idempotent associative product and $GL_n(q)$ is an ideal.

The "$q = 1$ case" consists of ordered strings from $1, 2, \ldots n$, without repeated values and the ideal becomes the symmetric group S_n. Thus if $s = (3, 5)$ and $t = (23145)$ $st = 35214$ and we see the move to the front chain.

Example 3. The free idempotent semigroup F_n on $1, 2, \ldots n$, may be described as the equivalence class of finite strings under the equivalence relations $w^2 = w$ for all subwords. For example, when $n = 2$, we get the six strings

$$S = \{1, 2, 12, 21, 121, 212\}$$

Brown (following Green and Reees (1952)) shows that F_3 has order 159 and F_n has order $\sum_{i=1}^{n} \binom{n}{i} \prod_{j=1}^{i} (i - j + 1)^{2^j}$.

Let I be the ideal of all words having each of $\{1, 2, \ldots, n\}$ appearing at least once (for $n = 2, I = \{12, 21, 121, 212\}$). Any probability measure on S induces a Markov chain on I by left multiplication.

Return now to the general case of an idempotent semigroup S. Brown introduces a support map supp : $S \to L$ with L an explicitly constructed semilattice. The support map is a subjection satisfying supp $(xy) =$ supp $x \vee$ supp y and supp $x \geq$ supp y if and only if $x = xyx$. The set L indexes the eigenvalues of the walk. For hyperplane walks, L is the intersection lattice. For matrices, L is the subspace spanned by the columns. For the free idempotent semigroup L is the collection of subsets of $\{1, 2, \ldots, n\}$ under union. The natural ideal I is the two sided ideal. $\{x : \text{supp } x = \hat{1}\}$. This specializes to the ideals given in the three examples above.

Brown gives a version of theorems one and two of Section 4A: For each $X \in L$ there is an eigenvalue $\lambda_X = \sum_{\text{supp } x \leq X} w(x)$ with a neat way of computing multiplicities. If the product of x with $w_x \neq 0$ is in I then there is a unique stationary distribution π which may be described as the distribution of the random element $x_1 x_2 \ldots c_0$ with $x_1, x_2, \ldots$ sampled *without* replacement from $w(x)$. Finally, for any starting state $C_0 \in I$,

$$\|K_{C_0} - \pi\| \leq \sum_H \lambda_H^\ell$$

where H ranges over the maximal elements of L.

The key to the analysis is a surprising, complete character theory. (Most semigroups do not have a reasonable character theory.) Brown shows that all representations of S are one dimensional and that the representations are indexed by L; the 'Fourier transform' of the random walk now yields the eigenvalues.

One aspect of Theorem 1 that needn't go through: the Markov chain needn't be diagonalizable. To help the reader navigate, Brown first worked in idempotent semigroups satisfying the additional identity $xyx = xy$. These are called left regular bands in the semigroup literature; most of the examples considered above are left regular bands. Under this condition the chain *is* diagonalizable. In later work, Brown showed that nearly everything goes through in the general case. There is a tantalizing extension to a walk on the chambers of a building. Here, while a product is well defined, it is not associative. This creates a mess but there are some positive results as well.

5. Some Open Problems

1. Almost none of the walks presented here have good lower bounds available. Examples include riffle shuffles with the deck cut exactly in two (see Section 3A) or any of Fulman's shuffles (Section 3C). It would be nice to have a lower bound in some generality for the general hyperplane walks of Section 4A. Usually, reasonable lower bounds are easier to prove than upper bounds. See [Diaconis, 1988] or [Saloff-Coste 1997] for the usual techniques. One idea for a systematic approach: Brown's method (Section 4C) finds a representation theoretic interpretation. With characters available, perhaps David Wilson's [2001] approach may be pushed through.

2. It should be the case that essentially all the walks discussed here show a sharp cutoff in their approach to stationarity; proving this requires sharp upper bounds as well as sharp lower bounds. The general upper bounds (e.g., Theorem 2 of Section 4A) are often slightly off in the few cases where sharp answers are known. For example, for ordinary riffle shuffles, the general approach shows $2\log_2 n + C$ shuffles suffice for randomness while Theorem 1 of Section 2A shows the right answer is $\frac{3}{2}\log_2 n + C$. The original paper of Bidigare-Hanlon-Rockmore gives a potentially sharper upper bound. It would be very instructive to compare the two variations. In preliminary work, Brown-Diaconis [1998] found them similar but Uyemura-Reyes [2002] found examples where the BHR bound is a genuine improvement. It may be that the bounds of BHR or Theorem 2 of Section 4A are sharp for some other metric; this happens for ordinary riffle shuffles with separation distance as discussed in Section 2C. At a more abstract level, it may be possible to prove

the existence of a sharp threshold without being able to locate it along the line of concentration inequalities see Ledoux [2000].

3. A very clear set of problems is to give any kind of upper bounds for Lalley's f-shuffles of Section 3C. Presumably, these all mix n cards in order $\log n$ steps but at present we don't know that order 2^n steps suffice.

4. For practical reasons it is natural to seek models of riffle shuffling cards that result in neater shuffles than the GSR shuffles. This arises in studying the way Las Vegas dealers shuffle; they drop cards in close to perfect alternation while the GSR method has packet sizes geometrically distributed. Here is a suggestion whose analysis is completely open: the *Markovian Model* is driven by a 2-state Markov chain with transition matrix.

$$\begin{pmatrix} p_{00} & p_{01} \\ p_{10} & p_{11} \end{pmatrix}$$

To shuffle a deck of n cards, run the chain starting in stationarity to produce $x_1, x_2, \ldots, x_n$ a binary sequence. If this sequence has k zeros and $n-k$ ones, cut off the top k cards as a left hand pile, the $n-k$ remaining as a right hand pile. Use the zeros and ones (from right to left say) to dictate if the next drop is from left or right.

For example, with $n = 10$ cards the sequence 0101100110 results in 5 cards being cut off and the final arrangement $1, 6, 2, 7, 8, 3, 4, 9, 10, 5$. This includes the GSR model by taking $p_{ij} = 1/2$ and perfect shuffles by taking $p_{01} = p_{10} = 1$. It is natural to begin with a symmetric cut, so $p_{01} = p_{10}$ and $p_{00} = p_{11}$. There is every hope that this model will produce neat and useful analyses. It must be the case that (for symmetric shuffles with $0 < p_{01} < 1$) there is a sharp threshold at $\theta \log n + C$ with $\theta = \theta(p)$. For practical purposes one could estimate $\theta(p)$ from computer experiments and also estimate p by watching dealers shuffle. This would allow one to derive reasonable ways of exploiting the structure if the dealers do not shuffle enough. The following seems clear: Since $p_{ij} = 1/2$ requires seven shuffles and this is the fastest method, most neat shufflers will require a good many more shuffles and there will be plenty of structure to take advantage of! Incidentally, the case of 'random perfect shuffles', each time choosing randomly to do a perfect in or out shuffle, has been analyzed by Uyemura-Reyes (2002). For perfect shuffles, not all permutations are possible so the walk is random on a subgroup. For decks of size 2^k, he shows order k^2 steps are necessary and suffice. The case of general decks remains open.

5. I want to record empirical work which yields a quite different natural model for riffle shuffling. In joint work with my student Arnab Chakraborty we studied commonly available machines for shuffling cards. These machines have the user cut the deck in two halves (placed into the left and right sides of the machine). Then a button is pushed which activates rubber wheels touching the bottoms of the two packets. These spin cards off the bottoms into a central region where they drop onto a collecting place. At the end of one shuffle the user retrieves the deck, cuts it into two and the process continues.

In our empirical work we found that an "opposite" GSR model seemed to fit the data. In the GSR model, if at some stage there are A cards in the left half and B cards in the right half, the chance of dropping the next card from the left half is $A/(A+B)$. In the mechanical shuffler, the chance seemed to be $B/(A+B)$; it was more likely for a card to be dropped from the smaller half. Of course, once all the cards in a half are used up, the remaining cards are dropped on top. This seems like a natural candidate for careful study.

One may interpolate between models with a one-parameter family of the following form. If at some stage there are A cards in the left half and B cards in the right half, the chance of dropping the next card from the left is $A^\theta B^{1-\theta}/(A^\theta B^{1-\theta} + B^\theta A^{1-\theta})$. Again, for practical purposes θ could be fit from data and a cutoff parameter could be estimated by simulation.

6. The GSR is the most uniform single riffle shuffle in the sense that, given the cut, it makes all shuffles equally likely. It is not clear (though it seems plausible) that it is also the probability on shuffles so that $Q * Q$ is closest to uniform (as well as Q^{*k} for all k). This may be a simple problem but it seems worth clarifying. A similar case that would shed some light: on $\mathbf{Z}/m\mathbf{Z}$ (the integers mod m), consider all probability measures with support in $[-a, a]$. Find the probability P in this set such that P^{*k} is closest to uniform (say in entropy or total variation distance). Is this uniform?

7. A beautiful set of conjectures has arisen from thesis work of J.C. Uyemura-Reyes. To describe them, consider first the random to top shuffle. This has eigenvalues $0, 1/n, 2/n, \dots (n-2)/n$, 1 with multiplicity of j/n the number of permutations in S_n with j fixed points. This was proved by Phatarphod and independently by Wallach (1986) and follows from Theorem 1 of Section 4A. Next consider the multiplicative reversibilization of random to top. This is random to top followed by top to random: It may also be described as: remove a random card and insert it in a random position.

   Numerical work shows that the eigenvalues are all of the form quadratic function of $(j)/n^2$. For some cases this can be proved. For example, zero occurs with multiplicity the number of derrangements and the eigenvalues in representations near the trivial representation (or alternating representation) can be proved of this form. There must be a way to understand these! Work of Phil Hanlon and Patricia Hersh indicates that this question fits very neatly into algebra, along the lines of Section 3A.

   Using the available results one may conjecture where the cutoff occurs for mixing. For either random to top or top to random, $n \log n$ is the cutoff. It seems that random to random must be faster but perhaps not by more than a factor of two. We conjecture that $3/4n \log n$ is the cutoff here. This is what is required to kill the eigenvalues from the $n-1$ dimensional representation. Uyemura-Reyes [2002] proves a lower bound of form $\frac{1}{2} n \log n$ and an upper bound of $4n \log n$. At present writing we do not know that $n^2 \times$ eigenvalue is an integer.

   Again, in preliminary work, it seems as if the eigenvalues of the multiplicative symmetrization of any hyperplane walk from a Coxeter group with symmetric face weights generated by a finite reflection group will be "nice" in the same sense. To be specific, consider the permutation group S_n. Fix a composition $\mu = (\mu_1, \mu_2, \ldots, \mu_r)$ of n. A symmetric μ shuffle removes a uniformly chosen subset of μ_1 cards (keeping them in their same relative order, then, from the remaining $n - \mu_1$ cards, a random subset of size μ_2, and so on. With a final packet of size μ_r. These r packets are shuffled together by a GSR shuffle.

8. A very simple to state conjecture: After k GSR shuffles of n cards consider turning up cards from the top one at a time. What is the optimal guessing strategy to maximize the expected number of correct guesses? A conjectured optimal strategy due to McGrath is described in Bayer-Diaconis [1992]. There is related work in Ciucu (1998). Prove that McGrath's strategy is optimal. An easier version (still open) asks the same question following k top to random shuffles with k fixed and known.

9. Less of a conjecture than a suggestion; many of the semigroup walks in Brown [2000] seem worthwhile studying in depth. To take one example; Brown [2000, Section 6.3] introduced a fascinating family of walks on phylogenetic trees. Walks on such trees are currently an active area of study. See Diaconis-Holmes [2002] for pointers to work by Aldous and to the currently very active work in biology. Brown's walks are driven by weights w_{ij}. A first natural problem is to study the stationary distribution of Brown's walk as a natural family of non-uniform distributions on trees. They carry over to trees the Luce model which has been very actively studied for permutations. One might even contemplate estimating Brown's parameters w_{ij} from data. It is also natural to carry out some careful analyses of rates of convergence for natural families of weights: randomly chosen i.i.d. uniform weights, Zipf type weights, or w_{ij} the distance from i to j in some natural geometric structure.

10. A most annoying problem: find some use for the eigenvalues of the many walks in the section above. For reversible Markov chains there are good bounds on the rate of convergence based on eigenvalues. Are there *any* explicit bounds on, e.g., L^2 distances for non-symmetric chains? Going further, are there bounds on the multiplicative symmetrization of a chain based on knowledge of the eigenvalues of the original chain? This would allow the wealth of eigenvalue information reported above to be used for comparison purposes as explained in Saloff-Coste [1997].

REFERENCES

Aguiar, M. and Sottile, F. (2002), Structure of the Malvenuto-Reutenauer Hopf algebra of permutations. Technical Report, Dept. of Mathematics, University of Massachusetts, Amherst.

Aldous, D. (1983), Random walks on finite groups and rapidly mixing Markov chains, in *Springer Lecture Notes in Mathematics* **986**.

Aldous, D. and Diaconis, P. (1986), Shuffling cards and stopping times, *Amer. Math. Monthly* **93**, 333-348.

Aldous, D. and Fill, J. (2002, *Reversible Markov Chains*, Available at http://www.stat.berkeley.edu/users/aldous.

Bayer, D. and Diaconis, P. (1992), Trailing the Dovetail shuffle to its lair, *Ann. Appl. Probab* **2**, 294-313.

Berger, P. (1973), On the distribution of hand patterns in bridge: Man-dealt versus computer dealt, *Canadian Jour. Statist.* **1**, 261-266.

Bergeron, F., Bergeron, W. and Garsia, H. (1989), *The Free Lie Algebra and q-Enumeration.* In D. Stanton (ed.), *Invariant Theory and Tableaux*, Springer, New York, 166-190.

Bergeron, W. (1991), A hyperoctahedral analogue of the free Lie algebra, *Jour. Combin. Th. A.* **55**, 80-92.

Bergeron, F. and Bergeron, N. (1992), Orthogonal idempotents in the descent algebra of B_n and applications, *Jour. Pure Appl. Alg.* **79**, 109-129.

Bergeron, F., Bergeron, N., Howlett, R. and Taylor, D. (1992), A decomposition of the descent algebra of a finite Coxeter group, *Jour. Alg. Combinatorics* **1**, 23-44.

Bergeron, N. (1994), *On Hochschild homology and Vassikiev invariants*, DIMACS Conference, 1-10.

Bergeron, N. (1995a), Hyperoctahedral operations on Hochschild homology, *Adv. Math.* **110**, 255-276.

Bergeron, N. (1995b), Décomposition hyperoctahédrale de l'homologie de Hochschild, *Discrete Math.* **139**, 33-48.

Bergeron, W. and Wolfgang, H. (1995), The decomposition of Hochschild cohomology and Gerstenhaber operations, *Jour. Pure Appl. Alg.* **104**, 243-265.

Bidigare, P. (1997), Hyperplane arrangements face algebras and their associated Markov chains. Ph.D. thesis, University of MIchigan, Department of Mathematics.

Bidigare, P., Hanlon, P. and Rockmore, D. (1999), A combinatorial description of the spectrum for the Tsetlin library and its generalization to hyperplane arrangements, *Duke Math. Journal* **99**, 135-174.

Billera, L., Brown, K. and Diaconis, P. (1999), Random walks and plane arrangements in three dimensions, *Amer. Math. Monthly* **106**, 502-524.

Billera, L. Hsiao, S. and Van Willigenburg (2001), Peak quasi-symmetric functions and eulerian enumeration. Preprint, Dept. of Mathematics, Cornell University.

Billera, L. and Sarangarajan (1996), The combinatorics of permutation polytopes. In L. Billera et al. (eds.). *Formal Power Series and Algebraic Combinatorics*, Amer. Math. Soc. Providence.

Borel, E. and Chéron, A. (1940), Théorie mathématique du bridge a la portée de tous. Gauthier-Villars, Paris. English translation by Alec Traub, (N.D.), Monna Lisa Publishing, Taiwan.

Buhler, J., Eisenbud, D. Graham, R. and Wright, C. (1994), Juggling drops and descents, *Amer. Math. Monthly* **101**, 507-519.

Brown, K. (2000), Semigroups, rings, and Markov chains, *Jour. Theoretical Probability* **13**, 871-930.

Brown, K. (2001), Notes on bands. Preprint, Department of Mathematics, Cornell University.

Brown, K. and Diaconis, P. (1998), Random walks and hyperplane arrangements, *Ann. Probab.* **26**, 1813-1854.

Cartan, H. (1967), *Formes Différantielles*, Hermann, Paris.

Carter, R. (2002), Semisimple conjugacy classes and classes in the Weyl Group. Preprint Mathematics Institute, University of Warwick.

Ciucu, M. (1998), No-feedback card guessing for dovetail shuffles, *Ann. Appl. Probab.* **8**, 1251-1269.

Diaconis, P. (1988), *Group Representations in Probability and Statistics*, IMS, Hayward, CA.

Diaconis, P. (1996), The cutoff phenomenon in finite Markov chains, *Proc. Nat. Acad. Sci. USA* **93**, 1659-1664.

Diaconis, P. (1998), From shuffling cards to walking around the building: An introduction to modern Markov chain theory. In *Proc. Int. Congress Math.* **1**, 187-204.

Diaconis, P., Graham, R. and Kantor, W. (1983), The mathematics of perfect shuffles, *Adv. Appl. Math.* **4**, 175-193.

Diaconis, P., Fill, J. and Pitman, J. (1992), Analysis of top to random shuffles, *Combinatorics Probability and Computing* **1**, 135-155.

Diaconis, P., McGrath, M. and Pitman, J. (1995), Riffle shuffles, cycles and descents, *Combinatorica* **15**, 11-20.

Diaconis, P. and Holmes, S. (2000), Analysis of a card mixing scheme. Unpublished report.

Diaconis, P. and Holmes, S. (2001), Random walk on trees and matchings, *Electronic Jour. Probab.* **7**.

Dobrow, R. and Fill, J. (1995), *The Move to the Front Rule for Self-organizing Lists with Markov Dependence.* In D. Aldous et al. Ed. *Discrete Probability and Algorithms*, Springer, New York, pg. 51-80.

Doob, J. (1954), *Stochastic Processes*, Wiley, N.Y.

Dupont, J. (2001), *Scissors Congruences, Group Homology and Characteristic Classes*, World Scientific, Hong Kong.

Fill, J. (1996), An Exact formula for the move-to-front rule for self-organizing lisis, *Jour. Theoret. Prob.* **9**, 113-160.

Fulman, J. (1998), The combinatorics of biased riffle shuffles, *Combinatorics* **18**, 173-174.

Fulman, J. (1999), Counting semisimple orbits of finite Lie algebras by genus, *Jour. Algebra* **217**, 170-179.

Fulman, J. (2000A), Semisimple orbits of Lie algebras and card-shuffling measures on Coxeter groups, *Jour. Algebra* **224**, 151-165.

Fulman, J. (2000B), Affine shuffles, shuffles with cuts, the Whitehouse module, and patience sorting, *Jour. Algebra* **231**, 614-639.

Fulman, J. (2001A), Descent algebras, hyperplane arrangements, and shuffling cards, *Proc. Amer. Math. Soc.* **129**, 965-973.

Fulman, J. (2001B), Applications of the Brauer complex: Card shuffling, permutation statistics, and dynamical systems, *Jour. Algebra* **243**, 96-122.

Fulman, J. (2002), Applications of symmetric functions to cycle and increasing subsequence structure after shuffles. Preprint, Department of Mathematics, University of Pittsburgh. To appear, *Jour. Alg. Combinatorics.*

Fulton, W. (1997), *Young Tableaux*, Cambridge Press, Cambridge.

Garsia, A. and Remmel, J. (1985), Shuffles of permutations and the Kronecker product., *Graphs Combinatorics* **1**, 217-263.

Garsia, A. and Reutenauer, C. (1989), A decomposition of Solomon's descent algebra, *Adv. in Math.* **77**, 189-262.

Garsia, A. (1990), Combinatorics of the free Lie algebra and the symmetric group. In *Analysis ETC*, Jurgen Moser Festschrift. Academic Press, N.Y., pg. 309-82.

Garsia, A. (2002), *On the powers of top to random shuffling.* Typed notes, UCSD.

Garsia, A. and Wallach, N. (2002), Quasi-symmetric functions modulo symmetric functions are Cohen-Macually. Preprint, Dept. of Mathematics, UCSD.

Gerstenhaber, M. and Schack, S. (1987), A hodge-type decomposition for commutative algebra cohomology, *Jour. Pure Apl. Alg.* **48**, 229-247.

Gilbert, E. (1955), Theory of Shuffling. Technical Report, Bell Laboratories.

Gessel, I. and Reutenauer, C. (1993), Counting permutations with given cycle structure and descent set, *Jour. Combin. Theory. Ser. A* **64**, 189-215.

Green, J. and Rees, D. (1952), On semigroups in which $x^r = x$, *Proc. Camb. Phil. Soc.* **48**, 35-40.

Greenbaum, A. (2002), Card shuffling and the polynomial numerical hull of degree k. To appear, *SIAM J. Sci. Comput.*

Hanlon, P. (1990), The action of S_n on the components of the Hodge decomposition of Hochschild homology, *Mich. Math. Jour.* **37**, 105-124.

Hanlon, P. (1992), Order and disorder in algebraic combinatorics, *Math. Intell* **14**, 20-25.

Hanlon, P. and Hersh, P. (2002), A hodge decomposition for the complex of injective words. Technical Report, Dept. of Mathematics, University of Michigan.

Hatcher, A. (2002), *Algebraic Topology*, Cambridge University Press, Cambridge.

Jonsson, G. and Trefehten, L. (1998), A numerical analyst looks at the cutoff phenomenon in card shuffling and other Markov chains, in *Numerical Analysis* **1997**, (D. Griffiths et al. (eds.), Addison Wesley.

Keller, J. (1995), How many shuffles to mix a deck? *SIAM Rev* **37**, 88-89.

Lalley, S. (1996), Cycle structure of riffle shuffles, *Ann. Probab.* **24**, 49-73.

Lalley, S. (1999), k-Riffle shuffles and their associated dynamical systems, *Jour. Theoret. Probab.* **12**, 903-932.

Lalley, S. (2000), On the rate of mixing for p-shuffles, *Ann. Appl. Prob.* **10**, 1302-1321.

Ledoux, M. (2000), *Concentration Measures*, American Math. Soc. Providence.

Loday, J. (1988), Partition Eulerienne et operations en homologie cyclique, *C.R. Acad. Sci. Paris Ser 1 Math.* **307**, 283-286.

Lovasz, L. and Winkler, P. (1995), Mixing of random walks and other diffusions on a graph. In *Surveys in Combinatorics* (P. Rowlinson ed.) London Math Soc. Lecture Notes, V. 218, pg. 119-154, Cambridge University Press, Cambridge.

Mann, B. (1995), How many times should you shuffle a deck of cards? In *Topics in Contemporary Probability and its Applications* (J.L. Snell, ed.) CRC Press, Boca Raton.

Macdonald, I. (1985), *Symmetric Functions and Hall Polynomials*, Clarendon Press, Oxford.

MacLane, S. (1950), Cohomology of Abelian groups, *International Congress of Mathematicians* **2**, 8-14.

MacLane, S. (1963), *Homology*, Springer, Heidelberg.

MacLane, S. (1970), The Milgram bar construction as a tensor product of functions, Springer Lecture Notes in Math, No. 168, pg. 135-152.

Mahajan, S. (2002), Shuffles on Coxeter groups. Preprint, Dept. of Math., Cornell University.

Markov, A. (1906), Extension of the law of large numbers to dependent events (Russian), *Bull. Soc. Math. Kazan* **2**, 155-156.

Milgram, J. (1967), The bar construction and abelian H-spaces, *Ill. Jour. Math.* **11**, 242-250.

Morris, S.B. (1998), *Magic Tricks, Card Shuffling and Dynamic Computer Memories*, M.A.A. Washington, D.C.

Pemantle, R. (1989), An analysis of the overhand shuffle, *Jour. Theoret. Probab.* **2**, 37-50.

Phatarfod, R. (1991), On the matrix occuring in a linear search problem, *Jour. Appl. Probab.* **28**, 336-346.

Poincare, H. (1912), Calcul des probabilités, 2nd ed. Gauthier Villars, Paris.

Reeds, J. (1981), Theory of shuffling, unpublished manuscript.

Reutenauer, C. (1993), *Free Lie Algebras*, Clarendon Press, Oxford.

Saloff-Coste, L. (1997), Lectures on finite Markov chains, *Springer Lecture Notes in Math* **1665**, 301-408.

Saloff-Coste, L. (2001), Probability on groups: Random walks and invariant diffusions, *Notices Amer. Math. Soc.* **48**, 968-977.

Saloff-Coste, L. (2002), Random walks on finite groups in H. Kesten (ed.), *Springer Encyclopedia of Mathematical Sciences, Discrete Probability Volume.* To appear.

Shahshahani, M. (2002), *Lecture Notes on Geometry,*

Shnider, S. and Sternberg, S. (1993), *Quantum Groups*, International Press, Cambridge, MA.

Solomon, S. (1976), A Mackey formula in the group ring of a Coxeter group, *Journal of Algebra* **41**, 255-268.

Stadler, J. (1997), Schur functions, juggling, and statistics on shuffled permutations, Ph.D. Dissertation, Dept. of Mathematics, Ohio State.

Stanley, R. (1972), *Ordered Structures and Partitions*, Memoirs Amer. Math Soc. **119**, Amer. Math Soc. Providence.

Stanley, R. (1997), *Enumerative Combinatorics*, Vol. I, 2nd ed., Cambridge University Press, Cambridge.

Stanley, R. (1999), *Enumerative Combinatorics*, Vol. II, Cambridge University Press, Cambridge.

Stanley, R. (2002), Generalized riffle shuffles and quasisymmetric functions. *Annals of Combinatorics* **5**, 479-491.

Stark, D., Ganesh, D. and O'Connell, N. (2002), Information loss in riffle-shuffling, *Combin. Probab. Comput.* **11**, 79-95.

Su, F. (1995), Methods for quantifying rates of convergence for random walks on groups. Ph.D. Thesis, Harvard University.

Steenrod, N. (1967), Milgrams classifying space of a topological group, *Topology* **7**, 319-368.

Thorpe, E., (1972), Non-random shuffling with applications to the game of faro, *Jour. Amer. Statist. Assoc.* **68**, 842-847.

Trefethen, L. and Trefethen, L. (2000), How many shuffles to randomize a deck of cards? *Proc. Roy. Soc. London A* **456**, 2561-2568.

Uyemura-Reyes, J.C. (2002), Random walk, semi-direct products, and card shuffling, Ph.D. Thesis, Dept. of Mathematics, Stanford University.

Wallach, N. (1988), Lie algebra cohomology and holomorphic continuation of generalized Jacquet integrals, *Adv. Studies Pure Math.* **14**. Representations of Lie groups, Hiroshima, 123-151.

Wilson, D. (2001), Mixing times of lozenge tiling and card shuffling. Preprint Microsoft Research, Seattle. To appear, *Ann. Appl. Probab.*

DERANGEMENTS IN SIMPLE AND PRIMITIVE GROUPS

JASON FULMAN AND ROBERT GURALNICK

ABSTRACT. We investigate the proportion of fixed point free permutations (derangements) in finite transitive permutation groups. This article is the first in a series where we prove a conjecture of Shalev that the proportion of such elements is bounded away from zero for a simple finite group. In fact, there are much stronger results. This article focuses on finite Chevalley groups of bounded rank. We also discuss derangements in algebraic groups and in more general primitive groups. These results have applications in questions about probabilistic generation of finite simple groups and maps between varieties over finite fields.

1. INTRODUCTION

Let G be a group and X a transitive G-set. An element of $g \in G$ is called a derangement on X if g has no fixed points on X. We are interested in showing that under certain hypotheses the set of derangements of G on X is large – in particular, we will mainly focus on the case where G is finite. We then define $\delta(G, X)$ to be the proportion of elements in G that are derangements acting on X. The rare situations when $\delta(G, X)$ is very small are also quite interesting and arise in the theory of permutation and exceptional polynomials, coverings of curves and graph theory.

The study of derangements goes back to the origins of permutation group theory. It is an elementary result of Jordan that if X is a finite transitive G-set of cardinality $n > 1$, then $\delta(G, X) > 0$. It is also one of the earliest problems in probability theory – the problem was considered by Montmort [?] in 1708. Diaconis pointed out to us that Frobenius in 1904 showed that $G \leq S_n$ is k-fold transitive if and only if the first k moments of the number of fixed points is equal to the first k moments of a Poisson(1) random variable. He used this to determine character tables of Mathieu groups.

Jordan's result fails if G is infinite. There are various constructions for example where any two nontrivial elements of G are conjugate. Then G contains no derangements in any transitive action with a nontrivial point stabilizer. Another example is the case of $GL(V)$ where V is a finite dimensional vector space over an algebraically closed field and X is the set of subspaces of fixed dimension or more generally X is the set of flags of a given type (every matrix is similar to an upper triangular matrix is the equivalent formulation). The same holds for any connected algebraic group over an algebraically closed field acting its on flag variety — every element is contained in a Borel subgroup and all Borel subgroups are conjugate.

Guralnick was partially supported by National Science Foundation grant DMS-9970305.

Derangements have proved to be very useful. In particular, they have applications to images of rational points for maps between curves over finite fields (and more generally to higher dimensional varieties as well). See [GW] for more details. They also are useful in studying probabilistic generation (see [GLSS]). ¿From Chapter 3D of [Dia], one sees that derangements are useful for bounding convergence rates of random walks on finite groups; page 43 gives applications to lower bounds and since character ratios are sometimes fixed point ratios, derangements are relevant to upper bounds as well. We will explore these ideas further in future articles.

Recall that G is called a Frobenius group of degree n if G acts transitively on a set of cardinality n such that no element in G fixes 2 points (and $|G| > n$). Surprisingly, it was only very recently that Cameron and Cohen proved:

Theorem 1.1. [CC] *If X is a transitive G-set of cardinality $n > 2$, then $\delta(G, X) \geq 1/n$ with equality if and only if G is a Frobenius group of cardinality $n(n-1)$ (and in particular, n is a prime power).*

The proof is quite elementary. Another proof of this is given in [GW] and the result is extended in various ways. In particular, it was shown that:

Theorem 1.2. [GW] *If X is a transitive G-set of cardinality $n > 6$, then $\delta(G, X) > 2/n$ unless G is a Frobenius group of cardinality $n(n-1)$ or $n(n-1)/2$ (and in particular, n is a prime power).*

The proof of this result seems to require the classification of almost simple 2-transitive groups (and so the classification of finite simple groups).

Note that when trying to produce lower bounds for the proportion of derangements, there is no loss in assuming that G acts faithfully on X. We will typically make that assumption.

We particularly want to focus on the case of primitive permutation groups and simple and almost simple groups. The Aschbacher-O'Nan-Scott theorem [AS] gives the structure of primitive permutation groups and reduces many questions about them to almost simple groups (groups which have a unique minimal normal subgroup which is nonabelian simple).

A primitive permutation group G of degree n is called affine if it preserves an affine structure on the set. This is equivalent to saying that G has a nontrivial normal elementary abelian p-subgroup N for some prime p. Necessarily, $|N| = n$ is a power of p. In particular, primitive Frobenius groups are always affine permutation groups.

The major part of this paper deals with $\delta(G, X)$ when G is a finite nonabelian simple group. In particular, we will discuss a recent result of the authors proving a conjecture that has been attributed to Shalev. This theorem is proved in a series of papers by the authors starting with this one – see also [FG1], [FG2] and [FG3].

Theorem 1.3. *There exists a positive number δ such that $\delta(G, X) > \delta$ for all finite nonabelian simple groups G and all nontrivial transitive G-sets X.*

Note that it suffices to prove the previous theorem when X is a primitive G-set (for if $f : Y \to X$ is a surjection of G-sets, then $\delta(G, Y) \geq \delta(G, X)$). We also note that as stated this is an asymptotic result – we only need to prove that there exists a $\delta > 0$ such that for any sequence G_i, X_i with $|X_i| \to \infty$ with $\delta(G_i, X_i) > \delta$ for all sufficiently large i. This result is known for G alternating essentially by [D] and [LuP] and for $G = PSL(d,q)$ for a fixed q [Sh] for many families of actions. Shalev's method used some difficult results about the order of a random matrix; we use simpler properties of random matrices.

We will prove much more specific theorems and obtain much better asymptotic results. Our proof shows that we can take δ to be roughly $1/25$ aside from finitely many exceptions (and it is likely that there are no exceptions).

This paper is a partially expository paper regarding variations on this theme in [FG1] and [FG2]. We will discuss in detail analogous results for algebraic groups and give the proof for finite Chevalley groups of bounded rank.

As in [LuP] and [Sh], we obtain results about families of subgroups as well. The following result is proved in [FG1], [FG2] and [FG3].

Theorem 1.4. *Let $X := X_n(q)$ be a classical group of dimension n over $\mathbb{F}_q$. Let $I(X)$ be the union of all proper irreducible subgroups of X except when $X = Sp_{2m}(q)$ we do not include the irreducible subgroups containing $\Omega_{2m}^{\pm}(q)$ with q even. Then* $\lim_{n\to\infty} |I(X)|/|X| = 0$.

In the theorem, we can take X to be a simple classical group or the full conformal subgroup or anything in between. Moreover, we can allow a center. Of course if X is not quasisimple, we only consider maximal subgroups which do not contain $F^*(X)$. In [FG4], we will use this result to obtain some new results about probabilistic generation.

One might think that Theorem 1.3 is valid for almost simple groups. However, examples constructed in [FGS] and [GMS] (coming from problems in coverings of curves) show that the result fails for almost simple groups. We give some examples of this phenomenon later in the paper. The presence of field automorphisms is critical in producing such examples.

However, we do prove that the result on simple groups does lead to a weaker bound for primitive groups which are not affine. See §8.

Theorem 1.5. *Let X be a primitive G-set with $|X| = n$. There exists a positive number δ such that either*

(1) *$\delta(G,X) > \delta/\log(n)$; or*
(2) *G preserves an affine structure on X.*

We will investigate the affine case in future work.

We will also consider a slight refinement of this problem:

Let G be a normal subgroup of A with A and G acting transitively on a finite A-set X. Assume that A/G is generated by the coset aG.

We wish to investigate the quantity $\delta(A,G,X)$, the proportion of derangements in the coset aG. We note the following easy facts:

(1) the quantity $\delta(A,G,X)$ does not depend on the choice of the generating coset aG; and
(2) $\delta(G,X) = \delta(G,G,X)$.

This quantity is important in studying maps of varieties over finite fields via a Cebotarev density theorem (see [GW] for more details). In contrast to the case $A = G$, there may be no derangements in a given coset. This turns out to be a very special and important case in the study of exceptional covers [FGS] and graph theory [GLPS]. See §6 and §7 for further discussion.

In [GW], the following was shown via an elementary proof:

Theorem 1.6. *Let G be a normal subgroup of A with A/G cyclic. Let X be a transitive A-set of cardinality $n > 2$. Either $\delta(A, G, X) = 0$ or $\delta(A, G, X) \geq 1/n$. Moreover, equality holds if and only if $A = G$ is a Frobenius group of order $n(n-1)$.*

One of the eventual goals of this project is to greatly improve this result.

We now give a brief sketch of the contents and ideas of the paper.

Let G be a group acting primitively and faithfully on the transitive G-set X and let H be the stabilizer of a point $x \in X$. Set

$$\mathcal{C}_G(H) := \cup_{u \in G} uHu^{-1}.$$

An element $g \in G$ is a derangement if and only if $g \notin \mathcal{C}_G(H)$.

The proofs of many of these results are heavily dependent upon the classification of finite simple groups – both in the fact that we are assuming the complete list of finite simple groups and in using information about subgroups of the finite simple groups. Since Theorem 1.3 is really an asymptotic result, we are considering the following situation – we have a sequence (G_i, X_i) where G_i is a finite nonabelian simple group and X_i is a primitive G_i-set of cardinality of n_i. We may assume that $|G_i|$ (or n_i) tends to infinity. We need to show that $\liminf \delta(G_i, X_i) \geq \delta$ for some positive δ (a single δ for all such sequences). This implies that $\delta(G, X) \geq \delta$ for all but finitely many simple G and primitive X, whence $\delta(G, X)$ is bounded away from 0 for all simple G and primitive X.

In [FG1], [FG2] and [FG3], we obtain much stronger results.

By passing to infinite subsequences, to prove Theorem 1.3, it suffices to assume that all the G_i are alternating groups (of increasing degree), are all Chevalley groups of a given type (and rank) over fields of size q_i with $q_i \to \infty$ or are classical groups of dimension d_i over fields of cardinality q_i with $d_i \to \infty$.

In the case of alternating (and symmetric groups), we can apply the work of Dixon [D] and Luczak-Pyber [LuP]. We improve some of these results in [FG1] and [FG2]. In the case of Chevalley groups of fixed type, we can use the theory of algebraic groups and algebraic geometry to obtain the desired results. Here the dichotomy is between subgroups that contain maximal tori and those that do not. See §2, §3 and §4.

Now consider the case that the G_i are classical groups of dimension d_i over a field of size q_i. We subdivide this case further. First of all, either the $q_i \to \infty$ or we may assume that $q_i = q$ is constant. In the first situation, by [GL], it suffices to consider only semisimple regular elements. We subdivide each case further using the idea of Aschbacher's classification of maximal subgroups of classical groups [A2]. In particular, we consider subspace stabilizers and show that we can reduce certain questions to the study of the Weyl group (and so to symmetric groups). We prove in [FG1] the following result about subspace stabilizers. For the next theorem, in the case of a linear group, all subspaces are considered to be totally singular.

Theorem 1.7. *Let G_i be a sequence of classical groups with the natural module of dimension d_i. Let X_i be a G_i-orbit of either totally singular or nondegenerate subspaces (of the natural module) of dimension $k_i \leq d_i/2$. If $k_i \to \infty$, then* $\lim \delta(G_i, X_i) = 1$. *If k_i is a bounded sequence, then there exists $0 < \delta_1 < \delta_2 < 1$ so that $\delta_1 < \delta(G_i, X_i) < \delta_2$.*

One of the key ingredients in the proof of Theorem 1.3 for fixed q is getting estimates for the number of conjugacy classes for finite Chevalley groups of rank r over a field of size q. We show that there is an explicit universal constant C so that the number of conjugacy classes of such a group is at most Cq^r – see §9. See also Gluck [Gl] and Liebeck-Pyber [LiP] for weaker estimates. These results are of independent interest and should be useful (see the above mentioned references for some applications). Two other important ideas in the proof are an upper bound for the maximum size of a conjugacy class and a result that says that most elements in a classical group are nearly regular semisimple (i.e. they are regular semisimple on a subspace of small codimension). Another ingredient we use in the proof of Shalev's conjecture (for q fixed) is precise estimates on proportions of regular semisimple elements (proved via generating functions). See [FNP]. Finally, we require results on random permutations.

This article is organized as follows:

We first discuss derangements in algebraic groups (in algebraic actions) – see §2. We then prove Theorem 1.3 for groups of bounded rank in §§3, 4 – the second of which focuses on subgroups containing a maximal torus. As a corollary (§5), we solve for bounded rank groups a problem studied by Dixon [D] and McKay (unpublished) for symmetric groups. The case of classical groups with rank going to ∞ is treated in [FG4]. In §§6, 7, we give some examples and mention the connection with so called exceptional permutation actions and give a short proof of Theorem 1.6. In §8, we then show how Theorem 1.5 follows as a corollary to Theorem 1.3. In the final section, we tabulate some of our results about conjugacy classes for classical groups.

2. Algebraic Groups

In this section, we investigate the existence of derangements in (algebraic) permutation actions for connected algebraic groups. We refer to [H1] for the basic results about algebraic groups.

We first make a simple observation that holds for solvable groups (not just algebraic groups).

Lemma 2.1. *Let G be a solvable group and H a proper subgroup of G. Then* $\cup_{g \in G} H^g \neq G$.

Proof. Let A be the last term in the derived series of G. If $HA \neq G$, we can pass to G/A and the result follows by induction on the derived length (the case of abelian groups being obvious). So assume that $G = HA$ and in particular, H does not contain A. Then $H \cap A$ is normal in G (since $H \cap A$ is normal in H and in the abelian group A). It follows that the only elements in A which have fixed points are the elements of $A \cap H$, a proper subgroup of A. □

We now consider connected algebraic groups. We restrict attention to semisimple groups.

Let G be a connected semisimple algebraic group and X a nontrivial faithful algebraic G-set G/H. In particular, H is a proper closed subgroup of G. Let $J = \cup_{g \in G} H^g$. Let J' denote the complement of J – thus, J' is precisely the set of derangements.

Let T denote a maximal torus of G and N its normalizer. Note that T is self centralizing and N/T is a finite group (the Weyl group of G). Moreover, any two elements of T are conjugate in G if and only if they are conjugate in N. We recall that any two maximal tori of G are conjugate.

Lemma 2.2. *Let G be a connected semisimple algebraic group over an algebraically closed field. Let H be a closed subgroup of G and $J = \cup_{g \in G} H^g$. Then the closure $\bar{J}$ of J is all of G if and only if T has a fixed point on X.*

Proof. Since the set of semisimple elements of G contains an open subvariety of G, the reverse implication is clear.

Assume that J is dense in G. Let S be a maximal torus of the connected component H_0 of H. Let $d = |H : H_0|$.

Since J is the image of the morphism $f : H \times G \to G$ with $f(h, g) = h^g$, it follows that J contains a dense open subset of its closure and so under this hypothesis an open subset of G.

Note that if $g \in G$ is semisimple regular, then the there are at most d^r solutions to $x^d = g$ (where r is the rank of G) – for $x^d = g$ implies that $x \in C_G(g)$, a maximal torus of rank r. This implies that the dth power map on G is dominant and so the set of dth powers of elements in J also contains an open subvariety of G. This implies that the union of the conjugates of H_0 contains an open subvariety of G. Since the union of the conjugates of S contains an open subvariety of H_0, we have that $\cup_{g \in G} S^g$ contains an open subvariety of G.

By conjugating, we may assume that $S \leq T$. We have the surjection from $G/T \times S \to \cup_{g \in G} S^g$ given by $(gT, s) \to s^g$, whence

$$\dim G = \dim \cup_{g \in G} S^g \leq \dim G + \dim S - \dim T,$$

and so $\dim S = \dim T$ and $S = T$. □

Of course, every element is contained in a Borel subgroup. So if H is a parabolic subgroup (i.e. an overgroup of a Borel subgroup), there are no derangements. We can give an easy proof that these are the only examples if H is connected.

Theorem 2.3. *Let G be a semisimple algebraic group over an algebraically closed field k of characteristic p. Let H be a closed proper subgroup of G. Assume that H is connected or that p does not divide the order of the Weyl group of G (this includes the case $p = 0$).*

(a) *If H contains a maximal torus of G and a regular unipotent element of G, then H is a parabolic subgroup of G.*
(b) *If $\cup_{g \in G} H^g = G$, then H is a parabolic subgroup.*

Proof. If $\cup_{g \in G} H^g = G$, the previous lemma implies that H contains a maximal torus. Clearly, it contains a regular unipotent element, whence (b) follows from (a).

We now prove (a). Let H_0 be the connected component of H. We will first show that H_0 contains a regular unipotent element. We can then reduce to the case that H is connected.

If $p = 0$, this is clear because all unipotent subgroups are connected. Let T be a maximal torus of H_0 (which is also a maximal torus of G). Since all maximal tori of H are H_0-conjugate, it follows that $H = N_H(T)H_0$, whence $|H : H_0| = |N_H(T) : N_{H_0}(T)|$ is a divisor of the order of the Weyl group of G. In particular, it has order prime to p and so H_0 contains all unipotent elements of H.

If H_0 is a parabolic subgroup, then so is H (and indeed $H = H_0$ as any overgroup of a Borel subgroup is a parabolic subgroup). So we may assume that H is connected.

Let B_H be a Borel subgroup of H containing T and let B be a Borel subgroup of G containing B_H. Let U be the unipotent radical of B. Since H is connected, B_H contains a regular unipotent element u as well (because every unipotent element of H is conjugate to an element of U). We can write $u = v \prod U_\alpha(t_\alpha)$ where the α are the simple roots relative to T, $t_\alpha \neq 0$ and $v \in [U, U]$. It follows that $u^T[U, U]$ contains all elements in U which have a nonzero entry in U_α for each simple root α. Thus, $[u, T][U, U] = U$. Since U is nilpotent, this implies that $U = [u, T]$ and so $B = TU \leq H$. Thus, $B_H = B$ and $H \geq B$ as required. □

There are only a handful of examples of proper closed nonconnected subgroups containing a conjugate of every element of G. This requires a result of Saxl and Seitz. We note that the result of [SaSe] has the unneeded hypothesis that the characteristic is good (their proof never uses this fact). We will use the following fact in the next result – any positive dimension closed subgroup of a simple algebraic group is contained in a maximal closed subgroup.

Theorem 2.4. *Let G be a simple algebraic group over an algebraically closed field k of characteristic p. Let H be a closed proper subgroup of G. Assume that H is not contained in a parabolic subgroup. The following are equivalent:*

(a) *H contains a maximal torus of G and a conjugate of every unipotent element of G;*
(b) *H contains a conjugate of every element of G;*
(c) *The characteristic of k is 2 and $(G, H) = (Sp(2m, k), O(2m, k))$ or $(G, H) = (G_2(k), A_2(k).2)$.*

Proof. Clearly (b) implies (a).

We next show that (a) implies (c). So assume that H satisfies (a). If H is maximal (among closed subgroups), then Theorem C of [SaSe] shows that (c) holds.

Let M be a maximal closed subgroup of G containing H. Then (G, M) satisfies the conclusion of (a) as well and so as noted, (G, M) satisfies (c). In particular, k has characteristic 2. Moreover, H must have maximal rank and is not connected.

If $G = G_2(k)$ and H is a proper (disconnected) rank 2 subgroup of M, then the only possibility is that H is contained in the normalizer of a maximal torus, which does not contain a conjugate of every unipotent element.

So we may assume that $G = Sp(2m, k)$. If $m = 1$, then M is the normalizer of a maximal torus T and M/T has order 2 and so clearly $H = M$. So consider the case that $G = Sp(2m, k), m \geq 2$. If H acts reducibly on the natural module V for G, then H is contained in the stabilizer of a proper subspace W. Take W of minimal dimension. Since H is not contained in a parabolic subgroup, it follows that this subspace is nondegenerate. The stabilizer of such a subspace is not contained in a conjugate of M, a contradiction.

Suppose that the connected component H_0 does not act irreducibly on V. Then either H is contained in a maximal subgroup not contained in $O(2m, k)$, a contradiction or $V = W_1 \oplus W_2$, where H_0 is irreducible on each W_i and W_i is a maximal totally singular subspace of dimension m. Thus, H is contained in the stabilizer of a pair of complementary totally singular subspaces. We claim that H contains no transvections. A transvection in H cannot swap the two spaces and so would have to stabilize each W_i. The action on W_1 is dual to that on W_2 and so the element is not a transvection.

Thus, H_0 acts irreducibly on V, whence H_0 is semisimple. Since H has rank m, this forces H_0 to contain the connected component of $O(2m, k)$, whence $H = O(2m, k)$.

All that remains is to verify that (c) implies (b).

This is well known for the first family (see [SaSe], Lemma 4.1) The latter case is an easy consequence of the first case (since $G_2(k) \leq Sp(6, k)$, $Sp(6, k) = O(6, k)A_2.2$ and $A_2.2 = G_2(k) \cap O_6(k)$). □

3. Groups of Bounded Rank I

In this section and the next, we consider the case where the groups have bounded rank. We will prove Theorem 1.3 in this case. The next section deals with subgroups containing a maximal torus. We deal with the other cases in this section.

As we have observed, as stated it is an asymptotic result. We only need to produce a δ so that the proportion of derangements is at least δ for all but finitely many cases. If this fails, there would be a sequence with the proportion of derangements all less than δ. Thus, Theorem 1.3 is an asymptotic result (as noted, eventually we want a non-asymptotic version). Since the groups have bounded rank, we may assume that they have fixed type $X(q_i)$ with X a simple algebraic group and only the field size is varying. We can use methods of algebraic geometry and algebraic groups to study this situation.

We recall that $F^*(H)$ is the generalized Fitting subgroup of H. See [A1]. In particular, $F^*(H)$ simple just means that $F^*(H) \leq H \leq \mathrm{Aut}(H)$. There is no harm in considering covering groups of almost simple groups since all the maximal subgroups will contain the center.

Fix a type of simple algebraic group X of rank r. Let σ be an endomorphism of X with fixed point group X_σ of finite order. We will typically write $X(q) = X_\sigma$ if q is the absolute value of the eigenvalues of σ on the character group of the maximal torus T of X. In the case of the Suzuki or Ree groups q will not be an integer. This will cause no problems. Indeed, in those cases, one knows all the maximal subgroups and it is quite easy to obtain our results. We may take X simply connected or of adjoint type or anything in between – this allows us to obtain results for Chevalley groups generated by inner-diagonal automorphisms.

The maximal subgroups H of $X(q)$ (which do not contain $F^*(X(q))$) are of four types:

(1) $|H| < N$ for some fixed $N = N(X)$;
(2) $H = N_{X(q)}(X(q'))$ for some q' dividing q (this includes the twisted forms, e.g., ${}^2E_6(q) \leq E_6(q^2)$);
(3) $H = Y_\sigma$ where Y is a proper σ-invariant algebraic subgroup of X of rank $s < r$ and the connected component of Y is semisimple;
(4) $H = Y_\sigma$ where Y is a proper σ-invariant algebraic subgroup of X of maximal rank r.

This is well known for the case of classical groups (see [A2]). If H is of exceptional type, this follows in a very precise way from results of Liebeck and Seitz [LS1]. See the remarkable paper of Larsen and Pink [LaP] for a classification free proof of the previous result. Note that if Y is a maximal σ-invariant positive dimensional algebraic subgroup, then either Y has maximal rank or the connected component of Y is semisimple (for if the unipotent radical of Y is nontrivial, Y is a parabolic subgroup by the Borel-Tits theorem and if it contains a normal torus, then Y contains a maximal torus).

We also note the following result.

Lemma 3.1. *If X is a simple connected algebraic group and Y is a proper positive dimensional σ-invariant subgroup , then Y is contained in a maximal proper closed σ-invariant subgroup. Moreover, there is a bound $m = m(X)$ for the number of connected components for any maximal σ-invariant closed subgroup.*

Proof. Let Y_1 be a proper closed σ-invariant subgroup of X containing Y that has maximal dimension. Let Y_0 be the connected component of Y_1. Then $N_X(Y_0)$ is maximal among σ-invariant closed subgroups (for any such overgroup would have the same dimension as Y_0 whence would have connected component Y_0).

All that remains is to prove the statement about the number of components. So we may assume that Y is a maximal proper closed σ-invariant subgroup. If Y_0 has a unipotent radical, then Y is contained in a parabolic subgroup and in particular is connected.

So assume Y_0 is reductive. If Y_0 is not semisimple, then the connected component of $Z(Y_0)$ is a nontrivial torus and Y is contained in the normalizer of this torus, whence we may take Y to be the normalizer of this torus. Thus, Y_0 contains a maximal torus and so contains its centralizer (which is contained in the maximal torus). By the Frattini argument, any closed subgroup containing a maximal torus has at most $|W|$ components, where W is the Weyl group of X.

So we may assume that Y_0 is semisimple. Let $C = C_X(Y_0)$. Then C is finite (since $C \cap Y_0$ is finite). If X is classical and Y_0 is not simple acting irreducibly on the natural module, then the result follows by [A2] which gives all possibilities for Y (although the fields are assumed to be finite, the proofs go through without change for the algebraic closure – see [LS3] for a treatment of the algebraic group case). If Y_0 is simple acting irreducibly, then $C \leq Y_0$ and Y/Y_0 is bounded by the size of the (algebraic) outer automorphism group and so has order at most 6.

If X is exceptional, all such maximal subgroups are classified (see [LS2], Corollary 2 – we only need to handle the case where $F^*(Y)$ is not quasisimple, such maximal subgroups were classified much earlier) and the result follows by inspection. □

Let $\mathcal{M}_i$ denote the maximal subgroups in the corresponding families $i = 1, 2, 3$ or 4 above.

We will deal with each of these families separately. In this section, we deal with the first three cases. In the next section, we deal with the remaining case. The purpose of this section is to prove:

Theorem 3.2. $\lim_{q\to\infty} |\cup_{i=1}^{3} \cup_{M\in\mathcal{M}_i} M|/|X(q)| = 0.$

This is not true for $\mathcal{M}_4$ (compare with the result Lemma 2.2 for algebraic groups). We first start with some general known results.

Lemma 3.3. *Let U be a unipotent connected group of dimension r defined over the finite field F_q. Let $\tau = g\sigma$ where g is an algebraic automorphism of U and σ is the q-Frobenius map. Then $|U_\tau| = q^r$.*

Proof. Suppose that Y is a connected τ invariant subgroup of U. The result would follow by induction and Lang's theorem (since $|U_\tau| = |Y_\tau||(U/Y)_\tau|$). So we may assume this is not the case. It follows that U is abelian of exponent p (where p is the prime dividing q). Then X is just a product of copies of the field and another application of Lang's theorem (applied to $\mathrm{Aut}(X)$) gives that τ and σ are conjugate via an element of $\mathrm{Aut}(X)$ and the result follows. □

Lemma 3.4. *Let $x \in X(q)$. Let C be the centralizer of x in $X(q)$.*

(1) *If x is unipotent, then $|C|$ is divisible by q^r.*
(2) $|C| \geq (q-1)^r$.

Proof. If x is unipotent, the result follows since all unipotent classes are known as well as their centralizers. Aside from the cases of Suzuki and Ree groups, this also follows from the previous lemma. Let B be a σ invariant Borel subgroup with unipotent radical U containing x and consider the connected component of $C_U(x)$. Since regular unipotent elements are dense in B, it follows that $\dim C_B(x) \geq r$ and so C has order divisible by the cardinality of the subgroup of fixed points in $C_U(x)$. By the previous lemma (applied to σ acting on U), this cardinality is divisible by q^r. A variation of the previous lemma could be applied to the case of Suzuki and Ree groups.

We note that the result holds for semisimple groups as well.

We prove the second statement more generally for reductive groups of rank r. Write $x = su = us$ with u unipotent and s semisimple. Pass to the connected component of D of $C_X(s)$. This is still reductive of rank r. Write $D = AB$ with A a central torus in C and $B = [C, C]$ semisimple with A of rank a and B of rank b. Since a torus of rank a over the field of q elements has at least size $(q-1)^a$ and $C_{B(q)}(u)$ has order divisible by q^b, we see that $|C| \geq (q-1)^a q^b$, whence the second statement holds. □

The following was originally proved by Steinberg in the case of simply connected X. See [Ca] or [H2].

Lemma 3.5. *The number of conjugacy classes of semisimple elements in $X(q)$ is at most q^r with equality if X is simply connected.*

The next result follows from [GL].

Lemma 3.6. *The proportion of regular semisimple elements in $X(q)$ is greater than $1 - 5/(q-1)$.*

The previous result indicates that the proportion of elements which are not semisimple regular goes to 0 linearly with $1/q$. The same is thus true for the set of derangements which are not semisimple regular. Thus, it suffices to consider the set of derangements which are semisimple (and indeed regular). We will do so in the next two sections without further comment.

Lemma 3.7. $|\cup_{M\in\mathcal{M}_1} M|/|X(q)| \to 0$ *as* $q \to \infty$.

Proof. $\cup_{M\in\mathcal{M}_1} M$ is a union of at most N' conjugacy classes of elements for some N' (that depends only on N and so only on X). Thus the union has order at most $|X(q)|N'/(q-1)^r$ and the result follows. □

Lemma 3.8. $|\cup_{M\in\mathcal{M}_2} M|/|X(q)| \to 0$ *as* $q \to \infty$.

Proof. Consider $X(q')$. The number of semisimple conjugacy classes in $X(q')$ is at most $(q')^r$. Let $S(q', q)$ denote the union of the semisimple conjugacy classes of $X(q)$ intersecting $X(q')$. Thus,

$$|S(q', q)| \leq |X(q)|(q')^r/(q-1)^r.$$

In the case of the Suzuki or Ree groups, we write $X = X(p^{2a+1})$ (this conflicts slightly with our notation above). The number of possible classes of subfield groups is the number of distinct prime divisors of $2a+1$, whence the estimate above shows that the union of the semisimple elements in any subfield group is certainly at most $\sum_b |X(q)|q^{r/b}/(q-1)^r$, where b ranges over prime divisors of $2a+1$. This yields the result.

Consider the remaining cases. Write $q = p^a$. Note that for each choice of q' (corresponding essentially to a prime divisor of a), there are at most $2c$ choices for $S(q, q')$ where c is the order of the group of outer diagonal automorphisms ($6c$ in case $X = D_4$). This is because we may take $\sigma = \alpha\tau f_{q'}$ where τ is a graph automorphism, α is a diagonal automorphism and f_q is the standard Frobenius (any two such elements in the coset with the same order are conjugate up to diagonal automorphisms – see I.7.2 [GoLy]). In fact as noted above, we only need to consider semisimple elements, the diagonal outer automorphisms will not make a difference, but we do not need to use this. □

Lemma 3.9. $|\cup_{M\in\mathcal{M}_3} M|/|X(q)| \to 0$ *as* $q \to \infty$.

Proof. It follows by the theory of high weights if X is classical [GKS] and by [LS2] if X is exceptional that there are only finitely many conjugacy classes (with a bound depending only upon X) in $\mathcal{M}_3$. Thus, it suffices to show the result for a fixed type of subgroup $Y < X$. Then Y_σ has at most cq^s conjugacy classes of semisimple elements (where c is the number of connected components of Y – note that c is bounded in terms of X). It follows that $|\cup_{g\in X_\sigma} Y_\sigma^g| \leq |X_\sigma|cq^s/(q-1)^r$, whence the result. □

This completes the proof of Theorem 3.2. The next section deals with $\mathcal{M}_4$.

4. Maximal Rank Subgroups

In this section we consider $\mathcal{M}_4$. It follows from the results on algebraic groups that the proportion of derangements will be positive in this case. For the Suzuki and Ree groups, one just inspects the maximal rank subgroups and the result about derangements follows quite easily. We assume for the rest of the section that we are not in any of those cases. We remark again that it suffices to consider only regular semisimple elements (since as $q \to \infty$, the proportion of regular semisimple elements is $1 + O(1/q)$).

Keep notation as in the previous section. Let Y be a σ-stable subgroup of X of maximal rank and $H = Y_\sigma$. The possibilities are that Y is a parabolic subgroup (maximal with respect to being σ-stable) or is reductive. Let Y_0 denote the connected component of Y. Let $H_0 = (Y_0)_\sigma$. This is a normal subgroup of H.

There exists a σ-stable maximal torus T contained in a Borel subgroup B of X. A maximal torus of X_σ is S_σ where S is a σ-stable maximal torus of X. There is a notion of nondegenerate maximal tori (for example, if $X = SL(n)$, then over the field of 2-elements, a maximal torus might be trivial, see §3.6 in [Ca] for details). We will just note that if the maximal torus contains a regular semisimple element, then $N_{X_\sigma}(S_\sigma) = N_X(S)_\sigma$ and so is nondegenerate – this follows since $S = C_X(S_\sigma)$. Moreover (for fixed X), if q is sufficiently large, all maximal tori contain regular semisimple elements (indeed almost all elements are regular semisimple).

Let W be the Weyl group of G (more precisely identify $W = N_X(T)/T$). Consider the semidirect product $W\langle\sigma\rangle$. There is a bijection between conjugacy classes of maximal tori in X_σ and W-classes of elements in the coset σW (see [SpSt] or [Ca]). In particular, if σ is a field automorphism, σ commutes with W and so the correspondence is with W-conjugacy classes (this latter fact is still true for all groups of type A and type D_n with n odd).

Let T_w denote a maximal torus of X_σ corresponding to σw. Let N_w be the normalizer in X_σ of T_w. Then $|N_w : T_w| = |C_W(\sigma w)|$. Let $f(w)$ be the size of W-class of σw. So $f(w) = |W : C_W(\sigma w)| = |W||T_w|/|N_w|$.

In particular, we see that

$$|\cup_{g \in X_\sigma} T_w^g|/|X_\sigma| < |T_w|/|N_w| = f(w)/|W|.$$

Since a semisimple regular element lies in a unique maximal torus, it follows that the union of all regular semisimple elements of X_σ that are conjugate to an element of T_w has cardinality at most $|X_\sigma|f(w)/|W|$.

Since the proportion of elements which are not semisimple regular tends to 0 as $q \to \infty$ and the same is true for each maximal torus, it follows that the inequality above becomes equality as $q \to \infty$.

We first show that the collection of elements which are conjugate to an element of H but not H_0 is small. We need the following result. A very easy result (see Proposition 4.3 below) gives an upper bound (always at least 1/2) for the proportion of derangements contained in H_0 (assuming that $H \neq H_0$).

Lemma 4.1. *Let G be a connected reductive algebraic group with σ an endomorphism of G such that G_σ is finite. Assume that all eigenvalues of σ on the character group of T have absolute value q. Let S and T be distinct σ-stable maximal tori of G. Then $|T_\sigma : (S \cap T)_\sigma| \geq (q-1)/2$.*

Proof. Consider a counterexample with $\dim G$ minimal. Since G is reductive, G is the central product of Z and H where H is semisimple and Z is the connected component of the center of G. Since Z is contained in every maximal torus, there is no loss in taking $G = H$ to be semisimple. We can replace G by its universal central extension (since the center will be contained in every maximal torus) and so assume that G is a direct product of simply connected simple algebraic groups.

If $S \cap T = Z(G)$, the result is clear (pass to the simple case). Otherwise, we can consider $H = C_G(x)$ with $x \in S \cap T \setminus Z(G)$. Then H is connected and reductive and S, T are maximal tori in H. □

Note that if $G = SL(2)$, we do have equality in the previous result.

Proposition 4.2.

$$\lim_{q \to \infty} |\cup_{g \in X_\sigma} (Y \setminus Y_0)^g_\sigma| / |X_\sigma| = 0.$$

Proof. It suffices to consider a single coset yY_0 for some element $y \in Y_\sigma \setminus Y_0$.

We will obtain an upper bound on the number of conjugacy classes of semisimple regular elements of X_σ that intersect yY_0. We will do this by bounding the number of Y_σ classes in that coset.

Suppose that $u \in yY_0 \cap X_\sigma$ is a semisimple regular element. Then the centralizer of u in the algebraic group is a σ-stable maximal torus T. Let S be a σ stable maximal torus of Y_0 containing $T \cap Y_0$. The number of $(Y_0)_\sigma$ conjugates of u is

$$|(Y_0)_\sigma : (S \cap T)_\sigma| \geq |(Y_0)_\sigma|(q-1)/2|S_\sigma|.$$

Since $|S_\sigma| \leq (q+1)^r$, it follows that the number of conjugates of u in the coset $u(Y_0)_\sigma$ is at least $|(Y_0)_\sigma|q^{r-1}/2$ (up to a small error term). This implies that there are at most $2q^{r-1}$ classes of semisimple regular elements in this coset (again up to a term of smaller order). Since each class has size approximately $O(|X_\sigma|)/q^r$, the union of these classes has size $O(|X_\sigma|/q)$ as required. □

We now consider the connected component Y_0 and its fixed points H_0. We first note that if $H_0 \neq H$, then we have the following easy estimate for derangements.

Lemma 4.3. *If $H \neq H_0$, then $|\cup_{g \in X_\sigma} H_0^g| / |X_\sigma| < 1/|H : H_0| \leq 1/2$.*

Proof. Since H normalizes H_0, $\cup_{g \in X_\sigma} H_0^g$ is the union where g ranges over a transversal of X_σ/H, whence the cardinality of this union is less than $|X_\sigma : H||H_0| = |X_\sigma|/|H : H_0|$. □

We just remark that Lang's theorem implies that $|H : H_0|$ is the number of σ stable cosets of Y_0 in Y.

Let S be a σ stable maximal torus of Y_0. Then $S = xTx^{-1}$ where $x^{-1}\sigma(x) \in N(T)$. Note that $x^{-1}N(S)x = N(T)$. So we have subgroups $T \leq x^{-1}N_H(S)x \leq x^{-1}N_Y(S)x \leq N(T)$ and this gives rise to corresponding subgroups $1 \leq W_0 \leq W_1 \leq W$ in W the Weyl group of T.

The σ-stable maximal tori of H (up to H_σ-conjugacy) are of the form $ySy^{-1} = yxT(yx)^{-1}$ where $v := y^{-1}\sigma(y) \in N_H(S)$. Moreover, we see that S is conjugate to T_w where

$$w = (yx)^{-1}\sigma(yx)T = x^{-1}y^{-1}\sigma(y)\sigma(x) \in \tau W_0,$$

where $\tau = x^{-1}\sigma(x)T \in W$.

Thus, setting R to be the union of all X_σ conjugates of maximal tori of H_σ, we see that $|R|/|X_\sigma| \leq \sum f(w)/|W|$ where the sum is a set of representatives Γ of conjugacy classes in W that are represented by elements in τW_0.

We note that R is precisely the set of conjugacy classes of regular semisimple elements conjugate to an element of H_σ (since the centralizer of such an element will be a maximal torus in H). Thus, we have an upper bound for the proportion of regular semisimple elements in X_σ that are not derangements in X_σ/Y_σ. By Proposition 4.2, we can replace H_σ by Y_σ (up to an $O(1/q)$ term) and we can consider all elements (not just regular semisimple elements) by introducing another such term (by [GL]) and so we see that $\delta(X_\sigma, Y_\sigma) \geq 1 - \sum_{w\in\Gamma} f(w)/|W| + O(1/q)$. We just need to bound $\sum_{w\in\Gamma} f(w)/|W|$ away from 1.

There is one very easy case – if σ does not involve a graph automorphism and $W_1 \neq W$, then $1 - \sum_{w\in\Gamma} f(w)/|W| \leq \delta(W, W/W_1)$. Note in this case $f(w)$ is just the size of the W-conjugacy class of w. This can be computed for the exceptional Weyl groups. In any case, for bounded rank, we can even use the Jordan bound to see this is bounded away from 1. For classical groups, using [D], [LuP], [FG1], [FG2] we see that this quantity will typically be at most 2/3.

If σ does involve a graph automorphism, then we consider the group Z defined above. Since σ stabilizes both W_1 and W_0, we can define Z_1 and Z_0 in an obvious manner. Note that in this case $f(w)$ is the size of the W-class of σw. Thus, we still have a bound $1 - \sum_{w\in\Gamma} f(w)/|W| \leq \delta(Z, W, Z/Z_1)$.

We note by inspection that unless $W = W_1$, there are always derangements in the coset σW. Since except for type D_4, W can be thought as of a subgroup of index 2 in Z (only the graph automorphism makes a difference), exceptionality would force $|W : W_1|$ to have odd index.

There are only a few cases where $W = W_1$. In all cases, we see that whenever this happens $H \neq H_0$ and so the upper bound of $1/2 + O(1/q)$ holds. One possibility is that H is a maximal torus. In that case, we note directly that $f(w)/|W| \leq 1/2$. Another possibility is $X = G_2$ and $H = A_2$. Similarly, there is the possibility of $(X, Y) = (F_4, D_4)$.

The only other such possibility for X classical is in characteristic 2 with X of type B_n and Y of type D_n. Then W_0 has index 2 in $W = W_1$. In this case, one sees that there are two possible forms of Y_σ (i.e. two conjugacy classes corresponding the single X_σ class of σ stable conjugates of Y) – the two forms of orthogonal groups. One form of the orthogonal group has maximal tori T_w with $w \in W_0$ and the other the complement (note a maximal torus is contained in a unique orthogonal group in the symplectic group). Thus, $\delta(Sp(2m, 2^a), O^\epsilon(2m, 2^a)) = 1/2 + O(1/2^a)$.

We note that our analysis works for any form of the Chevalley group and for any fixed coset in the group of inner-diagonal automorphisms.

Thus, we have proved:

Theorem 4.4. *Let r be a positive integer. Let S be a simple Chevalley group of rank at most r over the field of q elements and $S \leq G$ with G contained in the group of inner-diagonal automorphisms of S. Let X be a transitive faithful G-set. Then there exists $\delta > 0$ such that $\delta(G, S, V) \geq \delta + O(1/q)$ for any transitive G-set V.*

We will give an explicit δ in the sequel. Note that the error term depends only on r (and we do remove that dependence in [FG1], [FG2] and [FG3]).

If X is classical, then the possibilities for Y are rather limited. There is the special case in characteristic 2 when X is symplectic and Y is an orthogonal group. The remaining cases are essentially when Y_σ is the group preserving a decomposition of the space or Y_σ is an extension field group (both forms of $Y \leq C \wr S_m$ where C is a classical group on a subspace). In the bounded rank case, we have seen that we could work with the connected piece.

If the rank increases, there are two added complications. If q is fixed, then we can no longer deal with only semisimple regular elements (the error term may be larger than the main term). Even if q increases, the error term associated with reducing to the connected component may be increasing with the rank. Thus, the analysis is much more difficult. See [FG1] and [FG2]. We also want to produce an explicit δ that is valid for either all cases or all but a specified finite set of cases.

We close this section by considering a few examples.

(1) Let $G = PSL(n, q)$ and let H be the stabilizer of a k-dimensional vector space. In this case H is the set of fixed points of a connected subgroup and so we see from the analysis above that for a fixed n, $\lim_{q\to\infty} \delta(G, H) = \delta(S_n, Y_k)$ where $Y_k = S_k \times S_{n-k}$ is a Young subgroup. By [D], $\delta(S_n, Y_k) \geq 1/3$ and by [LuP] (and also by [FG1]), $\delta(S_n, Y_k) \to 1$ as $k \to \infty$ (for $k \leq n/2$). This example holds more generally for any parabolic subgroup – the limiting proportion of derangements is precisely the proportion of derangements of the Weyl group acting on the cosets of the corresponding parabolic subgroup.

(2) Let $G = Sp(2n, q)$ with q even. Let $H = O^\epsilon(2n, q)$. Then H is the set of σ fixed points on some σ invariant conjugate of $O(2n) < Sp(2n)$. The Weyl group of the connected component of $O(2n)$ has index 2 in the Weyl group of $Sp(2n)$ and so we see that $\lim_{q\to\infty} \delta(G, H) = 1/2$ (each type corresponds to maximal tori in one coset of the Weyl group of Ω). Note that a regular semisimple element is contained in precisely one orthogonal group.

(3) Let $G(q) = E_8(q)$ and $H(q) = D_8(q)$. Since the corresponding algebraic subgroup is connected, it follows that

$$\lim_{q\to\infty} \delta(G(q), G(q)/H(q)) = \delta(W(E_8), W(D_8)).$$

5. Generation and Derangements

In this section, we indicate how some generation results follow immediately from our results. See [FG4] for more results about probabilistic generation that follow from the results in this paper, [FG1], [FG2] and [FG3].

If G is a finite simple group, set $P_G(x)$ the probability that a random $y \in G$ satisfies $G = \langle x, y \rangle$. Let P_G be the minimum of $P_G(x)$ over all nontrivial x. It follows by [GK] that $P_G > 0$. One can easily deduce the following result (a special case of [GLSS]):

Theorem 5.1. *Let X be a type of simple algebraic group. Then one has that $\lim_{q\to\infty} P_{X(q)} = 1$.*

Recall that a group G is generated *invariably* by the elements $x_1, \cdots, x_m$ if the elements $y_1, \cdots, y_m$ generate G whenever y_i is conjugate to x_i for $i = 1, \cdots, m$. Luczak and Pyber [LuP] proved the following conjecture of McKay, useful in computational Galois theory. (We make the constants in Theorem 5.2 more explicit in forthcoming work).

Theorem 5.2. *([LuP]) There exists N so that for all $n \geq N$ and all $\epsilon > 0$, there is a constant $C = C(\epsilon)$ so that C permutations, chosen from S_n uniformly and independently, generate S_n invariably with probability at least $1 - \epsilon$.*

For Chevalley groups of bounded rank, we have the following result.

Theorem 5.3. *Let X be a type of simple algebraic group. For any $\epsilon > 0$, there is a constant $C = C(\epsilon)$ (not depending upon q) so that C elements, chosen from $X(q)$ uniformly and independently, generate $X(q)$ invariably with probability at least $1-\epsilon$.*

Proof. The probability that some $y_1, \ldots, y_m$ generate a maximal subgroup in $\mathcal{M}_i, i \leq 3$ tends to 0 as $q \to \infty$ by Theorem 3.2. Indeed our proof shows that this probability is $O(1/q^m)$. There are at most d (depending only on X) conjugacy classes of maximal subgroups in $\mathcal{M}_4$ and the probability that some conjugate of a random element $x \in G$ is contained in one is at at most $1 - \delta$ for some $\delta > 0$ (for some δ depending only on X). Thus, the probability that some collection of y_i are contained in one of these maximal subgroups is at most $d(1 - \delta)^m$. So for q sufficiently large, we can choose an m so with probability greater than $1 - \epsilon$, m random elements invariably generate $X(q)$. Note that we are ignoring the possibility that $X(q)$ may not be simple – however, this quotient is bounded in terms of X and so is not a problem. □

We will prove the analogous result for classical groups of unbounded rank in a future article.

6. Exceptionality and Derangements

In this section, we discuss the notion of exceptional permutation representations and its connection to curves. See [GMS] for a more elaborate discussion of these ideas.

Let G be a normal subgroup of A. Let X be a transitive A-set that is also transitive for G. We say that (A, G, X) is exceptional if A and G have no nontrivial common orbits on $X \times X$ (the trivial orbit being the diagonal). We note the following easy example. See [GMS] for more examples and some classification theorems.

Recall that a Hall subgroup H of a finite group G is a subgroup with $\gcd(|H|, |G : H|) = 1$.

Theorem 6.1. *Let A be a finite group and G a normal Hall subgroup. Then $A = GD$ for some complement D (by the Schur-Zassenhaus theorem). Let $H = N_A(D)$ and $X = A/H$. Then (A, G, X) is exceptional.*

Proof. Suppose not. We can identify X with the set of conjugates of D. Let $K = C_G(D) = G \cap H$. It is easy to see that exceptionality is equivalent to K and H having no common orbit (other than D itself). Suppose $D \neq E$ is in a common orbit. Then the length of this orbit divides $|K|$ and in particular has order prime to $|D|$. Thus, $|H : N_H(E)|$ has order prime to D, whence $N_H(E)$ contains a Hall π-subgroup D_1 of H (where π is the set of primes dividing $|D|$). Since D and D_1 are both Hall π-subgroups of H, they are conjugate in H (by the Schur-Zassenhaus theorem), whence D normalizes some K-conjugate of E. So we may assume that D normalizes E. Then DE is a π-subgroup, whence $D = E$, a contradiction. □

In particular, this result applies to the case G is a Chevalley group defined over the field of q^b elements, b is relatively prime to $|G|$ and we take D to be the cyclic group of order b of field automorphisms of G. Specifically, take $G = L(2, p^b)$ with b any prime not dividing $p(p^2 - 1)$. See [GMS]. In this family of examples, the degree of the permutation representation has size less than p^{3b} and the proportion of derangements is less than $1/b$ (since all derangements are contained in G and $|A : G| = b$). Thus, even for almost simple groups acting primitively, the proportion of derangements can be less than any given $\epsilon > 0$. The best general result one could hope for is $\delta(A, X) > C/\log|X|$. See §8 for such a result.

Some easy facts about exceptional triples are (see [FGS], [GW], [GMS]):

(1) If A/G is generated by the coset aG, then (A, G, X) is exceptional if and only if every element in the coset aG has a unique fixed point or equivalently $\delta(A, G, X) = 0$.
(2) If A/G is cyclic and (A, G, X) is exceptional, then so is (A, G, Y) where Y is the image of a morphism of A-sets from X is an A-morphism.

In particular, if A/G has prime order and (A, G, X) is exceptional, then $\delta(A, X) < |G|/|A|$ (since all derangements are contained in G).

So the analog of Shalev's conjecture for almost simple groups fails. In future work, we hope to obtain a result that says that Shalev's conjecture holds except for certain primitive actions (mostly related to the case where the point stabilizer is the set of fixed points of some Lang-Steinberg endomorphism of an algebraic group).

As we have remarked, $\delta(A, G, X)$ is related to images of rational points for maps between curves and higher dimensional varieties over finite fields. The connection is through the following estimate that follows from the Cebotarev density theorem (see [GTZ] or [GW] for more details).

We make this more precise.

Let U, V be smooth projective curves defined over $F := F_q$ the field of q elements. Let $U(q^a)$ denote the F_{q^a} rational points of U. Let $f : U \to V$ be a separable rational map of degree n also defined over F. Let $F(U)$ and $F(V)$ be the function fields of U and V over F. Let A be the arithmetic monodromy group of this cover (i.e. A is the Galois group of the Galois closure of $F(U)/F(V)$) and G the geometric monodromy group of the cover (the subgroup of A which acts trivially on the algebraic closure of F). Let H be the subgroup of A trivial on $F(U)$ (so $|A : H| = n$). Note that A/G is cyclic. Let xG be a generator for A/G. It follows from the Cebotarev density theorem (cf. [GTZ]) that:

Theorem 6.2.
$$|f(U(q_a))| = 1 - \delta(\langle x^a, G\rangle, G, H) + O(q^{a/2}).$$

The special case where $\delta(\langle x^a, G\rangle, G, H) = 0$ gives rise to exceptional covers. In this case, it is not difficult to show that f is in fact bijective on rational points. Of course, this cannot be the case if $x^a \in G$. See [FGS], [GMS] and [GSa] for more about exceptional covers. By [GSt], any group theoretic solution does give rise to some cover of curves with the appropriate property.

7. Derangements in a Coset

In this section, we present a proof of the Guralnick-Wan result – Theorem 1.6 that is a bit different than the one given in [GW]. It is more in the spirit of the proof in [CC] and an unpublished proof of Marty Isaacs (both for the case $A = G$).

Let G be a normal subgroup of A with A/G generated by aG. Suppose that A and G both act on the finite set X. Let $f(g)$ be the number of fixed points of g on X. We note the following well known easy result (cf. [GW]).

Lemma 7.1. *$\sum_{g \in G} f(ag) = |G|c$, where c is the number of common A, G-orbits on X.*

Now suppose that A and G are both transitive on X (and so in particular $c = 1$ in the previous result). Let H be the stabilizer of a point and set $K = H \cap G$.

Let Δ denote the derangements in the coset xG. There must be some element in the coset with a fixed point and so we may assume that $a \in H$.

We split xG into three disjoint sets, xK, Δ and Γ (the complement of the union of xK and Δ).

Breaking up the sum into the sum into two pieces, one over xK and the other the remaining terms, we see that

$$|G| = \sum_{g \in K} f(ag) + \sum_{xg \in \Gamma} f(ag) \geq c|K| + |G| - |K| - |\Delta|,$$

where d is the number of common H, K orbits on X (of course, $d \geq 1$).

This yields $|\Delta| \geq (d-1)|K|$ or $\delta(A, G, X) \geq (d-1)/n$.

If $d = 1$, then it is easy to see that Δ is empty (using the fact that the average number of fixed points is 1). So we obtain:

Theorem 7.2. *If (A, G, X) is not exceptional, then $\delta(A, G, X) \geq 1/n$.*

If $d \geq 3$, we see that $\delta(A, G, X) \geq 2/n$. It would be interesting to characterize those groups where $d = 2$ (this includes the case where G is 2-transitive) and classify the actions where $\delta(A, G, X) \leq 2/n$ (presumably only Frobenius groups and exceptional actions).

8. Primitive Groups

As we have seen in the previous section, we cannot hope to extend Shalev's conjecture to the almost simple case. There are many more examples in case of affine primitive groups and also diagonal actions (again related to exceptionality – see [GMS] for examples).

In this section we show how one can obtain a weaker result for primitive groups with no normal abelian subgroup (so in particular as long as the degree is not a prime power). The example in the previous section shows that one can do no better than this theorem. We do hope to classify which primitive representations have few derangements.

Theorem 8.1. *Let G be a primitive group of degree n and assume that G has no normal abelian subgroup. Then there exists a positive constant δ such that $\delta(G,X) > \delta/\log n$.*

We prove this by reducing to the almost simple case and then to the simple case.

We first need some auxiliary lemmas.

We will use the following result (which depends on the classification of finite simple groups – see [GMS] for a proof).

Lemma 8.2. *If h is an automorphism of a finite nonsolvable group J, then $C_J(h) \neq 1$.*

Lemma 8.3. *Let G be a transitive permutation group with a regular nonsolvable normal subgroup N acting on X. Then $\delta(G,X) \geq 1/2$.*

Proof. We can identify N with X. A point stabilizer H is a complement to N and the action on X is equivalent to the conjugation action of H. If $h \in H$, then the number of fixed points is just $|C_N(h)| > 1$ (by the previous result). Thus every element either has zero or at least 2 fixed points. Since the average number of fixed points is 1, this implies that $\delta(G,X) \geq 1/2$. □

We say that G preserves a product structure on X if X can be identified with $Y \times \ldots \times Y$ ($t > 1$ copies) and G embeds in $S_Y \wr S_t$ in its natural action (S_Y is the symmetric group on Y and each of the t copies acts on one copy of Y, the S_t permutes the coordinates). In particular, there is a homomorphism π from G into S_t. We assume that this image is transitive (which is always the case if G is primitive on X). Let G_1 denote the preimage of the stabilizer of 1 in $\pi(G)$. So G_1 acts on Y. If G is primitive, it follows that G_1 is as well [AS].

Lemma 8.4. $\delta(G,X) \geq \delta(G_1,Y)/t$.

Proof. The proportion of elements in G_1 is $1/t$. If $g \in G_1$, then g a derangement on Y implies that g is a derangement on X. □

Note in particular that $\log|X| = t\log|Y|$.

By examining the possibilities of primitive permutation groups (see [AS]) and using the two previous lemmas, there are only two cases remaining – G is almost simple or X is of full diagonal type (we explain this more fully below). Let H be a point stabilizer. In particular, G has a unique minimal normal subgroup N a direct product of t copies of a nonabelian simple L and either $t = 1$ or we may view $H \cap N \cong L$ as the diagonal subgroup of N (note that all diagonal subgroups are conjugate in $\mathrm{Aut}(N)$ so there is no loss of generality in assuming that $H \cap N$ is the canonical diagonal subgroup – alternatively, the arguments below are valid with $H \cap N$ any diagonal subgroup).

We next handle the diagonal case.

Lemma 8.5. *Let G be a finite group with a unique minimal normal subgroup $N = L^t$ with L a nonabelian finite simple group and $t > 1$. Let D be a diagonal subgroup of N and assume that $G = NH$ with $H = N_G(D)$. Then $\delta(G, G/H) > 1/\log_2 |G/H|$.*

Proof. Suppose that $g \in G$ has a unique fixed point on G/H. We claim that g is transitive on the t conjugates of L. Conjugating by an element of the transitive subgroup N allows us to assume that $g \in H$. Since g has a unique fixed point, it is invariant under $C_G(g)$ and so $C_G(g) \leq H$. In particular, $C_N(g) \leq D$. This implies the claim – for if g leaves invariant some proper factor N_1 of N, then $C_{N_1}(g) \neq 1$ (by Lemma 8.2) but $N_1 \cap D = 1$.

Now the proportion of elements in G that induce at t-cycle on the t conjugates of L is at most $1 - 1/t$ ($1/t$ of the elements normalize L). Thus, the proportion of elements with a unique fixed point is at most $1 - 1/t$, whence at least half the remaining elements must be derangements. Thus, the proportion of derangements is at least $1/2t$. Since $|G : H| = |L|^{t-1} \geq 60^{t-1}$, we have $2t < (t-1)\log_2 60 \leq \log_2 |G/H|$.

□

One can show that in most cases above, one can obtain an estimate not involving the log term. However, if the action of G on the t-conjugates of L is cyclic of order t and t does not divide the order of L, then in fact one can do no better than the previous result (this is another example of exceptionality).

If G is almost simple, then we can apply Theorem 1.3. Note that this implies the same result for almost simple groups as long as the socle of G has bounded index (with perhaps a different constant). Indeed, in the sequel we actually prove the result for all Chevalley groups contained in the group of inner diagonal automorphisms. Since the group of graph automorphisms always has order at most 6, we only need worry about field automorphisms. A simple inspection shows that the group of the field automorphisms has order at most $\log_2 n$ where n is the degree of the permutation representation. Thus, we have proved our result follows from Theorem 1.3. We have proved Theorem 1.3 in the bounded rank case. As we noted in the introduction, the complete proof of Theorem 1.3 is contained in [FG1], [FG2] and [FG3].

9. Numbers of conjugacy classes in finite classical groups

To conclude this paper we record some upper bounds on the number of conjugacy classes in the finite classical groups. These are treated fully in [FG2] where the results are used as a key ingredient in proving Theorem 1.3 and more. We mention that upper bounds on numbers of conjugacy classes are also of interest in random walks [Gl], [LiSh] and for computation of Fourier transforms on finite groups [MR].

The bounds we present are of the form cq^r where r is the rank and c is a small explicit constant. The paper [LiP] had previously established the bound $(6q)^r$ and the paper [Gl] had established bounds such as cq^{3r}. Thus our bounds in Theorem 9.1 are sharper for classical groups. For exceptional groups, one can compute precisely the number of classes as a monic polynomial in q (see [H2] for some discussion of this and references). In even characteristic $O(2n+1, q)$ is isomorphic to $Sp(2n, q)$

so we omit this case. Here we only state the results for a specific form of each group. This gives bounds for the simple groups using the fact that the number of conjugacy classes decreases when one takes homomorphic images and also using the lemma below to pass to a subgroup or overgroup of bounded index. In [FG2], we actually prove results for more forms of the groups.

Theorem 9.1. *Let $k(G)$ denote the number of conjugacy classes of a finite group G.*

(1) $k(SL(2,q)) \leq q+4$.
(2) $k(SL(3,q)) \leq q^2+q+8$.
(3) $k(SU(3,q)) \leq q^2+q+10$.
(4) *For* $n \geq 4$, $k(SL(n,q)) \leq \frac{q^n}{q-1} + q^{1+\frac{n}{2}}$.
(5) *For* $n \geq 4$, $k(SU(n,q)) \leq 11.5\left(\frac{q^n}{q+1} + \frac{q+1}{q-1}q^{n/2+1}\right)$.
(6) $k(Sp(2n,q)) \leq 12q^n$ *if q is odd.*
(7) $k(Sp(2n,q)) \leq 21.4q^n$ *if q is even.*
(8) $k(O^{\pm}(2n,q)) \leq 29q^n$ *if q is odd.*
(9) $k(O^{\pm}(2n,q)) \leq 19.5q^n$ *if q is even.*
(10) $k(SO(2n+1,q)) \leq 7.38q^n$ *if q is odd.*

Our proof of Theorem 9.1 uses generating functions for numbers of conjugacy classes in finite classical groups [Lu], [M], [MR], [W] and is largely inspired by the proof in [MR] that $GL(n,q)$ has at most q^n classes and that $GU(n,q)$ has at most $8.26q^n$ conjugacy classes. However some new ingredients (combinatorial identities) are required.

Let $k_p(G)$ denote the number of conjugacy classes of p'-elements of G. This is also the number of absolutely irreducible representations of G in characteristic p. If p does not divide G, then $k_p(G) = k(G)$ the number of conjugacy classes of G (and also the number of irreducible complex representations). We also employ the following useful lemma, which allows us to move between various forms of the finite classical groups–at least when $|G/H|$ is bounded. This is proved in [Ga] for $k(G)$. The modification for p'-classes is straightforward and we omit the proof.

Lemma 9.2. *Let H be a subgroup of G with G/H of order d. Fix a prime p. Then $k_p(G) \leq dk_p(H)$ and $k_p(H) \leq dk_p(G)$. If H is normal in G, then $k_p(G) \leq k_p(H)k_p(G/H)$.*

In fact using generating functions it is possible to understand the asymptotic behavior of the constant c in the bound cq^n of Theorem 9.1. More precisely, we establish in [FG2] the following result.

Theorem 9.3. (1) $lim_{n\to\infty}\frac{k(GL(n,q))}{q^n}=1$

(2) $lim_{n\to\infty}\frac{k(GU(n,q))}{q^n}=\prod_{i\geq 1}\frac{1+1/q^i}{1-1/q^i}$

(3) $lim_{n\to\infty}\frac{k(Sp(2n,q))}{q^n}=\prod_{i=1}^{\infty}\frac{(1+\frac{1}{q^i})^4}{(1-\frac{1}{q^i})}$ *if q is odd.*

(4) $lim_{n\to\infty}\frac{k(Sp(2n,q))}{q^n}=\prod_{i=1}^{\infty}\frac{1-1/q^{4i}}{(1-1/q^{4i-2})(1-1/q^i)^2}$ *if q is even.*

(5) $lim_{n\to\infty}\frac{k(O^{\pm}(2n,q))}{q^n}=\frac{1}{4\prod_{i=1}^{\infty}(1-1/q^i)}(\prod_{i=1}^{\infty}(1+1/q^{i-1/2})^4+\prod_{i=1}^{\infty}(1-1/q^{i-1/2})^4)$ *if q is odd.*

(6) $lim_{n\to\infty}\frac{k(O^{\pm}(2n,q))}{q^n}=\frac{1}{2}\frac{\prod_{i=0}^{\infty}(1-1/q^{2i+2})(1+1/q^{2i+1})^2}{\prod_{i=1}^{\infty}(1-1/q^i)^2}$ *if q is even.*

(7) $lim_{n\to\infty}\frac{k(SO(2n+1,q))}{q^n}=\prod_{i=1}^{\infty}\frac{(1-1/q^{4i})^2}{(1-1/q^i)^3(1-1/q^{4i-2})^2}$ *if q is odd.*

References

[A1] Aschbacher, M., Finite Group Theory, Cambridge University Press, Cambridge, 1986.

[A2] Aschbacher, M., On the maximal subgroups of the finite classical groups, Invent. Math. 76 (1984), 469–514.

[AS] Aschbacher, M. and Scott, L., Maximal subgroups of finite groups, J. Algebra 92 (1985), 44–80.

[CC] Cameron, P. and Cohen, A., On the number of fixed point free elements in a permutation group, Discrete Math. 106/107 (1992), 135–138.

[Ca] Carter, R., Finite groups of Lie type. Conjugacy classes and complex characters. Reprint of the 1985 original. John Wiley and Sons, Chichester, 1993.

[Co] Cohen, S., Permutation group theory and permutation polynomials, Algebras and combinatorics (Hong Kong, 1997), 133–146, Springer, Singapore, 1999.

[Dia] Diaconis, P., Group representations in probability and statistics, Lecture Notes Monograph Series 11, Institute of Mathematical Statistics, Hayward, CA 1988.

[D] Dixon, J., Random sets which invariably generate the symmetric group, *Discrete Math.* **105** (1992), 25-39.

[FGS] Fried, M., Guralnick, R., and Saxl, J., Schur covers and Carlitz's conjecture, Israel J. Math. 82 (1993), 157–225.

[FG1] Fulman, J. and Guralnick, R., Derangements in subspace actions of classical groups, preprint.

[FG2] Fulman, J. and Guralnick, R., Derangements in classical groups for nonsubspace actions, preprint.

[FG3] Fulman, J. and Guralnick, R., Derangements in simple and primitive groups II, in preparation.

[FG4] Fulman, J. and Guralnick, R., The probability of generating an irreducible subgroup, preprint.

[FNP] Fulman, J., Neumann, P.M. and Praeger, C.E., A generating function approach to the enumeration of matrices in finite classical groups, preprint.

[Ga] Gallagher, P., The number of conjugacy classes in a finite group, Math. Z. 118 (1970), 175–179.

[Gl] Gluck, D., Characters and random walks on finite classical groups, Adv. Math. 129 (1997), 46–72.

[GoLy] Gorenstein, D. and Lyons, R., The local structure of finite groups of characteristic 2 type. Mem. Amer. Math. Soc. 42 (1983), no. 276.

[GK] Guralnick, R. and Kantor, W., Probabilistic generation of finite simple groups, J. Algebra 234 (2000), 743–792.

[GKS] Guralnick, R., Kantor, W. and Saxl, J., The probability of generating a classical group, Comm. Algebra 22 (1994), 1395–1402.

[GLPS] Guralnick, R., Li, P., Praeger, C., and Saxl, J., Exceptional primitive group actions and partitions of orbitals, preprint.

[GLSS] Guralnick, R., Liebeck, M., Saxl, J, and Shalev, A., Random generation of finite simple groups, J. Algebra 219 (1999), 345–355.

[GL] Guralnick, R., Lübeck, F., On p-singular elements in Chevalley groups in characteristic p. Groups and computation, III (Columbus, OH, 1999), 169–182, Ohio State Univ. Math. Res. Inst. Publ., 8, de Gruyter, Berlin, 2001.

[GMS] Guralnick, R., Müller, P., and Saxl, J., The rational function analogue of a question of Schur and exceptionality of permutation representations, Memoirs of the Amer. Math. Soc., to appear.

[GSa] Guralnick, R. and Saxl, J., Exceptional polynomials over arbitrary fields, Proceedings of a Conference in honor of Abhyankar, to appear.

[GSt] Guralnick, R. and Stevenson, K., Prescribing ramification, 387–406 in Arithmetic fundamental groups and noncommutative algebra, Proceedings of Symposia in Pure Mathematics, 70 (2002) editors M. Fried and Y. Ihara, 1999 von Neumann Conference on Arithmetic Fundamental Groups and Noncommutative Algebra, August 16-27, 1999 MSRI.

[GTZ] Guralnick, R. Tucker, T. and Zieve, M., Exceptional covers and bijections of rational points, preprint.

[GW] Guralnick, R., and Wan, D., Bounds for fixed point free elements in a transitive group and applications to curves over finite fields, Israel J. Math. 101 (1997), 255–287.

[H1] Humphreys, J., Linear algebraic groups. Graduate Texts in Mathematics, No. 21. Springer-Verlag, New York-Heidelberg, 1975

[H2] Humphreys, J., Conjugacy classes in semisimple algebraic groups, Mathematical Surveys and Monographs, 43, American Mathematical Society, Providence, RI, 1995.

[LaP] Larsen, M. and Pink, R., Finite subgroups of algebraic groups, preprint.

[LiP] Liebeck, M. and Pyber, L., Upper bounds for the number of conjugacy classes of a finite group, J. Algebra 198 (1997), 538–562.

[LS1] Liebeck, M. and Seitz, G., On the subgroup structure of exceptional groups of Lie type, Trans. Amer. Math. Soc. 350 (1998), 3409–3482.

[LS2] Liebeck, M. and Seitz, G., The maximal subgroups of positive dimension in exceptional algebraic groups, Memoirs of the Amer. Math. Soc., to appear.

[LS3] Liebeck, M. and Seitz, G., On the subgroup structure of classical groups, Invent. Math. 134 (1998), 427–453.

[LiSh] Liebeck, M. and Shalev, Diameters of finite simple groups: sharp bounds and applications, Annals of Math. 154 (2001), 383-406.

[LuP] Luczak, T. and Pyber, L., On random generation of the symmetric group, Combin. Probab. Comput. 2 (1993), 505–512.

[Lu] Lusztig, G., Irreducible representations of the finite classical groups, *Invent. Math* **43** (1977), 125-176.

[M] Macdonald, I., Numbers of conjugacy classes in some finite classical groups, Bull. Austral. Math. Soc. 23 (1981), 23-48.

[MR] Maslen, D. and Rockmore, D., Separation of variables and the computation of Fourier transforms on finite groups, I., J. Amer. Math. Soc. 10 (1997), 169-214.

[Mo] de Montmort, P. R., Essay d'Analyse sur les Jeux de Hazard, 1708.

[No] Nori, M., On subgroups of $GL_n(F_p)$, Invent. Math. 88 (1987), 257–275.

[SaSe] Saxl, J. and Seitz, G., Subgroups of algebraic groups containing regular unipotent elements, J. London Math. Soc. (2) 55 (1997), 370–386.

[Sh] Shalev, A., A theorem on random matrices and some applications, J. Algebra 199 (1998), 124–141.

[SpSt] Springer, T. and Steinberg, R., Conjugacy classes. 1970 Seminar on Algebraic Groups and Related Finite Groups (The Institute for Advanced Study, Princeton, N.J., 1968/69) pp. 167–266, Lecture Notes in Mathematics, Vol. 131, Springer, Berlin.

[W] Wall, G. E., On the conjugacy classes in the unitary, symplectic, and orthogonal groups, *J. Aust. Math. Soc.* **3** (1963), 1-63.

Computing with matrix groups*

William M. Kantor and Ákos Seress

1 Introduction

A group is usually input into a computer by specifying the group either using a presentation or using a generating set of permutations or matrices. Here we will emphasize the latter approach, referring to [Si3, Si4, Ser1] for details of the other situations. Thus, the basic computational setting discussed here is as follows: a group is given, specified as $G = \langle X \rangle$ in terms of some generating set X of its elements, where X is an arbitrary subset of either S_n or $\mathrm{GL}(d, q)$ (a familiar example is the group of Rubik's cube). The goal is then to find properties of G efficiently, such as $|G|$, the derived series, a composition series, Sylow subgroups, and so on.

When G is a group of permutations there is a very well-developed body of literature and algorithms for studying its properties (see Section 2). The matrix group situation is much more difficult, and is the focus of the remaining sections of this brief survey. Sections 4 and 5 discuss the case of simple groups, and section 6 uses these to deal with general matrix groups. We will generally emphasize the group-theoretic aspects of the subject, rather than ones involving implementation in the computer systems GAP [GAP4] or MAGMA [BCP]. Thus, the word "efficiently" used above will usually mean for us "in time polynomial in the input length of the problem" rather than "works well in practice".

One can ask for the relevance of such questions to finite group theory. Certainly computers have been involved in the construction of sporadic simple groups, as well as in the study of these and other simple groups. We will make a few comments concerning the expected uses in GAP and MAGMA of the results presented here. However, our point of view includes a slightly different aspect: the purely mathematical questions raised by computational needs have led to new points of view and new questions concerning familiar groups.

2 Permutation groups

We begin with a brief discussion of the case of permutation groups. Here, X is a set of permutations of $\{1, 2, \ldots, n\}$, and then the word "efficiently" will mean "in time

2000 Mathematics Subject Classification: Primary 20D06; Secondary: 20-04, 20P05, 68W30.
* This research was supported in part by the National Science Foundation.

polynomial in the input length $|X|n\log n$ of the problem" ($|X|n\log n$ is roughly the number of keystrokes needed to input X into a computer). The problem is that a small generating set X can specify a very large group G, so large that it is absurd (both from the theoretical and practical points of view) to imagine listing the elements of G.

The development of efficient computer algorithms for permutation groups was begun by Sims [Si1, Si2]. If $G^{[i]}$ is the pointwise stabilizer of $\{1,\ldots,i-1\}$, then

$$G = G^{[1]} \geq G^{[2]} \geq \cdots \geq G^{[n]} = 1 \tag{2.1}$$

where $|G| = \prod_1^{n-1} |G^{[i]}\colon G^{[i+1]}|$ and $|G^{[i]}\colon G^{[i+1]}|$ is the length of the orbit $\mathcal{O}_i$ of i under $G^{[i]}$ and hence is at most n. Sims developed a data structure to find (generators for) all of these subgroups $G^{[i]}$ and orbits $\mathcal{O}_i$ simultaneously and efficiently. This yielded $|G|$ using only elementary group theory: it did not involve structural properties of groups.

The ideas behind the point stabilizer chain construction can also be used for finding many other properties of G, such as the derived series, solvability, and nilpotence, in polynomial time. The application most important from an algorithmic point of view is a *Membership Test: given $h \in S_n$, decide whether or not $h \in G$; and if it is, obtain h from the generating sets of the $G^{[i]}$*.

The above ideas have been implemented in GAP and MAGMA. A detailed description of point stabilizer constructions, and of many other permutation group algorithms, can be found in [Ser2].

3 Matrix groups

We now turn to the case of a group $G = \langle X\rangle$ in which X is an arbitrary set of invertible matrices over some finite field $\mathbb{F}_q$. The questions remain the same: efficiently find properties of G, such as $|G|$, solvability, a composition series, etc. If $X \subset GL(d,q)$ then the input length is $|X|d^2\log q$ (since $\log q$ bits are required to write each of the d^2 entries of a matrix)[1]. Once again a small generating set X can specify a very large group G.

These problems seem to be very hard. The fundamental difference from the permutation group setting is that there is no longer, in general, a decreasing sequence of subgroups from G to 1 in which all successive indices are "tiny" (as was the case in (2.1)), even with the very generous definition of "tiny" meaning "polynomial in the input length". However, under reasonable additional conditions, and allowing probabilistic components to the algorithms, this has become an actively studied area. Some of the results use the exact representation of G on $\mathbb{F}_q^d$ implicit in the above description (such as eigenvalues, minimal polynomials and so on), but most of those we survey below avoid trying to deal with the exact representation.

First we need to know that random elements can be found. According to an amazing result of Babai [Ba2], *in polynomial time, with high probability one can find independent, nearly uniformly distributed*[2] *random elements of* $G = \langle X\rangle \leq$

[1]Logs are always to the base 2.

[2]Meaning: for all $g \in G$, $(1-\varepsilon)/|G| \leq \mathrm{Prob}(x=g) \leq (1+\varepsilon)/|G|$ for some fixed $\varepsilon \leq 1/2$.

GL(V). This tour de force involves combinatorial methods but nothing about the structure of G; note that $|G|$ is never known here. The results presented in this survey all involve probabilistic estimates, and it is straightforward to have these estimates take the "nearly" part of these "nearly uniform" elements into account; therefore it is convenient to make believe that we actually have uniformly distributed random elements when discussing later results. In practice, a heuristic algorithm from [CLMNO] is used for finding random group elements, and that method is adequate for algorithms in which correctness of the output is ultimately verified (cf. [Ba3]). The points of view in [Ba2] and [CLMNO] are merged in the recent paper [Pak]. Moreover, a new algorithm for random element generation is described in [Co].

A second important tool in almost all of the results below involves the order of an element $g \in \mathrm{GL}(d, q)$. In general, we do not want to assume that we can find $|g|$; for example, testing that an element has order $q^d - 1$ seems to require at least having the prime factorization of the rather large integer $q^d - 1$ (however, compare Theorem 4.3 below and the remarks following it). Nevertheless, it is possible to determine qualitative properties of $|g|$ without actually computing the order. There are algorithms in [NeP1, NiP, KS2] that can be used to decide whether or not $|g|$ is divisible by some *primitive prime divisor*[3] of $p^k - 1$ for a given prime p and given exponent k.

4 Nonconstructive recognition of simple groups

In the matrix group setting, the problem of *recognizing* simple groups began with the following groundbreaking result:

Theorem 4.1 [NeP1] *There is a randomized polynomial-time algorithm which, when given* $0 < \varepsilon < 1$ *and* $G = \langle X \rangle \leq \mathrm{GL}(V)$, *outputs either* "$G$ definitely contains SL(V)" *or* "G does not contain SL(V)*, and the probability that the latter assertion is incorrect, given that G does contain* SL(V)*, is less than* ε*.* [4]

Thus, the algorithm gives an answer guaranteed to be correct if it is "Yes", but there is a small probability that the answer "No" will be incorrect. Randomized algorithms that may return an incorrect answer, where the probability of an incorrect output can be controlled by the user, are called *Monte Carlo algorithms.* A special case of Monte Carlo algorithms is the class of *Las Vegas algorithms*: in this case, an incorrect answer can be recognized, so we can achieve that the output is always correct; however, the algorithm may report failure.

The proof of Theorem 4.1 relies heavily on CFSG[5]: the algorithm searches for certain matrices in G that occur with high probability in SL(V), and then uses nonalgorithmic consequences of CFSG to determine the subgroups of GL(d, q) containing such elements. This theorem was followed by others [NiP, CLG1] that

[3]This means that $|g|$ is divisible by a prime divisor of $p^k - 1$ that does not divide $p^i - 1$ whenever $1 \leq i < k$. Such prime divisors exist for all but a very limited type of pairs p, k [Zs].

[4]In the future we will avoid ε and merely say that such an algorithm succeeds "with arbitrarily high probability".

[5]The classification of the finite simple groups.

decide, similarly, whether or not a given subgroup $G = \langle X \rangle \leq \mathrm{GL}(d,q)$ contains a classical group defined on V as a normal subgroup.

These are *nonconstructive* recognition algorithms, outputting either "G contains a normal classical group", or "G probably does not contain any classical group of $d \times d$ matrices as a normal subgroup". They do *not* tell us how to "get" any given elements of the classical group from the given generating set (e.g., elementary matrices in the situation of the above theorem).

Of course, there is no reason to expect that a quasisimple subgroup G of $\mathrm{GL}(d,q)$ appears in the most natural representation. Even if we have an irreducible representation of $\mathrm{SL}(d,q)$ on some vector space V, the characteristic and dimension of V may very well be quite different from those of the more familiar d-dimensional representation. In order to handle arbitrary matrix groups, this possibility must be taken into account; and when dealing with an unknown representation some of the more standard tools of linear algebra (minimal and characteristic polynomials, eigenvalues and eigenvectors) do not appear to be sufficiently helpful in identifying composition factors of the group being studied.

In general, it would be especially nice to be able to recover the more natural representations from the given "arbitrary" one; we will return to this in the next section. For now, we note that the name of a simple factor can be determined under suitable additional conditions (Theorems 4.2 and 4.3).

Theorem 4.2 [BKPS] *There is a polynomial-time Monte Carlo algorithm which, when given $G = \langle X \rangle \leq \mathrm{GL}(V)$ such that $G/Z(G)$ is isomorphic to a simple group of Lie type of known characteristic p, finds the* name *of $G/Z(G)$.*

Note that the name gives at least one additional piece of information about G, namely $|G/Z(G)|$. The proof of this theorem in [BKPS] handles all situations except for distinguishing the pairs $\mathrm{PSp}(2m,q), \Omega(2m+1,q)$ with q odd and $m \geq 3$, where entirely different techniques were needed [AB]. Our proof is relatively simple (using information already obtained while proving Theorem 4.3 below). We start with a sample of independent (nearly) uniformly distributed random elements of G. We then find the three largest integers $v_1 > v_2 > v_3$ such that a member of the list has order divisible by a primitive prime divisor of one of the integers $p^v - 1$ for $v = v_1, v_2$ or v_3; our sample is chosen large enough so that, with high probability, these are the three largest v such that $|G|$ is divisible by a primitive prime divisor of $p^v - 1$. In a lot of cases, the triple $\{v_1, v_2, v_3\}$ determines the name of G. In the remaining cases, we investigate the occurrence of element orders divisible by two appropriately chosen primitive prime divisors. While this idea is simple enough, it becomes more awkward and detailed if p is a Fermat or Mersenne prime and no such primitive prime divisor greater than 2 occurs. Nevertheless, the algorithm is not complicated, and has already been implemented in MAGMA by Malle and O'Brien.

While the assumption that p is known is a natural one (cf. Section 6), it would be better to be able to avoid this. A result that preceded the previous theorem attempts to do this:

Theorem 4.3 [KS3] *There is a polynomial-time Monte Carlo algorithm which, when given $G = \langle X \rangle \leq \mathrm{GL}(V)$ such that $G/Z(G)$ is isomorphic to a simple group* *of Lie type of unknown characteristic and such that the order of any given element of G can be computed, finds the* name *of $G/Z(G)$.*

The proof of this theorem rests on a nonalgorithmic property of groups G of Lie type in characteristic p. Define a graph $\Gamma(G)$ whose vertices are the prime powers r^a that occur as orders of elements of G, for all primes $r \neq p$ and integers $a > 0$. Prime powers r^a, s^b are joined if and only if G has an element of order $\mathrm{lcm}(r^a, s^b)$ (thus, every vertex of $\Gamma(G)$ has a loop). We say that two vertices of $\Gamma(G)$ are *equivalent* if they have the same neighbors, and denote by $\Delta(G)$ the quotient graph with respect to this equivalence relation, with vertex set $V(\Delta(G))$. We view $\Delta(G)$ as a simple graph (i.e., without loops and multiple edges) and as a *weighted graph*: the *weight* of $v \in V(\Delta(G))$ is the least common multiple of the prime powers in the equivalence class v. This weighted graph usually determines G:

Theorem 4.4 [KS3] *Let G and G^* be finite simple groups of Lie type such that $\Delta(G) \cong \Delta(G^*)$. Then $G \cong G^*$ with some specific exceptions.*

These exceptions include, of course, the pairs $\mathrm{PSp}(2m, q), \Omega(2m+1, q)$ for q odd and $m \geq 3$; additional exceptions are $\mathrm{PSp}(4, q), \mathrm{PSL}(2, q^2)$; $\mathrm{PSp}(6, q), \mathrm{P}\Omega^+(8, q), \Omega(7, q)$; $\mathrm{PSp}(8, q), \mathrm{P}\Omega^-(8, q)$; and $\mathrm{PSL}(3, 2), G_2(2)'$.

Since p is involved in the definition of $\Delta(G)$, how can the above theorem be used to prove Theorem 4.3? This requires additional properties of G:

(i) [GL] If G has characteristic p and is defined over $\mathbb{F}_q$, then the proportion of elements of order divisible by p is at most $5/q$. (We note that a *lower* bound of $2/5q$ for this proportion was proved in [IKS], also motivated by uses in Computational Group Theory.)

(ii) [KS3, Lü] If $r, s \neq p$ are primes such that G has an element of order divisible by $\mathrm{lcm}(r^a, s^b)$, then the proportion of such elements is large (at least $c/(\text{Lie-rank}(G))^3$, for an absolute constant c).

Now the proof of Theorem 4.3 starts by testing all "small" primes p ("small" means bounded from above by an explicit function of the input length) using [KS1, KM] (cf. Theorem 5.3 and the remarks following it) in order to try to find the characteristic of G. (Note that Theorem 4.2 does not quite apply: it is at least conceivable that that theorem could output an answer even if p is not the characteristic of G; and we do not know the probability that this strange possibility might occur.) If this fails then we find a set of suitably many independent random elements of G, and find their orders. This number is chosen so that, by (i), with high probability none of these orders is divisible by p. This number is also chosen so that, by (ii), for every pair r^a, s^b arising in the definition of $\Gamma(G)$, with high probability one of our elements has order divisible by $\mathrm{lcm}(r^a, s^b)$. Using this we determine $\Delta(G)$, and then the name of G.

According to E. O'Brien, in actual computations with matrix groups G using MAGMA it is standard to find exact orders of elements of G using extensive tables of prime power factorizations of integers of the form $p^k - 1$ for suitable p and k. Therefore, we expect that there will be a version of the above theorem of more than theoretical importance.

Theorem 4.4 leads to a question already alluded to that might make the theorem even more useful in our computational setting. Consider a group H of Lie type over a field of characteristic $r \neq p$. Define a weighted graph $\Delta^p(H)$ for H using the same description as above but with p in place of the correct characteristic r (so that, for example, r is one of the vertices of $\Delta^p(H)$). Then we conjecture that, if $\Delta^p(H) \cong \Delta(G)$ for a group G of Lie type in characteristic p, then $H \cong G$.

Once again we emphasize that the results in this section do not provide any means of calculating with the given matrix group G using its more familiar representations.

5 Constructive recognition of simple groups

As suggested in the preceding section, there is a need for *constructive* recognition algorithms, which allow us to get from our generating set X to any given element of G.[6] These are of fundamental importance when simple groups are used to handle general groups (see the next section).

In the situation of Theorem 4.1, constructive recognition means the following:

Theorem 5.1 [CLG2] *There is a Las Vegas algorithm which, when given $G = \langle X \rangle$ such that $\mathrm{SL}(V) \leq G \leq \mathrm{GL}(V)$, with arbitrarily high probability outputs a new generating set X^* (in terms of X) such that there is a polynomial-time procedure that gets from X^* to any given $g \in G$.*

However, the algorithm in [CLG2] producing X^* does not quite run in polynomial time: there is a factor q in the timing, where V is a vector space over $\mathbb{F}_q$. The corresponding result has also been proved for all classical groups: given $G = \langle X \rangle \leq \mathrm{GL}(d, q)$ having a normal classical subgroup C defined on V, algorithms in [Ce, Bro1, Bro2] output, with high probability, a new generating set X^* such that there is a polynomial-time procedure that gets from X^* to any given $g \in G$. The version of this theorem in [Bro2] handles all symplectic, orthogonal and unitary groups simultaneously in a more or less uniform manner.

It is not known how to get around the factor of q in the timing indicated above without some other type of assumption. In [CoLG] a lovely idea was introduced to avoid this factor: assume the availability of a way to handle *Discrete Logarithms.* Given $\mathbb{F}_q^* = \langle \rho \rangle$ and $\alpha \in \mathbb{F}_q^*$, the Discrete Log Problem asks for an exponent i such that $\alpha = \rho^i$. There are procedures for accomplishing this that are significantly faster than the $O(q)$ time approach that tests all integers with $0 \leq i < q$. Discrete Logs led to the next

Theorem 5.2 [CoLG] *There is a Las Vegas algorithm which, when given $G = \langle X \rangle$ such that $\mathrm{SL}(V) \leq G \leq \mathrm{GL}(V) = \mathrm{GL}(2, q)$, and also given a way to handle Discrete Logs in $\mathbb{F}_q^*$, with arbitrarily high probability outputs a new generating set X^* such that there is a polynomial-time procedure that gets from X^* to any given $g \in G$. The*

[6] More precisely, such that we can find a *straight-line program* from X to any given $g \in G$: a sequence $g_1, \ldots, g_k = g$ with each term either a member of X, the product of two previous terms or the inverse of a previous term.

time requirement is a polynomial of the input length plus the time of polynomially many calls to the Discrete Log subroutine.

This result has been extended in [LGO] to deal with *arbitrary irreducible* representations of $\mathrm{SL}(2, q)$. This extension is fundamental for Theorems 5.5 and 6.1 below.

We next turn to arbitrary representations of classical groups. While we could assume irreducibility, this does not seem to provide useful information about the general situation.

Theorem 5.3 [KS2] *There is a Las Vegas algorithm which, when given $G = \langle X \rangle \leq \mathrm{GL}(V)$ with $G = G'$ and $G/Z(G)$ isomorphic to some (unknown) classical simple group of given characteristic, with arbitrarily high probability finds the classical group C, and outputs a new generating set X^* (in terms of X) together with an injective map $X^* \to C$ that extends to a* constructive *isomorphism $\Psi: G/Z(G) \to C$.*

This means that there is a polynomial-time procedure to get to any given $g \in G$ from X^*, and polynomial-time procedures which take any given $g \in G$ or $c \in C$ and find $(gZ(G))\Psi$ or $c\Psi^{-1}$; moreover, it means that if a set X^* and map $X^* \to C$ are output then they are guaranteed to behave as just indicated.

Versions of this theorem are in [CFL], where it was first shown that this type of result could be proved (in the case $G \cong \mathrm{PSL}(d, 2)$), and later in [Bra] when $G/Z(G) \cong \mathrm{PSL}(d, q)$ with $d \geq 4, q > 4$. As in Theorem 5.1, the previous theorem does not quite run in polynomial time: once again there is a factor of at least q in the timing. The case of the exceptional groups of Lie type other than ${}^2F_4(q)$ (also assuming a given characteristic) has been close to completion for a few years [KM]; once again the algorithm has an undesirable factor of q in its timing.

Remark 5.4 We stated Theorem 5.3 in its simplest form. It can be extended to handle matrix groups G that have an almost simple classical factor group G/N of given characteristic, *provided that we can test membership in N.* This extension will play an important role in Section 6. So will the fact that there are similar extensions for the exceptional groups [KM] and for the alternating groups [BLNPS]. These and related results are discussed in [Ka2].

The characteristic assumption in the preceding theorem can be removed using Theorem 4.3. When the characteristic is known, the idea behind the theorem is to try to construct an element in a large conjugacy class, one of whose powers is a (long) root element of G (but usually not a long root element of the underlying group $\mathrm{GL}(V)$!); with reasonably high probability[7], an element of G has this property. These root elements and their random conjugates are then used to generate larger subgroups, eventually leading to a subgroup of rank one less than that of G (if G does not already have rank 1).

Combining the Discrete Log results in Theorem 5.2 and its sequel [LGO] with ideas from the proof of Theorem 5.3 and some new ideas (in [Bro2]) has led to algorithms for many classical groups:

[7] But much less than $1/q$, requiring many more that q selections to make it likely that an element of the desired sort is obtained; this is a principal cause of the timing not being polynomial.

Theorem 5.5 [BK, Bro2] *There is an algorithm which, when given* $G = \langle X \rangle \leq \mathrm{GL}(V)$ *such that* $G/Z(G) \cong C = \mathrm{PSL}(d,q)$, $\mathrm{PSp}(2m,q)$ *or* $\mathrm{PSU}(d,q)$ *and* $(q, |V|) \neq 1$, *and also given a way to handle Discrete Logarithms in* $\mathbb{F}_q^*$, *with arbitrarily high probability outputs a constructive isomorphism* $\Psi: G/Z(G) \to C$. *The time requirement is a polynomial of the input length plus the time of polynomially many calls to the Discrete Log subroutine.*

The orthogonal groups present additional difficulties, but should be completed in the near future. Analogous constructive isomorphisms for alternating groups are in [BB1, BLNPS, BP]. There are only a bounded number of sporadic groups, so these do not enter into our asymptotic timing questions.

The algorithms announced in Theorems 5.1–5.5 can also be used as Monte Carlo algorithms to decide whether a given group G is such that $G/Z(G)$ is simple of a type indicated in these theorems. As in the case of nonconstructive recognition, the correctness of a "Yes" answer can be verified, but the verification is much more complicated than in the cases covered by Theorem 4.1 and its extensions. Namely, we have to compute a generating set X^{**} and a short presentation in terms of X^{**}, and prove that $G = \langle X^{**} \rangle$ by expressing the original generators of G in terms of X^{**}. A presentation for a quasisimple group G is called *short* if its length[8] is $O(\log^2 |G|)$. Such short presentations are known for almost all simple groups:

Theorem 5.6 [BGKLP, Suz, HS] *For all simple groups except, perhaps,* ${}^2G_2(q)$, *there is a presentation of length* $O(\log^c |G|)$, *where* $c = 2$; *in fact* $c = 1$ *for most* G.

The proof in [BGKLP] uses simple tricks to adapt the usual Curtis-Steinberg-Tits presentations for these groups when the rank is at least 2, while the cases ${}^2B_2(q)$ and $\mathrm{PSU}(3,q)$ require different ideas to modify the standard presentations for these groups [Suz, HS]. Short presentations have the following nonalgorithmic consequence needed in the proof of Theorem 6.1:

Theorem 5.7 [BGKLP] *Every finite group* G *with no composition factor of the form* ${}^2G_2(q)$ *has a presentation of length* $O(\log^3 |G|)$.

The exponent 3 here is best possible.

Although the primary use of constructive recognition algorithms is in computations with matrix groups, they are useful for computing with permutation groups as well. For example, all modern Sylow subgroup algorithms for permutation groups reduce to the case of simple groups [Ka1, Mo, KLM, CCH]. For any given simple permutation group one first determines an explicit isomorphism with a known simple group, and afterwards studies Sylow subgroups of the concrete simple groups. Deterministic algorithms producing such isomorphisms are in [Ka1, KLM]. In the matrix group setting, finding Sylow subgroups should not be difficult, but conjugating one to another may present some difficulties. Another application of constructive recognition algorithms is the computation of maximal subgroups of permutation groups [EH].

[8] The *length* of a presentation $\langle X \mid R \rangle$ is $|X| + \sum_{r \in R} l_X(r)$.

6 General matrix groups

Given $G = \langle X \rangle \leq \mathrm{GL}(d, q)$, there are two basic approaches to exploring properties of G. One of these is a geometric approach, led by Leedham-Green, and commonly called the "The computational matrix group project". This approach uses Aschbacher's classification [Asch] of subgroups of $\mathrm{GL}(d, q)$. (It was first suggested in [Pr] to use Aschbacher's theorem as a guide in the design of what amounted to nonconstructive algorithms for the study of matrix groups.) There are nine categories in this classification, and the goal is to find at least one category to which G belongs. Eight of these categories describe geometric subgroups of $\mathrm{GL}(d, q)$, which means that G preserves some structure associated with the action of G on the vector space $V = \mathbb{F}_q^d$. Moreover, in seven categories, the kernel N of the action on this structure enables us to consider N and G/N acting in smaller dimension, or over a smaller field, or as a permutation group on a small domain. For example, one category consists of irreducible but imprimitive matrix groups. This means that the dimension d can be written as a product $d = ab$, and there is a decomposition $V = V_1 \oplus \cdots \oplus V_a$ into subspaces of dimension b such that G transitively permutes the set $\{V_1, \ldots, V_a\}$; the normal subgroup N is the kernel of this permutation action, and G/N is a transitive permutation group of degree a.

If we can recognize the action of G on the appropriate structure then handling G can be reduced to recursively handling both N and G/N. This reduction bottoms out when a group is a classical group in its natural action (which is the eighth geometric subgroup category of the Aschbacher classification), or G modulo the scalar matrices is almost simple (the ninth category). These two cases are handled by the constructive recognition algorithms for almost simple groups described in Section 5. Note that, even if we have the images of generators of G under a homomorphism φ defined by the action on some geometric structure where $\mathrm{Im}(\varphi)$ is almost simple, usually constructive recognition of $\mathrm{Im}(\varphi)$ is needed in order to obtain generators for $\mathrm{Ker}(\varphi)$.

As a result of extensive research summarized in [LG], there are practical algorithms for recognizing most categories of the Aschbacher classification.

By contrast, the other approach, initiated by Babai and Beals [BB1], tries to determine the abstract group-theoretic structure of G. Every finite group G has a series of characteristic subgroups $1 \leq O_\infty(G) \leq \mathrm{Soc}^*(G) \leq \mathrm{Pker}(G) \leq G$, where $O_\infty(G)$ is the largest solvable normal subgroup of G; $\mathrm{Soc}^*(G)/O_\infty(G)$ is the socle of the factor group $G/O_\infty(G)$, so that $\mathrm{Soc}^*(G)/O_\infty(G)$ is isomorphic to a direct product $T_1 \times \cdots \times T_k$ of nonabelian simple groups that are permuted by conjugation in G; and $\mathrm{Pker}(G)$ is the kernel of this permutation action. Given $G = \langle X \rangle \leq \mathrm{GL}(d, p^e)$, Babai and Beals [BB2] construct subgroups $H_1, \ldots, H_k$ such that $H_i/O_\infty(H_i) \cong T_i$. Having these H_i at hand, it is possible to construct the permutation group $G/\mathrm{Pker}(G) \leq S_k$, which then can be handled by permutation group methods.

The Babai–Beals algorithm is Monte Carlo, and runs in polynomial time in the input length. Contrary to the geometric approach, it does not use the geometry associated with the matrix group action of G. The fact that $G \leq \mathrm{GL}(d, p^e)$ is only used when appealing to a simple consequence of [LS, FT]: if T_i is of Lie type in characteristic different from p, then T_i has a permutation representation of degree polynomial in d.

Combining the Babai–Beals method with constructive recognition algorithms for simple groups, we obtain the following result.

Theorem 6.1 [KS4] *Given $G = \langle X \rangle \leq GL(d, p^e)$, there is a Las Vegas algorithm that computes the following.*

(*i*) *The order of G.*

(*ii*) *A series of subgroups $1 = N_0 \lhd N_1 \lhd \cdots \lhd N_{m-1} \lhd N_m = G$, where N_i/N_{i-1} is a nonabelian simple group or a cyclic group for all i.*

(*iii*) *A presentation of G.*

(*iv*) *Given any $g \in GL(d, p^e)$, the decision whether $g \in G$, and if $g \in G$, then a straight-line program from X, reaching g.*

The algorithm uses an oracle to compute discrete logarithms in fields of characteristic p and size up to p^{ed}. In the case when all of those composition factors of Lie type in characteristic p are constructively recognizable with a Discrete Log oracle, the running time is a polynomial in the input length $|X|d^2 e \log p$, plus the time requirement of polynomially many calls to the Discrete Log oracle.

The current list of groups recognizable with a Discrete Log oracle is given in Theorem 5.5.

In part (ii) of Theorem 6.1, we construct a series of subgroups that is "almost" a composition series of G. However, some of the cyclic factor groups may not be simple, since we do not assume that we can factor large integers. Using discrete logs seems to be necessary, since already for 1×1 matrix groups $G \leq \mathrm{GL}(1, q)$, finding $|G|$ amounts to solving a version of the discrete log problem in $\mathbb{F}_q^*$. Also, finding and identifying the composition factors, or at least the nonabelian composition factors, seems to be unavoidable, even if the goal is only to compute the order of the input group.

The special case of Theorem 6.1, when the input group is solvable, was already covered a decade ago by the following remarkable theorem of Luks:

Theorem 6.2 [Lu] *Theorem* 6.1 *holds for solvable matrix groups. In fact, there is a* deterministic *algorithm that computes the required output.*

We sketch the proof of Theorem 6.1. Given $G = \langle X \rangle \leq GL(d, p^e)$, the algorithm announced in Theorem 6.1 starts by appealing to the results of [BB2] to compute a composition series for $G/\mathrm{Pker}(G)$, generators for $\mathrm{Pker}(G)$, and generators for some subgroups $H_i \leq \mathrm{Pker}(G)$, $i = 1, 2, \ldots, k$, such that $H_i/O_\infty(H_i) \cong T_i$ for the simple groups T_i involved in $\mathrm{Soc}^*(G)/O_\infty(G) \cong T_1 \times \cdots \times T_k$. Next, we use an extension of Theorem 4.2 to find polynomially many possibilities for the name of the T_i. Given any $g \in H_i$, we can test whether $g \in O_\infty(H_i)$, by testing the solvability of $\langle g^{H_i} \rangle$. This implies that the primitive prime divisor computations necessary for the algorithm in Theorem 4.2 can be carried out in $\overline{T_i} := H_i/O_\infty(H_i)$. If $\overline{T_i}$ is of Lie type then its characteristic is either p or a prime less than d^2 [LS, FT], so we have only polynomially many possibilities for this name.

After that, we replace each H_i by its normal closure in $\mathrm{Pker}(G)$. This step maintains the property that $H_i/O_\infty(H_i) \cong T_i$, but also makes H_i invariant under the conjugation action of $\mathrm{Pker}(G)$.

We now deal with the subgroups $H_1, \ldots, H_k$ sequentially. The conjugation action of $\mathrm{Pker}(G)$ on H_1 also defines a homomorphism $\varphi_1 \colon \mathrm{Pker}(G) \to \mathrm{Aut}(\overline{T_1})$. Using again our ability to test membership in $O_\infty(H_1)$, if $\overline{T_1}$ is *not* of Lie-type in characteristic p, or if $\overline{T_1}$ is defined in characteristic p but over a field of size $q \leq d^2$, then the extension of Theorem 5.3, mentioned in Remark 5.4, can be used to construct the kernel $K_1 := \mathrm{Ker}(\varphi_1)$ of this action.

The only remaining possibility is that $\overline{T_1}$ is of Lie type of characteristic p, and the size q of the field of definition is greater than d^2. In this case, the crucial observation is that $H_1/O_\infty(H_1)$ cannot act nontrivially on any elementary abelian section of $O_\infty(H_1)$ that is not a p-group, since then we would have a cross-characteristic representation of $H_1/O_\infty(H_1)$ of degree not allowed by [LS, FT]. Hence the solvable residual H_1^∞ (the last term of the derived series of H_1) is an extension of a p-group by a simple group isomorphic to $\overline{T_1}$. Therefore, in an appropriate basis, which can be found by the Meat-Axe [HR, IL, NeP2], the matrices for the elements of H_1^∞ have the following form:

$$\begin{pmatrix} I & * & * \\ 0 & A & * \\ 0 & 0 & * \end{pmatrix}$$

The blocks in position $(2,2)$ of these matrices define $\overline{T_1}$ modulo scalars. Hence, concentrating on these blocks, we can perform a constructive recognition with a Discrete Log oracle (see Theorem 5.5). After that, as we outlined for the other possibilities for the isomorphism type of T_1, we obtain generators for K_1.

The same procedure is repeated for the conjugation action of K_1 on H_2, constructing its kernel K_2, and so on. Eventually the kernel K_k is a solvable group, which is handled by Luks's methods (see Theorem 6.2).

As the very last step of the algorithm, we construct a presentation for G. This presentation verifies the correctness of the output.

References

[AB] C. Altseimer and A. V. Borovik, Probabilistic recognition of orthogonal and symplectic groups, pp. 1–20 in: Groups and Computation III (eds. W. M. Kantor and Á. Seress), The Ohio State Univ. Math. Res. Inst. Publ. 8, Walter deGruyter, Berlin–New York 2001.

[Asch] M. Aschbacher, On the maximal subgroups of the finite classical groups, Invent. Math. 76 (1984) 469–514.

[Ba1] L. Babai, Local expansion of vertex-transitive graphs and random generation in finite groups, pp. 164–174 in: Proc. 23rd ACM Symp. on Theory of Computing 1991.

[Ba2] L. Babai, Randomization in group algorithms: conceptual questions, pp. 1–17 in: Groups and Computation II (eds. L. Finkelstein and W. M. Kantor),

DIMACS Series in Discrete Math. and Theoretical Computer Science, vol. 28, AMS 1997.

[BB2] L. Babai and R. Beals, A polynomial-time theory of black-box groups I, pp. 30–64 in: Groups St Andrews 1997 in Bath, I (eds. C. M. Campbell, E. F. Robertson, N. Ruskuc, and G. C. Smith), LMS Lecture Note Series 260, Cambridge U. Press 1999.

[BGKLP] L. Babai, A. J. Goodman, W. M. Kantor, E. M. Luks and P. P. Pálfy, Short presentations for finite groups. J. Algebra 194 (1997) 79–112.

[BKPS] L. Babai, W. M. Kantor, P. P. Pálfy, and Á. Seress, Black-box recognition of finite simple groups of Lie type by statistics of element orders, J. Group Theory (to appear).

[BB1] R. Beals and L. Babai, Las Vegas algorithms for matrix groups, pp. 427–436 in: Proc. 34th IEEE Symp. on Found. of Comp. Science 1993.

[BLNPS] R. Beals, C. Leedham-Green, A. Niemeyer, C. Praeger, and Á. Seress, A black-box group algorithm for recognizing finite symmetric and alternating groups, Trans. Amer. Math. Soc. (to appear).

[BCP] W. Bosma, J. Cannon, and C. Playoust, The Magma algebra system I: The user language, J. Symbolic Comput. 24 (1997), 235–265.

[Bra] S. Bratus, Recognition of finite black-box groups, Ph.D. Thesis, Northeastern University 1999.

[BP] S. Bratus and I. Pak, Fast constructive recognition of a black-box group isomorphic to S_n or A_n using Goldbach's Conjecture. J. Symbolic Comput. 29 (2000), 33–57.

[Bro1] P. A. Brooksbank, A constructive recognition algorithm for the matrix group $\Omega(d,q)$, pp. 79–93 in: Groups and Computation III (eds. W. M. Kantor and Á. Seress), The Ohio State Univ. Math. Res. Inst. Publ. 8, Walter deGruyter, Berlin–New York 2001.

[Bro2] P. A. Brooksbank, Constructive recognition of the finite simple classical groups. Ph.D. thesis, University of Oregon 2001.

[BK] P. A. Brooksbank and W. M. Kantor, On constructive recognition of a black-box $\mathrm{PSL}(d,q)$, pp. 95–111 in: Groups and Computation III (eds. W. M. Kantor and Á. Seress), The Ohio State Univ. Math. Res. Inst. Publ. 8, Walter deGruyter, Berlin–New York 2001.

[CCH] J. Cannon, B. Cox and D. F. Holt, Computing Sylow subgroups in permutation groups. Computational algebra and number theory (London, 1993). J. Symbolic Comput. 24 (1997), 303–316.

[Ce] F. Celler, Matrixgruppenalgorithmen in GAP. Ph. D. thesis, RWTH Aachen 1997.

[CLG1] F. Celler and C. R. Leedham-Green, A non-constructive recognition algorithm for the special linear and other classical groups, pp. 61–67 in: Groups and Computation II (eds. L. Finkelstein and W. M. Kantor), DIMACS Series in Discrete Math. and Theoretical Computer Science, vol. 28, AMS 1997.

[CLG2] F. Celler and C. R. Leedham-Green, A constructive recognition algorithm for the special linear group, pp. 11–26 in: The atlas of finite groups: ten years on (eds. R. T. Curtis and R. A. Wilson), LMS Lecture Note Series 249, Cambridge U. Press 1998.

[CLMNO] F. Celler, C. R. Leedham-Green, S. H. Murray, A. C. Niemeyer and E. A. O'Brien, Generating random elements of a finite group, Comm. in Alg. 23 (1995) 4931–4948.

[CoLG] M. Conder and C. R. Leedham-Green, Fast recognition of classical groups over large fields, pp. 113–121 in: Groups and Computation III (eds. W. M. Kantor and Á. Seress), The Ohio State Univ. Math. Res. Inst. Publ. 8, Walter deGruyter, Berlin–New York 2001.

[Co] G. Cooperman, Towards a theoretically sound algorithm for random generation in finite groups, preprint.

[CFL] G. Cooperman, L. Finkelstein and S. Linton, Recognizing $GL_n(2)$ in non-standard representation, pp. 85–100 in: Groups and Computation II (eds. L. Finkelstein and W. M. Kantor), DIMACS Series in Discrete Math. and Theoretical Computer Science, vol. 28, AMS 1997.

[EH] B. Eick and A. Hulpke, Computing the maximal subgroups of a permutation group I, pp. 155–168 in: Groups and Computation III (eds. W. M. Kantor and Á. Seress), The Ohio State Univ. Math. Res. Inst. Publ. 8, Walter deGruyter, Berlin–New York 2001.

[FT] W. Feit and J. Tits, Projective representations of minimum degree of group extensions, Can. J. Math. 30 (1978), 1092–1102.

[GAP4] The GAP Group, Aachen, St Andrews, GAP – Groups, Algorithms, and Programming, Version 4.3, 2002 (`http://www-gap.dcs.st-and.ac.uk/~gap`).

[GL] R. M. Guralnick and F. Lübeck, On p-singular elements in Chevalley groups in characteristic p, pp. 169–182 in: Groups and Computation III (eds. W. M. Kantor and Á. Seress), The Ohio State Univ. Math. Res. Inst. Publ. 8, Walter deGruyter, Berlin–New York 2001.

[HR] D. F. Holt and S. Rees, Testing modules for irreducibility, J. Austral. Math. Soc. (Ser. A) 57 (1994), 1–16.

[HS] A. Hulpke and Á. Seress, Short presentations for three-dimensional unitary groups, J. Algebra 245 (2001), 719–729.

[IKS] I. M. Isaacs, W. M. Kantor and N. Spaltenstein, On the probability that a group element is p–singular, J. Algebra 176 (1995), 139–181.

[IL] G. Ivanyos and K. Lux, Treating the exceptional cases of the MeatAxe, Experiment. Math. 9 (2000), 373–381.

[Ka1] W. M. Kantor, Sylow's theorem in polynomial time, J. Comp. Syst. Sci. 30 (1985), 359–394.

[Ka2] W. M. Kantor, Simple groups in computational group theory, pp. 77–86 in: Proc. International Congress of Mathematicans, Berlin 1998, Vol. II.

[KLM] W. M. Kantor, E. M. Luks and P. D. Mark, Parallel algorithms for Sylow subgroups, J. Algorithms 31 (1999), 132–195.

[KM] W. M. Kantor and K. Magaard, Black-box exceptional groups of Lie type (in preparation).

[KS1] W. M. Kantor and Á. Seress, Permutation group algorithms via black box recognition algorithms, pp. 436–446 in: Groups St Andrews 1997 in Bath (Eds. C. Campbell et al.), LMS Lectures Notes Series 261, Cambridge U. Press 1999.

[KS2] W. M. Kantor and Á. Seress, Black box classical groups, Memoirs of the Amer. Math. Soc., 149 (2001), No. 708.

[KS3] W. M. Kantor and Á. Seress, Prime power graphs for groups of Lie type, J. Algebra 247 (2002), 370–434.

[KS4] W. M. Kantor and Á. Seress, Algorithms for finite linear groups (in preparation).

[KL] P. B. Kleidman and M. W. Liebeck, The subgroup structure of the finite classical groups, LMS Lecture Note Series 129, Cambridge U. Press 1990.

[LS] V. Landazuri and G. M. Seitz, On the minimal degrees of projective representations of the finite Chevalley groups, J. Algebra 32 (1974), 418–443.

[LG] C. R. Leedham-Green, The computational matrix group project, pp. 229–247 in: Groups and Computation III (eds. W. M. Kantor and Á. Seress), The Ohio State Univ. Math. Res. Inst. Publ. 8, Walter deGruyter, Berlin–New York 2001.

[LGO] C. R. Leedham-Green and E. A. O'Brien, Constructive recognition of SL$(2,q)$ (in preparation).

[Lu] E. M. Luks, Computing in solvable matrix groups, pp. 110–120 in: Proc. 33rd IEEE Symp. on Found. of Comp. Science 1992.

[Lü] F. Lübeck, Finding p'-elements in finite groups of Lie type, pp. 249–255 in: Groups and Computation III (eds. W. M. Kantor and Á. Seress), The Ohio State Univ. Math. Res. Inst. Publ. 8, Walter deGruyter, Berlin–New York 2001.

[Mo] P. Morje, A nearly linear algorithm for Sylow subgroups of permutation groups, Ph.D. thesis, The Ohio State University 1995.

[NeP1] P. M. Neumann and C. E. Praeger, A recognition algorithm for special linear groups, Proc. London Math. Soc. 65 (1992), 555–603.

[NeP2] P. M. Neumann and C. E. Praeger, Cyclic matrices and the Meataxe, pp. 291–300 in: Groups and Computation III (eds. W. M. Kantor and Á. Seress), The Ohio State Univ. Math. Res. Inst. Publ. 8, Walter deGruyter, Berlin–New York 2001.

[NiP] A. C. Niemeyer and C. E. Praeger, A recognition algorithm for classical groups over finite fields, Proc. London Math. Soc. (3) 77 (1998), 117–169.

[Pak] I. Pak, The product replacement algorithm is polynomial, Proc. 41st IEEEE Symp. on Found. Comp. Science 2000, 476–485.

[Pr] C. E. Praeger, Computation with matrix groups over finite fields, pp. 189–195 in: Groups and Computation (eds. L. Finkelstein and W. M. Kantor), DIMACS Series in Discrete Math. and Theoretical Computer Science, vol. 11, AMS 1993.

[Ser1] Á. Seress, An introduction to computational group theory, Notices AMS 44 (1997), 671–679.

[Ser2] Á. Seress, Permutation Group Algorithms, Cambridge University Press, 2002.

[Si1] C. C. Sims Computational methods in the study of permutation groups, pp. 169–183 in: Computational problems in abstract algebra (ed. J. Leech), Pergamon 1970.

[Si2] C. C. Sims, Computation with permutation groups, pp. 23–28 in: Proc. Symp. Symb. Alg. Manipulation (ed. S. R. Petrick), ACM 1971.

[Si3] C. C. Sims, Group-theoretic algorithms, a survey, pp. 979–985 in: Proc. ICM, Helsinki 1978.

[Si4] Computing with Finitely Presented Groups, Cambridge University Press, 1994.

[Suz] M. Suzuki, On a class of doubly transitive groups, Ann. Math. 75 (1962) 105–145.

[Zs] K. Zsigmondy, Zur Theorie der Potenzreste, Monatsh. Math. Phys. 3 (1892), 265-284.

A survey of maximal subgroups of exceptional groups of Lie type

Martin W. Liebeck and Gary M. Seitz

The object of this survey is to bring the reader up to date with recent results concerning the maximal subgroups of finite and algebraic groups of exceptional Lie type. The first section deals with algebraic groups, and the second with finite groups.

1 Maximal subgroups of exceptional algebraic groups

Let G be a simple algebraic group of exceptional type G_2, F_4, E_6, E_7 or E_8 over an algebraically closed field K of characteristic p. The analysis of maximal subgroups of exceptional groups has a history stretching back to the fundamental work of Dynkin [3], who determined the maximal connected subgroups of G in the case where K has characteristic zero. The flavour of his result is that apart from parabolic subgroups and reductive subgroups of maximal rank, there are just a few further conjugacy classes of maximal connected subgroups, mostly of rather small dimension compared to $\dim G$. In particular, G has only finitely many conjugacy classes of maximal connected subgroups.

The case of positive characteristic was taken up by Seitz [15], who determined the maximal connected subgroups under some assumptions on p, obtaining conclusions similar to those of Dynkin. If $p > 7$ then all these assumptions are satisfied. This result was extended in [7], where all maximal closed subgroups of positive dimension in G were classified, under similar assumptions on p.

In the years since [15, 7], the importance of removing the characteristic assumptions in these results has become increasingly clear, in view of applications to both finite and algebraic group theory (see for example Section 2 below for some such applications). This has finally been achieved in [11]. Here is a statement of the result.

Theorem 1 ([11]) *Let M be a maximal closed subgroup of positive dimension in the exceptional algebraic group G. Then one of the following holds:*

(a) *M is either parabolic or reductive of maximal rank;*

(b) *$G = E_7$, $p \neq 2$ and $M = (2^2 \times D_4).Sym_3$;*

(c) *$G = E_8$, $p \neq 2, 3, 5$ and $M = A_1 \times Sym_5$;*

(d) *M^0 is as in Table 1 below.*

The subgroups M in (b), (c) *and* (d) *exist, are unique up to conjugacy in* Aut(G), *and are maximal in* G.

Table 1

G	M^0 simple	M^0 not simple
G_2	$A_1\,(p \geq 7)$	
F_4	$A_1\,(p \geq 13)$, $G_2\,(p = 7)$,	$A_1G_2\,(p \neq 2)$
E_6	$A_2\,(p \neq 2,3)$, $G_2\,(p \neq 7)$, $C_4\,(p \neq 2)$, F_4	A_2G_2
E_7	A_1 (2 classes, $p \geq 17, 19$ resp.), $A_2\,(p \geq 5)$	$A_1A_1\,(p \neq 2,3)$, $A_1G_2\,(p \neq 2)$, A_1F_4, G_2C_3
E_8	A_1 (3 classes, $p \geq 23, 29, 31$ resp.), $B_2\,(p \geq 5)$	$A_1A_2\,(p \neq 2,3)$, $A_1G_2G_2\,(p \neq 2)$, G_2F_4

For notational convenience in the table, we set $p = \infty$ if K has characteristic zero; thus, for example, the condition $p \geq 7$ includes the characteristic zero case.

In fact [11] has a somewhat more general version of Theorem 1, which allows the presence of Frobenius and graph morphisms of G.

A few remarks are in order concerning the subgroups occurring in the conclusion of Theorem 1.

The subgroups of G of type (a) in the theorem are well understood. Maximal parabolic subgroups correspond to removing a node of the Dynkin diagram. Subgroups which are reductive of maximal rank are easily determined. They correspond to various subsystems of the root system of G, and a complete list of those which are maximal in G can be found in [11, Table 10.3].

The subgroups under (b) and (c) of Theorem 1 were constructed in [2, 7]: in (b), the connected component $M^0 = D_4$ lies in a subsystem A_7 of G, and in (c), $M^0 = A_1$ lies in a subsystem A_4A_4, with restricted irreducible embedding in each factor.

The subgroups in Table 1 are constructed in [15, 16, 17], apart from a few cases in small characteristic which can be found in [11].

Theorem 1 has a number of consequences. The first is the following, which applies to all types of simple algebraic groups, both classical and exceptional.

Corollary 2 *If H is a simple algebraic group over an algebraically closed field, then H has only finitely many conjugacy classes of maximal closed subgroups of positive dimension.*

Another major consequence of Theorem 1 is that sufficiently large maximal subgroups of finite exceptional groups of Lie type are known. We shall discuss this in the next section.

Also determined in [11] are the precise actions of maximal subgroups X in Table 1 on the adjoint module $L(G)$, as a sum of explicit indecomposable modules. An interesting feature of these actions is that very few types of indecomposables arise. Indeed, with one exception, each restriction $L(G) \downarrow X$ is the sum of indecomposables of one of the following three types: an irreducible module $V(\lambda)$; an indecomposable tilting module $T(\lambda)$; or an indecomposable module $\Delta(\lambda;\gamma)$ of shape $\mu|(\lambda \oplus \gamma)|\mu$ arising in the following way: suppose λ, γ, μ are dominant weights for X such that $T(\lambda) = \mu|\lambda|\mu$ and $T(\gamma) = \mu|\gamma|\mu$ (where μ denotes the irreducible $V(\mu)$, etc.). Then $\Delta(\lambda;\gamma)$ denotes an indecomposable module of shape $\mu|(\lambda \oplus \gamma)|\mu$ with socle and cosocle both of type μ, and which is obtained as a section of $T(\lambda) \oplus T(\gamma)$, by taking a maximal submodule and then factoring out a diagonal submodule of the socle. The one exception to the above is $X = G_2 < E_6$ with $p = 3$, in which case $L(G)' \downarrow X$ is uniserial with series $10|01|11|01|10$.

Finally, we mention that as a consequence of Theorem 1, together with work on finite subgroups of exceptional groups described in the next section, all closed subgroups of G which act irreducibly on either the adjoint module for G, or on one of the irreducible modules of dimension $26 - \delta_{p,3}$, 27 or 56 for $G = F_4, E_6$ or E_7 respectively, have been determined in [12].

2 Maximal subgroups of finite exceptional groups

In this section let G be an adjoint simple algebraic group of exceptional type over $K = \bar{\mathbb{F}}_p$, the algebraic closure of the prime field $\mathbb{F}_p$, where p is a prime, and let σ be a Frobenius morphism of G. Denote by G_σ the fixed point group $\{g \in G : g^\sigma = g\}$. Then $G_0 := G'_\sigma$ is a finite simple exceptional group (exclude the cases $G_2(2)' \cong U_3(3)$ and ${}^2G_2(3)' \cong L_2(8)$).

Throughout the section, let H be a maximal subgroup of G_σ; all the results below apply more generally to maximal subgroups of any almost simple group with socle G_0, but we restrict ourselves to G_σ for notational convenience. The ultimate aim is of course to determine completely all the possiblities for H up to conjugacy. This task is by no means finished, but there has been a great deal of recent progress, and our aim is to bring the reader up to date with this.

First, we present a "reduction theorem", reducing considerations to the case where H is almost simple. In the statement reference is made to the following *exotic* local subgroups of G_σ (one G_σ-class of each):

$$
\begin{array}{lcl}
2^3.SL_3(2) & < & G_2(p)\ (p > 2) \\
3^3.SL_3(3) & < & F_4(p)\ (p \geq 5) \\
3^{3+3}.SL_3(3) & < & E_6^\epsilon(p)\ (p \equiv \epsilon \bmod 3, p \geq 5) \\
5^3.SL_3(5) & < & E_8(p^a)\ (p \neq 2,5; a = 1 \text{ or } 2, \text{ as } p^2 \equiv 1 \text{ or } -1 \bmod 5) \\
2^{5+10}.SL_5(2) & < & E_8(p)\ (p > 2)
\end{array}
$$

Note that these local subgroups exist for $p = 2$ in lines 2, 3 and 4, but are non-maximal because of the containments $3^3.SL_3(3) < L_4(3) < F_4(2)$, $3^{3+3}.SL_3(3) < \Omega_7(3) < {}^2E_6(2)$ and $5^3.SL_3(5) < L_4(5) < E_8(4)$ (see [2]).

Theorem 3 ([7, Theorem 2]) *Let H be a maximal subgroup of G_σ as above. Then one of the following holds:*

(i) *H is almost simple;*

(ii) $H = M_\sigma$, *where* M *is a maximal* σ*-stable closed subgroup of positive dimension in* G *as in Theorem* 1*;*

(iii) H *is an exotic local subgroup;*

(iv) $G = E_8, p > 5$ *and* $H = (Alt_5 \times Alt_6).2^2$.

A version of this theorem was also proved by Borovik [1], who in particular discovered the interesting maximal subgroup in part (iv).

In view of this result attention focusses on the case where H is an almost simple maximal subgroup of G_σ. Let H be such, and write $H_0 = F^*(H)$, a simple group. The analysis falls naturally into two cases: $H_0 \in \mathrm{Lie}(p)$, and $H_0 \notin \mathrm{Lie}(p)$, where $\mathrm{Lie}(p)$ denotes the set of finite simple groups of Lie type in characteristic p. We call these *generic* and *non-generic* subgroups, respectively.

We first discuss non-generic subgroups. Here we have the following result, which determines the possibilities for H_0 up to isomorphism; however the problem of determining them up to conjugacy remains open.

Theorem 4 ([10]) *Let* S *be a finite simple group, some cover of which is contained in the exceptional algebraic group* G *, and assume* $S \notin \mathrm{Lie}(p)$*. Then the possibilities for* S *and* G *are given in Table* 2.

Table 2

G	S
G_2	$\mathrm{Alt}_5, \mathrm{Alt}_6, L_2(7), L_2(8), L_2(13), U_3(3)$, $\mathrm{Alt}_7 (p=5), J_1(p=11), J_2(p=2)$
F_4	above, plus: $\mathrm{Alt}_{7-10}, L_2(17), L_2(25), L_2(27), L_3(3), U_4(2), Sp_6(2), \Omega_8^+(2), {}^3D_4(2), J_2$ $\mathrm{Alt}_{11}(p=11), L_3(4)(p=3), L_4(3)(p=2), {}^2B_2(8)(p=5), M_{11}(p=11)$
E_6	above, plus: $\mathrm{Alt}_{11}, L_2(11), L_2(19), L_3(4), U_4(3), {}^2F_4(2)', M_{11}$, $\mathrm{Alt}_{12}(p=2,3), G_2(3)(p=2), \Omega_7(3)(p=2), M_{22}(p=2,7)$, $J_3(p=2), Fi_{22}(p=2), M_{12}(p=2,3,5)$
E_7	above, plus: $\mathrm{Alt}_{12}, \mathrm{Alt}_{13}, L_2(29), L_2(37), U_3(8), M_{12}$, $\mathrm{Alt}_{14}(p=7), M_{22}(p=5), Ru(p=5), HS(p=5)$
E_8	above, plus: $\mathrm{Alt}_{14-17}, L_2(16), L_2(31), L_2(32), L_2(41), L_2(49)$, $L_2(61), L_3(5), PSp_4(5), G_2(3), {}^2B_2(8)$, $\mathrm{Alt}_{18}(p=3), L_4(5)(p=2), Th(p=3), {}^2B_2(32)(p=5)$

This is actually a condensed version of the main result of [10], which also determines precisely which simple groups (rather than just covers thereof) embed in adjoint exceptional groups.

We now move on to discuss generic maximal subgroups H of G_σ - namely, those for which $H_0 = F^*(H)$ lies in $\mathrm{Lie}(p)$. The expectation in this case is that in general H is of the form M_σ, where M is a maximal closed σ-stable subgroup of positive dimension in G, given by Theorem 1. This is proved in the next result, under some assumptions on the size of the field over which H_0 is defined. In [9], a certain constant $t(G)$ is defined, depending only on the root system of G; and R. Lawther has computed the values of $t(G)$ for all exceptional groups except E_8: we have $t(G) = u(G) \cdot (2, p-1)$, where $u(G)$ is as follows

G	G_2	F_4	E_6	E_7
$u(G)$	12	68	124	388

Theorem 5 ([9]) *Let H be a maximal subgroup of the finite exceptional group G_σ such that $F^*(H) = H(q)$, a simple group of Lie type over $\mathbb{F}_q$, a field of characteristic p. Assume that*

$$\begin{array}{ll} q > t(G), & \text{if } H(q) = L_2(q),\ {}^2B_2(q) \text{ or } {}^2G_2(q) \\ q > 9 \text{ and } H(q) \neq A_2^\epsilon(16), & \text{otherwise.} \end{array}$$

Then one of the following holds:

(i) *$H(q)$ has the same type as G (possibly twisted);*

(ii) *$H = M_\sigma$ for some maximal closed σ-stable subgroup M of positive dimension in G (given by Theorem 1).*

Writing $G_\sigma = G(q_1)$, the subgroups in (i) are subgroups of the form $G(q)$ or a twisted version, where $\mathbb{F}_q$ is a subfield of $\mathbb{F}_{q_1}$; they are unique up to G_σ-conjugacy, by [8, 5.1].

One of the points of this result is that it excludes only finitely many possibilities for $F^*(H) = H(q)$, up to isomorphism. Since there are also only finitely many non-generic simple subgroups up to isomorphism, the following is an immediate consequence.

Corollary 6 *There is a constant c, such that if H is a maximal subgroup of G_σ with $|H| > c$, then either $F^*(H) = H(q)$ has the same type as G, or $H = M_\sigma$ where M is maximal closed σ-stable of positive dimension in G.*

This is all very well, but in practice one needs more information concerning the generic almost simple maximal subgroups which are not covered by Theorem 5. A useful result in this direction is the following, which determines generic maximal subgroups of rank more than half the rank of G. For a simple group of Lie type $H(q)$, let $\mathrm{rk}(H(q))$ denote the untwisted Lie rank of $H(q)$.

Theorem 7 ([5, 13]) *Suppose H is a maximal subgroup of G_σ such that $F^*(H) = H(q)$, a simple group of Lie type in characteristic p, with $rk(H(q)) > \frac{1}{2}rk(G)$. Then either $H(q)$ has the same type as G, or $H = M_\sigma$ where M is maximal closed σ-stable of positive dimension in G. In the latter case, the possibilities are as follows:*

(i) *M is a subgroup of maximal rank (possibilities determined in [6]);*

(ii) *$G_\sigma = E_6^\epsilon(q)$ and $H(q) = F_4(q)$ or $C_4(q)$ (q odd);*

(iii) *$G'_\sigma = E_7(q)$ and $H(q) = {}^3D_4(q)$ (with M as in Theorem 1(b)).*

This is proved in [5, Theorem 3] assuming that $q > 2$, and in [13] for $q = 2$. The maximal subgroups in part (iii) were omitted in error in [5]. They arise when M is the maximal closed subgroup $(2^2 \times D_4).Sym_3$ in Theorem 1(b) and σ acts on M as $\sigma_q w$, where σ_q is the standard field morphism and $w \in Sym_3$ has order 3 (so that $M_\sigma = {}^3D_4(q).3$).

For the reader's convenience, we now present a compendium result which summarises almost all of the work above on maximal subgroups of G_σ.

Theorem 8 *Let H be a maximal subgroup of the finite exceptional group G_σ over $\mathbb{F}_q$, $q = p^a$. The one of the following holds:*

(I) *$H = M_\sigma$ where M is maximal closed σ-stable of positive dimension in G; the possibilities are as follows:*

(a) *M (and H) is a parabolic subgroup;*

(b) *M is reductive of maximal rank: the possibilities for H are determined in* [6]*;*

(c) *$G = E_7$, $p > 2$ and $H = (2^2 \times P\Omega_8^+(q).2^2).Sym_3$ or ${}^3D_4(q).3$;*

(d) *$G = E_8$, $p > 5$ and $H = PGL_2(q) \times Sym_5$;*

(e) *M is as in Table* 1*, and $H = M_\sigma$ as in Table* 3 *below.*

(II) *H is of the same type as G;*

(III) *H is an exotic local subgroup;*

(IV) *$G = E_8$, $p > 5$ and $H = (Alt_5 \times Alt_6).2^2$;*

(V) *$F^*(H) = H_0$ is simple, and not in* Lie(p)*: the possibilities for H_0 are given up to isomorphism by* [10] *(see also Theorem* 4 *above);*

(VI) *$F^*(H) = H(q_0)$ is simple and in* Lie(p)*; moreover $rk(H(q_0)) \leq \frac{1}{2}rk(G)$, and one of the following holds:*

(a) *$q_0 \leq 9$;*

(b) *$H(q_0) = A_2^\epsilon(16)$;*

(c) *$q_0 \leq t(G)$ and $H(q_0) = A_1(q_0)$, ${}^2B_2(q_0)$ or ${}^2G_2(q_0)$.*

In cases (I)-(IV)*, H is determined up to G_σ-conjugacy.*

Table 3

G'_σ	possibilities for $F^*(M_\sigma), M$ in Table 1
$G_2(q)$	$A_1(q)\ (p \geq 7)$
$F_4(q)$	$A_1(q)\ (p \geq 13),\ G_2(q)\ (p = 7),\ A_1(q) \times G_2(q)\ (p \geq 3, q \geq 5)$
$E_6^\epsilon(q)$	$A_2^\epsilon(q)\ (p \geq 5),\ G_2(q) (p \neq 7),\ C_4(q)\ (p \geq 3),\ F_4(q),$ $A_2^\epsilon(q) \times G_2(q)\ ((q,\epsilon) \neq (2,-))$
$E_7(q)$	$A_1(q)\ (2 \text{ classes}, p \geq 17, 19),\ A_2^\epsilon(q)\ (p \geq 5),\ A_1(q) \times A_1(q)\ (p \geq 5),$ $A_1(q) \times G_2(q)\ (p \geq 3, q \geq 5),\ A_1(q) \times F_4(q)\ (q \geq 4),\ G_2(q) \times C_3(q)$
$E_8(q)$	$A_1(q)\ (3 \text{ classes}, p \geq 23, 29, 31),\ B_2(q)\ (p \geq 5),\ A_1(q) \times A_2^\epsilon(q)\ (p \geq 5),$ $G_2(q) \times F_4(q),\ A_1(q) \times G_2(q) \times G_2(q)\ (p \geq 3, q \geq 5),$ $A_1(q) \times G_2(q^2)\ (p \geq 3, q \geq 5)$

We remind the reader that, as mentioned before, the above results apply more generally to maximal subgroups of all almost simple groups whose socle is a finite exceptional group of Lie type.

Bounds for the orders of maximal subgroups of finite groups of Lie type have proved useful in a variety of applications. For exceptional groups, the first such bounds appeared in [4]; using some of the above results, these were improved as follows in [14].

Theorem 9 ([14, 1.2]) *Let H be a maximal subgroup of the finite exceptional group G_σ over $\mathbb{F}_q$, $q = p^a$. Assume that $|H| \geq 12aq^{56}$, $4aq^{30}$, $4aq^{28}$ or $4aq^{20}$, according as $G = E_8, E_7, E_6$ or F_4, respectively. Then H is as in conclusion* (I)(a),(b) or (e) *of Theorem* 8.

It should be possible to improve these bounds substantially.

Despite the progress reported above, there remain some substantial problems to tackle in the theory of maximal subgroups of finite exceptional groups. The most obvious ones are the determination of the conjugacy classes of non-generic simple subgroups (Theorem 8(IV)), and of generic simple subgroups over small fields (Theorem 8(V)). Of these, perhaps the most challenging and important is to reduce substantially the $t(G)$ bound for subgroups of rank 1 in Theorem 8(V(c)), especially for $G = E_8$, where $t(G)$ is currently unknown (and in any case is known to be quite large).

References

[1] A.V. Borovik, "The structure of finite subgroups of simple algebraic groups" (Russian), *Algebra i Logika* **28** (1989), no. 3, 249-279; translation in *Algebra and Logic* **28** (1989), no. 3, 163-182 (1990).

[2] A.M. Cohen, M.W. Liebeck, J. Saxl and G.M. Seitz, "The local maximal subgroups of exceptional groups of Lie type, finite and algebraic", *Proc. London Math. Soc.* **64** (1992), 21-48.

[3] E.B. Dynkin, "Semisimple subalgebras of semisimple Lie algebras", *Translations Amer. Math. Soc.* **6** (1957), 111-244.

[4] M.W. Liebeck and J. Saxl, "On the orders of maximal subgroups of the finite exceptional groups of Lie type", *Proc. London Math. Soc.* **55** (1987), 299-330

[5] M.W. Liebeck, J. Saxl and D. Testerman, "Subgroups of large rank in groups of Lie type", *Proc. London Math. Soc.* **72** (1996), 425-457.

[6] M.W. Liebeck, J. Saxl and G.M. Seitz, "Subgroups of maximal rank in finite exceptional groups of Lie type", *Proc. London Math. Soc.* **65** (1992), 297-325.

[7] M.W. Liebeck and G.M. Seitz, "Maximal subgroups of exceptional groups of Lie type, finite and algebraic", *Geom. Dedicata* **36** (1990), 353-387.

[8] M.W. Liebeck and G.M. Seitz, "Subgroups generated by root elements in groups of Lie type", *Annals of Math*, **139**, (1994), 293-361.

[9] M.W. Liebeck and G.M. Seitz, "On the subgroup structure of exceptional groups of Lie type", *Trans. Amer. Math. Soc.* **350** (1998), 3409-3482.

[10] M.W. Liebeck and G.M. Seitz, "On finite subgroups of exceptional algebraic groups", *J. reine angew. Math.* **515** (1999), 25-72.

[11] M.W. Liebeck and G.M. Seitz, "The maximal subgroups of positive dimension in exceptional algebraic groups", submitted to *Mem. Amer. Math. Soc.*.

[12] M.W. Liebeck and G.M. Seitz, "Subgroups of exceptional algebraic groups which are irreducible on an adjoint or minimal module", to appear.

[13] M.W. Liebeck and G.M. Seitz, "Subgroups of large rank in exceptional groups of Lie type", to appear.

[14] M.W. Liebeck and A. Shalev, "The probability of generating a finite simple group", *Geom. Dedicata* **56** (1995), 103-113.

[15] G.M. Seitz, "Maximal subgroups of exceptional algebraic groups", *Mem. Amer. Math. Soc.* Vol. 90, No. 441, 1991.

[16] D.M. Testerman, "A construction of certain maximal subgroups of the algebraic groups E_6 and F_4", *J. Algebra* **122** (1989), 299-322.

[17] D.M. Testerman, "The construction of the maximal A_1's in the exceptional algebraic groups", *Proc. Amer. Math. Soc.* **116** (1992), 635-644.

Bases of primitive permutation groups

Martin W. Liebeck and Aner Shalev

1 Introduction

Let G be a permutation group on a finite set Ω of size n. A subset of Ω is said to be a *base* for G if its pointwise stabilizer in G is trivial. The minimal size of a base for G is denoted by $b(G)$. Bases have been studied since the early years of permutation group theory, particularly in connection with orders of primitive groups and, more recently, with computational group theory. In this paper we survey some of the recent developments in this area, with particular emphasis on some well known conjectures of Babai, Cameron and Pyber.

We begin with a number of examples.

(1) Obviously $b(S_n) = n - 1$ and $b(A_n) = n - 2$.

(2) At the other extreme, $b(G) = 1$ if and only if G has a regular orbit on Ω.

(3) Let $G = S_k$ acting on the set Ω of pairs in $\{1, \ldots k\}$. Write $k = 3l + r$ with $0 \leq r \leq 2$, and define B to be the subset of Ω consisting of the pairs $\{1,2\}, \{2,3\}, \{4,5\}, \{5,6\}, \ldots, \{3l-2, 3l-1\}, \{3l-1, 3l\}$ (adding also $\{3l, 3l+1\}$ if $r = 2$). It is easy to see that B is a base so that $b(G) \leq \frac{2}{3}k + 1$ in this example.

(4) If $G = PGL_d(q)$ acting on the set Ω of 1-spaces in the underlying vector space $V_d(q)$, then $b(G) = d + 1$, a minimal base being $\{\langle v_1 \rangle, \ldots, \langle v_d \rangle, \langle v_1 + \ldots + v_d \rangle\}$, where $v_1, \ldots, v_d$ is a basis for $V_d(q)$.

(5) Let G be the affine group $AGL_d(q)$ acting on $V_d(q)$, of degree q^d. Then $b(G) = d + 1$.

(6) Let $G = S_2 \wr C_k$ in its natural imprimitive, transitive representation of degree $2k$ having k blocks of imprimitivity of size 2. Then $b(G) = k$.

If $\{\omega_1, \ldots, \omega_b\}$ is a base for G of size $b = b(G)$, then

$$|G| = |G : G_{\omega_1 \ldots \omega_b}| = |G : G_{\omega_1}|\,|G_{\omega_1} : G_{\omega_1\omega_2}| \ldots |G_{\omega_1 \ldots \omega_{b-1}} : G_{\omega_1 \ldots \omega_b}|.$$

Each term on the right hand side is at most n and at least 2, and so we have

Proposition 1.1 *We have $2^{b(G)} \leq |G| \leq n^{b(G)}$. Consequently*

$$\log_2 |G| \geq b(G) \geq \frac{\log |G|}{\log n}.$$

In examples (1), (3), (4) and (5) above we see that $b(G) \sim \frac{\log|G|}{\log n}$ (where $f \sim g$ means that f/g is bounded between two positive constants); whereas in example (6), $b(G) \sim \log|G|$.

Despite the elementary nature of Proposition 1.1, the connection it gives between the order of G and the value of $b(G)$ has been much exploited, leading to a number of important results and conjectures which we shall discuss below.

2 General bounds

The problem of bounding the order of a primitive permutation group of degree n not containing A_n is one of the oldest in permutation group theory, going back to the 19th century. One of the principal methods is to bound $b(G)$ and use Proposition 1.1. The most striking early result is due to Bochert:

Theorem 2.1 (Bochert [5]) *If G is a primitive permutation group of degree n not containing A_n, then $b(G) \leq \frac{n}{2}$.*

Using Proposition 1.1 it follows from this that $|G| \leq n^{n/2}$.

Almost a hundred years later, Babai proved the first substantial improvement of Bochert's result:

Theorem 2.2 (Babai [1, 2]) *Let G a primitive permutation group of degree n not containing A_n.*

(i) *If G is not 2-transitive then $b(G) < 4\sqrt{n}\log n$.*

(ii) *If G is 2-transitive then $b(G) < c^{\sqrt{\log n}}$, where c is an absolute constant.*

The 2-transitive case was improved by Pyber:

Theorem 2.3 (Pyber [16]) *If G is a 2-transitive group of degree n not containing A_n, then $b(G) < c\log^2 n$, where c is an absolute constant.*

Note that Example (3) in the Introduction gives a primitive, not 2-transitive group of base size $c\sqrt{n}$; and Examples (4), (5) give 2-transitive groups of base size $c\log n$. This shows that the bounds in Theorems 2.2(i) and 2.3 are not far off best possible.

The above results were proved using combinatorial methods, in particular not using the classification of finite simple groups.

The first general result on base sizes using the classification was the following:

Theorem 2.4 (Liebeck [11]) *If G is a primitive group of degree n, then either*

(i) *$b(G) < 9\log_2 n$, or*

(ii) *G is a subgroup of $S_m \wr S_r$ containing $(A_m)^r$, where the action of S_m is on k-sets and the wreath product has the product action of degree $\binom{m}{k}^r$.*

Using this one can easily deduce a sharp result of the form $b(G) \leq c\sqrt{n}$ for primitive groups and $b(G) < c\log n$ for 2-transitive groups (where $G \not\geq A_n$), somewhat improving the classification-free results 2.2 and 2.3.

3 Conjectures of Babai and Cameron

In this section we discuss some conjectures and results concerning base sizes of some important classes of primitive permutation groups. The first conjecture stems from the following well known result.

Theorem 3.1 (Babai-Cameron-Pálfy [3]) *Let d be a positive integer, and let G be a primitive group of degree n not involving A_d as a section. Then $|G| < n^{f(d)}$, where $f(d)$ depends only on d.*

The function $f(d)$ obtainable from the proof in [3] is of the form $O(d \log d)$. A new proof by Pyber (unpublished) shows that $f(d)$ can be chosen to be linear in d.

Seeking a structural explanation of this result in the light of Proposition 1.1, Babai conjectured that any group G as in the statement has a base of size bounded in terms of d alone. The first indication that this might be true came from analysis of the solvable case:

Theorem 3.2 (Seress [18]) *If G is a solvable primitive permutation group, then $b(G) \leq 4$.*

This corresponds very closely to the rather tight bound $|G| < 24^{-1/3}n^{3.244}$ on the order of a primitive solvable group G obtained in [15, 20]; indeed, the bound of 4 in Theorem 3.2 is best possible, since there are primitive solvable groups of order larger than n^3.

Babai's conjecture was finally proved:

Theorem 3.3 (Gluck-Seress-Shalev [10]) *There exists a function $g(d)$ such that if G is a primitive group not involving A_d, then $b(G) < g(d)$.*

The proof in [10] shows that $g(d)$ can be chosen as a quadratic function of d. This is improved to a linear function in [13, 1.4].

The other class of primitive groups we shall discuss in this section are the almost simple primitive groups. Here again there is a result on orders:

Theorem 3.4 (Liebeck [11]) *If G is an almost simple primitive permutation group of degree n, then one of the following holds:*

(i) $|G| < n^9$;

(ii) $F^*(G) = A_m$ *acting on k-subsets or an orbit of partitions of* $\{1,\ldots,m\}$;

(iii) $F^*(G)$ *is a classical group in a subspace action.*

In (iii), a *subspace* action of a classical group $G_0 = F^*(G)$ with natural module V is a primitive action on an orbit of subspaces of V, or pairs of subspaces of complementary dimensions (when $G_0 = PSL(V)$ and G contains a graph automorphism), or on the cosets of an orthogonal subgroup $O^{\pm}_{2m}(q) < G_0 = Sp_{2m}(q)$ with q even.

A version of this result with n^c in (i) (c unspecified) appeared in [6, 6.1]; and an improvement with n^5 replacing n^9 in (i), allowing also the exceptions $G = M_n$ with $n \in \{23, 24\}$, appears in [12, Proposition 2].

Definition We call primitive actions of groups as in (ii) or (iii) of Theorem 3.4 *standard* actions.

It is natural to ask whether there is a base-size analogue of Theorem 3.4, and indeed the following conjecture was posed by Cameron.

Conjecture 3.5 (Cameron [7, 3.4]) *There is a constant c such that if G is an almost simple primitive group in a non-standard action, then $b(G) < c$.*

In [8], Cameron and Kantor suggested a probabilistic strengthening of this conjecture: if G is as above, then almost every c-tuple is a base for G. This has now been established:

Theorem 3.6 (Liebeck-Shalev [13, 1.3]) *There is a constant c such that if G is an almost simple primitive group in a non-standard action, then the probability that a random c-tuple of points from the permutation domain forms a base for G tends to 1 as $|G| \to \infty$. In particular, Cameron's conjecture holds.*

For G an alternating or symmetric group, Theorem 3.6 was proved by Cameron and Kantor [8] with $c = 2$.

The proofs of Theorems 3.3 and 3.6 use results on fixed point ratios as a main tool, as we shall now discuss. For a permutation group G on a set Ω of size n, and an element $x \in G$, define $\mathrm{fix}(x)$ to be the number of fixed points of x and $\mathrm{rfix}(x) = \mathrm{fix}(x)/n$. Thus $\mathrm{rfix}(x)$ is the probability that a random point of Ω is fixed by x. Therefore the probabilibity that a random k-tuple is fixed by x is $\mathrm{rfix}(x)^k$. If a given k-tuple is not a base, then it is fixed by some element $x \in G$ of prime order. Hence if $Q(G, k)$ is the probability that a random k-tuple is not a base for G, then

$$Q(G,k) \leq \sum \mathrm{rfix}(x)^k, \qquad (\dagger)$$

the sum being over elements $x \in G$ of prime order.

Now assuming G is primitive and M is a point stabilizer, we have $\mathrm{rfix}(x) = |x^G \cap M|/|x^G|$. In the crucial case where G is an almost simple group of Lie type in a non-standard action, it is established in [13, Theorem (*)] that this ratio is bounded above by $|x^G|^{-\epsilon}$, where ϵ is a positive constant. Plugging this into (†) and choosing k large enough (greater than $11/\epsilon$ will do), it is possible to deduce that $Q(G, k) \to 0$ as $|G| \to \infty$, which is enough to prove Theorem 3.6.

As for Theorem 3.3, the proof starts with a far from straightforward reduction to the cases where G is almost simple, or of affine type with a point stabilizer G_0 being a primitive linear group. Define $\mathrm{rfix}(G)$ to be the maximum value of $\mathrm{rfix}(x)$ for $1 \neq x \in G$. Then (†) gives

$$Q(G,k) \leq |G|\mathrm{rfix}(G)^k.$$

In the two cases above, it is shown in [10] that the right hand side tends to 0 as $|G| \to \infty$ for a suitable choice of $k = g(d)$. Thus in fact a stronger, probabilistic form of Theorem 3.3 holds for these types of primitive groups.

4 Pyber's Conjecture

A conjecture for arbitrary primitive groups which generalizes the conjectures in the previous section was formulated by Pyber in [17]:

Conjecture 4.1 (Pyber) *There is a constant c, such that if G is a primitive permutation group of degree n, then*

$$b(G) < c\frac{\log |G|}{\log n}.$$

Note that this conclusion does not hold for all transitive groups G, as is shown by Example (6) of the Introduction.

Seress [19] has shown that to prove Pyber's conjecture, it is sufficient to establish it in the cases where G is either almost simple or of affine type.

Suppose first that G is almost simple. If the action is non-standard, then by Theorem 3.6, $b(G)$ is bounded above by a constant, and so Pyber's conjecture holds in this case. For standard actions, Benbenishty [4] has verified that Pyber's conjecture holds. Hence we have

Theorem 4.2 *Pyber's conjecture holds for almost simple groups.*

Now suppose that G is affine. Here $G \leq AGL(V)$, where V is a finite vector space of order $n = p^d$ (p prime); identifying V with the group of translations we have $G = VH$, where the point stabilizer $H = G_0$ is an irreducible subgroup of $GL(V)$, and $b(G) = 1 + b(H)$ (where $b(H)$ is the minimal size of a base for H in its action on vectors).

A couple of special cases of the problem have appeared: the solvable case (see Theorem 3.2), and the case where H is a p'-group. In the latter case Gluck and Magaard [9] show that $b(H) \leq 95$.

Recently we have solved the case in which H acts primitively as a linear group on V (in other words, H does not preserve any non-trivial direct sum decomposition of V).

Theorem 4.3 (Liebeck-Shalev [14]) *There is a constant c such that if $H \leq GL(V)$ is an irreducible, primitive linear group on a finite vector space V, then either*

(i) $b(H) < c$, *or*

(ii) $b(H) < 18\frac{\log |H|}{\log |V|} + 27$.

In proving this result, we study the structure of primitive linear groups which have unbounded base sizes. The first step is to analyse quasisimple linear groups. Here are some obvious examples of such groups having unbounded base sizes.

(1) Let $H = Cl_d(q)$, a classical group with natural module V of dimension d over $\mathbb{F}_q$. Then in its action on V, we have $b(H) \sim d$.

(2) Let $H = Cl_d(q^{1/r})$, where $\mathbb{F}_{q^{1/r}}$ is a subfield of $\mathbb{F}_q$, and take H to act naturally on $V = V_d(q)$. If $v_1, \ldots, v_d$ is an $\mathbb{F}_q$-basis of V, and $\lambda_1, \ldots, \lambda_r$ is a basis

for $\mathbb{F}_q$ over $\mathbb{F}_{q^{1/r}}$, then $\sum_1^r \lambda_i v_i, \sum_1^r \lambda_i v_{r+i}, \ldots$ is a base for H, and hence we see that $b(H) \sim d/r$ in the unbounded case.

(3) Let $H = A_{d+\delta}$ ($\delta = 1$ or 2) acting on its irreducible deleted permutation module $V = V_d(q)$ over $\mathbb{F}_q$. It is straightforward to see that $b(H) \sim \log d/\log q$ in the unbounded case.

An important intermediate result in [14] shows that these are the only examples of quasisimple groups with unbounded base sizes:

Proposition 4.4 ([14, 2.2]) *If $H \leq GL_d(q)$ with $E(H)$ quasisimple and absolutely irreducible on $V_d(q)$, then either*

(i) *$b(H) < c$ for some absolute constant c, or*

(ii) *$E(H) = Cl_d(q^{1/r})$ or $A_{d+\delta}$ as in Examples* (2), (3) *above.*

In the statement, $E(H)$ as usual denotes the product of all quasisimple subnormal subgroups of H.

Note that Proposition 4.4 does not require the assumption of primitivity of H as a linear group.

The next step in the proof involves analysis of tensor products, and we present another couple of examples.

(4) Let $V = V_m(q) \otimes V_m(q)$, and let $H = GL_m(q) \otimes GL_m(q)$ acting naturally on V (where $GL_m(q) \otimes GL_m(q)$ denotes the image of $GL_m(q) \times GL_m(q)$ in $GL(V)$). We claim that $b(H) \leq 3$. To see this, identify V with $M_m(q)$, the space of all $m \times m$ matrices over $\mathbb{F}_q$, with H-action $(g,h) : A \to g^T A h$ for $g, h \in GL_m(q)$, $A \in V$. Then the stabilizer of the identity matrix, $H_I = \{(h^{-T}, h) : h \in GL_m(q)\}$, and (h^{-T}, h) sends A to $h^{-1}Ah$. It is well known that $SL_m(q)$ is 2-generated, say $SL_m(q) = \langle C, D \rangle$. Then $H_{I,C,D} = 1$, proving the claim.

(5) Extending the previous example, it can be shown that if $V = V_a(q) \otimes V_b(q)$ with $a \leq b$, and $H = Cl_a(q) \otimes Cl_b(q^{1/r})$ acting naturally on V, then either $b(H)$ is bounded or $b(H) \sim b/ar$.

Here is our structure theorem, on which the proof of Theorem 4.3 is based. It is a simplified version of [14, Theorem 2].

Theorem 4.5 ([14]) *Suppose $H \leq GL_d(q)$ is primitive and absolutely irreducible. Then one of the following holds:*

(i) *$b(H) < c$ for some absolute constant c;*

(ii) *$H \leq GL_a(q) \otimes Cl_b(q^{1/r})$ $(d = ab)$, H contains the factor $Cl_b(q^{1/r})'$, and $b(H) \sim b/ar$;*

(iii) *$H \leq GL_a(q) \otimes S_{b+\delta}$ $(d = ab, \delta = 1$ or $2)$, H contains the factor $A_{b+\delta}$, and $b(H) \sim \log b/(a \log q)$.*

In view of the above results, to complete the proof of Pyber's conjecture it remains to handle the affine case where the linear group $H = G_0$ acts imprimitively on V.

References

[1] L. Babai, On the order of uniprimitive permutation groups, *Ann. of Math.* **113** (1981), 553–568.

[2] L. Babai, On the order of doubly transitive permutation groups, *Invent. Math.* **65** (1982), 473–484.

[3] L. Babai, P.J. Cameron and P. Pálfy, On the orders of primitive groups with restricted nonabelian composition factors, *J. Algebra* **79** (1982), 161–168.

[4] C. Benbenishty, Base sizes of standard actions of alternating and classical groups, to appear.

[5] A. Bochert, Über die Zahl verschiedener Werte, die eine Funktion gegebener Buchstaben durch Vertauschung derselben erlangen kann, *Math. Ann.* **33** (1889), 584-590.

[6] P.J. Cameron, Finite permutation groups and finite simple groups, *Bull. London Math. Soc.* **13** (1981), 1–22.

[7] P.J. Cameron, Some open problems on permutation groups, in *Groups, combinatorics and geometry* (eds. M. Liebeck and J. Saxl), pp.340–350, London Math. Soc. Lecture Note Ser., 165, Cambridge Univ. Press, Cambridge, 1992.

[8] P.J. Cameron and W.M. Kantor, Random permutations: some group-theoretic aspects, *Combin. Probab. Comput.* **2** (1993), 257–262.

[9] D. Gluck and K. Magaard, Base sizes and regular orbits for coprime affine permutation groups, *J. London Math. Soc.* **58** (1998), 603–618.

[10] D. Gluck, A. Seress and A. Shalev, Bases for primitive permutation groups and a conjecture of Babai, *J. Algebra* **199** (1998), 367-378.

[11] M.W. Liebeck, On minimal degrees and base sizes of primitive permutation groups, *Arch. Math.* **43** (1984), 11-15.

[12] M.W. Liebeck and J. Saxl, Maximal subgroups of finite simple groups and their automorphism groups, *Contemp. Math.* **131** (1992), 243-259.

[13] M.W. Liebeck and A. Shalev, Simple groups, permutation groups, and probability, *J. Amer. Math. Soc.* **12** (1999), 497-520.

[14] M.W. Liebeck and A. Shalev, Bases of primitive linear groups, *J. Algebra* **252** (2002), 95-113.

[15] P.P. Pálfy, A polynomial bound for the orders of primitive solvable groups, *J. Algebra* **77** (1982), 127-137.

[16] L. Pyber, On the orders of doubly transitive permutation groups, elementary estimates, *J. Combin. Theory Ser. A* **62** (1993), 361–366.

[17] L. Pyber, Asymptotic results for permutation groups, in *Groups and Computation* (eds. L. Finkelstein and W. Kantor), DIMACS Series on Discrete Math. and Theor. Comp. Science, Vol. 11, pp.197-219, Amer. Math. Soc., Providence, 1993.

[18] A. Seress, The minimal base size of primitive solvable permutation groups, *J. London Math. Soc.* **53** (1996), 243-255.

[19] A. Seress, Bases for non-affine primitive groups, to appear.

[20] T.R. Wolf, Solvable and nilpotent subgroups of $\mathrm{GL}(n, q^m)$, *Canad. J. Math.* **34** (1982), 1097–1111.

Finite groups of local characteristic p
An Overview

Ulrich Meierfrankenfeld, Bernd Stellmacher, Gernot Stroth

Contents

Introduction

Let p be a fixed prime and H be a finite group whose order is divisible by p. A *p-local subgroup* of H is a subgroup of the form $N_H(U)$, where U is a non-trivial p-subgroup of H.

We say that H has *characteristic* p if $C_H(O_p(H)) \leq O_p(H)$, where $O_p(H)$ is the largest normal p-subgroup of H. If all the p-local subgroup of H have characteristic p, we say that H has *local characteristic* p.

In this paper we describe the current status of a project whose goals are

- to understand the p-local structure of finite simple groups of local characteristic p, and
- to classify the finite simple groups of local characteristic 2.

The generic examples of groups of local characteristic p are the groups of Lie type defined over fields of characteristic p. Also some of the sporadic groups have local characteristic p, for example J_4, M_{24} and Th for $p = 2$, McL for $p = 3$, Ly for $p = 5$, and $O'N$ for $p = 7$.

But also every group with a self-centralizing cyclic Sylow p-subgroup, like $Alt(p)$, is of local characteristic p. These latter groups are particular examples of groups with a strongly p-embedded subgroup. Because of such groups we used the word "understand" rather than "classify" in the first item.

We hope to obtain information that allows to understand why, apart from groups with a strongly p-embedded subgroup, p-local subgroups of groups of local characteristic p look like those in the above examples.

For $p = 2$ Bender's fundamental classification of groups with a strongly 2-embedded subgroup puts us in a much better situation. In this case the information collected about the 2-local structure actually suffices to classify the finite simple groups of local characteristic 2. This then can be seen as part of a third generation proof of the classification of the finite simple groups.

At this point we should also justify another technical hypothesis we have not mentioned yet. We will assume that the simple sections (i.e., the composition factors of subgroups) of p-local subgroups are "known" simple groups, a property that surely holds in a minimal counterexample to the Classification Theorem of the finite simple groups.

One final word about a possible third generation proof of the classification and its relation to existing proofs. In 1954 R. Brauer [Br] suggested to classify the finite simple groups by the structure of the centralizers of their involutions. In principle the classification went this way, based on the epoch-making Theorem of Feit-Thompson [FeTh] which shows that every non-abelian finite simple group possesses involutions. Of course, a priori, there are as many centralizers as there are finite groups, so one of the main steps in the proof is to give additional information about the possible structure of centralizers of involutions in finite simple groups (this corresponds to the first item of our project).

In a given simple group the centralizers of involutions are particular 2-local subgroups, and there are basically two cases: Either every such centralizer has characteristic 2, in which case the group is of local characteristic 2, or this is not the case.

In the latter case, with a great amount of work, one can prove that there exists a centralizer of an involution that has a certain standard form. There is a well established machinery that can then be used to classify the corresponding groups.

The situation is more complicated if the simple group has local characteristic 2. The actual classification then works with a suitably chosen odd prime p and centralizers of elements of order p rather than involutions. For example, in the groups $L_n(2^m)$, which are of local characteristic 2, one would choose an element of order p in a standard torus, or an element of order 3 if $m = 1$. The idea is then to prove that there exists a p-element whose centralizer is again in some standard form. This needs very delicate signalizer functor and uniqueness group arguments, moreover, the classification of quasi-thin groups has to be done separately.

If successful, our classification of groups of local characteristic 2 would give an alternative proof that does not need the above described switch to another prime and also does not need the separate treatment of quasi-thin groups.

In fact, in view of the part of the classification that deals with groups that are not of local characteristic 2, it might be desirable to classify groups of parabolic characteristic 2 rather than of local characteristic 2. Here a *parabolic subgroup* of H is a subgroup of H which contains a Sylow p-subgroup of H. And H has *parabolic characteristic* p if all p-local, parabolic subgroups of H have characteristic p. The remaining simple groups would then have a 2-central involution whose centralizer is not of characteristic 2, a condition which seems to be fairly strong. We hope that our methods also work in the more general situation of groups of parabolic characteristic p, but have not spent much time on it.

Notation and Hypothesis

Let p be a fixed prime and H be a finite group whose order is divisible by p.

The largest normal p-subgroup of H, $O_p(H)$, is called the *p-radical* of H.

H is *p-minimal* if every Sylow p-subgroup S of H is contained in a unique maximal subgroup of H and $S \neq O_p(H)$. The p-minimal parabolic subgroups of H are called *minimal parabolic subgroups.*

If every simple section of H is a known finite simple group, then H is a $\mathcal{K}$-group. If every p-local subgroup of H is a $\mathcal{K}$-group, then H is a $\mathcal{K}_p$-group.

A proper subgroup M of H is called *strongly p-embedded* if p divides $|H|$, but does not divide $|H \cap H^g|$ for any $g \in G \setminus H$.

$F_p^*(H)$ is defined by $F_p^*(H)/O_p(H) = F^*(H/O_p(H))$.

For any set $\mathcal{T}$ of subgroups of H and $U \leq H$ we set

$$\mathcal{T}_U := \{T \in \mathcal{T} \mid T \leq U\} \text{ and } \mathcal{T}(U) := \{T \in \mathcal{T} \mid U \leq T\}.$$

We further set

$$\mathcal{L} := \{L \leq H \mid C_H(O_p(L)) \leq O_p(L)\} \text{ and } \mathcal{P} := \{P \in \mathcal{L} \mid P \text{ is } p\text{-minimal}\},$$

and denote the set of maximal elements of $\mathcal{L}$ by $\mathcal{M}$. Observe that in the case when H has local characteristic p and $S \in Syl_p(H)$,

$\mathcal{L}$ contains every p-local subgroup of H,

$\mathcal{M}$ is the set of maximal p-local subgroups of H,

$\mathcal{L}(S)$ is the set of parabolic subgroups containing S with a non-trivial p-radical,

$\mathcal{P}(S)$ is the set of p-minimal parabolic subgroups containing S with a non-trivial p-radical.

Let $1 = D_0 < D_2 < \dots D_{n-1} < D_n = H$ be a chief-series for H and put $V_i = D_i/D_{i-1}$. The *shape* of H is define to be the ordered tuple $(H/C_H(V_i), V_i)_{1 \leq i \leq n}$. Isomorphisms between the shapes of two groups are defined in the canonical way. Note that by the Jordan Hölder Theorem the shape of H is unique up to isomorphism. Abusing language we will say that two groups have the same shape if they have isomorphic shapes.

From now on we assume

Main Hypothesis *G is a finite $\mathcal{K}_p$-group of local characteristic p with trivial p-radical.*

In the following we will discuss the principal steps and subdivisions in the investigation of G. It splits into three major parts:

- Modules
- Local Analysis
- Global Analysis.

In the first part we collect information about pairs (H, V), where H is a finite $\mathcal{K}$-group and V is a faithful $\mathbb{F}_pH$-module fulfilling certain assumptions, like quadratic action or $2F$. The results of this part serve as an invaluable background for the local analysis.

The local analysis generates information about the structure of the p-local subgroups of G, and in the global analysis this information is used to identify G up to isomorphism.

1 The Modules

In this part we collect some theorems about finite groups and their $\mathbb{F}_p$-modules that are needed in the local analysis of groups of local characteristic p. Some of these theorems are known, others are not. Proofs for the theorems in this section will appear in [BBSM].

Let H be a finite group, V an $\mathbb{F}_pH$ module and $A \leq H$. We say that A acts *quadratically* on V if $[V, A, A] = 0$. Let i be a positive real number. We say that A is an *iF-offender* provided that $|V/C_V(A)| \leq |A/C_A(V)|^i$. A is an *offender* if A is an $1F$-offender. If in addition $[V, A] \neq 0$, A is called a *non-trivial iF-offender.* If there exists a non-trivial iF-offender in H then V is called an *iF-module.* An *FF-module* is a $1F$-module.

We say that H is a $\mathcal{CK}$-group if every composition-factor of H is isomorphic to one of the known finite simple groups.

1.1 Results

Theorem 1.1.1 (Quadratic Module Theorem) *Let H be a finite $\mathcal{CK}$-group with $F^*(H)$ quasi-simple, V be a faithful irreducible $\mathbb{F}_pH$-module and $A \leq G$ such that*

(i) $[V, A, A] = 0$.

(ii) $H = \langle A^H \rangle$.

(iii) $|A| > 2$.

Then one the following holds:

1. $p = 2$, $H \cong Alt(n)$ *or* $Sym(n)$ *and V is the natural module.*

2. $p = 2$, $H \cong Alt(n)$ *and V is the spin-module.*

3. $p = 2$, $H \cong 3.Alt(6)$, $Alt(7)$, $3.U_4(3)$, M_{12}, $\mathrm{Aut}(M_{12})$, $\mathrm{Aut}(M_{22})$, $3.M_{22}$, M_{24}, J_2, Co_1, Co_2 *or* $3.Sz$ *and V is known.*

4. $p = 2$, $H \cong O^{\pm}_{2n}(2)$ *and V is the natural module.*

5. $p = 3$, $H \cong 2.Alt(n)$ *and V is the spin module.*

6. $p = 3$, $H \cong PGU_n(2)$ *and V is the Weil-module.*

7. $p = 3$, $H \cong 2.Sp_6(2)$, $2.\Omega_8(2)$, $2.J_2$, $2.G_2(4)$, $2.Sz$, $2.Co_1$ *and V is known.*

8. $G = F^*(H) \cong {}^{\sigma}G_{\Phi}(\mathbb{F})$ *is a group of Lie type over the field $\mathbb{F}$ with* $\mathrm{char}\,\mathbb{F} = p$. *Moreover, if $|A| > |\mathbb{F}|$ or if there exists a root subgroup R of H with $A \cap R \neq 1$ and $A \not\leq R$, then $V = V(\lambda_i)$ where λ_i is a fundamental weight with $\lambda_i(\alpha) = 1$ for the highest long root $\alpha \in \Phi$.*

Theorem 1.1.2 (FF-Module Theorem) *Let H be a finite $\mathcal{CK}$-group and V a faithful, irreducible FF-module for H over $\mathbb{F}_p$. Suppose that $F^*(H)$ is quasi-simple and that H is generated by the quadratic offenders on V. Then one of the following holds (where q is a power of p):*

1. $H \cong SL_n(q)$, $n \geq 2$; $Sp_{2n}(q)$, $n \geq 2$; $SU_n(q)$, $n \geq 4$; $\Omega^+_{2n}(q)$, $n \geq 3$; $\Omega^-_{2n}(q)$, $n \geq 4$; *or* $\Omega_n(q)$, $n \geq 7$, *n and q odd; and V is the corresponding natural module.*

2. *$H \cong SL_n(q)$, $n \geq 3$ and V is the exterior square of a natural module.*
3. *$H \cong \Omega_7(q)$, and V is the spin-module.*
4. *$H \cong \Omega_{10}^{+}(q)$ and V is one of the two half-spin modules.*
5. *$H \cong O_{2n}^{\pm}(q)$, $p = 2$, $n \geq 3$ and V is the natural module.*
6. *$H \cong G_2(q)$, $p = 2$ and $|V| = q^6$.*
7. *$H \cong Alt(n)$ or $Sym(n)$, $p = 2$ and V is the natural module.*
8. *$H \cong Alt(7)$, $p = 2$ and $|V| = 2^4$.*
9. *$H \cong 3.Alt(6)$, $p = 2$ and $|V| = 2^6$.*

Let V be an $\mathbb{F}_pH$-module and $S \in Syl_p(H)$. The group $O^{p'}(C_H(C_V(S)))$ is called a point stabilizer for H on V. V is called p-reduced if $O_p(H/C_H(V)) = 1$.

Lemma 1.1.3 (Point Stabilizer Theorem) *Let H be a finite $\mathcal{CK}$- group, V a $\mathbb{F}_pH$-module, L a point stabilizer for H on V and $A \leq O_p(L)$.*

(a) *If V is p-reduced, then $|V/C_V(A)| \geq |A/C_A(V)|$.*

(b) *Suppose V is faithful and irreducible for H, $F^*(H)$ is quasi-simple, $H = \langle A^H \rangle$ and A is a non-trivial offender on V. Then $H \cong SL_n(q)$, $Sp_{2n}(q)$, $G_2(q)$ or $Sym(n)$, where $p = 2$ in the last two cases, $n = 2, 3 \mod (4)$ in the last case, and q is a power of p. Moreover, V is the corresponding natural module.*

Theorem 1.1.4 *Let H be a finite $\mathcal{CK}$-group with $F^*(H)$ quasi-simple. Let V be a faithful irreducible $\mathbb{F}_pH$ module. Suppose there exists $1 \neq A \leq T \in \mathrm{Syl}_p(H)$ such that $|A| > 2$ and $\langle A^L \rangle$ acts quadratically on V for all $S \leq L < H$. Then*

(a) *$F^*(H)A \cong SL_n(q), Sp_{2n}(q), SU_n(q), G_2(q)'$ or $Sz(q)$, where $p = 2$ in the last two cases.*

(b) *Let I be an irreducible $F^*(H)A$ submodule of V. Then one of the following holds:*

   1. *I is a natural module for $F^*(H)A$.*
   2. *$p = 2$, $F^*(H)A \cong L_3(q)$, H induces a graph automorphism on $F^*(H)$ and I is the adjoint module.*
   3. *$p = 2$, $F^*(H)(A) \cong Sp_6(q)$ and I is the spin-module.*

(c) *Either A is contained in a long root subgroup of $F^*(H)A$, or $p = 2$, $F^*(H)A \cong Sp_4(q)$, $A \leq Z(S \cap F^*(H)A)$ and H induces a graph automorphism on $F^*(H)$.*

The information given in the above theorem can be used to prove the following corollary, which is of great help in the local analysis.

Corollary 1.1.5 (Strong L-Lemma) *Let L be a finite $\mathcal{CK}$-group with $O_p(L) = 1$ and V a faithful $\mathbb{F}_pL$-module. Suppose that there exists $1 \neq A \leq S \in Syl_p(L)$ such that*

(∗) *$\langle A^P \rangle$ acts quadratically on V for every proper subgroup $P < L$ satisfying $A \leq P$ and $S \cap P \in Syl_p(P)$.*

Then

(a) *$L \cong SL_2(p^m)$, $Sz(2^m)$ or D_{2r}, where $p = 2$ in the last two cases and r is an odd prime.*

(b) *$[V, L]C_V(L)/C_V(L)$ is a direct sum of natural modules for L.*

Let H be a finite group, V a $\mathbb{F}_pH$-module and $A \leq H$. We say that A is *cubic* on V if $[V, A, A, A] = 0$. We say that V is a *cubic $2F$-module* if H contains a non-trivial cubic $2F$-offender. The following theorem is due to R. Guralnick and G. Malle [GM]:

Theorem 1.1.6 (The $2F$-Module Theorem,I) *Let H be a finite $\mathcal{CK}$-group and V a faithful irreducible cubic $2F$-module for H. Suppose that $F^*(H)$ is quasi-simple, but $F^*(H)$ is not a group of Lie-type in characteristic p. Then one of the following holds:*

1. *$F^*(H)/Z(F^*(H)) \cong Alt(n)$, $p = 2$ or 3 and one of the following holds.*
   1. *V is the natural module.*
   2. *$H \cong Alt(n)$, $p = 2$, $n = 7$ or 9 and V is a half-spin module.*
   3. *$H \cong Sym(7)$, $p = 2$ and V is the spin-module.*
   4. *$F^*(H) \cong 2.Alt(5)$, $p = 3$ and V is the half spin module.*
   5. *$F^*(H) \cong 3.Alt(6)$ and $|V| = 2^6$.*
2. *$F^*(H) \cong G_2(2)'$, $p = 2$ and $|V| = 2^6$.*
3. *$F^*(H) \cong 3.U_4(3)$, $p = 2$ and $|V| = 2^{12}$.*
4. *$F^*(H) \cong 2.L_3(4)$, $p = 3$ and $|V| = 3^6$.*
5. *$F^*(H) \cong Sp_6(2)$, $p = 3$ and $|V| = 3^7$.*
6. *$F^*(H) \cong 2.Sp_6(2)$, $p = 3$ and $|V| = 3^8$.*
7. *$F^*(H) \cong 2.\Omega_8^+(2)$, $p = 3$ and $|V| = 3^8$.*
8. *$F^*(H) \cong M_{12}, M_{22}, M_{23}, M_{24}$, $p = 2$ and V is a non-trivial composition factor of dimension $10, 10, 11, 11$ resp. of the natural permutation module.*
9. *$F^*(H) = 3.M_{22}$, $p = 2$ and $|V| = 2^{12}$.*
10. *$F^*(H) = J_2$, $p = 2$ and V is the 12-dimensional module which arises from the embedding into $G_2(4)$.*

11. $F^*(H) \cong Co_2$ *or* Co_1, $p = 2$ *and* V *is 22– resp. 24-dimensional module arising from the Leech Lattice*

12. $F^*(H) \cong M_{11}$ *or* $2.M_{12}$, $p = 3$ *and* $|V| = 3^5$ *and* 3^6 *respectively.*

It is not known whether Case 11 in the preceding theorem really occurs. We tend to believe it does not.

1.2 An Example

To get an idea how these theorems are used in the local analysis we now discuss briefly a particular but fairly general situation.

Let G be as in the Main Hypothesis, that is a finite $\mathcal{K}_p$-group of local characteristic p with trivial p-radical. Fix $S \in Syl_p(G)$ and put $Z := \Omega_1 Z(S)$. Let $M_1, M_2 \in \mathcal{L}(S)$ and put $F_i = F_p^*(M_i)$. Suppose that

(i) $F_i/O_p(F_i)$ is quasisimple, $i = 1, 2$,

(ii) $O_p(\langle F_1, F_2\rangle) = 1$,

(iii) $M_i = SF_i$, $i = 1, 2$.

Let $Z_i := \langle Z^{M_i}\rangle$ and $V_i := \langle Z_j^{M_i}\rangle$ for $i \neq j$. Note first that $Z_j \leq Z(O_p(M_j)) \leq S \leq M_i$, so Z_i and V_i are normal subgroups of M_i.

As an elementary consequence of (i) we get:

(1) Let $U \leq M_i$ and $F_i \leq N_{M_i}(U)$. Then either $F_i \leq U$, or $U \cap O_p(M_i) \in Syl_p(U)$.

This property (1) together with (iii) applied to $U = C_{M_i}(F_i/O_p(M_i))$ and $U = C_{M_i}(V)$, V a non-central M_i-chief factor in $O_p(M_i)$, gives:

(2) Suppose that V is a non-central M_i-chief factor in $O_p(M_i)$. Then

$$C_S(V) = O_p(M_i) = C_S(F_i/O_p(M_i)).$$

Next we show that one of the following cases holds:

(I) There exist $g \in G$ and $i \in \{1, 2\}$, say $i = 1$, such that

$$[Z_1, Z_1^g] \neq 1,\ [Z_1, Z_1^g] \leq Z_1 \cap Z_1^g \text{ and } Z_1 Z_1^g \leq M_1 \cap M_1^g.$$

(II) There exists an $i \in \{1, 2\}$, say $i = 1$, such that $Z_1 \not\leq O_p(M_2)$, and (I) does not hold.

(III) V_1 and V_2 are elementary abelian, and (I) does not hold.

To see this, assume that (I) and (II) do not hold. Then $V_1V_2 \leq O_p(M_1) \cap O_p(M_2)$, and either (III) holds, or for some $i \in \{1,2\}$, say $i = 2$, V_2 is not abelian. In the latter case, there exists $g \in M_2$ such that $[Z_1, Z_1^g] \neq 1$. Since $\langle Z_1, Z_1^g \rangle \leq O_p(M_2) \leq M_1 \cap M_1^g$ we also have $[Z_1, Z_1^g] \leq Z_1 \cap Z_1^g$. This gives (I) contrary to our assumption. We now discuss these three cases separately.

Assume case (I). We can choose the notation such that

$$|Z_1^g / C_{Z_1^g}(Z_1)| \geq |Z_1 / C_{Z_1}(Z_1^g)|,$$

so Z_1^g is a quadratic offender on Z_1.

Clearly $[Z_1, M_1] \neq 1$ since $[Z_1, Z_1^g] \neq 1$, so the definition of Z_1 implies that $M_1 \neq C_{M_1}(Z_1)S$. Thus $[Z_1, O^p(M_1)] \neq 1$ and there exists a non-central M_1-chief factor $V = U/W$ of Z_1. From (2) we conclude that $C_{Z_1^g}(Z_1) = C_{Z_1^g}(V) = Z_1^g \cap O_p(M_1)$. It follows that

$$|V/C_V(Z_1^g)| \leq |U/C_U(Z_1^g)| \leq |Z_1/C_{Z_1}(Z_1^g)| \leq |Z_1^g/C_{Z_1^g}(Z_1)| = |Z_1^g/C_{Z_1^g}(V)|.$$

Hence Z_1^g is a non-trivial quadratic offender on V, and the FF-Module Theorem gives the structure of $F_1/O_p(M_1)$ and V.

Assume case (II). Then

$$[O_p(M_2), Z_1, Z_1] \leq [O_p(M_2) \cap Z_1, Z_1] = 1,$$

so Z_1 is quadratic on every M_2-chief factor V of $O_p(M_2)$. Hence (unless it is the case that $|Z_1 O_p(M_1)/O_p(M_1)| = 2$), the Quadratic Module Theorem applies to $M_2/C_{M_2}(V)$ and $A = Z_1 C_{M_2}(V)/C_{M_2}(V)$. But in this case one also gets information about M_1:

Among all subgroups $U \leq M_2$ with $Z_1 \leq U$, $U \cap S \in Syl_p(U)$ and $Z_1 \not\leq O_p(U)$ choose U minimal and set $\overline{U} = U/O_p(U)$. Then for every proper subgroup $O_p(U) \leq P < U$ with $S \cap P \in Syl_p(P)$ and $Z_1 \leq P$ we get that $Z_1 \leq O_p(P)$. But this implies, since we are not in case (I), that $X := \langle Z_1^P \rangle$ is abelian. Hence as above, since X is normal in $S \cap P$, $[O_p(P), X, X] = 1$. This shows that the Strong L-Lemma 1.1.5 applies with $L = \overline{U}$, $V = O_p(U)/\Phi(O_p(U))$ and $A = \overline{Z}_1$.

Set $\widetilde{B} := Z_1 \cap O_p(U)$ and $B := \widetilde{B}^x$ for some $x \in U \setminus N_U(S \cap U)$. Note that $U = \langle Z_1, Z_1^x \rangle$ and so U normalizes $\widetilde{B}B$ and $\widetilde{B} \cap B \leq Z(U)$. By 1.1.5, $|Z_1/\widetilde{B}| \leq |B/\widetilde{B} \cap B|$ and $C_B(y) = \widetilde{B} \cap B$ for every $y \in Z_1 \setminus \widetilde{B}$. It follows that

$$|Z_1/C_{Z_1}(B) \leq |Z_1/\widetilde{B} \cap B| = |Z_1/\widetilde{B}||\widetilde{B}/\widetilde{B} \cap B| = |Z_1/\widetilde{B}||B/\widetilde{B} \cap B| \leq |B/\widetilde{B} \cap B|^2,$$

so $B/\widetilde{B} \cap B$ is a $2F$-offender on Z_1. Using (ii) we see that the $2F$-Module Theorem applies to $M_1/C_{M_1}(Z_1)$ and a non-central M_1-chief factor of Z_1.

Assume case (III). Note that $\langle C_{M_i}(V_i), M_i \rangle \leq N_G(Z_i)$ and so by condition (ii)

$$F_i \not\leq C_{M_i}(V_i) \text{ for every } i \in I.$$

In particular by (1) $C_S(V_i) \leq O_p(M_i)$ for every $i \in I$. Since by property (ii) $\mathrm{J}(S) \not\leq O_p(M_1) \cap O_p(M_2)$ we may assume that

(**) $F_1 \not\leq C_{M_1}(V_1)$ and $J(S) \not\leq C_{M_1}(V_1)$.

Let D be the inverse image of $O_p(M_1/C_{M_1}(V_1))$. Pick $A \in \mathcal{A}(S)$ such that $A \not\leq C_{M_1}(V_1)$. According to the Thompson Replacement Theorem we may assume that A acts quadratically on V_1. The maximality of A gives

$$|V_1||C_A(V_1)||V_1 \cap A|^{-1} = |V_1 C_A(V_1)| \leq |A|$$

and thus $|V_1/C_{V_1}(A)| = |V_1/V_1 \cap A| \leq |A/C_A(V_1)|$, so A is a quadratic offender on V_1. This looks promising, but $A_0 := A \cap D$ might not centralize V_1. This is an obstacle for the application of the FF-Module Theorem to $F_1 A$ and non-central $F_1 A$-chief factors of V_1.

Evidently $|A_0/C_A(V_1)| \leq |A/A_0|$ or $|A/A_0| \leq |A_0/C_A(V_1)|$. In the first case

$$|V_1/C_{V_1}(A_0)| \leq |V_1/V_1 \cap A| \leq |A/C_A(V_1)| = |A/A_0||A_0/C_A(V_1)| \leq |A/A_0|^2,$$

so in this case, using again (2), A/A_0 is a quadratic 2F-offender on the non-central $F_1 A$-chief factors of V_1.

In the second case

$$|V_1/C_{V_1}(A_0)| \leq |V_1/V_1 \cap A| \leq |A/C_A(V_1)| = |A/A_0||A_0/C_{A_0}(V_1)| \leq |A_0/C_{A_0}(V_1)|^2,$$

so A_0 is a quadratic 2F-offender on V_1. An elementary calculation then shows that there exists a quadratic 2F-offender on Z_2.

This concludes the discussion of the cases (I) – (III). In all cases the module theorems from 1.1 reveal the structure of $F_1/O_p(F_1)$ or $F_2/O_p(F_2)$.

2 The Local Analysis

In this part we discuss the p-local structure of G, where G is, according to our Main Hypothesis, a finite $\mathcal{K}_p$-group of local characteristic p with trivial p-radical. We fix

$$S \in Syl_p(G), \ Z := \Omega_1 Z(S).$$

For further notation see the introduction.

The basic idea is to study the structure of $L \in \mathcal{L}$ by its action on elementary abelian normal subgroups contained in $Z(O_p(L))$ and by its interaction with other elements of $\mathcal{L}$ having a common Sylow p-subgroup. It is here where the module results of Part 1 are used.

The appropriate candidates for such normal subgroups in $Z(O_p(L))$ are the *p-reduced* normal subgroups, i.e. elementary abelian normal subgroups V of L with $O_p(L/C_L(V)) = 1$. Note that an elementary abelian normal subgroup V is p-reduced iff any subnormal subgroup of L that acts unipotently on V already centralizes V. Here are the basic properties of p-reduced normal subgroups. They include the fact that there exists a unique maximal p-reduced normal subgroup of L which we always denote by Y_L.

Lemma 2.0.1 *Let L be a finite group of characteristic p and $T \in \mathrm{Syl}_p(L)$.*

(a) *There exists a unique maximal p-reduced normal subgroup Y_L of L.*

(b) *Let $T \le R \le L$ and X a p-reduced normal subgroup of R. Then $\langle X^L \rangle$ is a p-reduced normal subgroup of L. In particular, $Y_R \le Y_L$.*

(c) *Let $T_L = C_T(Y_L)$ and $L_T = N_L(T_L)$. Then $L = L_T C_L(Y_L)$, $T_L = O_p(L_T)$ and $Y_L = \Omega_1 Z(T_L)$.*

(d) $Y_T = \Omega_1 \mathrm{Z}(T)$, $Z_L := \langle \Omega_1 \mathrm{Z}(T)^L \rangle$ *is p-reduced for L and $\Omega_1 \mathrm{Z}(T) \le Z_L \le Y_L$.*

(e) *Let V be p-reduced normal subgroup of L and K a subnormal subgroup of L. Then $[V, O^p(K)]$ is a p-reduced normal subgroup of K.*

Of course, the action of L on Y_L might be trivial, whence $Y_L = \Omega_1 Z(T)$, $T \in$ *$Syl_p(L)$*. This leads to another notation. Let H be any finite group and $T \in$ *$Syl_p(H)$*. Then $P_H(T) := O^{p'}(C_H(\Omega_1 Z(T)))$ is called a *point stabilizer* of H. In the above situation trivial action on Y_L implies that $O^{p'}(L) = P_L(T)$. Here are some basic (but not entirely elementary) facts about point stabilizers.

Lemma 2.0.2 *Let H be a finite group of local characteristic p, $T \in \mathrm{Syl}_p(H)$ and L a subnormal subgroup of H. Then*

(a) (Kieler Lemma) $C_L(\Omega_1 \mathrm{Z}(T)) = C_L(\Omega_1 \mathrm{Z}(T \cap L))$

(b) $P_L(T \cap L) = O^{p'}(P_H(T) \cap L)$

(c) $C_L(Y_L) = C_L(Y_H)$

(d) *Suppose $L = \langle L_1, L_2 \rangle$ for some subnormal subgroups L_1, L_2 of H. Then*

  (da) $P_L(T \cap L) = \langle P_{L_1}(T \cap L_1), P_{L_2}(T \cap L_2) \rangle$.

  (db) *For $i = 1, 2$ let P_i be a point stabilizer of L_i. Then $\langle P_1, P_2 \rangle$ contains a point stabilizer of L.*

It is evident that all elements of $\mathcal{L}(S)$ having a normal point stabilizer are contained in $N_G(Z)$. Therefore, controlling $N_G(Z)$, or better a maximal p-local subgroup containing $N_G(Z)$, means controlling all elements of $L \in \mathcal{L}(S)$ with trivial action on Y_L. This point of view leads to the next definition and subdivision.

Let $\widetilde{C}$ be a fixed maximal p-local subgroup of G containing $N_G(Z)$. Put

$$E := O^p(F_p^*(C_{\widetilde{C}}(Y_{\widetilde{C}}))), \quad Q := O_p(\widetilde{C})$$

The major subdivision is:

Non-E-Uniqueness $(\neg E!)$: E is contained in at least two maximal p-local subgroups of G.

E-Uniqueness $(E!)$: $\widetilde{C}$ is the unique maximal p-local subgroup containing E.

Another subdivision refers to the rank of G. Define the *rank* of G to be the minimal size of a non-empty subset Σ of $\mathcal{P}(S)$ with $\langle \Sigma \rangle \notin \mathcal{L}$. If no such subset exists we define the rank to be 1. Note that $\operatorname{rank} G = 1$ if and only if $|\mathcal{M}(S)| = 1$. The cases $\operatorname{rank} G = 1$ and $\operatorname{rank} G \geq 2$ are treated separately, so in the $E!$-case we will assume, in addition, that G has rank at least 2.

The subgroup $\langle \mathcal{M}(S) \rangle$ is called the *p-core* of G (with respect to S). Note that G has a proper p-core if G has rank 1, so the rank 1 case can be treated in this more general context.

2.1 Pushing Up

At various times in the local analysis we encounter a p-local subgroup L of G and a parabolic subgroup H of L such that $N_G(O_p(H))$ and L are not contained in a common p-local subgroup of G. In other words $O_p(\langle L, N_G(O_p(H)) \rangle) = 1$. In this section we provide theorems that allow us, under additional hypotheses, to determine the shape of L.

For a p-group R we let $\mathcal{PU}_1(R)$ be the class of all finite $\mathcal{CK}$-groups L containing R such

(a) L is of characteristic p,

(b) $R = O_p(N_L(R))$

(c) $N_L(R)$ contains a point stabilizer of L.

Let $\mathcal{PU}_2(R)$ be the class of all finite $\mathcal{CK}$- groups L containing R such that L is of characteristic p and

$$L = \langle N_L(R), H \mid R \leq H \leq L, H \in \mathcal{PU}_1(R) \rangle.$$

Let $\mathcal{PU}_3(R)$ be the class of all finite $\mathcal{CK}$-groups L such that

(a) L is of characteristic p.

(b) $R \leq L$ and $L = \langle R^L \rangle$

(c) $L/C_L(Y_L) \cong SL_n(q), Sp_{2n}(q)$ or $G_2(q)$, where q is a power of p and $p = 2$ in the last case.

(d) $Y_L/C_{Y_L}(L)$ is the corresponding natural module.

(e) $O_p(L) < R$ and $N_L(R)$ contains a point stabilizer of L.

(f) If $L/C_L(Y_L) \ncong G_2(q)$ then $R = O_p(N_L(R))$.

Let $\mathcal{PU}_4(R)$ be the class of all finite $\mathcal{CK}$- groups L containing R such that L is of characteristic p and

$$L = \langle N_L(R), H \mid R \leq H \leq L, H \in \mathcal{PU}_3(R) \rangle.$$

For a finite p-group T let $\mathcal{A}(T)$ be the set of elementary abelian subgroups of maximal order in T, $\mathrm{J}(T) = \langle \mathcal{A}(T) \rangle$, the *Thompson subgroup* of T, and $\mathrm{B}(T) = C_R(\Omega_1 \mathrm{Z}(\mathrm{J}(T)))$, the *Baumann subgroup* of T. Recall that a finite group F is p-closed if $O^{p'}(F) = O_p(F)$. The following lemma is a generalization of a well known lemma of Baumann, also the proof is similar to Baumann's.

Lemma 2.1.1 (Baumann Argument) *Let L be a finite group, R a p-subgroup of L, $V := \Omega_1 \mathrm{Z}(O_p(L))$, $K := \langle \mathrm{B}(R)^L \rangle$, $\tilde{V} = V/C_V(O^p(K))$, and suppose that each of the following holds:*

(i) *$O_p(L) \leq R$ and $L = \langle \mathrm{J}(R)^L \rangle N_L(\mathrm{J}(R))$.*

(ii) *$C_K(\tilde{V})$ is p-closed.*

(iii) *$|\tilde{V}/C_{\tilde{V}}(A)| \geq |A/C_A(\tilde{V})|$ for all elementary abelian subgroups A of R.*

(iv) *If U is an FF-module for $L/O_p(L)$ with $\tilde{V} \leq U$ and $U = C_U(B(R))\tilde{V}$, then $U = C_U(O^p(K))\tilde{V}$.*

Then $O_p(K) \leq \mathrm{B}(R)$.

Using the Point Stabilizer Theorem 1.1.3 and the Baumann Argument 2.1.1 one can prove

Lemma 2.1.2 *Let R be a p-group. Then $\mathcal{PU}_2(R) \subseteq \mathcal{PU}_4(\mathrm{B}(R))$.*

Similarly,

Lemma 2.1.3 *Let L be a finite p-minimal $\mathcal{CK}$- group of characteristic p. Let $T \in$ $\mathrm{Syl}_p(L)$. Then either L centralizes $\Omega_1 \mathrm{Z}(T)$ (and so $P_L(T)$ is normal in L) or $L \in \mathcal{PU}_4(\mathrm{B}(T))$.*

If R is a group and Σ is a set of groups containing R we define

$$O_R(\Sigma) = \langle T \leq R \mid T \trianglelefteq L,\ \forall L \in \Sigma \rangle$$

So $O_R(\Sigma)$ is the largest subgroup of R which is normal in all $L \in \Sigma$.

Theorem 2.1.4 *Let R be a finite p-group with $R = \mathrm{B}(R)$ and Σ a subset of $\mathcal{PU}_3(R)$. Suppose $O_R(\Sigma) = 1$. Then there exists $L \in \Sigma$ such that $O^p(L)$ has one of the following shapes: (where q is a power of p.)*

1. $q^n SL_n(q)'$;
2. $q^{2n} Sp_{2n}(q)', p$ *odd;*
3. $q^{1+2n} Sp_{2n}(q)', p = 2$;
3. $2^6 G_2(2)', p = 2$;

4. $q^{1+6+8}Sp_6(q), p = 2$;

5. $2^{1+4+6}L_4(2), p = 2$; *or*

6. $q^{1+2+2}SL_2(q)', p = 3$.

Examples for above configurations can be found in $SL_{n+1}(q)$, $L_{p^n}(r)$(with $q = p \mid r-1$), $Sp_{2n+2}(q)$, Ru, $F_4(q)$, Co_2 and $G_2(q)$, respectively.

We are currently working on determining the shapes of all $L \in \Sigma$, not only of one. We expect all elements $L \in \Sigma$ to have one of the structures of the previous theorem, except for one additional possibility namely $L/O_p(L) \cong SL_2(q)$ and all non-central chief-factors for L on $O_p(L)$ are natural. For a given R, the number of such chief-factors is bounded. But as R varies it cannot be bounded.

About the proof: Using elements $A \in \mathcal{A}(R)$ and their interaction with the Y_L's, $L \in \Sigma$ one shows that there exist $L, M \in \Sigma$ such that $\langle Y_L^M \rangle$ is not abelian. The fact that $\langle Y_L^M \rangle$ is not abelian allows us to pin down the structure of L and M. (Compare this with the cases (I) and (II) in 1.2).

Theorem 2.1.5 (The Pushing Up Theorem) *Let R be a finite p-group, $1 \leq i \leq 4$, and Σ a subset of $\mathcal{PU}_i(R)$ with $O_R(\Sigma) = 1$. If $i = 3$ or 4 suppose that $R = \mathrm{B}(R)$. Then the shape of $\langle \mathrm{B}(R)^L \rangle$ will be known for all $L \in \Sigma$.*

Given 2.1.2, the Pushing Up Theorem should be a straightforward but tedious consequence of 2.1.4. The details still need to be worked out.

2.2 Groups with a Proper p-Core

Recall from the introduction that a proper subgroup $M < G$ is strongly p-embedded if M is not a p'-group but $M \cap M^g$ is a p'-group for every $g \in G \setminus M$. The following lemma is well known and elementary to prove:

Lemma 2.2.1 *Let H be a finite group, T a Sylow p-subgroup of H and M a proper subgroup of H with p dividing $|M|$. Put $K := \langle N_G(A) \mid 1 \neq A \leq T \rangle$. Then M is strongly p-embedded iff $N_G(A) \leq M$ for all non-trivial p-subgroups A of M and iff M contains a conjugate of K. In particular, H has a strongly p-embedded subgroup if and only if p divides $|H|$ and K is a proper subgroup of M.*

Note that the group K from the preceding lemma contains the p-core of H with respect to T. Thus if our G has a strongly p-embedded subgroup then G also has a proper p-core. We say that G satisfies CGT if G has proper p-core but no strongly p-embedded subgroups.

2.2.1 Strongly p-embedded subgroups

Suppose that G has a strongly p-embedded subgroup. If $p = 2$, we can apply Bender's theorem [Be]:

Theorem 2.2.2 (Bender) *Let H be a finite group with a strongly 2-embedded subgroup. Then one of the following holds:*

1. *Let t be an involution in H. Then $H = O(H)C_H(t)$ and t is the unique involution in $C_H(t)$.*

2. $O^{2'}(H/O(H)) \cong L_2(2^k), U_3(2^k)$ *or* $Sz(2^k)$.

If $p \neq 2$ we end our analysis without a clue.

2.2.2 CGT

Suppose that G satisfies CGT. Let $M := \langle \mathcal{M}(S) \rangle$ be the p-core with respect to S. According to CGT, M is a proper subgroup of G, but M is not strongly p-embedded. Thus there exists $g \in G \setminus M$ such that $|M \cap M^g|_p \neq 1$. Evidently we can choose g such that $M \cap S^g$ is a Sylow p-subgroup of $M \cap M^g$. Thus $|S^g \cap M|_p \neq 1$. If $S^g \leq M$, then $S^{gm} = S$ for some $m \in M$. Since $N_G(S) \leq M$ we obtain the contradiction to $g \notin M$. Thus $S^g \not\leq M$. Also $S^g \in \mathcal{L}$.

Among all $L \in \mathcal{L}$ satisfying $L \not\leq M$ we choose L such that $|L \cap M|_p$ is maximal. Then $|L \cap M|_p \geq |S^g \cap M|_p \neq 1$. Let $T \in Syl_p(L \cap M)$ and without loss $T \leq S$.

If $T = S$ we get that $L \in \mathcal{L}(S)$ and so by the definition of M, $L \leq M$, a contradiction. Thus $T \neq S$. Let C be a non-trivial characteristic subgroup of T. Then $N_S(T) \leq N_G(C)$ and so $|M \cap N_G(C)|_p > |M \cap L|$. Hence the maximal choice of $|M \cap L|_p$ implies $N_G(C) \leq M$. In particular, $N_L(C) \leq M \cap L$. For $C = T$ we conclude that $T \in \mathrm{Syl}_p(L)$. We can now apply the following theorem with L in place of H:

Theorem 2.2.3 (Local C(G,T)-Theorem) *Let H be a finite $\mathcal{K}_p$-group of characteristic p, T a Sylow p-subgroup of H, and suppose that*

$$C(H,T) := \langle N_H(C) \mid 1 \neq C \text{ a characteristic subgroup of } T \rangle$$

is a proper subgroup of H. Then there exists an H-invariant set $\mathcal{D}$ of subnormal subgroups of H such that

(a) $H = \langle \mathcal{D} \rangle C(H,T)$

(b) $[D_1, D_2] = 1$ *for all* $D_1 \neq D_2 \in \mathcal{D}$.

(c) *Let $D \in \mathcal{D}$, then $D \not\leq C(H,T)$ and one of the following holds:*

   1. *$D/Z(D)$ is the semi-direct product of $SL_2(p^k)$ with a natural module for $SL_2(p^k)$. Moreover $O_p(D) = [O_p(D), D]$ is elementary abelian.*
   2. *$p = 2$ and D is the semi-direct product of $Sym(2^k + 1)$ with a natural module for $Sym(2^k + 1)$.*
   3. *$p = 3$, D is the semi-direct product of $O_3(D)$ and $SL_2(3^k)$, $\Phi(D) = Z(D) \leq O_3(D)$ has order 3^k, and both $[Z(O_3(D)), D]$ and $O_3(D)/Z(O_3(D))$ are natural modules for $D/O_p(D)$.*

For $p = 2$ the local $C(G,T)$-theorem was proved by Aschbacher in [Asch] without using the $\mathcal{K}_2$-hypothesis. For us it will be a consequence of 2.1.5. Using the local $C(G,T)$ theorem and that G is of local characteristic p it is not difficult to show:

Theorem 2.2.4 *Suppose that G fulfills CGT. Let M be a p-core for G and $L \in \mathcal{L}$ such that $|L \cap M|_p$ is maximal with respect to $L \not\leq M$. Then there exists a normal subgroup D of L such that $D/Z(D) \cong q^2 SL_2(q)$ and $C_L(D) \leq O_p(L)$.*

Using the preceding theorem, A. Hirn is currently trying to show that for $p = 2$, G cannot fulfill CGT.

2.3 ¬E!

In this section we assume that we are in the ¬E!-case, so E is contained in $\widetilde{C}$ and at least one other maximal p-local subgroup of G. To illustrate this situation we look at a few examples.

Let $p = 2$, $q = 2^k$ and $G = F_4(q)\langle\sigma\rangle$ where σ induces a graph automorphism of order 2. (Yes, G is not of local characteristic 2, only of parabolic characteristic 2. But as we mostly look at subgroups containing a Sylow p-subgroup, or at least a large part of the Sylow p-subgroup, it is difficult for us to detect that G is not of local characteristic p.)

Note that G' is a group of Lie-type with Dynkin-diagram

Also S is only contained in two parabolic subgroups, namely the σ-invariant B_2- and $A_1 \times A_1$-parabolic. Trying to treat this amalgam would not be easy. To determine E, note that $Z(S)$ has order q and is contained in the product of the highest long root group and the highest short root group. It follows that $E \leq G'$ and E is essentially the B_2-parabolic. So E is contained in the B_3- and C_3-parabolic, and G will be identified by the $(Sp_6(q), Sp_6(q))$-amalgam.

As a second example consider $G = E_8(q) \wr Sym(p^k)$ (Again this is a group of parabolic characteristic p, but not of local characteristic p.) Here E helps us to find p-local subgroups which are not of characteristic p. Let H be the normalizer of a root subgroup in $E_8(q)$, i.e. the E_7-parabolic. Then $\widetilde{C}$ is $H \wr Sym(p^k)$, and E is essentially a direct product of p^k copies of H. Hence, E is contained in a p-local subgroup L which is a direct product of $p^k - 1$ copies of H and $E_8(q)$, so L is not of characteristic p.

As a final example consider $p = 2$ and $G = M_{24}$. Then $\widetilde{C} = 2^4 L_4(2)$ and so $C_{\widetilde{C}}(Y_{\widetilde{C}}) = O_2(\widetilde{C})$ and $E = 1$. It seems that E is not of much use in this case, but $E = 1$ can only occur if $\widetilde{C}/O_p(\widetilde{C})$ acts faithfully on $Y_{\widetilde{C}}$. Together with the fact that $\widetilde{C}$ contains $N_G(Z)$, the $E = 1$-situation can be handled with the amalgam method.

To summarize, the $\neg E!$-case detects situations which allow a treatment via the amalgam method. The general idea is to find a p-subgroup R and a set Σ of subgroups of G containing R such that we can apply the Pushing Up Theorem 2.1.5 to (R, Σ).

To get started we choose a subgroup X of $\widetilde{C}$ such that X is the point stabilizer of some subnormal subgroup $\widetilde{X}$ of $\widetilde{C}$ and such that X is maximal with respect to $\mathcal{M}(EX) \neq \{\widetilde{C}\}$. By assumption $\mathcal{M}(E) \neq \widetilde{C}$ so such a choice is possible. For $L \in \mathcal{L}(EX)$ let $S_{\widetilde{C}}(L)$ be the largest subnormal subgroup of $\widetilde{C}$ contained in L. We choose L such that in consecutive order

$L1$. $L \in \mathcal{L}(EX)$ with $L \not\leq \widetilde{C}$.

$L2$. $|\widetilde{C} \cap L|_p$ is maximal.

$L3$. $S_{\widetilde{C}}(L)$ is maximal.

$L4$. $\widetilde{C} \cap L$ is maximal.

$L5$. L is minimal.

Let $R = O_p(\widetilde{C} \cap L)$. Consider the following two conditions:

(PU-L) $\quad N_{\widetilde{C}}(R) \not\leq L \cap \widetilde{C}$.

$\neg$ (PU-L) $\quad N_{\widetilde{C}}(R) = L \cap \widetilde{C}$

If (PU-L) holds we define $H := N_{\widetilde{C}}(R)$. Note here that $L \cap \widetilde{C} < H$.

If $\neg$ (PU-L) holds we choose a $\widetilde{C} \cap L$-invariant subnormal subgroup N of $\widetilde{C}$ minimal with respect to $N \not\leq L$ and put $H = N(L \cap \widetilde{C})$.

Note that in both cases $H \not\leq L$ and $H \cap L = \widetilde{C} \cap L$, since $\widetilde{C} \cap L \leq H \leq \widetilde{C}$. Let T be a Sylow p-subgroup of $H \cap L$ such that $T \cap X$ is a Sylow p-subgroup of X. Without loss $T \leq S$.

Lemma 2.3.1 (a) $O_p(\langle H, L \rangle) = 1$.

(b) $N_G(\Omega_1 Z(T)) \leq \widetilde{C}$.

(c) *T is a Sylow p-subgroup of L and $H \cap L$ contains a point stabilizer of L.*

(d) *If $\neg$(PU-L) holds, then $O_p(N_H(R)) = R$ and $Q \leq R$. In particular, H is of characteristic p.*

Proof: Suppose (a) is false. Then there exists a p-local subgroup L^* of G with $\langle H, L \rangle \leq L^*$. Since $L \leq L^*$, L^* fulfills all the assumptions on L (except for the minimality of L, our last choice). But $H \leq L^*$ and so $\widetilde{C} \cap L < \widetilde{C} \cap L^*$ contradicting (L4). This proves (a).

We claim that $EX \leq N_G(\Omega_1 Z(T))$. Since $\tilde{X}$ is subnormal in $\widetilde{C}$, and T contains a Sylow p-subgroup of X and so of $\widetilde{X}$, we conclude that T is a Sylow p-subgroup of $\langle \widetilde{X}, T \rangle$. Thus by the Kieler Lemma 2.0.2, $X \leq C_{\widetilde{X}}(\Omega_1 Z(T \cap X)) \leq C_G(\Omega_1 Z(T)) \leq N_G(\Omega_1 Z(T))$. Similarly $E < N_G(\Omega_1 Z(T))$.

By the choice of $\widetilde{C}$, $N_G(\Omega_1 Z(S)) \leq \widetilde{C}$. Thus to prove (b) we may assume $T \neq S$. Since $N_S(T) \leq N_{\widetilde{C}}(\Omega_1 Z(T))$ we conclude that $|\widetilde{C} \cap N_G(\Omega_1 Z(T))|_p > |\widetilde{C} \cap L|_p$. If (L1) holds for $N_G(\Omega_1 Z(T))$ we obtain a contradiction to (L2). Thus $N_G(\Omega_1 Z(T)) \leq \widetilde{C}$ and (b) holds.

By (b) $N_L(T) \leq H$ and so T is a Sylow p-subgroup of L. Hence (c) follows from (b).

Suppose $\neg$ (PU-L) holds. Then

$$L \cap \widetilde{C} \leq N_H(R) \leq N_{\widetilde{C}}(R) = L \cap \widetilde{C}, \text{ and}$$

$$N_Q(R)) \leq O_p(N_{\widetilde{C}}(R)) = O_p(\widetilde{C} \cap L) = R,$$

so $R = O_p(N_H(R))$ and $Q \leq R$. This is (d). □

Proposition 2.3.2 *Suppose $\neg E!$ and that (PU-L) holds. Set $\Sigma = L^H$. Then $\Sigma \subseteq \mathcal{PU}_1(R)$ and $O_R(\Sigma) = 1$.*

Proof: By 2.3.1(c) and since R is normal in H, $\Sigma \subseteq \mathcal{PU}_1(R)$. As H and L both normalize $O_R(\Sigma)$ we get from 2.3.1(a) that $O_R(\Sigma) = 1$. □

In view of the preceding proposition the (PU-L)-case can be dealt with via the Pushing Up Theorem 2.1.5. The $\neg$ (PU-L)-case is more complicated. As a first step we show

Lemma 2.3.3 *Suppose $\neg E!$ and $\neg$ (PU-L). Put $\Sigma = \{H, L\}$. Then $O_R(\Sigma) = 1 = O_{\mathrm{B}(R)}(\Sigma)$ and $L \in \mathcal{PU}_1(R) \subseteq \mathcal{PU}_4(\mathrm{B}(R))$. If $H \cap L$ contains a point stabilizer of H, then $\Sigma \subseteq \mathcal{PU}_1(R) \subseteq \mathcal{PU}_4(\mathrm{B}(R))$.*

Proof: By 2.3.1(a) $O_R(\Sigma) = 1$ and by 2.3.1(c), $L \in \mathcal{PU}_1(R)$. If $H \cap L$ contains a point stabilizer of H, then by 2.3.1(d) $H \in \mathcal{PU}_1(R)$. By 2.1.2, $\mathcal{PU}_1(R) \subseteq \mathcal{PU}_4(\mathrm{B}(R))$. Also $O_{\mathrm{B}(R)}(\Sigma) \leq O_R(\Sigma) = 1$ and all parts of the lemma have been verified. □

The preceding lemma is the main tool in the proof of:

Proposition 2.3.4 *Suppose $\neg E!$, $\neg$ (PU-L) and $Y_H \leq O_p(L)$. Then $H \in \mathcal{PU}_4(\mathrm{B}(R))$.*

Outline of a Proof: Suppose that $H \notin \mathcal{PU}_4(\mathrm{B}(R))$. Note that by 2.3.1(d) $Q \leq O_p(H) \leq R$, so $Y_H \leq Q$. Since $H \notin \mathcal{PU}_4(\mathrm{B}(R))$, $\mathrm{B}(R)$ is not normal in H and so $\mathrm{B}(R) \not\leq O_p(H)$. The definition of H (and the minimal choice of N) shows that $N = [N, B(R)]$. As $H \notin \mathcal{PU}_1(R)$, we get $[Y_H, N] \neq 1$ and thus also $[Y_H, B(R)] \neq 1$. It follows that Y_H is an FF-module for H. Let $\overline{H} := H/C_H(Y_H)$. Then there exist subnormal subgroups $K_1, \dots, K_m$ of N such that $\overline{K}_i$ is quasi-simple and

$$\overline{N} = \overline{K}_1 \times \cdots \times \overline{K}_m \text{ and } [Y_H, N] = \bigoplus_{i=1}^{m} [Y_H, K_i].$$

(Note here that $H \notin \mathcal{PU}_1(R)$ rules out the case where $\overline{N}$ is solvable.) We show next:

(∗) Let $1 \neq x \in C_S((S \cap K_i)C_S(K_i/O_p(K_i))$. Then $\mathcal{M}(C_G(x)) = \{\widetilde{C}\}$.

Since K_i and $\widetilde{X}$ are subnormal in $\widetilde{C}$ and $K_i \not\leq \widetilde{X}$ we get $[K_i, \widetilde{X}] \leq O_p(K_i)$. By 2.0.2 and the choice of x, $C_G(x)$ contains a point stabilizer of $\langle K_i, \widetilde{X}\rangle$. Suppose that $C_G(x) \leq L^*$ for some $L^* \in \mathcal{L}$ with $L^* \not\leq \widetilde{C}$. Then the maximal choice of X implies that X contains a point stabilizer of K_i. But then by 2.0.2(d), $H \cap L$ contains a points stabilizer of H, which contradicts 2.3.3 and $H \notin \mathcal{PU}_1(R)$. So (∗) holds.

We apply the amalgam method to (H, L) using the standard notation as it is given in [DS]. For $\alpha = Hg$ put $K_{\alpha i} = K_i^g$ and $\widetilde{C}_\alpha = \widetilde{C}^g$.

Suppose that b is even and (α, α') is a critical pair with $\alpha = H$. Typically we will find $1 \neq x \in [Y_\alpha, Y_{\alpha'}]$ such that x is centralized by a Sylow p-subgroup of $K_{\alpha i}$ and $K_{\alpha' j}$. Thus $(*)$ implies $\widetilde{C}_\alpha = \widetilde{C}_{\alpha'}$. But this contradicts $Y_\alpha \leq O_p(\widetilde{C}_\alpha)$ and $Y_\alpha \not\leq O_p(G_{\alpha'})$.

A typical case where one cannot find such an x is, when $\overline{K}_i \cong \Omega_{2n}^{\pm}(2)$ and $A := [[Y_\alpha, K_{\alpha i}], Y_{\alpha'}]$ has order 2 (for some i) and $\mathcal{M}(C_G(A)) \neq \{\widetilde{C}\}$. Then there exists $C_G(A) \leq L^* \in \mathcal{L}$ with $L^* \not\leq \widetilde{C}$. It is easy to see that $K_i R^* \in \mathcal{PU}_4(R^*)$. Using the Pushing Up Theorem 2.1.5 one derives a contradiction.

Suppose that b is even, but $\alpha \neq H$ for every critical pair (α, α'). One then proves that $Y_H Y_L$ is normal in L and $O_p(\langle O_p(H)^L \rangle) \leq O_p(H)$. Another application of 2.1.5 gives a contradiction.

So b is odd and without loss $\alpha = H$. Let $\alpha' + 1 \in \Delta(\alpha')$ with $Y_\alpha \not\leq G_{\alpha'+1}$. One usually gets that $Y_\alpha \cap Q_{\alpha'} \cap Q_{\alpha'+1}$ contains an element x as in $(*)$. This forces $Y_{\alpha'+1} \leq \widetilde{C}_{\alpha'} \cap G_{\alpha+1} \leq G_\alpha$. This allows us to find $y \in Y_{\alpha'+1} \cap Q_\alpha$ with $C_G(y) \leq \widetilde{C}_{\alpha'+1}$. Hence $Y_\alpha \leq G_{\alpha'+1}$, a contradiction. □

The propositions in this section together with the Pushing Up Theorem leave us with the following open problem:

2.3.1 The open "$\neg E!$, $b = 1$"-Problem

Suppose $\neg E!$, $\neg$(PU-L), $Y_H \not\leq O_p(L)$ and $H \notin \mathcal{PU}_4(\mathrm{B}(R))$. Determine the shapes of H and L.

2.4 E!

The way we usually use $E!$ is through an intermediate property called *Q-Uniqueness*.

$$(Q!) \quad C_G(x) \leq \widetilde{C} \text{ for all } 1 \neq x \in C_G(Q)$$

Lemma 2.4.1 *$E!$ implies $Q!$.*

Proof: Since $\widetilde{C}$ is a maximal p-local subgroup, $N_G(Q) = \widetilde{C}$. Thus $x \in C_G(Q) = C_{\widetilde{C}}(Q)$. Since $\widetilde{C}$ is of characteristic p we conclude $x \in Z(Q)$. Without loss $|x|$ has order p and thus $x \in \Omega_1 Z(Q)$. Note that EQ/Q has no p-chief-factors and so $\Omega_1 Z(Q) = Y_{EQ}$. By 2.0.2(c)

$$C_E(Y_{EQ}) = C_E(Y_E) = C_E(Y_{\widetilde{C}}) = E$$

Thus $E \leq C_G(x)$ and $E!$ implies $C_G(x) \leq \widetilde{C}$. □

The reader might want to verify that $L_n(q)$ is an example of a group which fulfills $Q!$ but not $E!$.

In this section we assume $Q!$ and that G has rank at least two. For $L \in \mathcal{L}$ define $L^\circ = \langle Q^g \mid g \in G, Q^g \leq L \rangle$.

Lemma 2.4.2 *Suppose $Q!$.*

(a) $\widetilde{C}^\circ = Q$, *in particular, any p-subgroup of G contains at most one conjugate of Q.*

(b) *If $L \in \mathcal{L}$ with $Q \leq O_p(L)$, then $L \leq \widetilde{C}$. In particular, if $1 \neq X \leq Z(Q)$ then $N_G(X) \leq \widetilde{C}$.*

(c) *If $Q_1, Q_2 \in Q^G$ with $Z(Q_1) \cap Z(Q_2) \neq 1$, then $Q_1 = Q_2$.*

(d) *Let $L \in \mathcal{L}$ with $Q \leq L$. Then*

(da) $L^\circ = \langle Q^{L^\circ} \rangle$

(db) $L = L^\circ(L \cap \widetilde{C})$.

(dc) $[C_L(Y_L), L^\circ] \leq O_p(L)$.

(dd) *If L acts transitively on $Y_L^\sharp$, then $L^\circ = N_G(Y_L)^\circ$.*

(de) *If $L^\circ \neq Q$, then $C_{Y_L}(L^\circ) = 1$.*

Proof: (a) Let $g \in G$ with $Q^g \leq \widetilde{C}$. We may assume that $Q^g \leq S$. Then $Z(S) \leq C_G(Q^g)$ and thus $S \leq C_G(x) \leq \widetilde{C}^g$ for $1 \neq x \in Z(S)$. Since $N_G(S) \leq N_G(\Omega_1 Z(S)) \leq \widetilde{C}$ we conclude that S is in a unique conjugate of $\widetilde{C}$, so $\widetilde{C} = \widetilde{C}^g$ and $Q = Q^g$.

(b) By (a) $Q = O_p(L)^\circ \trianglelefteq L$ and so $L \leq N_G(Q) = \widetilde{C}$. By $Q!$ we have $C_G(X) \leq \widetilde{C}$, so $Q \leq O_p(C_G(X)) \leq O_p(N_G(X))$ and we are done.

(c) As $\langle Q_1, Q_2 \rangle \leq C_G(Z(Q_1) \cap Z(Q_2))$, we get from $Q!$ and (a) that $Q_1 = Q_2$.

(d) By (a) each Sylow p-subgroup of L° contains a unique G-conjugate of Q. Thus Sylow's Theorem gives

$$\{Q^g \mid Q^g \leq L\} = Q^{L^\circ} = Q^L,$$

in particular (da) holds and by the Frattini argument $L = L^\circ N_L(Q)$. Then also (db) holds since $N_L(Q) \leq \widetilde{C}$. Note that $C_{Y_L}(Q) \neq 1$, so $C_L(Y_L) \leq L \cap \widetilde{C}$ by $Q!$. Thus

$$[C_L(Y_L), Q] \leq C_L(Y_L) \cap Q \leq O_p(C_L(Y_L)) \leq O_p(L),$$

and (dc) follows from (da).

Let $Q^g \leq N_G(Y_L)$. Then there exists $1 \neq x \in C_{Y_L}(Q^g)$. If L is transitive on $Y_L^\#$, then x is also centralized by an L-conjugate of Q. On the other hand, by $Q!$ and (a) $C_G(x)$ contains a unique conjugate of Q. Hence $Q^g \leq L$ and $N_G(Y_L)^\circ = L^\circ$.

(de) follows immediately from $Q!$ and (a). $\square$

2.4.1 The Structure Theorem

In this section we assume $Q!$ and that G has rank at least two. Our goal is to determine the action of L on Y_L for all $L \in \mathcal{L}(S)$ with $L \not\leq \widetilde{C}$.

For this let $\mathcal{M}^{\ddagger}(S)$ be the set of all $M \in \mathcal{M}(S)$ such that

$$\mathcal{M}(L) = \{M\} \text{ for all } L \in \mathcal{L}_M(S) \text{ with } M = LC_M(Y_M)$$

To explain the relevance of this set we define a partial ordering on a certain subset of $\mathcal{L}(S)$. For $L \in \mathcal{L}$ define $L^{\dagger} = LC_G(Y_L)$ and so $L = L^{\dagger}$ iff $C_G(Y_L) \leq L$. Then clearly Y_L is a p-reduced normal subgroup of $L^{\dagger}$ and so $Y_L \leq Y_{L^{\dagger}}$. Thus $C_G(Y_{L^{\dagger}}) \leq C_G(Y_L) \leq L^{\dagger}$. We conclude that every $L \in \mathcal{L}$ is contained in a member of

$$\mathcal{L}^{\dagger} = \{L \in \mathcal{L} \mid C_G(Y_L) \leq L\}$$

For $L_1, L_2 \in \mathcal{L}^{\dagger}(S)$ we define

$$L_1 \ll L_2 \Leftrightarrow L_1 = (L_1 \cap L_2)C_G(Y_{L_1})$$

The following lemma has an elementary proof:

Lemma 2.4.3 (a) *$\ll$ is a partial ordering on $\mathcal{L}^{\dagger}(S)$.*

(b) *$\mathcal{M}^{\ddagger}(S)$ is precisely the set of maximal elements in $\mathcal{L}^{\dagger}(S)$ with respect to $\ll$.*

(c) *If $L, H \in \mathcal{L}(S)$ with $L^{\dagger} \ll H^{\dagger}$, then $Y_L \leq Y_{H^{\dagger}}$ and $L^{\circ} \leq H^{\circ}$.* □

Let $L \in \mathcal{L}(S)$ with $L \not\leq \widetilde{C}$. As we have said earlier, we want to determine the action of L on Y_L. This will be done using a particular point of view based on the following elementary observations.

By the preceding lemma $L^{\dagger} \ll M$ for some $M \in \mathcal{M}^{\ddagger}(S)$, so

$$L = (L \cap M)C_L(Y_L) = (L \cap M)(L \cap \widetilde{C})$$

since $C_L(Y_L) \leq \widetilde{C}$; in particular, also $M \not\leq \widetilde{C}$.

It is easy to see that

$$L = \langle \mathcal{P}_L(S) \rangle N_L(S) = \langle \mathcal{P}_L(S) \rangle (L \cap \widetilde{C}),$$

so there exists $P \in \mathcal{P}_L(S)$ with $P \not\leq \widetilde{C}$.

According to these observations it suffices to study the action of M on Y_M, where $M \in \mathcal{M}^{\ddagger}(P)$ for a given $P \in \mathcal{P}(S)$ with $P \not\leq \widetilde{C}$. This point of view allows a case subdivision that requires another definition:

For $L \in \mathcal{L}(S)$ we write $gb(L) = 1$ if $Y_M \not\leq Q$ for some $M \in \mathcal{L}(L)$, and $gb(L) > 1$ otherwise. In the above discussion we now distinguish the cases $gb(P) > 1$ and $gb(P) = 1$. These two cases are treated in the next two sections. We remark that of the actual groups have $gb(P) = 1$. Indeed among the groups of Lie Type in characteristic p, only ${}^2F_4(2^k)$, ${}^3D_4(q)$ and (for $p \neq 3$) $G_2(q)$ fulfill $E!$, $\operatorname{rank} G > 1$ and $gb(P) > 1$.

We further set

$$\mathcal{L}^\circ = \{L \in \mathcal{L} \mid O^p(L) \leq L^\circ\} \text{ and } \mathcal{P}^\circ = \mathcal{P} \cap \mathcal{L}^\circ.$$

Note that for $P \in \mathcal{P}(S)$, $P \in \mathcal{P}^\circ$ iff $P \not\leq \widetilde{C}$.

2.4.1.1 The Structure Theorem for $Y_M \leq Q$

In this section we discuss a proof of the following theorem:

Theorem 2.4.4 (M-Structure Theorem for $Y_M \leq Q$) *Suppose $Q!$ and that $P \in \mathcal{P}^\circ(S)$ with $gb(P) > 1$. Let $M \in \mathcal{M}^\ddagger(P)$. Then one of the following two cases holds for $\overline{M} := M/C_M(Y_M)$ and $M_0 := M^\circ C_S(Y_M)$:*

(a) (aa) $\overline{M_0} \cong SL_n(p^k)$ *or* $Sp_{2n}(p^k)$ *and* $C_{\overline{M}}(\overline{M_0}) \cong C_q$, $q | p^k - 1$, *or* $\overline{M} \cong Sp_4(2)$ *and* $\overline{M_0} \cong Sp_4(2)'$ *(and* $p = 2$*),*

(ab) $[Y_M, M^\circ]$ *is the corresponding natural module for* $\overline{M_0}$*,*

(ac) $C_{M_0}(Y_M) = O_p(M_0)$*, or* $p = 2$ *and* $M_0/O_2(M_0) \cong 3Sp_4(2)'$.

(b) (ba) $P = M_0 S$, $Y_M = Y_P$*, and there exists a unique normal subgroup* P^* *of* P *containing* $O_p(P)$ *such that*

(bb) $\overline{P^*} = K_1 \times \cdots \times K_r$, $K_i \cong SL_2(p^k)$, $Y_M = V_1 \times \cdots \times V_r$*, where* $V_i := [Y_M, K_i]$ *is a natural* K_i*-module,*

(bc) Q *permutes the subgroups* K_i *of (bb) transitively,*

(bd) $O^p(P) = O^p(P^*) = O^p(M_0)$*, and* $P^* C_M(Y_P)$ *is normal in* M*,*

(be) *either* $C_{M^\circ}(Y_P) = O_p(M_0)$*, or* $p = 2$, $r > 1$, $K_i \cong SL_2(2)$*, and* $C_{M_0}(Y_P)/O_2(M_0) = Z(M_0/O_2(M_0))$ *is a 3-group.*

A second look at the situation discussed in section 1.2 (with M_1 corresponding to M) might help the reader to appreciate the conclusion of the Structure Theorem. In section 1.2 we have assumed that $F^*(M_1/O_p(M_1))$ is quasisimple. Here we get a similar statement as a conclusion in part (a), and part (b) shows that only for "small groups" it is not true (in fact, this case later will be ruled out in the $P!$-Theorem).

In section 1.2 we found that Y_{M_1} is an FF-module or a $2F$-module for M_1, where the second case is basically ruled out here by the hypothesis $Y_M \leq Q$. But in the FF-module case a glance at the FF-Module Theorem 1.1.2 shows that by far not all possible groups actually occur in the conclusion of the Structure Theorem. In the following we want to demonstrate, using the groups $Sym(I)$ and $G_2(2^k)$ as examples, how these additional groups are ruled out.

Suppose that $\overline{M} \cong Sym(I)$, $|I| \geq 9$, $p = 2$, and $Y := [Y_M, M]$ is the non-central irreducible constituent of the natural permutation module for $Sym(I)$. To describe the action of M on Y let V be a $GF(2)$-vector space with basis v_k, $k \in I$, and set

$v_J = \sum_{k\in J} v_k$ for every $J \subseteq I$. Then $Sym(I)$ acts on V via $v_k \mapsto v_{kx}$, $x \in Sym(I)$. Let $V_e := \{v_J \mid J \subseteq I,\ |J| \text{ even}\}$ and $\overline{V}_e = V_e + \langle v_I \rangle / \langle v_I \rangle$. Then $\overline{V}_e$ is the irreducible constituent meant above, so $Y = \overline{V}_e$.

Assume first that Q does not act transitively on I. Then there exists a proper Q-invariant subset J of I with $|J| \leq |I \backslash J|$ and $\overline{v}_J \in C_Y(Q)$. Hence $Q!$ gives $C_M(\overline{v}_J) \leq \widetilde{C}$ and $Q \leq O_2(C_M(\overline{v}_J))$. Note that $C_{\overline{M}}(a_J) \cong Sym(J) \times Sym(I \setminus J)$ (respectively $Sym(I) \wr C_2$, if $|J| = |I \setminus J|$). By 2.4.2(b) $\overline{Q} \neq 1$, we conclude that $O_2(Sym(J)) \neq 1$ or $O_2(Sym(I \setminus J)) \neq 1$. Since $|I| \geq 9$, $|J| \leq |I \setminus J|$, and $O_2(Sym(n)) = 1$ for all $n \geq 5$, we get that $|J| \in \{2, 4\}$ and $O_2(Sym(I \setminus J)) = 1$. Thus $\overline{Q} \leq Sym(J)$, and Q centralizes every $\overline{v}_{J^*}$ for $J^* \subseteq I \setminus J$. Choose such an J^* with $|J^*| = 2$. Then $C_{\overline{M}}(\overline{v}_{J^*}) = Sym(J^*) \times Sym(I \setminus J^*)$ and $\overline{Q} \leq O_2(C_{\overline{M}}(\overline{v}_{J^*})) \cap O_2(Sym(J))$. We conclude that $\overline{Q} = 1$ since $O_2(Sym(I \setminus J^*)) = 1$. But this is impossible by 2.4.2(b).

Assume now that Q is transitive on I. Let J be an orbit of a maximal subgroup of Q that contains the stabilizer of a point. Then $|J| = \frac{1}{2}|I|$ and $\overline{Q}$ centralizes $\overline{v}_J$ since $\overline{v}_J = \overline{v}_{I \setminus J}$. Now a similar argument as above leads to a contradiction.

As a second example let $p = 2$, $\overline{M} \cong G_2(q)$, $q = 2^k$, and $Y := [Y_M, M]$ be the module of order q^6. In addition, suppose that there exists $g \in G$ with $YY^g \leq M \cap M^g$ and $[Y, Y^g] \neq 1$. Then it is easy to see that $|\overline{Y^g}| = q^3$ and $|[Y, Y^g]| = q^3$. Let $1 \neq x \in [Y, Y^g]$. Since M act transitively on Y, there exist $h \in M$ such that $[x, Q^h] = 1$ From $Q!$ applied to Q^h, $C_G(x) \leq \widetilde{C}^h$. From the hypothesis $Y_M \leq Q$ we get $Y \leq Q^h$ and so $Y \leq O_2(C_G(x))$; in particular

$$Y \leq O_2(C_{M^g}(x)) \text{ for all } 1 \neq x \in [Y, Y^g].$$

This contradicts the action of M^g on Y^g.

Outline of a proof for 2.4.4: Let H be minimal in M with $S \leq H$ and $M = HC_M(Y_M)$. Then by definition of $\mathcal{M}^\ddagger(S)$, M is the unique maximal p-local subgroup containing H. Let $Y = Y_H (= Y_M)$.

We consider the following cases:

(a) *[The Orthogonal Case]* $p = 2$, $\overline{H} \cong O^\epsilon_{2n}(2)$, $[Y, H]$ is the natural module and $C_H(y) \not\leq M$ for every non-singular element $y \in [Y, H]$.

(b) *[The Symmetric Case]* (a) does not hold and there exists $g \in G$ with $YY^g \leq H \cap H^g$ and $[Y, Y^g] \neq 1$.

(c) *[The Non Abelian Asymmetric Case]* Neither (a) nor (b) holds and there exists $L \in \mathcal{L}$ with $O_p(H) \leq L$ and $Y \not\leq O_p(L)$.

(d) *[The Abelian Asymmetric Case]* None of (a),(b) or (c) holds.

In the following we show how these cases arise from the amalgam method and how they are dealt with.

Choose $P_1 \in \mathcal{P}_{\widetilde{C}}(S)$ with $P_1 \not\leq M$ and P_1 minimal. Since M is the unique maximal p-local containing H, $O_p(\langle H, P_1 \rangle) = 1$ and we can apply the amalgam method to the pair (H, P_1). For notation see [DS].

Assume that b is even. Let (α, α') be a critical pair. Then $Q!$ shows that $G_\alpha \sim H$, and we obtain $g \in G$ with $YY^g \leq H \cap H^g$ and $[Y, Y^g] \neq 1$. Hence either the Symmetric Case or the Orthogonal Case holds. Note that the symmetry in H and H^g allows to assume that Y^g is an offender on Y.

Suppose that there exists $1 \neq x \in [Y, Y^g]$ that is p-central in both, H and H^g. Then our hypothesis $Y \leq Q$ and $Q!$ imply that $Y^g \leq O_p(C_H(x))$, and (after a technical reduction to one component of $H/C_H(Y)$) the Point Stabilizer Theorem 1.1.3 applies. This gives the desired conclusion since the preceding discussion already ruled out $Sym(n)$ and $G_2(2^k)$.

Suppose now that $[Y, Y^g]$ does not contain such an element. Then (again omitting a reduction to components) the FF-Module Theorem 1.1.2 shows that Case (b) of the Structure Theorem holds, or that $|[Y, Y^g]| = 2$. The latter possibility leads to the Orthogonal Case.

Assume now that b is odd and (α, α') is a critical pair. Then again $G_\alpha \sim H$. If there exists $1 \neq x \in Y_\alpha$ with $[x, O^p(G_{\alpha'})] = 1$, then $Y_\alpha \not\leq O_p(C_{G_{\alpha'}}(x))$ and the Non-Abelian Asymmetric Case (or (a) or (b)) hold.

Suppose that $[x, O^p(G_{\alpha'})] \neq 1$ for all $x \in Y_\alpha^\#$ (in the actual proof we do not use $O^p(G_{\alpha'})$ but a possibly smaller subgroup of $G_{\alpha'}$). Using the action of Y_α on $V_{\alpha'}$ one can show the existence of a strong offender on Y_α. Here an offender A on a module V is called strong, if $C_V(a) = C_V(A)$ for all $a \in A \setminus C_A(V)$. This rules out most of the cases of the FF-module Theorem 1.1.2, and we get what we want, (except that it does not rule out $SL_n(q)$ on a direct sum of natural modules, a case which we will not discuss here).

This leaves us with the Orthogonal Case or the Non-Abelian Asymmetric Case. In the Orthogonal Case we choose L minimal with $C_H(x) \leq (L \cap H)C_H(Y)$ and $L \not\leq M$, where x is a non-singular vector (i.e. a non-p-central element) in $[Y, H]$. Let z be a non-zero singular vector in $[Y, H]$ perpendicular to x, so z is p-central in H. Then $[z, Q^h] = 1$ for some $h \in H$. Let $Q_z := Q^h$. We now show that $O_p(\langle Q_z, L\rangle) = 1$, $[Q_z, C_L(z)] \leq Q_z \cap L$, and that z and y are not conjugate in G. Then 2.1.5 gives the shape of L, and one obtains a contradiction.

It remains to discuss the Non-Abelian Asymmetric Case. Let $U \in \mathcal{L}(O_p(H))$ with $Y \not\leq O_p(U)$ such that first $|U \cap H|_p$ is maximal and then U is minimal. Let $T \in Syl_p(U \cap H)$. If $N_G(T) \not\leq M$, then considering the amalgam $(H, N_G(T))$ we obtain $g \in G$ with $YY^g \in H \cap H^g$ and $[Y, Y^g] \neq 1$. But this contradicts the assumptions of the Non-Abelian Asymmetric Case. Hence $N_G(T) \leq M$, in particular T is a Sylow p-subgroup of U. If $Q \not\leq U$ we can apply 2.1.5 and get a contradiction. So $Q \leq U$. Since $Y \leq Q$ but $Y \not\leq O_p(U)$, we have $U \not\leq \widetilde{C}$, and 2.4.2(de) implies $C_{Y_U}(U) = 1$.

Let $T \leq X < U$. Then by minimality of U, $Y \leq O_p(X)$. Since $O_p(X) \leq T \leq H$ we get $\langle Y^X \rangle \leq H$. Hence $\langle Y^X \rangle$ is abelian since we are not in the symmetric case. So 1.1.4 gives the structure of $U/O_p(U)$ and Y_U. Moreover, in most cases we can conclude that Y_U is a strong dual offender on Y and in all cases we get some strong dual offender on Y. Here a group A is called a strong dual offender on a module V if A acts quadratically on V and $[v, A] = [V, A]$ for all $v \in V \setminus C_V(A)$. The existence of a strong dual offender on Y together with the FF-Module Theorem 1.1.2 gives the desired conclusion. □

2.4.1.2 The Structure Theorem for $Y_M \not\leq Q$

In this section we outline a proof of the following theorem. (It might be worthwhile to mention that given $E!$ we do not need to assume in this section that G is of local characteristic p but only that G is of parabolic characteristic p.)

Theorem 2.4.5 (M-Structure Theorem for $Y_M \not\leq Q$) *Let $M \in \mathcal{M}(S)$ with M° maximal and put $\overline{K} = F^*(M^\circ/C_{M^\circ}(Y_M))$. Suppose $E!$, $Y_M \not\leq Q$ and that $M^\circ S$ is not p-minimal, then one of the following holds*

1. *$\overline{K}$ is quasisimple and isomorphic to $SL_n(q)$, $\Omega_n^\pm(q)$, or $E_6(q)$. In case of $\overline{K} \cong SL_n(q)$, or $E_6(q)$ no element in M induces a diagram automorphism.*
2. *$\overline{K} \cong SL_n(q)' \circ SL_m(q)'$.*
3. *$p = 2$ and $\overline{K} \cong Alt(6)$, $3Alt(6)$, $Sp_8(2)$, M_{22}, or M_{24}*
4. *$p = 3$ and $\overline{K} \cong M_{11}$ or M_{12}*

Moreover, the module Y_M is a 2F–module with quadratic or cubic offender and contains a module V as in the table below.

K	prime	module	example
$SL_n(q)$	p	ext. square	$\Omega_{2n}(q)$
$SL_n(q)$	p odd	sym. square	$Sp_{2n}(q)$
$SL_n(q^2)$	p	$V(\lambda_1) \otimes V(\lambda_1^\sigma)$	$SU_{2n}(q)$
$SL_3(2)$	2	natural	$G_2(3).2$
$Alt(6)$	2	natural	Suz
$3Alt(6)$	2	6-dim	M_{24}
$Sp_8(2)$	2	8-dim	F_2
$\Omega_n^\pm(q)$	p	natural	$\Omega_{n+2}^\pm(q)$
$\Omega_{10}^\pm(q)$	p	half spin	$E_6(q)$
$E_6(q)$	p	$V(\lambda_1)$	$E_7(q)$
M_{11}	3	5-dim	Co_3
$2M_{12}$	3	6-dim	Co_1
M_{22}	2	10-dim	$M(22)$
M_{24}	2	11-dim	$M(24)$

The proof of the above theorem corresponds to the discussion of the Cases (I) and (II) in section 1.2. Let $L \in \mathcal{L}_{\widetilde{C}}$ be minimal with $Y_M \leq S \cap L \in Syl_p(L)$ and $Y_M \not\leq O_p(L)$. Note that such a choice is possible since $Y_M \not\leq Q$. Let $Y_M \leq P < L$ and $S \cap P \in Syl_p(P)$. Then by the minimal choice of L, $Y_M \leq O_p(P)$ and so $\langle Y_M^P \rangle \leq O_p(P) \leq S \leq M$. We now consider the following two cases separately:

(1F) There exists $g \in G$ such that $1 \neq [Y_M, Y_M^g] \leq Y_M \cap Y_{M^g}$ and $Y_M Y_M^g \leq M \cap M^g$.

(2F) $\langle Y_M^P \rangle$ is abelian for all $Y_M \leq P < L$ with $S \cap P \in Syl_p(P)$.

In the 1F-Case, possibly after replacing g be g^{-1}, we may assume that $A := Y_M^g$ is a quadratic offender on Y_M.

In the 2F-Case 1.1.5 can be used to get a cubic $2F$-offender on Y_M as in case (III) of 1.2.

The FF-module Theorem1.1.2 and the 2F-module Theorem 1.1.6 now allow us to identify the components (or solvable variants of components) of $M^\circ/C_{M^\circ}(Y_M)$ which are not centralizes by A. In the iF-case one can show that A centralizes all but i of the components. Let K be the product of the components not centralized by A. By 2.4.2(de), $M^\circ/C_{M^\circ}(Y_M)$ acts essentially faithful on $[Y_M, K]$. This allows also to obtain information about all of $M^\circ/C_{M^\circ}(Y_M)$.

2.4.2 The P!-Theorem

In this section we assume $Q!$ and that G has rank at least 2. Note that this implies that $\mathcal{P}^\circ(S) \neq \emptyset$. We investigate the members of $\mathcal{P}^\circ(S)$, and distinguish the two cases $\langle \mathcal{P}^\circ(S)\rangle \notin \mathcal{L}$ and $\langle \mathcal{P}^\circ(S)\rangle \in \mathcal{L}$. Detailed proofs for the following two theorems can be found in [PPS].

Theorem 2.4.6 (The P! Theorem,I) *Suppose $Q!$ hold and $\langle \mathcal{P}^\circ(S)\rangle \notin \mathcal{L}$. Then*

(a) *p is odd.*

(b) *$Q = \mathrm{B}(S)$, $\widetilde{C} = N_G(\mathrm{B}(S))$ and $|Q|$ has order q^3, q a power of p.*

(c) *$P^\circ \sim q^2SL_2(q)$ for all $P \in \mathcal{P}^\circ(S)$*

Outline of a Proof: Let $L = N_G(B(S))$. By our assumption not every element of $\mathcal{P}^\circ(S)$ is in L. We first investigate an element $P \in \mathcal{P}^\circ(S)$ with $P \not\leq L$. Observe that $Q!$ implies that $\Omega_1 Z(X) = 1$ for every $X \in \mathcal{P}^\circ(S)$, so by 2.1.3 $P \in \mathcal{P}U_4(B(S))$; i.e.

(*) $P = \langle N_P(B(S)), P_0 \mid P_0 \leq P, P_0 \in \mathcal{P}U_3(B(S))\rangle$.

An application of 2.1.4 and a short argument show that for the groups P_0 in (*):

(1) $Y_P = O_p(P_0) = O_p(P)$,

(2) P_0/Y_P, Y_P is a natural $SL_2(q)$-module for P_0/Y_P ($q = p^m$), and $|\,\mathrm{B}(S)| = q^3$.

(3) P_0 is normal in P, and $P = SP_0$.

Suppose that $p = 2$. Then $|\mathcal{A}(S)| = 2$, so $L = SO^2(L) \leq N_G(A)$ for all $A \in \mathcal{A}(S)$. It is now easy to see that there exist exactly two maximal 2-local subgroups containing S. One of them is $\widetilde{C}$ and so $\langle \mathcal{P}^\circ(S)\rangle$ is contained in the other. But this contradicts our hypothesis.

So p is odd. Suppose that $Q \not\leq \mathrm{B}(S)$. Then (ii) shows that $q = p^{pk}$ for some integer $k \geq 1$. Moreover, $[Y_P, Q]$ has order at least $p^{(2p-1)k}$. Let $V = \langle [Y_P, Q]^{\widetilde{C}}\rangle$.

Note that an elementary abelian p-subgroup of S not contained in $\mathrm{B}(S)$ has order at most p^{2k+1}. Since $p > 2$, $(2p-1)k > 2k+1$ and so $V \leq \mathrm{B}(S)$.

In particular, $Z(V) \leq C_{\mathrm{B}(S)}([Y_P, Q]) \leq Y_P$. It follows that either $V \leq Y_P$ or $[V, Y_P] = Z(V) = Z(\mathrm{B}(S))$. In both cases $\langle Y_P^{\widetilde{C}} \rangle$ acts trivially on the series $1 \leq Z(V) \leq V \leq Q$, so $Y_P \leq Q$ since $\widetilde{C}$ has characteristic p. As Y_P is not normal in $\widetilde{C}$ we get $\mathrm{B}(S) = \langle Y_P^{\widetilde{C}} \rangle \leq Q$. In particular $\mathrm{B}(S) = \mathrm{B}(Q)$, so $\widetilde{C} = L$ and $Q \leq O_p(N_P(\mathrm{B}(S))) = \mathrm{B}(S)$.

We have proved that $Q \leq \mathrm{B}(S)$, so by 2.4.2(b) $L \leq \widetilde{C}$. In particular $P \not\leq L$ for every $P \in \mathcal{P}^\circ(S)$, and (1) – (3) hold for every $P \in \mathcal{P}^\circ(S)$. It remains to prove that $Q = \mathrm{B}(S)$.

Suppose that $Q \neq \mathrm{B}(S)$. Again as $\widetilde{C}$ is of characteristic p, we get that $Z(\mathrm{B}(S)) < Q$. Note that $N_P(\mathrm{B}(S))$ acts irreducibly on $Y_P/Z(\mathrm{B}(S))$ and $\mathrm{B}(S)/Y_P$. It follows that either $Y_P \leq Q$ or $Y_P \cap Q = Z(\mathrm{B}(S))$. The first case gives $Q = Y_P$ contrary to our assumption. The second case shows, with an argument as above using the series $1 \leq \Omega_1 Z(Q) \leq Q$, that $\Omega_1 Z(Q) \neq Z(\mathrm{B}(S))$, so Q is elementary abelian of order q^2.

For every $P \in \mathcal{P}^\circ(S)$ let t_P be an involution in P that maps onto the central involution of P_0/Y_P. Then t_P normalizes $\mathrm{B}(S)$ and so also Q. We conclude that t_P inverts $Z(\mathrm{B}(S))$ and $\mathrm{B}(S)/Q$ and centralize $Q/Z(\mathrm{B}(S))$. There exists $X \in \mathcal{P}^\circ(S)$ with $Y_X \neq Y_P$. Let $u = t_P t_X$. Then u centralizes $Z(\mathrm{B}(S))$, $\mathrm{B}(S)/Q$ and $Q/Z(\mathrm{B}(S))$. So u induces a p-element on $\mathrm{B}(S)$ and since $N_G(\mathrm{B}(S))$ has characteristic p, $\mathrm{B}(S)\langle u \rangle$ is a p-group. By (1)-(3) we conclude that $u \in B(S)$ and $t_P \mathrm{B}(S) = t_X \mathrm{B}(S)$. But then $Y_P = [\mathrm{B}(S), t_P] = [\mathrm{B}(S), t_X] = Y_X$, a contradiction.

We have shown that $Q = \mathrm{B}(S)$, and the lemma is proved. □

We say that $P!$ holds in G provided that:

(P!-1) There exists a unique $P \in \mathcal{P}^\circ(S)$.

(P!-2) $P^\circ/O_p(P^\circ) \cong SL_2(q)$, q a power of p.

(P!-3) Y_P is a natural module for P°.

(P!-4) $C_{Y_P}(S \cap P^\circ)$ is normal in $\widetilde{C}$.

Theorem 2.4.7 (The P! Theorem,II) *Suppose that*

(i) *$Q!$ holds and G has rank at least 2.*

(ii) *P is a maximal element of $\mathcal{P}^\circ(S)$ and $gb(P) = 1$.*

(iii) *$M := \langle \mathcal{P}^\circ(S) \rangle \in \mathcal{L}$*

Then $P!$ holds in G.

Outline of a Proof: Applying the Structure Theorem 2.4.4 to some $\widetilde{M} \in \mathcal{M}^\ddagger(M)$ it is fairly easy to see that $P = M$. In case (a) of the Structure Theorem

2.4.4 $P^\circ/O_p(P^\circ) \cong SL_2(q)$ and Y_p is the natural $P^\circ/O_p(P^\circ)$-module. In this case we define $Z_0 := C_{Y_P}(P^\circ \cap S)$ $(= \Omega_1 Z(S \cap P^\circ))$. In case (b) of the Structure Theorem we define $Z_0 := C_{Y_P}(S \cap P^*)$, where P^* is as given there.

The main step in the proof of the P!-Theorem is to show that Z_0 is normal in $\widetilde{C}$. Suppose not and let $\widetilde{P} \in \mathcal{P}_{\widetilde{C}}(S)$ be minimal with $Z_0 \ntrianglelefteq \widetilde{P}$. Another application of the Structure Theorem shows that $O_p(\langle P, \widetilde{P}\rangle) = 1$. So we can apply the amalgam method to the pair $(P, \widetilde{P})$.

For $\gamma = \widetilde{P}g$ put $\widetilde{C}_\gamma = \widetilde{C}^g$. Let (α, α') be a critical pair. Suppose that $\alpha \sim \widetilde{P} \sim \alpha'$. Then both Q_α and $Q_{\alpha'}$ contain a conjugate of Q. Since $1 \neq [Z_\alpha, Z_{\alpha'}] \leq Z(Q_\alpha) \cap Z(Q_{\alpha'})$ we conclude from 2.4.2(c) that $\widetilde{C}_\alpha = \widetilde{C}_{\alpha'}$ and so $Z_\alpha Z_{\alpha'} \leq Z_{\widetilde{C}_\alpha}$. Thus $[Z_\alpha, Z_{\alpha'}] = 1$, a contradiction.

So we may assume that $\alpha = P$. Since $Y_P \leq Q \leq O_p(\widetilde{P})$ we have $b > 1$. Suppose that $b = 2$. By the Structure Theorem Q (and so also Q_β) acts transitively on the "components" of G_α/Q_α. Hence $Z_0 = [Z_\alpha, Z_{\alpha'}]$. This is used to show that $Z_0 \trianglelefteq G_\beta$, a contradiction.

Thus $b \geq 3$. A lengthy amalgam argument now leads to contradiction.

We have established that Z_0 is normal in $\widetilde{C}$. In Case (a) of the Structure Theorem we are done. So suppose that Case (b) of the Structure Theorem holds. Since $N_P(Z_0) \leq \widetilde{C}$, $Q \leq O_p(N_P(Z_0))$. Since Q acts transitively on the components we conclude that $q = p = 2$.

Note that $\widetilde{M}$ is the unique maximal 2-local subgroup of G containing P. Suppose that $N_G(\mathrm{B}(S)) \nleq \widetilde{M}$. Then $O_2(\langle P, N_G(\mathrm{B}(S))\rangle) = 1$ and 2.1.4 gives a contradiction. Hence $N_G(\mathrm{B}(S)) \leq \widetilde{M}$. Since $\mathrm{B}(S) \leq C_G(Z_0)$ and Z_0 is normal in $\widetilde{C}$, the Frattini argument implies $\widetilde{C} = (\widetilde{C} \cap \widetilde{M})C_G(Z_0)$.

Let K be the one of the $Sym(3)$-components of $P/O_p(P)$, T a subgroup of index 2 in S with $N_S(K) \leq T$, $X = \langle(([Y_P, K] \cap Z_0)^T\rangle$ and $L = N_G(X)$. Then $\langle K, T, C_G(Z_0)\rangle \leq L$. Since Q acts transitively on the components of $P/O_p(P)$, $Q \nleq T$ and $P = \langle L \cap P, Q\rangle$. Thus $O_2(\langle Q, L\rangle) = 1$. Suppose that T is not a Sylow 2-subgroup of L. Since T is of index 2 in a Sylow 2-subgroup of G, $N_L(T)$ contains a Sylow 2-subgroup of L and G. But $N_L(T) \leq N_G(\mathrm{B}(S)) \leq \widetilde{M}$ and so $N_L(T)$ contains a Sylow 2-subgroup of $\widetilde{M}$. One concludes that $P \leq L$, a contradiction.

Thus T is a Sylow 2-subgroup of L. Since $C_L(\Omega_1\mathrm{Z}(T)) \leq C_G(\Omega_1\mathrm{Z}(S)) \leq \widetilde{C}$ and $|Q/Q \cap T| = 2$ we get $C_L(\Omega_1\mathrm{Z}(T)) \trianglelefteq QC_L(\Omega_1\mathrm{Z}(T))$. So we can apply 2.1.5 to $\Sigma = L^Q$ and $R = O_p(C_L(\Omega_1\mathrm{Z}(T)))$. A little bit of more work gives a contradiction. □

2.4.3 The $\widetilde{P}$! Theorem

Suppose that G fulfills $Q!$ and $P!$. We say that $\widetilde{P}!$ holds in G provided that

($\widetilde{P}$!-1) There exists at most one $\widetilde{P} \in \mathcal{P}(S)$ such that $\widetilde{P}$ does not normalize P° and $M := \langle P, \widetilde{P}\rangle \in \mathcal{L}$.

($\widetilde{P}$!-2) If such a $\widetilde{P}$ exists then,

(a) $M \in \mathcal{L}^\circ$.

(b) $M^\circ/C_{M^\circ}(Y_M) \cong SL_3(q), Sp_4(q)$ or $Sp_4(2)'$

(c) Y_M is a corresponding natural module.

In this section we outline a proof of the following theorem from [MMPS]:

Theorem 2.4.8 (The $\widetilde{P}!$ Theorem) *Suppose $Q!$ and $gb(P) > 1$ for some $P \in \mathcal{P}^\circ(S)$. Then one of the following is true:*

1. *G fulfills $\widetilde{P}!$.*
2. *Let $\widetilde{P} \in \mathcal{P}(S)$ with $\widetilde{P} \not\leq N_G(P^\circ)$ and $M := \langle P, \widetilde{P}\rangle \notin \mathcal{L}$. Then*

   (a) *$p = 3$ or 5.*

   (b) *$M/O_p(M) \cong SL_3(p)$*

   (c) *$O_p(M)/Z(O_p(M))$ and $Z(O_p(M))$ are natural $SL_3(p)$-modules for $M/O_p(M)$ dual to each other.*

Outline of a Proof: We may assume that $\widetilde{P}!$ does not hold. Then there exists $P_1 \in \mathcal{P}(S)$ such that $M_1 = \langle P, P_1\rangle \in \mathcal{L}$ and $P_1 \not\leq N_G(P^\circ)$. The Structure Theorem 2.4.4 shows that $M_1/O_p(M_1) \cong SL_3(q)$ or $Sp_4(q)$ (or some variant of $Sp_4(2)$) and that Y_{M_1} is a corresponding natural module. In particular, if P_1 were unique $\widetilde{P}!$ would hold. Hence we can choose P_2 having the same properties as P_1 and $P_1 \neq P_2$. Define $M_2 = \langle P, P_2\rangle$. The Structure Theorem also implies that $\langle M_1, M_2\rangle \notin \mathcal{L}$ and so we can apply the amalgam method to (M_1, M_2). Fairly short and elementary arguments show that $b \leq 2$. In the $b = 1$ case one easily gets $M_i' \cong 2^4Sp_4(2)'$ and then obtains a contradiction to $Y_{M_i} \leq Q$. Fairly routine arguments in the $b = 2$ case show that $M_i \sim q^{3+3}SL_3(q)$ or $q^{3+3+3}SL_3(q)$. A little extra effort rules out the second of this possibilities. But the proof that $q = 3$ or 5 in the remaining case currently is a rather tedious commutator calculation. □

The next lemma collects some information about $\widetilde{C}/O_p(\widetilde{C})$ which can be easily obtained using $Q!$ and $\widetilde{P}!$:

Lemma 2.4.9 *Suppose $Q!$, $P!$, $\widetilde{P}!$ and that G has rank at least three. Let $L = N_G(P^\circ)$. Then*

(a) *$N_G(T) \leq L \cap \widetilde{C}$ for all $O_p(\widetilde{C} \cap L) \leq T \trianglelefteq S$.*

(b) *There exists a unique $\widetilde{P} \in \mathcal{P}_{\widetilde{C}}(S)$ with $\widetilde{P} \not\leq L$.*

(c) *$\widetilde{P}/O_p(\widetilde{P}) \sim SL_2(q).p^k$.*

(d) *$\widetilde{C}/Q$ has a unique component K/Q. Moreover, $\widetilde{P} \leq KS$.*

(e) *$\widetilde{C} = K(L \cap \widetilde{C})$, $L \cap \widetilde{C}$ is a maximal subgroup of $\widetilde{C}$ and $O_p(\widetilde{C} \cap L) \neq Q$.*

(f) *Let $Z_0 = C_{Y_P}(S \cap P^\circ)$ and $V = \langle Y_P^{\widetilde{C}}\rangle$. Then $Z_0 \trianglelefteq V$ and $V \leq Q$.*

(g) *Let* $D = C_{\widetilde{C}}(K/O_p(K))$. *Then* D *is the largest normal subgroup of* $\widetilde{C}$ *contained in* L *and* D/Q *is isomorphic to a section of the Borel subgroup of* $\mathrm{Aut}(SL_2(q))$.

(h) *Let* $\overline{V} = V/Z_0$. *Then*

(ha) $[\overline{V}, Q] = 1$

(hb) $C_{\widetilde{C}}(\overline{V}) \leq D$ *and* $C_{\widetilde{C}}(\overline{V}) \cap C_{\widetilde{C}}(Z_0) = Q$.

(hc) *Let* $1 \neq X \leq Y_P/Z_0$. *Then* $N_{\widetilde{C}}(X) \leq \widetilde{C} \cap L$.

(hd) $\widetilde{C} \cap L$ *contains a point-stabilizer for* $\widetilde{C}$ *on* $\overline{V}$.

(g) $\langle \widetilde{C}, L \rangle \notin \mathcal{L}$.

2.4.4 The Small World Theorem

Given $Q!$ and $P \in \mathcal{P}^\circ(S)$. We say that $gb(P) = 2$ if $gb(P) > 1$ and $\langle Y_P^E \rangle$ is not abelian. If neither $gb(P) = 1$ nor $gb(P) = 2$ for P we say that $gb(P)$ is at least three.

Theorem 2.4.10 (The Small World Theorem) *Suppose* $E!$ *and let* $P \in \mathcal{P}^\circ(S)$. *Then one of the following holds:*

1. G *has rank 1 or 2.*

2. $gb(P) = 1$ *or* $gb(P) = 2$.

Outline of a Proof: Assume that G has rank at least three and that $gb(P)$ is at least three . In the exceptional cases of the P!-theorems (2.4.7, 2.4.6) one easily sees that $gb(P) = 1$. Thus $P!$ holds. Also in the exceptional case of the $\widetilde{P}!$ -Theorem 2.4.8 one gets $gb(P) = 1$ or $gb(P) = 2$. Thus $\widetilde{P}!$ holds. We proved

Step 1 $P!$ *and* $\widetilde{P}!$ *hold.*

2.4.9 gives us a good amount of information about E. We use the notation introduced in 2.4.9.

Since $\langle \widetilde{C}, L \rangle \notin \mathcal{L}$, we can apply the amalgam method to the pair $(\widetilde{C}, L)$. A non-trivial argument shows

Step 2 *One of the following holds:*

1. $O_p(\widetilde{C} \cap L)/Q$ *contains a non-trivial quadratic offender on* $\overline{V}$.

2. *There exists a non-trivial normal subgroup* A *of* $\widetilde{C} \cap L/Q$ *and normal subgroups* $Y_P \leq Z_2 \leq Z_3 \leq V$ *of* $\widetilde{C} \cap L$ *such that:*

   (a) A *and* V/Z_3 *are isomorphic as* $\mathbb{F}_p C_{\widetilde{C} \cap L}(Y_P)$*-modules.*

   (b) $|Z_3/Z_2| \leq |A|$.

(c) $[\overline{V}, A] \le \overline{Z_2} \le C_{\overline{V}}(A)$. *In particular, A is a quadratic $2F$-offender.*

(d) $[\overline{x}, A] = \overline{Y_P}$ *for all* $x \in Z_3 \setminus Z_2$.

(e) *Let* $Z_1 = \langle Y_P^{\widetilde{P}} \rangle$. *Then* $Y_1 \le Z_2$ *and* $\overline{Z_1}$ *is a natural* $SL_2(q)$*-module for* $\widetilde{P} \cap C_{\widetilde{C}}(Z_0)$.

We remark that 1. and 2. of Step 2 correspond to the $b > 3$- and $b = 3$-Case for the amalgam $(\widetilde{C}, L)$.

Let $X = C_V(O^p(K))$. Using 1.1.1, 1.1.3 and 2.4.9 (and the Z^*-theorem [Gl]) to deal with the case $|A| = 2$) it is not too difficult to derive

Step 3 *$K/O_p(K) \cong SL_n(q)$, $(n \ge 3)$, $Sp_{2n}(q)'$,$(n \ge 2)$ or $G_2(q)'$,$(p = 2)$. Moreover, V/X is the natural module for $K/O_p(K)$ and $\widetilde{C} \cap L$ contains a point-stabilizer for $\widetilde{C}$ on V/W.*

An amalgam argument now shows that $X = Z_0$. In particular, K acts transitively on $\overline{V}$. Hence all elements in V are conjugate under K to an element of Y_P. From this it is not to difficult to show that $b = 3$ in the amalgam $(\widetilde{C}, L)$. Finally also the $b = 3$ case leads to a contradiction.

□

We finish this section with

2.4.4.1 The open "$gb = 2$"-Problem

Suppose $P \in \mathcal{P}^\circ(S)$, $gb(P) = 2$ and that G has rank at least three. Determine the shape of $\widetilde{C}$ and P

Note that by the definition of $gb(P) = 2$, $Y_P \le Q$ and $\langle Y_P^{\widetilde{C}} \rangle$ is not abelian. So it should be possible to treat the $gb = 2$ problem with the methods of Parker/Rowley from [PR].

2.4.5 Rank 2

In this section we consider the case where $Q!$ holds and G has rank 2. The general idea is to show that $\langle P, \widetilde{P} \rangle$ is a weak BN-pair and then apply the Delgado-Stellmacher weak BN-pair Theorem [DS]. More precisely we try to characterize the situations where no weak BN-pair can be found. The following theorem has been proved in [Ch1] and [Ch2]

Theorem 2.4.11 (The Rank 2 Theorem) *Suppose $Q!$, $P!$, and $\ddot{P}!$ and that G has rank 2. Choose $\widetilde{P} \in \mathcal{P}_{\widetilde{C}}(S)$ such that*

(i) $\langle P, \widetilde{P} \rangle \notin \mathcal{L}$.

(ii) $H := \langle P \cap \widetilde{C}, \widetilde{P} \rangle$ *is minimal with respect to (i).*

(iii) $\widetilde{P}$ *is minimal with respect to (i) and (ii).*

Then one of the following holds:

1. $Y_P \not\leq O_p(\widetilde{P})$
2. $(N_H(P^\circ)P^\circ, H)$ *is a weak BN-pair.*
3. (H, P) *has the same shape as a suitable pair of parabolic subgroups in one of the following groups.*
    1. *For* $p = 2$, $U_4(3).2^e$, $G_2(3).2^e$, $D_4(3).2^e$, $HS.2^e$, F_3, $F_5.2^e$ *or* Ru.
    2. *For* $p = 3$, $D_4(3^n).3^e$, Fi_{23} *or* F_2.
    3. *For* $p = 5$, F_2.
    4. *For* $p = 7$, F_1.

We will not go into the details of this proof. It is a rather technical application of the amalgam method applied to the pair $(N_H(P^\circ)P^\circ, H)$.

The Rank 2 Theorem leaves as in the rank 2 case with the following open problem.

2.4.5.1 The open "Rank 2, gb=1"-Problem

Suppose Q! holds and there exists $P, \widetilde{P} \in \mathcal{P}(S)$ *such that* $\langle P, \widetilde{P} \rangle \notin \mathcal{L}$, $P \in \mathcal{P}^\circ(S)$ *and* $gb(P) = 1$. *Determine the structure of* P.

2.4.6 $gb = 1$

In this section we assume $E!$ and that G has rank at least 3. We investigate the case where $Y_M \not\leq Q$ for some $M \in \mathcal{M}(S)$ with M° maximal. Put $M_0 = M^\circ S$. The Structure Theorem 2.4.5 tells us the action of $M^\circ/O_p(M^\circ)$ on Y_M.

But we can get a lot more information. Let us consider one example. Suppose $\overline{M_0} = F^*(M_0/O_p(M_0)) \cong SL_n(q)$ and Y_M is the natural module. Then $\overline{M_0}$ has the following Dynkin diagram

We have that $C_{\overline{M_0}}(\Omega_1(Z(S)) \cap Y_M)$ is a maximal parabolic, which then by $E!$ is in $\widetilde{C}$. Hence there is a unique minimal parabolic P in M_0 which is not in $\widetilde{C}$. Notice that most of our groups we aim at are groups of Lie type in which $\widetilde{C}$ is a maximal parabolic. So there is a unique $P \in \mathcal{P}^\circ(S)$. Hence we are going to approach this situation.

But this unique P does not exist in general, as one can see in Case 2 of the structure theorem 2.4.5, in Case 3 with $M^\circ/O_2(M^\circ) \cong 3A_6$ and Y_M a 6-dimensional module, and in Case 4 with $M^\circ/O_3(M^\circ) \cong M_{11}$ and Y_M a 5-dimensional module. Hence one cannot expect a similar theorem as 2.4.7 for "$gb = 1$".

To be able to state the theorems in this section we need to introduce some notation:

Let H^* be a finite group. We say the G is of *identification type* H^* provided that:

(I1) There exist $T^* \in \mathrm{Syl}_p(H^*)$ and $I^* \subseteq \mathcal{P}_{H^*}(T^*)$ with $H^* = \langle I^* \rangle$.

(I2) There exists $H \leq G$ with $C_G(H) = 1 = O_p(H)$ and $M_0 \leq N_G(H)$.

(I3) Let $T = S \cap H$. Then there exists $I \subseteq \mathcal{P}_H(T)$ with $H = \langle I \rangle$.

(I4) There exists a bijection $I \to I^*, L \mapsto L^*$ such that for all $J \subset I$,

$$\langle J \rangle / O_p(\langle J \rangle) \cong \langle J^* \rangle / O_p(\langle J^* \rangle).$$

(I5) There exist $M^*, C^* \in \mathcal{L}_{H^*}(T^*)$ such that $M_0 \cap H$ has the same structure as M^* and $\widetilde{C} \cap H$ has the same structure as $\widetilde{C}^*$.

Theorem 2.4.12 *Suppose $E!$, $gb(P) = 1$, $\operatorname{rank} G > 2$ and $\mathcal{P}^\circ(S) \neq \{P\}$. If $p = 2$ then G is of identification typ M_{24}, He or $L_n(q)$.*

So suppose from now on that $\mathcal{P}^\circ(S) = \{P\}$.

Here is another observation. Let $P \in \mathcal{P}^\circ_{M_0}(S)$. Then in our example P corresponds to an end node of the Dynkin diagram of M_0. Hence (in most cases) there is a unique $\widetilde{P}$ in $\mathcal{P}_{M_0}(S)$ with $\widetilde{P} \not\leq N_G(P^\circ)$. Let us consider the group G we aim at, a group of Lie type. Then again in most cases P corresponds to an end node of the Dynkin diagram of G and there exists a unique minimal parabolic in $\mathcal{L}(S)$ not normalizing P°.

Unfortunately this is not true in general, for example if Y_M is the exterior square of the natural $SL_n(q)$–module. To analyze this situation, we consider $P_1 \neq P_2$ in $\mathcal{P}(S)$ such that P_i does not normalize P° for $i = 1, 2$. Let $L = \langle P_1, P, P_2 \rangle$. The case $O_p(L) = 1$ should be approachable with the amalgam method, (see the open problem at the end of the section).

So suppose that $L \in \mathcal{L}$. From the structure theorem we conclude that $L^\circ/C_{L^\circ}(Y_L) \cong SL_n(q)$, $n \geq 4$ (on the exterior square), M_{24} (on a 11-dimensional module) or M_{22} (on a 10-dimensional module.) These cases lead to the different groups in our next theorem.

Theorem 2.4.13 *Suppose $E!$, $\operatorname{rank} G > 2$, $\mathcal{P}(S) = \{P\}$ and $gb(P) = 1$. Furthermore, assume that there exist $P_1 \neq P_2 \in \mathcal{P}(S)$ with $P_i \not\leq N_G(P^\circ)$ and $\langle P_1, P, P_2 \rangle \in \mathcal{L}$. Then G is of identification-type $\Omega^\pm_n(q)$ or (for $p = 2$) Co_2, $M(22)$, Co_1, J_4, or $M(24)'$*

From now on we can assume that there is a unique P in $\mathcal{P}^\circ(S)$ and a unique $\widetilde{P} \in \mathcal{P}_{M_0}(S)$ which does normalize P°.

Theorem 2.4.14 *Suppose $E!$, $\operatorname{rank} G > 2$, $\mathcal{P}(S) = \{P\}$, $gb(P) = 1$ and that there exists a unique $\widetilde{P} \in \mathcal{P}(S)$ with $\widetilde{P} \not\leq N_G(P^\circ)$. If $p = 2$, then G is of identification type $U_n(q)$, ${}^2E_6(q)$, $E_6(q)$, $E_7(q)$, Sz, F_2 or F_1.*

In the remainder of this section, we will illustrate in some examples the basic ideas of the proof of the theorems. All the examples will be for $p = 2$.

Example 2.4.15 *Let $K = F^*(M_0/O_2(M_0)) \cong M_{24}$ and assume that Y_M contains an 11-dimensional submodule V with $|Y_M : V| \leq 2$. Assume further that V is the module in which $L = C_K(C_V(S)) \cong 2^6 3Sym(6)$. Then G is of identification type J_4 or $M(24)'$.*

For L we have the following series in V

$$1 < V_1 < V_2 < V,$$

where $|V_1| = 2$, V_2/V_1 is the 6–dimensional $3Sym(6)$–module and V/V_1 is the 4–dimensional $Sym(6)$–module. As $\widetilde{C} \cap M/O_2(M_0)$ contains L, we see that $QO_2(M_0)/O_2(M_0)$ is the elementary abelian subgroup of order 2^6 in L.

Suppose $V \leq Q$. By 1.1.2 V is not an FF–module and we conclude that $W = \langle (Y_M \cap Q)^{\widetilde{C}} \rangle$ is elementary abelian. Hence $W \leq O_2(M_0)$, i.e. $[Y_M, W] = 1$. But as $Y_M \not\leq Q$ and $[Q, Y_M] \leq W$ this contradicts $C_{\widetilde{C}}(Q) \leq Q$.

So we have $V \not\leq Q$. This now gives $V_2 = [V, Q]$ and $V_1 = [V_2, Q]$. Define $W := \langle V_2^{\widetilde{C}} \rangle$. Then V acts quadratically and nontrivially on W. Further from M_0 we see that for any $x \in V$ we have $|[W/V_1, x]| \leq 2^4$. Let L_1 be the pre-image of L in M_0. Then $L_1/O_2(L_1) \cong 3Sym(6)$. Let $U = \langle V^{\widetilde{C}} \rangle Q$. Then $C_U(W)Q/Q \leq Z(U/Q)$. Hence as $|[W/V_1, x]| \leq 2^4$ for $x \in V$, we see that $[F(U/Q), V] = 1$. So there is some component U_1 of U/Q containing L_1'/Q. If U_1 is a group of Lie type defined over a field of characteristic 2 we see that it has to be $F_4(2)$ or $Sp_{2n}(2)$, for some n. But in both cases the $Sp_4(2)$–parabolic has no elementary abelian normal subgroup of order 16. So U_1 is not a group of Lie type defined over a field of characteristic 2. As VQ/Q acts quadratically an application of 1.1.1 yields that $U_1 \cong 3U_4(3)$ or $3M_{22}$. This now tells us that V_1 is normal in U and that W/V_1 involves exactly one nontrivial irreducible module, which is 12–dimensional. As $[V, Q] \leq W$, we see that $[U, Q] \leq W$. This shows that $L_1/O_2(L_1)$ possesses exactly three nontrivial chief-factors in $O_2(L_1)$, two of them 6–dimensional and one 4-dimensional. Since L_1 has a a 4–dimensional and a 6-dimensional factor in V and a 6-dimensional factor in $QO_2(M_0)/O_2(M_0)$, we get that $[F^*(M_0/O_2(M_0)), O_2(M_0)] \leq V$. This shows that $O_2(M_0) = Y_M$ and so $O_2(M_0) = V$ or $|O_2(M_0) : V| = 2$.

In both cases we get that Q is extraspecial of order 2^{13} and that $\widetilde{C}/Q$ is an automorphism group of $3M_{22}$ or $3U_4(3)$. In the former case we have $\widetilde{C}/Q \cong 3Aut(M_{22})$ and so G is of identification-type J_4. So assume the latter case. Then we have that $\widetilde{C}/Q \cong 3U_4(3).2$, or $3U_4(3).4$. Now M_0 has a geometry with diagram and $\widetilde{C}$ has one with diagram

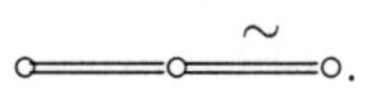

The intersection is the geometry for L_1. Let $P \in \mathcal{P}^{\circ}_{M_0}(S)$. Then P centralizes the foursgroup on which P_0 acts nontrivially. Hence $\langle P_0, P\rangle = P_0P$. This shows that we have a geometry with diagram

and that G is of identification-type $M(24)'$.

Example 2.4.16 *Let $K = F^*(M_0/O_2(M_0)) \cong \Omega^+_{10}(q)$, q a power of 2, and assume that Y_M contains V the half spin module. Then G is of identification-type $E_6(q)$.*

Let $L = C_K(C_V(S))$. Then $L \sim q^{\binom{n}{2}}L_n(q)$ and $QO_2(M_0)/O_2(M_0) = O_2(L)$. Note that V has the following L–series

$$1 < V_1 < V_2 < V,$$

where $|V_1| = q$, $|V_2/V_1| = q^{10}$, $|V/V_2| = q^5$. As in 2.4.15 $V \not\leq Q$, and so $V_2 = [V,Q] = V \cap Q$. Now $|V/V \cap Q| = q^5$ and $V/V \cap Q$ is a natural module for $L_1/O_2(L_1)$, where L_1 is the pre-image of L. We can now proceed as in 2.4.15. Let U be as before, then we again see that $[F(U/Q), V] = 1$. Let U_1 be a component of U/Q containing $O^2(L_1/Q)$. Because of quadratic action and the fact that $|VQ/Q| \geq 32$, we get with 1.1.1 that U_1 is a group of Lie type in characteristic two and the list of possible groups U_1 and the corresponding modules W in Q. As L_1 induces in some W on $C_W(O_2(O^2(L_1)))$ the 10 - dimensional module, we see that W is not a $V(\lambda)$ where λ belongs to an end node of the Dynkin diagram of U_1. Hence the possible groups U_1 are $SL_n(q)$, $Sp_{2n}(q)$ or $U_n(q)$. Further for $t \in V$ we have $[W,t] \leq C_W(O_2(O^2(L_1)))$ and $C_{L_1}(t)$ has to act on this group. This shows that $|[W,t]| = q^6$ or q^4. This in the first place shows that $U_1 \cong SL_n(q)$ and then that $W = V(\lambda_2)$ or $V(\lambda_3)$. In both cases we have $U_1 \cong SL_6(q)$, as $|[W,t]| \leq q^6$. Moreover, $[W,t]$ is not the natural $C_L(t)$–module in the case $|[W,t]| = q^6$. If we have $W = V(\lambda_2)$, then $W/C_W(V)$ is a 5-dimensional L_1–module, but there is no such module in $O_2(L)$. So we have that $W = V(\lambda_3)$. Now we see that $L/O_2(L)$ induces in $QO_2(M_0)$ exactly two 10-dimensional and one 5-dimensional module, as $[V,Q] \leq \langle V_2^{\widetilde{C}}\rangle$. But in V this group induces one 10-dimensional module and one 5-dimensional one. Further in $QO_2(M_0)/O_2(M_0)$ we see another 10-dimensional module. This shows $[K, O_2(M_0)] = V$. Again $Y_M = O_2(M_0)$ and so $Y_M = V$. Now we see that $M^{\circ} \cong q^{16}\Omega^+_{10}(q)$ and $U \cong q^{1+20}SL_6(q)$. The intersection is the $SL_5(q)$-parabolic. Now in this case we are in the situation of 2.4.14, so any minimal parabolic not in M_0 normalizes P°.

We try to show that $H = \langle M^{\circ}, U\rangle$ has a parabolic system with an E_6 – diagram.

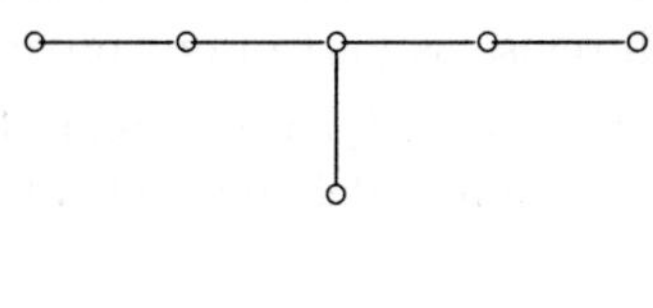

We have that $M_0 = M^\circ S$ and so there might be some field automorphism involved. But these field automorphisms are also field automorphisms on L, so they induce field automorphisms on $U/O_2(U)$. This shows that U and M^0 have a common Sylow 2–subgroup, and so G is of identification-type $E_6(q)$.

2.4.6.1 The open "$\widetilde{P}$!,gb=1"-Problem 2

Suppose $E!$, $\operatorname{rank} G > 2$, $\mathcal{P}^\circ(S) = \{P\}$, $gb(P) = 1$ and that there exist $P_1, P_2 \in \mathcal{P}(S)$ such that for $i = 1$ and 2:

(i) $P_i \not\leq N_G(P^\circ)$.

(ii) $M_i := \langle P, P_i \rangle \in \mathcal{L}$

(iii) $\langle M_1, M_2 \rangle \notin \mathcal{L}$.

Determine the shape of M_1 and M_2.

As a starting point towards a solution of the preceding problem we observe

Lemma 2.4.17 *Suppose $E!$ and $\mathcal{P}^\circ(S) = \{P\}$. Let $\widetilde{P} \in \mathcal{P}(S)$ with $\widetilde{P} \neq P$, $L := \langle P, \widetilde{P} \rangle \in \mathcal{L}$ and $\widetilde{P} \not\leq N_G(P^\circ)$. Then $L \in \mathcal{L}^\circ$ and $L^\circ/C_{L^\circ}(Y_L) \cong SL_3(q)$, $Sp_4(q)$, $\Omega_5(q)$ (and p odd), $Alt(6)$ (and $p = 2$), or $2.M_{12}$ (and $p = 3$).*

Note that in all cases of the preceding lemma $L/O_p(L)$ has a weak BN-pair of rank 2. Hence [StTi] provides a solution to the above open problem. But we believe that our stronger assumptions allow for a shorter solution.

3 The Global Analysis

We have not yet devoted much time to this part of the project, but here are some thoughts.

The main tool to identify the group G is via a diagram geometry for a non-local parabolic subgroup H of G. Usually we will not only know the diagram but also the group induced on each of the residues and so the isomorphism type of each of the residues. This allows to identify the geometry and then the group H.

For example if the diagram is the diagram of a spherical building of rank at least four, then the isomorphism type of the residues uniquely determines the building. This follows from the classification of spherical buildings, but can actually be proved using only a small part of the theory of buildings.

For many of the diagrams which we encounter, classification results are available in the literature. At this time we have not decided which of these results we will quote and which ones we will revise as part of our program.

The situation when $\mathcal{M}(S) = \{M_1, M_2\}$ is different. If (M_1, M_2) is a weak BN-pair associated to a BN-pair of rank 2 defined over a field which is not too small, one tries to recover the Weyl group. For p odd, this probably requires a $\mathcal{K}$-group assumptions not only for the p-local subgroups but also for some 2-locals. Once the Weyl group has been identified, H can be recognized as a group of Lie-type, see [BS].

Suppose (M_1, M_2) is not associated to a BN-pair of rank 2. If $p = 2$, knowledge of the parabolic subgroups often allows to determine the order of G by counting involutions. The actual identification will be done by some ad hoc methods depending on the group. If p is odd, the group is probably better left unidentified.

After the group H is identified, one still needs to deal with situations where $H \neq G$. Usually our choice of the group H will allow us to show that H is the p-core with respect to S, but some exceptions will have to be dealt with. The strongly p-embedded situation has been discussed before. If G has rank 1 the CGT-theorem 2.2.3 will limit the structure of H. For $p = 2$ this hopefully will lead to a contradiction, while for $p \neq 2$ we might not be able to identify G.

References

[Asch] M. Aschbacher, A Factorization Theorem for 2-constrained Groups, Proc. London. Math. Soc. (3) 43 (1981), 450-477.

[BBSM] B. Baumeister, U. Meierfrankenfeld et.al, The big book of small modules, in preparation.

[Be] H. Bender, Transitive Gruppen gerader Ordnung, in denen jede Involution genau einen Punkt festläßt, J. Algebra 17 (1971) 527 - 554.

[Br] R. Brauer, On the structure of groups of finite order, in : Proc. Internat. Congr. Math. Vol 1, 1954, 209 - 217.

[BS] C.D. Bennett, S. Shpectorov, A remark on a theorem of J. Tits, Proc. Amer. Math. Soc. 129, (2001) 2571-2579.

[Ch1] A. Chermak, Finite groups generated by a pair of minimal parabolic subgroups: Part I, preprint.

[Ch2] A. Chermak, Finite groups generated by a pair of minimal parabolic subgroups: Part II, draft.

[DS] A. Delgado, B. Stellmacher Weak BN-pairs of rank 2, in A. Delgado, D. Goldschmidt, B. Stellmacher, *Groups and graphs: new results and methods*, DMV Seminar, 6. Birkhäuser Verlag, Basel, (1985) 244 pp.

[FeTh] W. Feit, J. Thompson, Solvability of groups of odd order, Pacific J. Math. 13 (1963) 775–1029.

[Gl] G. Glauberman, Central elements in core-free groups, J. Algebra 4 (1966) 403–420.

[Gor] D. Gorenstein,*Finite Groups*, Chelsea (1980) New York.

[GM] R. Guralnick, G. Malle, Classification of 2F-Modules, I, submitted.

[MMPS] M. Mainardis, U. Meierfrankenfeld, G. Parmeggiani, B. Stellmacher, The $\widetilde{P}$!-Theorem, preprint.

[PR] C. Parker, P. Rowley, *Symplectic Amalgams*, Springer Monographs in Mathematics, Springer Verlag (2002) London. 360 pp.

[PPS] C. Parker, G. Parmeggiani, B. Stellmacher, The P!-Theorem, submitted.

[StTi] B. Stellmacher, F. Timmesfeld,Rank 3 amalgams, Mem. Amer. Math. Soc. 136, 1998, 123pp.

Modular subgroup arithmetic

Thomas W. Müller

1. Introduction

For a group $\mathfrak{G}$ denote by $s_n(\mathfrak{G})$ the number of subgroups of index n in $\mathfrak{G}$. If, for instance, $\mathfrak{G}$ is finitely generated or of finite subgroup rank, then $s_n(\mathfrak{G})$ is finite for all n. Groups $\mathfrak{G}$ enjoying the latter property will be referred to as FSG-groups. *Modular subgroup arithmetic* deals with divisibility properties of the sequence $\{s_n(\mathfrak{G})\}_{n\geq 1}$ or related subgroup counting functions and their connection with the algebraic structure of the underlying FSG-group $\mathfrak{G}$. The natural context for these studies is the theory of *subgroup growth*, which has evolved over the last two decades in the work of Grunewald, Lubotzky, Mann, Segal, and others including the present author; cf. Lubotzky's Galway notes [29], [30], as well as the forthcoming book [31] by Lubotzky and Segal.

In [29, § 1] particular mention is made of the problem to understand divisibility properties of the function $s_n(\mathfrak{G})$, and Lubotzky explicitly expresses his hope there that a very interesting theory might be awaiting its discovery. This prediction has turned out to be remarkably far–sighted and insightful; until recently however, modular subgroup arithmetic could hardly have been described as a subject area, since with the exception of Stothers' formulae for the classical modular group not much was known in this direction. For a finitely generated group $\mathfrak{G}$ define

$$\Pi(\mathfrak{G}) := \{n \in \mathbb{N} :\ s_n(\mathfrak{G}) \equiv 1 \bmod 2\}.$$

Moreover, for $\mathfrak{G}$ a finitely generated virtually free group, let

$$\Pi^*(\mathfrak{G}) := \{\lambda \in \mathbb{N} :\ f_\lambda(\mathfrak{G}) \equiv 1 \bmod 2\}.$$

Here, $f_\lambda(\mathfrak{G})$ is the number of free subgroups in $\mathfrak{G}$ of index $\lambda m_{\mathfrak{G}}$, where $m_{\mathfrak{G}}$ denotes the least common multiple of the orders of the finite subgroups in $\mathfrak{G}$. The sets $\Pi(\mathfrak{G})$ and $\Pi^*(\mathfrak{G})$ are called the *parity pattern* respectively *free parity pattern* of the group $\mathfrak{G}$. In this notation, the main result of Stothers [50] is that

$$\Pi(PSL_2(\mathbb{Z})) = \{2^{\sigma+1} - 3\}_{\sigma\geq 1} \cup 2\{2^{\sigma+1} - 3\}_{\sigma\geq 1}. \tag{1}$$

In the course of his proof, Stothers also shows that

$$\Pi^*(PSL_2(\mathbb{Z})) = \{2^{\sigma} - 1\}_{\sigma\geq 1}. \tag{2}$$

The latter pattern has been shown to occur for a larger class of virtually free groups of free rank 2, including free products $\mathfrak{G} = G_1 *_S G_2$ of two finite groups G_i with an amalgamated subgroup S of odd cardinality, whose indices $(G_i : S)$ satisfy $\{(G_1 : S), (G_2 : S)\} = \{2,3\}$ or $= \{2,4\}$; cf. [33, Prop. 6]. [1] One of the side results in the present author's 1989 dissertation [32] provides a somewhat surprising analogue of (1) for the special linear group $SL_2(\mathbb{Z})$:

$$\Pi(SL_2(\mathbb{Z})) = \Pi(PSL_2(\mathbb{Z})) \cup \{4\} = \{2^{\sigma+1} - 3\}_{\sigma\geq 1} \cup 2\{2^{\sigma+1} - 3\}_{\sigma\geq 1} \cup \{4\}; \tag{3}$$

[1] The free rank $\mu(\mathfrak{G})$ of a finitely generated virtually free group $\mathfrak{G}$ is defined as the rank (in the ordinary sense) of a free subgroup of index $m_{\mathfrak{G}}$ in $\mathfrak{G}$; cf. section 2.2.

cf. [32, Satz 2]. During the early 1990's, Grady and Newman in a series of papers established certain periodicity results for $s_n(\mathfrak{G})$ in the case when $\mathfrak{G}$ is a finitely generated free product; their results are summarized in Corollary 3. In this context, Grady and Newman also stated a number of interesting conjectures, which we list below.

Conjecture 1 ([15]). *If $\mathfrak{G}$ is a free product of $r \geq 2$ copies of the cyclic group of order 2, then $\Pi(\mathfrak{G}) = \mathbb{N}$.*

Conjecture 2 ([16, Conj. 1]). *If $\mathfrak{G}$ is a free product of finitely many cyclic groups of prime orders, containing in its free decomposition at least two copies of the cyclic group of order 2, then $\Pi(\mathfrak{G}) = \mathbb{N}$.*

Conjecture 3 ([16, Conj. 2]). *If $\mathfrak{G} = C_p^{*2} * C_{p_1} * \cdots * C_{p_k}$ with primes $p, p_1, \ldots, p_k$ satisfying $p_i \geq p$ for all i, then*

$$s_n(\mathfrak{G}) \equiv -\sum_{i=1}^{p-1} \frac{s_{n-p+i}(\mathfrak{G})}{(p-i)!} \quad \mathrm{mod}\ p, \quad n > p-1.$$

Conjecture 4 ([17, § 6]). *If p is a prime and $\mathfrak{G}$ is a free product of finitely many finite groups containing in its free decomposition a factor G^{*2}, where G contains a subgroup of index p, then $s_n(\mathfrak{G})$ is periodic modulo p.*

Clearly, Conjecture 3 $\Rightarrow$ Conjecture 2 $\Rightarrow$ Conjecture 1. [16, Theorem 1] implies the truth of Conjecture 1 for $r \geq 4$; an unconditional proof of Conjecture 1 is found in section 2.1. Otherwise, apart from numerical evidence, not much appears to be known concerning these conjectures.

Stothers' striking result (1), which had been conjectured for some time on the basis of numerical evidence, has, for more than 20 years, stood out as an indication that a fascinating chapter of subgroup arithmetic might be waiting to be explored; in particular, it had been a long standing open problem, whether Stothers' formulae generalize to Hecke groups, i.e., groups of the form $\mathfrak{G} = \mathfrak{H}(\mathfrak{q}) \cong C_2 * C_\mathfrak{q}$ for some $\mathfrak{q} \geq 3$. Nevertheless, despite the effort of a number of experts over the next 20 years, the only new result which eventually emerged in this direction was the author's discovery in 1998 of an analogue of Stothers' formula (1) for the Hecke group $\mathfrak{H}(5) \cong C_2 * C_5$, namely

$$\Pi(\mathfrak{H}(5)) = \left\{\frac{2^{2\sigma+1}-5}{3}\right\}_{\sigma\geq 1} \cup 2\left\{\frac{2^{2\sigma+1}-5}{3}\right\}_{\sigma\geq 1};$$

cf. [35]. The situation changed with the occurrence of the papers [37], [38], and [40]. Introducing a number of powerful new ideas, these papers have led to a massive breakthrough, and, at the time of writing, a substantial body of knowledge has formed concerning modular properties of the function $s_n(\mathfrak{G})$, as well as the beginnings of a systematic theory for the function $f_\lambda(\mathfrak{G})$ counting free subgroups of finite index. Moreover, some of the most recent results (like the description of the parity pattern for surface groups and other one–relator groups with its representation-theoretic background) seem to open up completely new and exciting horizons yet to be explored. The aim of this paper is to describe the present state of knowledge concerning these questions, to point out open problems and to state a number of conjectures. If these pages convey some of the complexity and delicate beauty of the subject matter, the joy and excitement of finding new results in this area, and perhaps inspire further development in the field, the effort of writing them will have been more than worthwhile.

As it presents itself at the moment, the theory falls naturally into five chapters.

(i) A descent principle.

(ii) A criterion for periodicity of the function $s_n(\mathfrak{G})$.

(iii) A systematic investigation of the mod p behaviour of $s_n(\mathfrak{G})$ for free products $\mathfrak{G}$ of the form

$$\mathfrak{G} = \mathfrak{H}(G, q_1, q_2) = C_{q_1} * \underbrace{G * \cdots * G}_{q_2}, \tag{4}$$

where p is a prime, G an arbitrary finite group, and q_1, q_2 are p–powers such that $q_1 q_2 > 1$.

(iv) Results concerning the function $f_\lambda(\mathfrak{G})$ counting torsion-free (i.e. free) subgroups of index $\lambda m_\mathfrak{q}$ in $\mathfrak{G}$ in the case when $\mathfrak{G} = \mathfrak{H}(\mathfrak{q})$ is a Hecke group for some $\mathfrak{q} \geq 3$, and with $m_\mathfrak{q} := m_{\mathfrak{H}(\mathfrak{q})} = [2, \mathfrak{q}]$.

(v) The determination of the parity pattern for a class of one–relator groups in particular containing all surface groups.

The descent principle relates the mod p behaviour of $s_n(\mathfrak{G})$ to that of a normal subgroup $\mathfrak{N}$ of $\mathfrak{G}$ in the case when the quotient $\mathfrak{G}/\mathfrak{N}$ is cyclic of p-power order for some prime p. Even in this restricted form the existence of a stable connection mod p between the subgroup arithmetic of a group and that of a finite index subgroup is highly surprising, and has important applications; for instance, it leads to the explicit determination of the subgroup arithmetic modulo p for fundamental groups of trees of groups all of whose vertex groups are cyclic of p–power order for some given prime p (Theorem 2). These aspects as well as a possible generalization of the descent principle will be described in section 2, while section 3 is devoted to a result concerning periodicity of the function $s_n(\mathfrak{G})$ modulo a distinguished prime: if a finitely generated group $\mathfrak{G}$ contains as a free factor a free product of finite abelian groups whose p–Sylow subgroups are non–trivial and have ranks at most 2 (for a fixed prime p), then generically (i.e. up to a rather weak arithmetic condition) $s_n(\mathfrak{G})$ will be periodic modulo p (Theorem 3).

As a consequence of the descent principle, the functions $s_n(\mathfrak{G})$ encountered in subsection 2.2 are periodic modulo p, since a tree group $\mathfrak{G}$ as described above can be shown to contain a free normal subgroup of index $m_\mathfrak{G}$, and the subgroup numbers of finitely generated free groups are periodic modulo every prime. Furthermore, it turns out that two finite abelian factors usually suffice in Theorem 3 to ensure periodicity, which places groups of the form (4) within a tight and interesting borderline region. Indeed, as we shall see in section 5, the generic picture for these groups is one of a peculiar type of fractal behaviour, contrasting with explicit closed formulae of Stothers' type for the set

$$\mathfrak{N}_{G,q_1,q_2} := \left\{n \in \mathbb{N} : s_n(\mathfrak{H}(G, q_1, q_2)) \not\equiv 0 \mod p\right\}$$

in certain arithmetic breakdown situations intimately related to Fermat primes. The key to our results for the groups $\mathfrak{H}(G, q_1, q_2)$ takes the form of a rather intricate and deep lying identity for the mod p projection $X_{G,q}(z)$ of the generating series

$\sum s_{n+1}(\mathfrak{H}(G,q))z^n$, where $\mathfrak{H}(G,q) := \mathfrak{H}(G,1,q) = G^{*q}$. Despite its complicated nature, this identity leads to a beautiful and quite comprehensive theory for the groups $\mathfrak{H}(G,q)$, and subsequently, via the descent principle, for groups of the more general form (4). A sketch of the proof of this functional equation for the $GF(p)$–series $X_{G,q}(z)$ is provided in section 4, while section 5 is devoted to some of its major consequences.

In section 6, we discuss the present state of affairs concerning modular properties of the function $f_\lambda(\mathfrak{G})$ counting free subgroups of index $m_{\mathfrak{G}}\lambda$ in a finitely generated virtually free group $\mathfrak{G}$. Here, a well developed theory exists only in the case when $\mathfrak{G} = \mathfrak{H}(\mathfrak{q})$ is a Hecke group for some $\mathfrak{q} \geq 3$, and $p = 2$. The main features of the latter theory parallel the behaviour modulo 2 of the function $s_n(\mathfrak{G})$ discussed in section 5; in particular, Fermat primes again play an important special role in determining the number–theoretic conditions for a breakdown of what is generically a peculiar type of fractal behaviour, this breakdown leading to patterns that can be described in explicit closed form. This area is clearly in need of further research.

One of the main results of [42] is that for $\mathfrak{G}$ ranging over a certain class of one-relator groups containing in particular all surface groups, the number $s_n(\mathfrak{G})$ of index n subgroups in $\mathfrak{G}$ is odd if and only if n is a square or twice a square, provided that $\mathrm{rk}(\mathfrak{G}) \geq 3$. This remarkable result relies on newly found parity properties of character values and multiplicities for symmetric groups, as well as on classical results of Gauß and Legendre concerning representation numbers of binary quadratic forms. Our last section is devoted to this topic.

Sections 4 and 6.1, which describe some of the key ideas and new methods underlying our recent progress in modular subgroup theory, are of a more technical nature; they may be skipped at first reading, or by readers only interested in an account of the main new results.

It is my pleasure to thank Peter Cameron for the enthusiasm he has shown for my work ever since I joined Queen Mary College in 1999, as well as for a number of stimulating questions and fruitful discussions, some of which are reflected in the results reported in these pages. Thanks are also due to Christian Krattenthaler for his valuable comments concerning an early version of some of the material presented here, and to Martin Liebeck for inviting this report.

2. The descent principle and some applications

2.1. **The descent principle.** In general, divisibility properties of subgroup counting functions appear to be rather peculiar to the particular group under investigation, and (unlike their growth behaviour) tend to react extremely sensitively to deformation of the underlying group within a commensurability class; in particular, when passing from a group to a subgroup of finite index, arithmetic structure of this kind is usually severely deformed if not completely destroyed. There is however one special situation, described in Theorem 1 below, where the existence of a stable connection modulo a distinguished prime between the subgroup arithmetic of a group and that of a finite index subgroup can be established. In order to state this as well as further results, it is convenient to fix some notation. For a prime p

and an FSG–group $\mathfrak{G}$ define the *p–pattern* $\Pi^{(p)}(\mathfrak{G})$ of $\mathfrak{G}$ to be the family of sets

$$\Pi^{(p)}(\mathfrak{G}) = \left\{\Pi_1^{(p)}(\mathfrak{G}), \Pi_2^{(p)}(\mathfrak{G}), \ldots, \Pi_{p-1}^{(p)}(\mathfrak{G})\right\},$$

where

$$\Pi_i^{(p)}(\mathfrak{G}) := \left\{n \in \mathbb{N}:\ s_n(\mathfrak{G}) \equiv i \bmod p\right\}, \quad 0 < i < p;$$

in particular, $\Pi_1^{(2)}(\mathfrak{G}) = \Pi(\mathfrak{G})$ is the parity pattern of $\mathfrak{G}$ introduced in the previous section. In this notation our result reads as follows.

Theorem 1 ([38, Theorem 1]). *Let $\mathfrak{G}$ be an* FSG*–group, p a prime, and let $\mathfrak{N} \trianglelefteq \mathfrak{G}$ be a normal subgroup of index p^r, with $\mathfrak{G}/\mathfrak{N}$ cyclic. Then the p-patterns $\Pi^{(p)}(\mathfrak{G})$ and $\Pi^{(p)}(\mathfrak{N})$ are related via the equations*

$$\Pi_i^{(p)}(\mathfrak{G}) = p^r\,\Pi_i^{(p)}(\mathfrak{N}) \cup \bigcup_{\rho=0}^{r-1} p^\rho\left(\Pi_i^{(p)}(\mathfrak{N}) \cap (\mathbb{N} - p\,\mathbb{N})\right), \quad 0 < i < p. \tag{5}$$

Equivalently, if $X_{\mathfrak{G}}(z)$ denotes the series $\sum_{n\geq 0} s_{n+1}(\mathfrak{G})z^n$ considered modulo p, and if $X_{\mathfrak{N}}(z)$ is the corresponding $GF(p)$–series for the group $\mathfrak{N}$, then under our assumptions

$$X_{\mathfrak{G}}(z) = \sum_{\rho=0}^{r} z^{p^\rho-1}\,X_{\mathfrak{N}}(z^{p^\rho}) + \sum_{\rho=0}^{r-1} z^{p^{\rho+1}-1}\,X_{\mathfrak{N}}^{(p-1)}(z^{p^\rho}). \tag{6}$$

The proof of Theorem 1, as given in [38], depends on the exploitation of various group actions as well as two facts from the theory of finite groups:

(i) Frobenius' generalization ([13, § 4, Theorem I]) of Sylow's third theorem, which states that

if a prime power p^s divides the order of a finite group G, then the number of subgroups in G of order p^s is congruent to 1 mod *p,*

(ii) a theorem due to Philip Hall ([19, Theorem 1.6]) to the effect that

for a finite group G, a positive integer n, and an automorphism $\alpha \in \mathrm{Aut}(G)$ whose order divides n, the number of solutions in G of the equation

$$x \cdot \alpha(x) \cdot \alpha^2(x) \cdot \ldots \cdot \alpha^{n-1}(x) = 1 \tag{7}$$

is divisible by the greatest common divisor of n and $|G|$.

For $\alpha = \mathrm{id}$, the identical automorphism of G, Hall's theorem reduces to the well–known result of Frobenius concerning the equation $x^n = 1$ in finite groups; cf. [13, § 2, Theorem II] and [22]. A short and elegant proof of Frobenius' version of Sylow's theorem, which is based on the idea of Wielandt [52], can be found in [21, Chap. I, § 7].

As a first illustration of Theorem 1 we give a proof of Conjecture 1.

Corollary 1. *Let $\mathfrak{G} = C_2^{*r}$ be a free product of $r \geq 2$ copies of the cyclic group of order 2. Then $\Pi(\mathfrak{G}) = \mathbb{N}$, i.e., the number $s_n(\mathfrak{G})$ of index n subgroups in $\mathfrak{G}$ is odd for all n.*

Proof. We have $m_{\mathfrak{G}} = 2$ and $\chi(\mathfrak{G}) = -\frac{r-2}{2}$, and hence $\mu(\mathfrak{G}) = 1 - m_{\mathfrak{G}}\chi(\mathfrak{G}) = r - 1$. Here, $\chi(\mathfrak{G})$ is the rational Euler characteristic of $\mathfrak{G}$ in the sense of Wall, and $\mu(\mathfrak{G})$ is defined as in Footnote 1. It follows that $\mathfrak{G}$ contains a finitely generated infinite free subgroup of index 2. On the other hand, M. Hall's recursion formula ([18, Theorem 5.2])

$$s_n(F_r) = n\,(n!)^{r-1} - \sum_{\mu=1}^{n-1} \left((n-\mu)!\right)^{r-1} s_\mu(F_r) \quad n \geq 1 \tag{8}$$

shows that a finitely generated infinite free group has all its subgroup numbers odd, and our claim follows from equation (5). □

Further applications of Theorem 1 will be presented in the next two subsections.

2.2. **Divisibility properties determined by free normal subgroups.** The category of graphs used here is described in Serre's book [47]. By Stallings' structure theorem on groups with infinitely many ends and the subsequent work of Karrass, Pietrowski, and Solitar, a finitely generated virtually free group $\mathfrak{G}$ can be presented as the fundamental group of a finite graph of groups $(\mathfrak{G}(-), Y)$ in the sense of Bass and Serre with finite vertex groups $\mathfrak{G}(v)$; cf. [49] and [26], or [11, Sect. IV.1.9]. The fact that, conversely, the fundamental group of a finite graph of finite groups is always virtually free of finite rank is more elementary, and can be found for instance in [47, Sect. II.2.6]. It follows in particular from this characterization and the universal covering construction in the category of graphs of groups that a torsion-free subgroup of a finitely generated virtually free group is in fact free (which was the original contribution of Stallings' work to the structure theory of virtually free groups).

If $\mathfrak{U}$ is a free subgroup of finite index in $\mathfrak{G}$ then, following an idea of Wall, one defines the rational Euler characteristic $\chi(\mathfrak{G})$ of $\mathfrak{G}$ as

$$\chi(\mathfrak{G}) = -\frac{\mathrm{rk}(\mathfrak{U}) - 1}{(\mathfrak{G} : \mathfrak{U})}. \tag{9}$$

This is well-defined in view of Schreier's theorem [46], and if $\mathfrak{G} \cong \pi_1(\mathfrak{G}(-), Y)$ is a decomposition of $\mathfrak{G}$ in terms of a graph of groups, then we have

$$\chi(\mathfrak{G}) = \sum_{v \in V(Y)} \frac{1}{|\mathfrak{G}(v)|} - \sum_{e \in E(Y)} \frac{1}{|\mathfrak{G}(e)|}, \tag{10}$$

where $V(Y)$ and $E(Y)$ denote respectively the set of vertices and (geometric) edges of Y. The latter formula reflects the fact that in our situation the Euler characteristic in the sense of Wall coincides with the equivariant Euler characteristic $\chi_T(\mathfrak{G})$ of $\mathfrak{G}$ relative to the tree T canonically associated with $\mathfrak{G}$ in the sense of Bass and Serre; cf. [6, Chap. IX, Prop. 7.3] or [48, Prop. 14]. Define the (free) rank $\mu(\mathfrak{G})$ of $\mathfrak{G}$ to be the rank of a free subgroup of index $m_{\mathfrak{G}} = [|\mathfrak{G}(v)| : v \in V(Y)]$ in $\mathfrak{G}$. The existence of such a subgroup follows from [47, Lemmas 8 and 10] or formulae (56) and (57) in section 6. Observe that, in view of (9), $\mu(\mathfrak{G})$ is connected with the Euler characteristic of $\mathfrak{G}$ via

$$\mu(\mathfrak{G}) + m_{\mathfrak{G}}\chi(\mathfrak{G}) = 1, \tag{11}$$

which shows in particular that $\mu(\mathfrak{G})$ is well-defined. The question is now: when is $\mathfrak{G}$ going to have a free normal subgroup $\mathfrak{F}$ of index $m_{\mathfrak{G}}$ and with quotient $\mathfrak{G}/\mathfrak{F}$ cyclic of p-power order? If such an $\mathfrak{F}$ exists, then every vertex group $\mathfrak{G}(v)$ must be cyclic of p-power order, since it embeds isomorphically into $\mathfrak{G}/\mathfrak{F}$. Conversely, if all $\mathfrak{G}(v)$

are cyclic of p–power order, then $m_{\mathfrak{G}} = p^r$, where $p^r = \max\{|\mathfrak{G}(v)| : v \in V(Y)\}$, and every normal free subgroup $\mathfrak{F}$ of index p^r satisfies $\mathfrak{G}/\mathfrak{F} \cong C_{p^r}$. Assume for the remainder of this subsection that all vertex groups $\mathfrak{G}(v)$ are cyclic of p–power order, and let $m_{\mathfrak{G}} = p^r$. Our next result provides a sufficient condition for $\mathfrak{G}$ to contain free normal subgroups of index $m_{\mathfrak{G}}$.

Lemma 1 ([38, Lemma 1]). *Let $(\mathfrak{G}(-), Y)$ be a finite tree of groups all of whose vertex groups $\mathfrak{G}(v)$ are cyclic of p–power order for some fixed prime p, and let $\mathfrak{G}$ be its fundamental group. Then $\mathfrak{G}$ contains precisely*

$$\frac{\prod_{v\in V(Y)} \varphi(|\mathfrak{G}(v)|)}{\prod_{e\in E(Y)} \varphi(|\mathfrak{G}(e)|)} \Big/ \varphi(m_{\mathfrak{G}}) \tag{12}$$

free normal subgroups of index $m_{\mathfrak{G}}$, where φ is Euler's totient function; in particular such subgroups exist.

Now let $\mathfrak{G}$ be as in Lemma 1, and suppose that $\chi(\mathfrak{G}) \leq 0$. Then $\mathfrak{G}$ contains a free normal subgroup $\mathfrak{F}$ of index $m_{\mathfrak{G}} = p^r$ with cyclic quotient and of rank $\mu(\mathfrak{G}) \geq 1$, and, by Theorem 1, the p–pattern of $\mathfrak{G}$ is determined via (5) by the p–pattern of $\mathfrak{F}$. Take for instance $p = 2$; then, since a finitely generated infinite free group has all its subgroup numbers odd, we find that the same is true for $\mathfrak{G}$. This yields a far reaching generalization of Corollary 1. For general p, the subgroup numbers $s_n(\mathfrak{G})$, when considered modulo p, will be periodic, as the same holds for every such free group and each prime. Both statements concerning free groups follow immediately from M. Hall's recursion formula (8). Let us consider the case $p = 3$ in more detail. Suppose first that $\mathrm{rk}(\mathfrak{F})$ is even. Then, by (8), the numbers $s_n(\mathfrak{F})$ satisfy $s_n(\mathfrak{F}) \equiv s_{n-2}(\mathfrak{F}) + 2s_{n-1}(\mathfrak{F}) \bmod 3$, with initial values $s_1(\mathfrak{F}) \equiv 1\ (3)$ and $s_2(\mathfrak{F}) \equiv 0\ (3)$. From this recurrence relation mod 3, one shows that $s_n(\mathfrak{F})$ is periodic mod 3 with period 8, and, more precisely, that

$$s_n(\mathfrak{F}) \equiv 1 \bmod 3 \text{ if and only if } n \equiv 0, 1, 3 \bmod 8$$

$$s_n(\mathfrak{F}) \equiv 2 \bmod 3 \text{ if and only if } n \equiv 4, 5, 7 \bmod 8.$$

If, on the other hand, $\mathrm{rk}(\mathfrak{F})$ is odd, then we find that $s_n(\mathfrak{F}) \equiv 1\ (3)$ for all $n \geq 1$. Using this information in (5), and summarizing the preceding discussion we obtain the following.

Theorem 2 ([38, Theorem 2]). *Let p be a prime, $(\mathfrak{G}(-), Y)$ a finite tree of groups all of whose vertex groups are cyclic of p–power order, and let $\mathfrak{G}$ be its fundamental group. Suppose that $m_{\mathfrak{G}} = p^r$ and that $\chi(\mathfrak{G}) \leq 0$. Then*

(i) *the function $s_n(\mathfrak{G})$ is periodic modulo p,*
(ii) *for $p = 2$ we have $\Pi(\mathfrak{G}) = \mathbb{N}$,*
(iii) *for $p = 3$ and $\mu(\mathfrak{G})$ odd we have $\Pi_1^{(3)}(\mathfrak{G}) = \mathbb{N}$,*
(iv) *for $p = 3$ and $\mu(\mathfrak{G})$ even, $s_n(\mathfrak{G})$ is periodic modulo 3 with period $8 \cdot 3^r$. More precisely, in this case $s_n(\mathfrak{G}) \equiv 1 \bmod 3$ if and only if n is congruent* mod $8 \cdot 3^r$ *to one of the 3^{r+1} numbers*
0, 3^{r-1}, 3^r, $8 \cdot 3^{r-1}$, 3^{r+1}, $11 \cdot 3^{r-1}$, $16 \cdot 3^{r-1}$, $17 \cdot 3^{r-1}$, $19 \cdot 3^{r-1}$, $3^\rho(1+24\lambda)$, $8 \cdot 3^\rho(1+3\lambda)$, $3^\rho(11+24\lambda)$, $8 \cdot 3^\rho(2+3\lambda)$, $3^\rho(17+24\lambda)$, $3^\rho(19+24\lambda)$
with $0 \leq \rho < r-1$ and $0 \leq \lambda < 3^{r-\rho-1}$;
and $s_n(\mathfrak{G}) \equiv 2 \bmod 3$ if and only if n is congruent mod $8 \cdot 3^r$ *to one of the 3^{r+1} numbers*
$4 \cdot 3^{r-1}$, $5 \cdot 3^{r-1}$, $7 \cdot 3^{r-1}$, $4 \cdot 3^r$, $13 \cdot 3^{r-1}$, $5 \cdot 3^r$, $20 \cdot 3^{r-1}$, $7 \cdot 3^r$, $23 \cdot 3^{r-1}$,

$4{\cdot}3^\rho(1+6\lambda)$, $4{\cdot}3^\rho(5+6\lambda)$, $3^\rho(5+24\lambda)$, $3^\rho(7+24\lambda)$, $3^\rho(13+24\lambda)$, $3^\rho(23+24\lambda)$ *with* $0 \leq \rho < r-1$ *and* $0 \leq \lambda < 3^{r-\rho-1}$.

2.3. **A reduction formula.** For a finite group G, a prime p, and p–powers q_1, q_2 with $q_1q_2 > 1$ form the group $\mathfrak{G} = \mathfrak{H}(G, q_1, q_2)$ as defined in (4). Write $\mathfrak{G} = \langle \zeta \mid \zeta^{q_1} = 1\rangle * \widetilde{\mathfrak{G}}$ with $\widetilde{\mathfrak{G}} = G^{*q_2}$, and consider the subgroup

$$\mathfrak{N} := \left\langle \widetilde{\mathfrak{G}},\, \widetilde{\mathfrak{G}}^{\zeta},\, \widetilde{\mathfrak{G}}^{\zeta^2},\, \ldots,\, \widetilde{\mathfrak{G}}^{\zeta^{q_1-1}} \right\rangle$$

generated in $\mathfrak{G}$ by all the conjugates $\widetilde{\mathfrak{G}}^{\zeta^j}$ for $0 \leq j < q_1$. By our construction and the normal form theorem in the group $\mathfrak{G} = \langle\zeta\rangle * \widetilde{\mathfrak{G}}$ we find that

(i) $\mathfrak{N} \trianglelefteq \mathfrak{G}$,

(ii) $\langle\zeta\rangle \cap \mathfrak{N} = 1$,

(iii) $\langle\zeta\rangle\,\mathfrak{N} = \mathfrak{G}$,

(iv) $\mathfrak{N} = \widetilde{\mathfrak{G}} * \widetilde{\mathfrak{G}}^{\zeta} * \cdots * \widetilde{\mathfrak{G}}^{\zeta^{q_1-1}} \cong \mathfrak{H}(G, q_1q_2)$.

In particular, $\mathfrak{G}$ and $\mathfrak{N}$ satisfy the hypotheses of Theorem 1, and (5) gives

$$\Pi_i^{(p)}(\mathfrak{H}(G, q_1, q_2)) = q_1\, \Pi_i^{(p)}(\mathfrak{H}(G, q_1q_2)) \cup \bigcup_{\substack{\sigma \mid q_1 \\ \sigma < q_1}} \sigma\Big(\Pi_i^{(p)}(\mathfrak{H}(G, q_1q_2)) \cap (\mathbb{N} - p\mathbb{N})\Big), \quad 0 < i < p. \tag{13}$$

In section 4 we shall sketch a rather deep lying method for analyzing the p–patterns of groups of the form G^{*q}, where $q > 1$ is a p–power. Equation (13) will then allow us to extend our results to the groups $\mathfrak{H}(G, q_1, q_2)$.

2.4. **A possible generalization of the descent principle.** In view of both its intrinsic interest as well as its applications, it would be important to obtain generalized versions of the descent principle. A detailed analysis of the proof of Theorem 1 carried out in [39] reveals that the existence of such a generalization hinges on a certain homological property of finite p–groups (to be of Frobenius type). We begin with the relevant definitions.

Definition 1. *Let p be a prime.*

(i) *A non–trivial finite p–group P is termed* **admissible** *if, for every finite group G with $p \mid |G|$ and every action $\alpha : P \to \mathrm{Aut}(G)$ of P on G, the corresponding set*

$$\mathrm{Der}(P, G) = \mathrm{Der}_\alpha(P, G)$$

of derivations $d : P \to G$ formed with respect to this action has cardinality a multiple of p.

(ii) *A finite p–group P is said to be of* **Frobenius type**, *if every subgroup $Q > 1$ of P is admissible.*

Definition 2. *A group $\mathfrak{G}$ is called quasi–hamiltonian (sometimes also a Dedekind group) if every subgroup of $\mathfrak{G}$ is normal; that is, if $\mathfrak{G}$ is abelian or hamiltonian.*

The structure of hamiltonian groups has been cleared up by Baer: a group $\mathfrak{G}$ is hamiltonian if and only if

$$\mathfrak{G} \cong \mathfrak{Q} \times \mathfrak{A} \times \mathfrak{B},$$

where $\mathfrak{Q}$ is the quaternion group of order 8, $\mathfrak{A}$ is an elementary abelian 2–group, and $\mathfrak{B}$ is a periodic abelian group with all its elements of odd order; cf. [4]. For finite groups, the latter result had already been proved by Dedekind; cf. [9]. It follows that a finite p–group P is quasi–hamiltonian if and only if it is either abelian, or $p = 2$ and $P \cong \mathfrak{Q} \times C_2^r$ for some $r \geq 0$. We can now state our (as it stands not very useful) generalization of Theorem 1.

Theorem 1'. *Let $\mathfrak{G}$ be an FSG–group, p a prime, and let $\mathfrak{N} \trianglelefteq \mathfrak{G}$ be a normal subgroup of index p^r with $\mathfrak{G}/\mathfrak{N}$ quasi–hamiltonian and of Frobenius type. Then equations* (5) *and* (6) *hold.*

In view of Theorem 1' and the above remarks, the existence of a useful generalization of Theorem 1 depends on the solution of the following.

Problem 1. *Is a non–trivial finite abelian p–group admissible? Are groups of the form $\mathfrak{Q} \times C_2^r$ with $r \geq 0$ of Frobenius type?*

In fact, the homological property described in Definition 1 appears to be quite deep and of some independent interest; hence, we might just as well ask the following more general question.

Problem 2. *Which finite p–groups are of Frobenius type?*

By Philip Hall's theorem explained in subsection 2.1, cyclic groups of prime power order are of Frobenius type. Indeed, an equivalent way of stating [19, Theorem 1.6] is as follows.

Proposition 1. *Let C be a finite cyclic group, G a finite group, and let $\alpha : C \to \mathrm{Aut}(G)$ be an action of C on G. Then*

$$|Der_\alpha(C, G)| \equiv 0 \mod \gcd(|C|, |G|).$$

Corollary 2. *A cyclic group of prime power order is of Frobenius type.*

Combining Theorem 1' with Corollary 2 we obtain Theorem 1. Unfortunately, no non–cyclic groups of Frobenius type are known so far. However, the main result of [54], which states that for a finite group G and a finite abelian group A, $|\mathrm{Hom}(A, G)|$ is divisible by $\gcd(|A|, |G|)$, may be viewed as a sign of hope that Problem 2 might have a positive answer, at least for abelian p–groups.

3. A criterion for periodicity modulo p

The functions $s_n(\mathfrak{G})$ encountered in Theorem 2 are all periodic modulo a distinguished prime p, since their p–divisibility is determined by a free normal subgroup. In fact, periodicity modulo some distinguished prime of the function $s_n(\mathfrak{G})$ appears to be a more general phenomenon. Results in this direction were first proved by Grady and Newman (see the corollary below). The purpose of this section is to briefly discuss a far reaching generalization of their results.

Theorem 3 ([38, Theorem 3]). *Let p be a prime, s a positive integer, and let $A_1, \ldots, A_s$ be finite abelian groups, whose orders are divisible by p, and whose p–Sylow subgroups $P_1, \ldots, P_s$ satisfy $rk(P_\sigma) \leq 2$. Put $P_\sigma \cong C_{p^{\ell_\sigma}} \times C_{p^{r_\sigma}}$ with $0 \leq r_\sigma \leq \ell_\sigma$ and $1 \leq \sigma \leq s$, let $\mathfrak{G}$ be a finitely generated group containing $A_1 * \cdots * A_s$ as a* *free factor, and suppose that*

$$s \geq \frac{p+2}{p} + \sum_{\sigma} (1 - r_\sigma)\, p^{-\ell_\sigma}. \tag{14}$$

Then the sequence $s_n(\mathfrak{G})$, considered modulo p, satisfies the linear recursion

$$s_n(\mathfrak{G}) \equiv - \sum_{0<j<p} \frac{|\mathrm{Hom}(\mathfrak{G}, S_{p-j})|}{(p-j)!}\, s_{n-p+j}(\mathfrak{G}) \quad \mathrm{mod}\ p, \qquad n \geq p, \tag{15}$$

of length $p-1$ over $GF(p)$; in particular, $s_n(\mathfrak{G})$ is periodic modulo p, and for $p = 2$ we have $\Pi(\mathfrak{G}) = \mathbb{N}$.

The proof of Theorem 3 is based on Dey's formula[2]

$$\sum_{\mu=1}^{n} \frac{|\mathrm{Hom}(\mathfrak{G}, S_{n-\mu})|}{(n-\mu)!}\, s_\mu(\mathfrak{G}) = \frac{|\mathrm{Hom}(\mathfrak{G}, S_n)|}{(n-1)!} \tag{16}$$

as well as the estimates

$$\nu_p(|\mathrm{Hom}(A_\sigma, S_n)|) \geq \sum_{j=1}^{\ell_\sigma} \left\lfloor \frac{n}{p^j} \right\rfloor - (\ell_\sigma - r_\sigma) \left\lfloor \frac{n}{p^{\ell_\sigma+1}} \right\rfloor, \quad 1 \leq \sigma \leq s,$$

for the p-adic norms of the arithmetic functions $|\mathrm{Hom}(A_\sigma, S_n)|$, which follow from [27, Theorem 5.1]. As an example, consider the case where $\mathfrak{G} = C_p^{*s} * \mathfrak{C}$ with some finitely generated group $\mathfrak{C}$ and a prime p. Then, for $1 \leq \sigma \leq s$, we have $A_\sigma = P_\sigma = C_p$, $r_\sigma = 0$, $\ell_\sigma = 1$, and equation (14) translates into the condition that $s \geq \frac{p+2}{p-1}$ or, equivalently,

$$s \geq \begin{cases} 4 & p = 2 \\ 3, & p = 3 \\ 2, & p \geq 5. \end{cases}$$

Applying Theorem 3, we find the following.

Corollary 3. *Let $\mathfrak{G}$ be a finitely generated free product.*

(i) *If $\mathfrak{G}$ contains in its free decomposition at least four copies of the cyclic group of order 2, then $\Pi(\mathfrak{G}) = \mathbb{N}$.*

(ii) *If $\mathfrak{G}$ contains in its free decomposition three or more copies of the cyclic group of order 3, then $s_n(\mathfrak{G})$, considered modulo 3, satisfies a two term linear recurrence relation, and hence is periodic modulo 3.*

(iii) *If $\mathfrak{G}$ contains in its free decomposition at least two copies of the cyclic group of prime order p with $p \geq 5$, then $s_n(\mathfrak{G})$ satisfies a $(p-1)$ term linear recurrence relation modulo p; in particular, $s_n(\mathfrak{G})$ is periodic modulo p.*

The results summarized in Corollary 3 are due to Grady and Newman; cf. [16, Theorem 1], [16, formulae (7), (8)], and [17, Theorem 2].

---

[2]Cf. [10, Theorem 6.10]. More general results in this direction can be found in [12] and [36].

4. A FUNCTIONAL EQUATION ASSOCIATED WITH THE GROUPS $\mathfrak{H}(G,q)$

As we already saw in section 2, the investigation of the p–patterns for groups of the form (4) can be reduced to studying the p–patterns associated with the symmetrized groups $\mathfrak{H}(G,q) = G^{*q}$; cf. formula (13). Here, we are going to sketch a method for analyzing the patterns $\Pi^{(p)}(\mathfrak{H}(G,q))$.

4.1. **A functional equation for $X_{G,q}(z)$.** Put $h_n(G) := |\operatorname{Hom}(G, S_n)|$, with the conventions that $h_0(G) = 1$ and that $h_n(G) = 0$ for $n < 0$. Our starting point is the recurrence relation[3]

$$h_n(G) = \sum_d s_d(G)\,(n-1)_{d-1}\,h_{n-d}(G), \qquad (n \geq 1,\ h_0(G) = 1), \tag{17}$$

which follows from the identity[4]

$$\sum_{n=0}^{\infty} h_n(G)\,\frac{z^n}{n!} = \exp\left(\sum_d s_d(G)\,\frac{z^d}{d}\right)$$

by differentiating and comparing coefficients. Raising (17) to the q–th power and separating terms gives

$$h_n^q(G) = \sum_{d|m} \Big(s_d(G)\,(n-1)_{d-1}\,h_{n-d}(G)\Big)^q$$

$$+ \sum_{\substack{||\underline{\nu}||=q \\ \ell(\underline{\nu})\geq 2}} \frac{q!}{\prod_{d|m} \nu_d!} \prod_{d|m} \Big(s_d(G)\,(n-1)_{d-1}\,h_{n-d}(G)\Big)^{\nu_d}. \tag{18}$$

Here, the $\underline{\nu}$'s are maps $\underline{\nu} : D \to \mathbb{N}_0$ defined on the set D of positive divisors of $m = |G|$, $\nu_d := \underline{\nu}(d)$, $||\underline{\nu}|| := \sum_{d|m} \nu_d$, and $\ell(\underline{\nu}) := |\{d \in D : \nu_d \neq 0\}|$. Multiply both sides of (18) by $z^{n-1}/(n-1)!$, sum over $n \geq 1$, and interchange summations to obtain

$$H'_{G,q}(z) = \sum_{d|m} s_d^q(G)\,\Sigma_d(z) + \sum_{\substack{||\underline{\nu}||=q \\ \ell(\underline{\nu})\geq 2}} \frac{q!}{\prod_{d|m} \nu_d!} \Big(\prod_{d|m} s_d^{\nu_d}(G)\Big)\,\overline{\Sigma}_{\underline{\nu}}(z), \tag{19}$$

where

$$H_{G,q}(z) := \sum_{n\geq 0} h_n^q(G) z^n/n! = \sum_{n\geq 0} |\operatorname{Hom}(\mathfrak{H}(G,q), S_n)|\, z^n/n!,$$

$$\Sigma_d(z) := \sum_{n\geq 1} \big((n \quad 1)_{d-1}\,h_{n-d}(G)\big)^q\, z^{n-1}/(n-1)!,$$

and

$$\overline{\Sigma}_{\underline{\nu}}(z) := \sum_{n\geq 1} \Big(\prod_{d|m} \big((n-1)_{d-1}\,h_{n-d}(G)\big)^{\nu_d}\Big)\, z^{n-1}/(n-1)!.$$

---

[3]For a ring R with identity element 1, an element $r \in R$, and an integer k, we set $(r)_k := r(r-1)\ldots(r-k+1)$, with the usual convention that an empty product should equal 1. This is the falling factorial r of order k. Its counterpart $\langle r\rangle_k := r(r+1)\ldots(r+k-1)$, the rising factorial r of order k, is sometimes called a Pochhammer symbol.

[4]Cf. for instance [12, Prop. 1].

We have

$$\Sigma_d(z) = \sum_{n\geq 0} \big((n+d-1)_{d-1}\big)^{q-1} h_n^q(G)\, z^{n+d-1}/n! \\ = z^{d-1}\big(D_z^{d-1} z^{d-1}\big)^{q-1}\big(H_{G,q}(z)\big), \tag{20}$$

where $D_z = \frac{d}{dz}$ is the differential operator. Next, consider the series $\overline{\Sigma}_{\underline{\nu}}(z)$ for some map $\underline{\nu} : D \to \mathbb{N}_0$, let $d_1 < d_2 < \cdots < d_\ell$ be the divisors $d \in D$ with $\nu_d \neq 0$, and write ν_i for ν_{d_i}. Proceeding as in the case of $\Sigma_d(z)$ one finds that

$$\overline{\Sigma}_{\underline{\nu}}(z) = z^{d_1-1}\prod_{i=1}^{\ell-1}\Big[\big(D_z^{d_i-1} z^{d_i-1}\big)^{\nu_i-1}\big(D_z^{d_i-1} z^{d_{i+1}-1}\big)\Big]\big(D_z^{d_\ell-1} z^{d_\ell-1}\big)^{\nu_\ell-1}\big(\overline{\mathcal{R}}_{G,\underline{\nu}}(z)\big), \tag{21}$$

where

$$\overline{\mathcal{R}}_{G,\underline{\nu}}(z) := \sum_{n\geq 0}\Big(\prod_{i=1}^{\ell} h_{n+d_\ell-d_i}^{\nu_i}(G)\Big) z^n/n!.$$

We now make use of two facts to be commented upon in the next subsection.

(A) *For every $\nu \in \mathbb{N}_0$ the series $H_{G,q}^{(\nu)}(z)/H_{G,q}(z)$ is an integral power series, and satisfies the congruence*

$$H_{G,q}^{(\nu)}(z)/H_{G,q}(z) \equiv \big(S_{G,q}(z) + S_{G,q}^{(p-1)}(z^{1/p})\big)^{\nu-k} \\ \times \sum_{\substack{\lambda_0,\lambda_1,\ldots,\lambda_k\geq 0\\ \lambda_0=\lambda_1=0}} \frac{(k)_{\sum_{\ell=1}^k \ell\lambda_\ell}}{\prod_{\ell=1}^k\big((\ell!)^{\lambda_\ell}\lambda_\ell!\big)}\big(S_{G,q}(z)\big)^{k-\sum_{\ell=1}^k \ell\lambda_\ell}\prod_{\ell=1}^k\big(S_{G,q}^{(\ell-1)}(z)\big)^{\lambda_\ell} \mod p, \tag{22}$$

where $\nu \equiv k\ (p)$, $0 \leq k < p$, and $S_{G,q}(z) := \sum_{n\geq 0} s_{n+1}(\mathfrak{H}(G,q))\, z^n$.

(B) *For every finite group G, each map $\underline{\nu} : D \to \mathbb{N}_0$ of norm $||\underline{\nu}|| = q$, and every integer $\mu \geq 0$ the series $\overline{\mathcal{R}}_{G,\underline{\nu}}^{(\mu)}(z)/H_{G,q}(z)$, viewed as a power series, has integral coefficients.*

Divide both sides of (19) by $H_{G,q}(z)$. Then, by Dey's formula (16), the left–hand side becomes $S_{G,q}(z)$. Moreover, using (21), assertion (B), and Leibniz's rule for the derivatives of a product function, we find that, for each finite group G and every map $\underline{\nu} : D \to \mathbb{N}_0$ of norm $||\underline{\nu}|| = q$, the series $\overline{\Sigma}_{\underline{\nu}}(z)/H_{G,q}(z)$ is an integral power series. Consequently, since a multinomial coefficient $\frac{q!}{\prod_{d|m}\nu_d!}$ with q a p–power and $\ell(\underline{\nu}) > 1$ is always divisible by p, the second sum on the right–hand side of (19), after division by $H_{G,q}(z)$, turns out to be an integral power series with coefficients divisible by p, and hence can be ignored modulo p. Similarly, using (20), the first part of assertion (A), Leibniz's rule, and Lucas' formula for binomial coefficients[5]

[5]Cf. for instance [7, Theorem 3.4.1].

we find that, for every divisor d of m, the series $\Sigma_d(z)/H_{G,q}(z)$ is integral, and that

$$\Sigma_d(z)/H_{G,q}(z) \equiv \sum_{0\le\sigma\le i(q-1)} c_{i,\sigma}^{(q)}\, z^{q(d-1)-\sigma}\, H_{G,q}^{((q-1)(d-1)-\sigma)}(z)/H_{G,q}(z) \mod p, \tag{23}$$

where $d \equiv i+1$ (p), $0 \le i < p$, and

$$c_{i,\sigma}^{(q)} := \sum_{\substack{0\le k_1,\dots,k_{q-1}\le i\\ k_1+\cdots+k_{q-1}=\sigma}} \prod_{j=1}^{q-1}\left[\binom{i}{k_j}\left(ji-\sum_{\ell=1}^{j-1}k_\ell\right)_{k_j}\right]. \tag{24}$$

From the previous discussion, formulae (19), (22), (23), and Fermat's Theorem, we obtain the following.

Theorem 4 ([40, Theorem 1]). *For each prime p, every finite group G, and each p–power $q>1$ the mod p projection $X_{G,q}(z)$ of the generating series $S_{G,q}(z)$ satisfies the functional equation*

$$X_{G,q}(z) = \sum_{0\le i<p}\sum_{0\le k<p}\sum_{d\equiv i+1\,(p)}\sum_{\substack{0\le\sigma\le i(q-1)\\ \sigma+i+k\equiv 0\,(p)}}\sum_{\substack{\lambda_0,\lambda_1,\dots,\lambda_k\ge 0\\ \lambda_0=\lambda_1=0}} s_d(G)\, c_{i,\sigma}^{(q)}\, \frac{(k)_{\sum_{\ell=1}^k \ell\lambda_\ell}}{\prod_{\ell=1}^k\left((\ell!)^{\lambda_\ell}\lambda_\ell!\right)}\, z^{q(d-1)-\sigma}$$
$$\times \left(X_{G,q}(z)\right)^{k-\sum_{\ell=1}^k \ell\lambda_\ell} \left(Y_{G,q}(z)\right)^{(q-1)(d-1)-\sigma-k} \prod_{\ell=1}^k \left(X_{G,q}^{(\ell-1)}(z)\right)^{\lambda_\ell}, \tag{25}$$

where

$$Y_{G,q}(z) := X_{G,q}(z) + X_{G,q}^{(p-1)}(z^{1/p}), \tag{26}$$

and the coefficients $c_{i,\sigma}^{(q)}$ are as in (24).

4.2. **Some remarks concerning assertions (A) and (B).** It turns out to be natural to establish assertion (A) in the following more abstract and general form.

Lemma 2 ([40, Lemma 3]). *Let p be a prime, and let $F(z)$ and $G(z)$ be power series with real coefficients. Suppose that* (i) $F(z)=G'(z)/G(z)$, (ii) $G(0)=1$, *and* (iii) *that $F(z)$ has integral coefficients. Then, for every $\nu\ge 0$, the series $G^{(\nu)}(z)/G(z)$ is integral, and satisfies the congruence*

$$G^{(\nu)}(z)/G(z) \equiv \left(F(z)+F^{(p-1)}(z^{1/p})\right)^{\nu-k}$$
$$\times \sum_{\substack{\lambda_0,\lambda_1,\dots,\lambda_k\ge 0\\ \lambda_0=\lambda_1=0}} \frac{(k)_{\sum_{\ell=1}^k \ell\lambda_\ell}}{\prod_{\ell=1}^k\left((\ell!)^{\lambda_\ell}\lambda_\ell!\right)} \left(F(z)\right)^{k-\sum_{\ell=1}^k \ell\lambda_\ell} \prod_{\ell=1}^k \left(F^{(\ell-1)}(z)\right)^{\lambda_\ell} \mod p, \tag{27}$$

where $0\le k<p$ and $\nu\equiv k$ (p).

The proof of Lemma 2 rests on an application to the equation

$$G(z)=\exp\left(\int F(z)\,dz\right)$$

of Bell's formula (61) for the derivatives of a composite function, and a detailed combinatorial analysis of the resulting expression for $G^{(\nu)}(z)/G(z)$. For a finitely generated group $\mathfrak{G}$ put

$$H_{\mathfrak{G}}(z) := \sum_{n\geq 0} |\operatorname{Hom}(\mathfrak{G}, S_n)|\, z^n/n! \quad \text{and} \quad S_{\mathfrak{G}}(z) := \sum_{n\geq 0} s_{n+1}(\mathfrak{G})\, z^n.$$

Noting that, in view of [33, Prop. 1], the series $S_{\mathfrak{G}}(z)$ and $H_{\mathfrak{G}}(z)$ satisfy the hypotheses of Lemma 2, we then deduce the following.

Corollary 4 ([40, Cor. 17]). *Let p be a prime, and let $\mathfrak{G}$ be a finitely generated group. Then, for every $\nu \in \mathbb{N}_0$, the series $H_{\mathfrak{G}}^{(\nu)}(z)/H_{\mathfrak{G}}(z)$ is integral, and satisfies the congruence*

$$H_{\mathfrak{G}}^{(\nu)}(z)/H_{\mathfrak{G}}(z) \equiv \big(S_{\mathfrak{G}}(z) + S_{\mathfrak{G}}^{(p-1)}(z^{1/p})\big)^{\nu-k}$$

$$\times \sum_{\substack{\lambda_0,\lambda_1,\ldots,\lambda_k\geq 0\\ \lambda_0=\lambda_1=0}} \frac{(k)_{\sum_{\ell=1}^k \ell\lambda_\ell}}{\prod_{\ell=1}^k \big((\ell!)^{\lambda_\ell}\lambda_\ell!\big)} \big(S_{\mathfrak{G}}(z)\big)^{k-\sum_{\ell=1}^k \ell\lambda_\ell} \prod_{\ell=1}^k \big(S_{\mathfrak{G}}^{(\ell-1)}(z)\big)^{\lambda_\ell} \mod p,$$

where $0 \leq k < p$ and $\nu \equiv k\ (p)$; in particular, assertion (A) *concerning the series $H_{G,q}(z)$ holds true.*

The proof of assertion (B) is more involved. For a non–negative integer σ define

$$\underline{\Lambda}_\sigma := \Big\{(\lambda_0, \lambda_1, \ldots, \lambda_\sigma) \in \mathbb{N}_0^{\sigma+1} :\ 0 = \lambda_0 \leq \lambda_1 \leq \ldots \leq \lambda_\sigma\Big\},$$

and put

$$\underline{\Lambda} := \bigcup_{\sigma\geq 0} \underline{\Lambda}_\sigma.$$

As usual, the norm $||\underline{\lambda}||$ of a vector $\underline{\lambda} = (\lambda_0, \ldots, \lambda_\sigma) \in \underline{\Lambda}$ is defined as $||\underline{\lambda}|| = \sum_{j=1}^{\sigma} \lambda_j$. Given a finite group G of order m, we define a system of polynomials $\mathfrak{p}_G^{\ell,\underline{\lambda}}(t) \in \mathbb{Z}[t]$ indexed by two extra parameters $\ell \in \mathbb{Z}$ and $\underline{\lambda} = (\lambda_0, \ldots, \lambda_\sigma) \in \underline{\Lambda}$ via the equations

$$\mathfrak{p}_G^{\ell,\underline{\lambda}}(t) = \delta_{\ell,0} \quad (\ell \in \mathbb{Z},\ ||\underline{\lambda}|| = 0), \tag{28}$$

$$\mathfrak{p}_G^{\ell,\underline{\lambda}}(t) = 0 \quad (l < 0,\ ||\underline{\lambda}|| > 0), \tag{29}$$

and

$$\mathfrak{p}_G^{\ell,\underline{\lambda}}(t) = \sum_{\substack{d|m\\ d>\lambda_\sigma}} s_d(G)\,(t+\lambda_\sigma-1)_{\lambda_\sigma-1}\, \mathfrak{p}_G^{\ell+\lambda_\sigma-d,\,(0,d-\lambda_\sigma+\lambda_0,\ldots,d-\lambda_\sigma+\lambda_{\sigma-1})}(t+\lambda_\sigma-d)$$

$$+\sum_{j=0}^{\sigma-1} \sum_{\substack{d|m\\ \lambda_\sigma-\lambda_{j+1}<d\leq\lambda_\sigma-\lambda_j}} s_d(G)\,(t+\lambda_\sigma-1)_{d-1}\, \mathfrak{p}_G^{\ell,\,(\lambda_0,\ldots,\lambda_j,\lambda_\sigma-d,\lambda_{j+1},\ldots,\lambda_{\sigma-1})}(t)$$

$$(\ell \geq 0,\ ||\underline{\lambda}|| > 0). \tag{30}$$

An immediate induction on ℓ, followed by induction on $||\underline{\lambda}||$, shows that (28) – (30) uniquely defines a system $\big\{\mathfrak{p}_G^{\ell,\underline{\lambda}}(t)\big\}_{(G,\ell,\underline{\lambda})}$ of integral polynomials indexed by the triples

$$(G, \ell, \underline{\lambda}) \in \mathbf{Fin} \times \mathbb{Z} \times \underline{\Lambda},$$

where **Fin** denotes the class of all finite groups. As our next result shows, these polynomials $\mathfrak{p}_G^{\ell,\underline{\lambda}}(t)$, for each finite group G and every $\sigma \geq 0$, relate the subdiagonals $\{\prod_{j=0}^{\sigma} h_{n+\lambda_j}(G)\}_{n\geq 0}$ with $\underline{\lambda} = (\lambda_0, \ldots, \lambda_\sigma) \in \underline{\Lambda}_\sigma$ of the $(\sigma+1)$–dimensional rectangular array $\big(h_{\mu_0}(G)\, h_{\mu_1}(G) \ldots h_{\mu_\sigma}(G)\big)_{\mu_0,\ldots,\mu_\sigma \geq 0}$ to the terms $h_n^{\sigma+1}(G)$ of its main diagonal. It is this important observation which underlies our proof of hypothesis (B).

Lemma 3 ([40, Lemma 4]). *For every finite group G, each vector $\underline{\lambda} = (\lambda_0, \lambda_1, \ldots, \lambda_\sigma) \in \underline{\Lambda}$, and every integer $n \geq 0$, and with $\mathfrak{p}_G^{\ell,\underline{\lambda}}(t)$ as defined above, we have*

$$\prod_{j=0}^{\sigma} h_{n+\lambda_j}(G) = \sum_{\ell=0}^{n} \mathfrak{p}_G^{\ell,\underline{\lambda}}(n)\, (n)_\ell\, h_{n-\ell}^{\sigma+1}(G). \tag{31}$$

For a finite group G, a vector $\underline{\lambda} = (\lambda_0, \lambda_1, \ldots, \lambda_\sigma) \in \underline{\Lambda}$, and an integer $\mu \geq 0$ consider the series

$$\mathcal{R}_G^{\underline{\lambda},\mu}(z) := \left\{ \sum_{n\geq 0} \left(\prod_{j=0}^{\sigma} h_{n+\lambda_j}(G) \right) z^n/n! \right\}^{(\mu)} \Big/ \left\{ \sum_{n \geq 0} h_n^{\sigma+1}(G)\, z^n/n! \right\}.$$

Expanding the polynomials $\mathfrak{p}_G^{\ell,\underline{\lambda}}(t)$ in terms of the basis $(t)_0, (t)_1, (t)_2, \ldots$,

$$\mathfrak{p}_G^{\ell,\underline{\lambda}}(t) = \sum_{\kappa=0}^{d_G^{\ell,\underline{\lambda}}} a_G^{\ell,\underline{\lambda}}(\kappa)\, (t)_\kappa, \quad (G, \ell, \underline{\lambda}) \in \mathbf{Fin} \times \mathbb{Z} \times \underline{\Lambda}, \tag{32}$$

where $d_G^{\ell,\underline{\lambda}} := \deg\big(\mathfrak{p}_G^{\ell,\underline{\lambda}}(t)\big)$, and with $a_G^{\ell,\underline{\lambda}}(\kappa) \in \mathbb{Z}$ for all $G, \ell, \underline{\lambda}$, and κ, and applying Lemma 3 as well as Leibniz's formula, one finds that

$$\mathcal{R}_G^{\underline{\lambda},\mu}(z) = \sum_{\ell \geq 0} \sum_{i \geq 0} \sum_{j \geq 0} \binom{\mu}{j} (\ell + i)_{\mu - j}\, A_{\ell,i}^{G,\underline{\lambda}}\, z^{\ell+i+j-\mu}\, H_{G,\sigma+1}^{(i+j)}(z)/H_{G,\sigma+1}(z),$$

where

$$A_{\ell,i}^{G,\underline{\lambda}} := \sum_{\kappa=0}^{d_G^{\ell,\underline{\lambda}}} \binom{\kappa}{i} (\ell)_{\kappa - i}\, a_G^{\ell,\underline{\lambda}}(\kappa).$$

Since the coefficients $A_{\ell,i}^{G,\underline{\lambda}}$ are well-defined integers, the family of series

$$\left\{ \binom{\mu}{j} (\ell + i)_{\mu-j}\, A_{\ell,i}^{G,\underline{\lambda}}\, z^{\ell+i+j-\mu}\, H_{G,\sigma+1}^{(i+j)}(z)/H_{G,\sigma+1}(z) \right\}_{\ell,i,j\geq 0}$$

is summable (because of the factor $z^{\ell+i+j-\mu}$), and the series $H_{G,\sigma+1}^{(i+j)}(z)/H_{G,\sigma+1}(z)$ are integral for all $i, j \geq 0$ by Corollary 4, we conclude that for every $G \in \mathbf{Fin}$, $\underline{\lambda} \in \underline{\Lambda}$, and $\mu \geq 0$ the series $\mathcal{R}_G^{\underline{\lambda},\mu}(z)$, considered as a power series, has integral coefficients; in particular, assertion (B) concerning the series $\overline{\mathcal{R}}_{G,\underline{\nu}}^{(\mu)}(z)/H_{G,q}(z)$ holds true.

4.3. **The coefficients $c_{i,\sigma}^{(q)}$.** In order to be able to exploit the functional equation (25), it is necessary to obtain information concerning the evaluation modulo p of the coefficients $c_{i,\sigma}^{(q)}$ occurring in this identity. Clearly, $c_{i,0}^{(q)} = 1$ for $i \geq 0$ and $c_{0,\sigma}^{(q)} = 0$ for $\sigma \geq 1$. No evaluation in closed terms is known for $c_{i,\sigma}^{(q)}$ with $i > 1$ and arbitrary σ. However, our next result provides such a description in terms of Stirling numbers of the second kind in the case when $i = 1$.

Lemma 4 ([40, Lemma 1]). *For $\sigma \geq 0$, we have $c_{1,\sigma}^{(q)} = S(q, q-\sigma)$.*
Moreover, the well-known identity[6]

$$\sum_{n\geq 0} S(n,k) z^n = \frac{z^k}{(1-z)(1-2z)\ldots(1-kz)}, \quad k \geq 0 \tag{33}$$

easily furnishes congruences for the $S(n,k)$ in terms of binomial coefficients, which in turn can be explicitly evaluated modulo a prime via Lucas' congruence. If, for instance, $p = 2$, equation (33) gives

$$\sum_{n\geq 0} S(n,k) z^n \equiv \frac{z^k}{(1+z)^{\lceil k/2 \rceil}} \mod 2,$$

which, after expanding $(1+z)^{-\lceil k/2 \rceil}$ and comparing coefficients, yields the elegant result that

$$S(n,k) \equiv \binom{\lceil k/2 \rceil + n - k - 1}{n-k} \mod 2. \tag{34}$$

For primes $p > 2$ the corresponding formulae are more involved. For instance, the result for $p = 3$ is that

$$S(n,k) \equiv \begin{cases} \binom{\frac{k}{3}+\frac{n-k}{2}-1}{\frac{n-k}{2}}, & k \equiv 0\ (3) \text{ and } n-k \equiv 0\ (2) \\ \sum_{0\leq\mu\leq\frac{n-k}{2}} \binom{\frac{k-1}{3}+\mu-1}{\mu}, & k \equiv 1\ (3) \\ \binom{\frac{k-2}{3}+\frac{n-k}{2}}{\frac{n-k}{2}}, & k \equiv 2\ (3) \text{ and } n-k \equiv 0\ (2) \\ 0, & \text{otherwise} \end{cases} \mod 3. \tag{35}$$

Combining Theorem 4 with Lemma 4, Congruence (34), and Lucas' formula one obtains a completely explicit functional equation for $X_{G,q}(z)$ in the case when $p = 2$.

Corollary 5 ([40, Cor. 1]). *For each finite group G and every 2–power $q > 1$ the $GF(2)$–series $X_{G,q}(z)$ satisfies the equation*

$$X_{G,q}(z) = \sum_{\substack{d\in\Pi(G)\\ d\equiv 1\,(2)}} z^{q(d-1)} \big(Y_{G,q}(z)\big)^{(q-1)(d-1)} + Z_{G,q}(z) \sum_{\substack{d\in\Pi(G)\\ d\equiv 0\,(2)}} z^{q(d-2)+1} \big(Y_{G,q}(z)\big)^{(q-1)(d-2)}, \tag{36}$$

where

$$Y_{G,q}(z) = X_{G,q}(z) + X'_{G,q}(z^{1/2})$$

and

$$Z_{G,q}(z) := 1 + X_{G,q}(z) \sum_{\substack{\mu | q\\ \mu>1}} z^{\mu-1} \big(Y_{G,q}(z)\big)^{\mu-2}.$$

In general, we have

$$c_{i,\sigma}^{(q)} = \sum_{0\leq r\leq \min(\sigma,q-1)} \;\sum_{\substack{\varphi:[r]\to\{0,1,\ldots,i-1\}\\ ||\varphi||=ri-\sigma}} \mathcal{A}^{(\varphi)}(r, q-r), \tag{37}$$

[6] Cf. for instance [53, formula (1.6.5)].

where

$$\mathcal{A}^{(\varphi)}(r,n) = \sum_{\nu=0}^{r} \alpha_\nu^{(\varphi)}(r,n)\, \mathcal{A}^{(\varphi)}(r-\nu, n-1), \quad n \geq 1, \tag{38}$$

and

$$\alpha_\nu^{(\varphi)}(r,n) := \prod_{\rho=r-\nu+1}^{r} \left[\binom{i}{\varphi(\rho)} \left(in + \sum_{\kappa=1}^{\rho-1} \varphi(\kappa) \right)_{i-\varphi(\rho)} \right]. \tag{39}$$

By definition, we also have

$$\mathcal{A}^{(\varphi)}(r,0) = \delta_{r,0}. \tag{40}$$

Formulae (37) – (40) yield a semi–recursive computation of the coefficients $c_{i,\sigma}^{(q)}$, which is particularly effective if σ is close to $i(q-1)$, since in that case $q-r$ is small, and the computation of the quantities $\mathcal{A}^{(\varphi)}(r, q-r)$ involved in equation (37) takes only few recursive steps. For instance, one finds in this way that

$$c_{i,i(q-1)}^{(q)} \equiv 1 \mod p, \quad i \geq 0, \tag{41}$$

$$c_{i,i(q-1)-1}^{(q)} \equiv \begin{cases} -1, & i = p-1 \\ 0, & i \leq p-2 \end{cases} \mod p, \quad i \geq 2, \tag{42}$$

and

$$c_{i,i(q-1)-2}^{(q)} \equiv \begin{cases} \frac{p+1}{2}, & i = p-1 \\ \frac{p-1}{2}, & i = p-2 \\ 0, & i \leq p-3 \end{cases} \mod p, \quad i \geq 2. \tag{43}$$

5. The patterns $\Pi^{(p)}(\mathfrak{H}(G, q_1, q_2))$

The purpose of this section is to explain some of the results for the p–patterns of groups of the form (4) which can be obtained via the approach to the patterns $\Pi^{(p)}(\mathfrak{H}(G,q))$ sketched in section 4 in combination with the reduction formula (13) coming from the descent principle. For further results and more details see [40, Sects. 4 - 8]. In what follows, p will be a prime, G a finite group, and q_1, q_2 will be p–powers such that $q_1 q_2 > 1$. Moreover, for $0 < j < p$, we put $\Pi_{G,q_1,q_2}^{(j)} := \Pi_j^{(p)}(\mathfrak{H}(G, q_1, q_2))$ and $\Pi_{G,q_1,q_2} := \Pi^{(p)}(\mathfrak{H}(G, q_1, q_2))$, and we write $X_{G,q_1,q_2}(z)$ for the mod p projection of the generating series $\sum_{n\geq 0} s_{n+1}(\mathfrak{H}(G, q_1, q_2))\, z^n$. Finally, we let $\mathcal{N}_{G,q_1,q_2} = \bigcup_{0<j<p} \Pi_{G,q_1,q_2}^{(j)}$.

5.1. The case where G is in $\mathbf{Fin}(p)$.

In order to be able to explain what is perhaps our most striking application of equation (25), we first have to introduce, for each prime p, a class $\mathbf{Fin}(p)$ of finite groups. Given p, a finite group G is said to be in the class $\mathbf{Fin}(p)$ if and only if G satisfies the (at first sight rather curious) condition that

$$\forall\, d \in \mathbb{N} \left(s_d(G) \not\equiv 0 \ (p) \rightarrow d \equiv 1 \ (p) \right).$$

Clearly, $\mathbf{Fin}(2)$ coincides with the class of all finite groups of odd order. For $p > 2$ no such description in structural terms is known, but it is easy to write down lower and upper bounds for $\mathbf{Fin}(p)$, showing in particular that these classes are fairly substantial. Indeed, if all prime divisors of the order of a finite group G

are congruent to 1 mod p, then G is in $\mathbf{Fin}(p)$ (note that, by Dirichlet's Theorem on primes in arithmetic progressions, there exist infinitely many primes congruent to 1 mod p). On the other hand, if G is in $\mathbf{Fin}(p)$, then the orders of all Sylow subgroups of G must be congruent to 1 mod p. For, if G is in $\mathbf{Fin}(p)$, then in particular $|G| \equiv 1\ (p)$, hence p does not divide $|G|$. Then p does not divide the number of Sylow $\mathfrak{q}$–subgroups (for any prime $\mathfrak{q}$), since they are all conjugate. But then the index (and hence the order) of a Sylow $\mathfrak{q}$–subgroup must be congruent to 1 mod p. The significance of these classes of groups $\mathbf{Fin}(p)$ stems from the fact that degeneration of the differential equation (25) is governed by a local–to–global principle: it can be shown that $X'_{G,q}(z) = 0$ (i.e., $\bigcup_{0<j<p} \Pi^{(j)}_{G,q} \subseteq 1 + p\mathbb{N}_0$) if and only if G is in $\mathbf{Fin}(p)$. In the latter case, the functional equation (25) reduces to the algebraic identity

$$X_{G,q}(z) = \sum_{d\geq 1} s_d(G)\, z^{q(d-1)} \left(X_{G,q}(z)\right)^{(q-1)(d-1)}, \tag{44}$$

which in turn can be explicitly solved by means of Lagrange inversion, to give

$$X_{G,q}(z) \equiv \sum_{\mu\geq 0} \left[\sum_{\substack{\underline{n}\in\mathbb{N}_0^r \\ \underline{d}_{G,p}\cdot\underline{n}=\mu}} \binom{1+p(q-1)\mu}{\underline{n},\, 1+p(q-1)\mu - ||\underline{n}||} \prod_{i=1}^{r} \left(s_{d_i}(G)\right)^{n_i} \right] z^{pq\mu} \quad \text{mod } p. \tag{45}$$

Here, the vector $\underline{d}_{G,p} \in \mathbb{N}^r$ attached to the group G and the prime p is defined as

$$\underline{d}_{G,p} := \left(\frac{d_1-1}{p}, \frac{d_2-1}{p}, \ldots, \frac{d_r-1}{p}\right),$$

where $1 = d_0 < d_1 < \cdots < d_r = |G|$ is the collection in increasing order of those positive integers d for which $s_d(G) \not\equiv 0\ (p)$.[7] Combining (45) with the reduction formula (13), we obtain the following explicit combinatorial description of the p–pattern Π_{G,q_1,q_2} in the case when G is in $\mathbf{Fin}(p)$.

Theorem 5 ([40, Theorem 12]). *Let p be a prime, let q_1 and q_2 be p–powers such that $q_1q_2 > 1$, and suppose that G is in $\mathbf{Fin}(p)$. Then we have*

$$\Pi^{(j)}_{G,q_1,q_2} = \bigcup_{\sigma|q_1} \sigma\,\Theta^{(j)}_{G,q_1,q_2}, \quad 0 < j < p,$$

where $\Theta^{(j)}_{G,q_1,q_2}$ consists of all positive integers $n \equiv 1$ mod pq_1q_2 such that the sum

$$\sum_{\substack{\underline{n}\in\mathbb{N}_0^r \\ \underline{d}_{G,p}\cdot\underline{n}=\frac{n-1}{pq_1q_2}}} \binom{1+(q_1q_2-1)(n-1)/(q_1q_2)}{\underline{n},\, 1+(q_1q_2-1)(n-1)/(q_1q_2) - ||\underline{n}||} \prod_{i=1}^{r} \left(s_{d_i}(G)\right)^{n_i}$$

is congruent to j modulo p.

The description of the patterns Π_{G,q_1,q_2} with G in $\mathbf{Fin}(p)$ just given simplifies considerably if the set $\bigcup_{0<j<p} \Pi^{(p)}_j(G)$ consists of precisely two elements.

---

[7] As usual, the norm of a vector $\underline{v} = (v_1, \ldots, v_r) \in \mathbb{N}_0^r$ is defined as $||\underline{v}|| = \sum_j v_j$, and if $\underline{u} = (u_1, \ldots, u_r)$ and $\underline{v} = (v_1, \ldots, v_r)$ are two such vectors, then their scalar product is given by $\underline{u}\cdot\underline{v} = \sum_j u_j v_j$.

Corollary 6 ([40, Cor. 13]). *If* $|G| = m \equiv 1\ (p)$ *and* $|\bigcup_{0<j<p} \Pi_j^{(p)}(G)| = 2$, *then, for* $0 < j < p$ *and any* p*–powers* q_1, q_2 *with* $q_1q_2 > 1$, *we have* $\Pi_{G,q_1,q_2}^{(j)} = \bigcup_{\sigma|q_1} \sigma\Theta_{G,q_1,q_2}^{(j)}$, *where*

$$\Theta_{G,q_1,q_2}^{(j)} = \left\{1+q_1q_2(m-1)\lambda :\ \lambda \in \mathbb{N}_0 \ \text{ and } \ \binom{1+(q_1q_2-1)(m-1)\lambda}{\lambda} \equiv j \bmod p\right\}.$$

5.2. **Rationality.** In this subsection we are going to focus on two related themes: the question when the series $X_{G,q_1,q_2}(z)$ is a rational function over $GF(p)$, and the more specific question under which conditions we will have that $X_{G,q_1,q_2}(z) = 1/(1-z)$, i.e., that $\Pi_{G,q_1,q_2}^{(1)} = \mathbb{N}$. Our first result shows that the latter condition on the function $s_n(\mathfrak{H}(G, q_1, q_2))$ is equivalent to a certain system of linear $GF(p)$–equations for the subgroup numbers $s_2(G), \ldots, s_p(G)$ involving the coefficients $c_{i,\sigma}^{(q_1q_2)}$, and for $p = 2, 3, 5$ these equations are translated into an equivalent system of structural conditions on the group G.

Theorem 6 ([40, Theorem 9]). *Let* p *be a prime, let* q_1 *and* q_2 *be* p*–powers such that* $q_1q_2 > 1$, *and let* G *be a finite group.*

(a) *We have* $\Pi_{G,q_1,q_2}^{(1)} = \mathbb{N}$ *if and only if the* $GF(p)$*–equations*

$$\sum_{0\le i<p}\ \sum_{\substack{0\le\sigma\le i(q_1q_2-1)\\ \sigma+p>i(q_1q_2-1)}} (-1)^{q_1q_2i+\sigma}\binom{p-i(q_1q_2-1)+\sigma-1}{p+i-\nu-1} \times ((q_1q_2-1)i-\sigma)!\, c_{i,\sigma}^{(q_1q_2)}\, s_{i+1}(G) = \binom{p-2}{\nu} \tag{46}$$

hold for $\nu = 1, 2, \ldots, 2(p-1)$.

(b) *For* $p = 2$, *we have* $\Pi_{G,q_1,q_2}^{(1)} = \mathbb{N}$ *if and only if* G *contains a subgroup of index* 2; *in particular, if* G *is nilpotent of even order, then* $\Pi_{G,q_1,q_2}^{(1)} = \mathbb{N}$ *for any* p*–powers* q_1 *and* q_2 *with* $q_1q_2 > 1$.

(c) *For* $p = 3$, *we have* $\Pi_{G,q_1,q_2}^{(1)} = \mathbb{N}$ *if and only if* $r_2(\overline{G})$, *the rank of the Sylow* 2*–subgroup* $S_2(\overline{G})$ *of the abelianized group* $\overline{G} = G/[G,G]$, *is odd, and* G *contains a normal subgroup of index* 3; *in particular, if* G *is nilpotent of order divisible by* 3, *then* $\Pi_{G,q_1,q_2}^{(1)} = \mathbb{N}$ *if and only if* $r_2(\overline{G}) \equiv 1\ (2)$.

(d) *For* $p = 5$, *we have* $\Pi_{G,q_1,q_2}^{(1)} = \mathbb{N}$ *if and only if* (i) $r_2(\overline{G}) \equiv 1\ (4)$, (ii) *the number* $\mathfrak{c}_3(G)$ *of conjugacy classes of self–normalizing subgroups of index* 3 *in* G *is connected with* $r_3(\overline{G})$ *(the rank of the Sylow* 3*–subgroup of* $\overline{G}$*) via* $\mathfrak{c}_3(G) + 3^{r_3(\overline{G})} \equiv 3\ (5)$, (iii) $\mathfrak{u}_4(G)$, *the number of non–normal subgroups of index* 4 *in* G, *satisfies* $\mathfrak{u}_4(G) + 2\cdot 3^{\nu_1} \equiv 2\ (5)$, *where* $S_2(\overline{G}) \cong \prod_{\rho\ge1} C_{2^\rho}^{\nu_\rho}$, *and* (iv) G *contains a normal subgroup of index* 5; *in particular, if* G *is nilpotent of order divisible by* 5, *then* $\Pi_{G,q_1,q_2}^{(1)} = \mathbb{N}$ *if and only if* $r_2(\overline{G}) \equiv r_3(\overline{G}) \equiv 1\ (4)$ *and* $\mathfrak{c}_{4,2}(G) + 3^{\nu_1} \equiv 1\ (5)$, *where* $\mathfrak{c}_{4,2}(G)$ *is the number of conjugacy classes of index* 4 *subgroups in* G *whose normalizer has index* 2.

Part (a) is first established in the case $q_1 = 1$ by means of the functional equation (25), and then generalized to groups of the form (4) via formula (13). Using the information concerning the coefficients $c_{i,\sigma}^{(q)}$ provided in subsection 4.3, one then shows that for $p = 2, 3, 5$ the system of equations (46) is equivalent to the requirement that

$$s_2(G) \equiv s_3(G) \equiv \cdots \equiv s_p(G) \equiv 1 \mod p. \tag{47}$$

Assertions (b), (c), and (d) follow from this equivalence, an explicit computation for the number of normal subgroups of index a prime or the square of a prime in a finitely generated group, and well–known properties of nilpotent groups. On the basis of these observations one might expect the following to be true.

Conjecture 5. *For a prime p, a finite group G, and p–powers q_1, q_2 such that $q_1q_2 > 1$, we have $\Pi^{(1)}_{G,q_1,q_2} = \mathbb{N}$ if and only if G satisfies (47).*

In fact, the proof of Conjecture 5 in the cases $p = 2, 3$, and 5 suggests the following strengthened version of Conjecture 5.

Conjecture 6. (i) *The congruences modulo p*

$$\sum_{0\le i<p}\ \sum_{0\le\sigma\le i(q-1)} (-1)^{qi+\sigma}\binom{p-i(q-1)+\sigma-1}{p+i-\nu-1}\,((q-1)i-\sigma)!\,c^{(q)}_{i,\sigma} \equiv \binom{p-2}{\nu} \tag{48}$$

hold for $1 \le \nu < p$ and every p–power $q > 1$.

(ii) *For $p \le \nu \le 2(p-1)$, all q, and $0 \le i < p$ we have*

$$\sum_{0\le\sigma\le i(q-1)} (-1)^{qi+\sigma}\binom{p-i(q-1)+\sigma-1}{p+i-\nu-1}\,((q-1)i-\sigma)!\,c^{(q)}_{i,\sigma} \equiv 0 \mod p. \tag{49}$$

We now turn to the related question, when the $GF(p)$–series $X_{G,q_1,q_2}(z)$ is rational. Here, we have the following result.[8]

Theorem 7. (a) *If one of the series $X_{G,q_1,q_2}(z)$ and $X_{G,q_1q_2}(z)$ is rational over $GF(p)$, then both are, and we have $\deg(X_{G,q_1,q_2}(z)) \ge 0$ if and only if $\deg(X_{G,q_1q_2}(z)) \ge 0$.*

(b) *Let p be a prime, G a finite group, and let $q > 1$ be a p–power. Suppose that $X_{G,q}(z)$ is rational over $GF(p)$, and write v for the (total) degree of $X_{G,q}(z)$.*

(i) *If $v \ge 0$, then $G = 1$.*

(ii) *If $v < 0$ and $|G| \equiv 0, 1 \mod p$, then $X_{G,q}(z)$ satisfies the functional equation*

$$X_{G,q}(z) = 1 + \sum_{0<i<p}\ \sum_{(q-1)i-p<\sigma\le(q-1)i}\ \sum_{\substack{\lambda_0,\lambda_1,\dots,\lambda_k\ge 0\\ \lambda_0=\lambda_1=0}} s_{i+1}(G)\,c^{(q)}_{i,\sigma}\,\frac{(k)_{\sum_{\ell=1}^k \ell\lambda_\ell}}{\prod_{\ell=1}^k (\ell!)^{\lambda_\ell}\lambda_\ell!}\,z^{qi-\sigma}$$
$$\times\,\big(X_{G,q}(z)\big)^{k-\sum_{\ell=1}^k \ell\lambda_\ell}\prod_{\ell=1}^k \big(X^{(\ell-1)}_{G,q}(z)\big)^{\lambda_\ell} \tag{50}$$

with $k := (q-1)i - \sigma$, as well as the equation

$$X_{G,q}(z) + X^{(p-1)}_{G,q}(z^{1/p}) = 0. \tag{51}$$

[8] Cf. [40, Theorems 6 and 10].

From formula (13) and Wilson's Theorem one easily finds that

$$X_{G,q_1,q_2}(z) = \sum_{\sigma|q_1} z^{\sigma-1} X_{G,q_1q_2}(z^\sigma) + \sum_{\substack{\sigma|q_1\\ \sigma>1}} z^{\sigma-1} X^{(p-1)}_{G,q_1q_2}(z^{\sigma/p}). \tag{52}$$

Hence, if $X_{G,q_1q_2}(z)$ is rational over $GF(p)$, then so is $X_{G,q_1,q_2}(z)$. To prove the converse, one shows (with somewhat more effort) that

$$X_{G,q_1q_2}(z) = X_{G,q_1,q_2}(z) + z^{p-1}\Big[X^{(p-1)}_{G,q_1,q_2}(z) + (-1)^{r+1}\mathcal{H}^r X^{(p-1)}_{G,q_1,q_2}(z)\Big], \tag{53}$$

where $q_1 = p^r$ and $\mathcal{H} : GF(p)^+[[z]] \to GF(p)^+[[z]]$ is the operator on the subalgebra

$$GF(p)^+[[z]] := \Big\{f(z) \in GF(p)[[z]] : \; f'(z) = 0\Big\}$$

of $GF(p)[[z]]$ given by $f(z) \mapsto \mathcal{H}f(z) := (f(z^{1/p}))^{(p-1)}$. It is not hard to see that $\mathcal{H}$ maps rational functions to rational functions, hence we find from (53) that rationality of $X_{G,q_1,q_2}(z)$ also implies rationality of $X_{G,q_1q_2}(z)$. The statement in (a) concerning degrees then follows from (52), while the assertions under (b) are deduced from the functional equation (25). As a consequence of Theorem 7 we have the following partial answer to the question when the series $X_{G,q_1,q_2}(z)$ will be rational over $GF(p)$.

Corollary 7 ([40, Cor. 11]). (i) *If $X_{G,q_1,q_2}(z)$ is rational and* $\deg(X_{G,q_1,q_2}(z)) \geq 0$, *then $G = 1$; in particular, the set $\mathcal{N}_{G,q_1,q_2}$ is infinite, provided that $G \neq 1$.*

(ii) *If G is in* **Fin**(p), *then $X_{G,q_1,q_2}(z)$ is rational over $GF(p)$ if and only if $G = 1$.*

(iii) *Let G be a finite group, and let q_1 and q_2 be 2–powers such that $q_1q_2 > 1$. Then $X_{G,q_1,q_2}(z)$ is rational over $GF(2)$ if and only if $G = 1$ or G contains a subgroup of index 2.*

5.3. **Divisibility patterns and Fermat primes.** To conclude this section, we briefly discuss the question, under which circumstances the sets $\mathcal{N}_{G,q_1,q_2}$, which capture much of the information contained in the p–patterns Π_{G,q_1,q_2}, allow for a characterization in terms of closed formulae, and, more precisely, under which conditions formulae of Stothers' type arise. As is already apparent from the explicit combinatorial description given in Theorem 5, the generic picture for the sets $\mathcal{N}_{G,q_1,q_2}$ – at least in the case when G is in **Fin**(p) – is very far from allowing for such a description in closed terms, but rather tends to display a peculiar kind of fractal behaviour typical of binomial and multinomial coefficients when considered modulo a prime. For $G \in$**Fin**(p) and $G \neq 1$, the only exceptions, where closed formulae are found, are in fact of Stothers' type, and are listed below. It seems rather unlikely that, apart from the case of periodic behaviour, other closed formulae will be found if $G \notin$ **Fin**(p), but the present state of knowledge does not allow us to completely rule out this possibility. The more precise question as to the existence of a (maximal) generalization of Stothers' formula is answered by the following result.

Theorem 8 ([40, Theorem 13]). *For a prime p, an integer $m > 1$, and a p-power $q > 1$ denote by $\Lambda_{m,q}$ the set of partial sums of the series $1 + q\sum_{\sigma\geq1}(m-1)^\sigma$. Then, for G a finite group and p–powers q_1, q_2 with $q_1q_2 > 1$, we have*

$$\mathcal{N}_{G,q_1,q_2} = \bigcup_{\sigma|q_1} \sigma\Lambda_{|G|,q_1q_2}$$

if and only if $q_1q_2 = 2$, *and* G *is either cyclic of order a Fermat prime, or* G *is an elementary abelian* 3*–group of rank* 2.

As is well known, Fermat primes, i.e., prime numbers of the form $2^{2^\lambda}+1$ with $\lambda \geq 0$, satisfy (or can even be characterized by) a number of curious regularity conditions; for instance, according to Gauß [14, § 366] a regular $\mathfrak{q}$–gon ($\mathfrak{q} > 2$ a prime) can be constructed by compass and ruler if and only if $\mathfrak{q}$ is a Fermat prime. Rather surprisingly, as our last result shows, Fermat primes also play an important special role in the context of subgroup growth theory. In fact, by specializing Theorem 8 to the case when $p = q_1 = 2$, $q_2 = 1$, and $G = C_\mathfrak{q}$ with $\mathfrak{q} \geq 3$, we obtain a new characterization of Fermat primes in terms of the subgroup arithmetic of Hecke groups.

Corollary 8 ([40, Cor. 16]). *Let* $\mathfrak{q} \geq 3$ *be an integer. Then* $\mathfrak{q}$ *is a Fermat prime if and only if*

$$\Pi(\mathfrak{H}(\mathfrak{q})) = \left\{\frac{2(\mathfrak{q}-1)^\sigma - \mathfrak{q}}{\mathfrak{q}-2}\right\}_{\sigma\geq 1} \cup 2\left\{\frac{2(\mathfrak{q}-1)^\sigma - \mathfrak{q}}{\mathfrak{q}-2}\right\}_{\sigma\geq 1}.$$

6. Free parity patterns in Hecke groups

In [37], results rather analogous to those for the p–patterns Π_{G,q_1,q_2} are established for the free parity pattern $\Pi^*(\mathfrak{G})$ in the Hecke group case, i.e., when

$$\mathfrak{G} = \mathfrak{H}(\mathfrak{q}) \cong C_2 * C_\mathfrak{q}, \quad \mathfrak{q} \geq 3.$$

The basis for these results is the functional equation

$$X_\mathfrak{q}^*(z) = 1 + z\left(X_\mathfrak{q}^*(z)\right)^{\mu_\mathfrak{q}} \tag{54}$$

for the mod 2 projection $X_\mathfrak{q}^*(z)$ of the generating function $1 + \sum_{\lambda\geq 1} f_\lambda(\mathfrak{H}(\mathfrak{q}))z^\lambda$, where $\mu_\mathfrak{q}$ is the free rank of $\mathfrak{H}(\mathfrak{q})$; equation (54) being arrived at by a method somewhat different from that leading to equation (25). We begin by recalling certain facts about virtually free groups leading up to a sketch of our approach to equation (54), before describing some of the rather striking consequences of this identity for the patterns $\Pi_\mathfrak{q}^* := \Pi^*(\mathfrak{H}(\mathfrak{q}))$.

6.1. The type $\tau_\mathfrak{q}$ and identity (54). The type $\tau(\mathfrak{G})$ of a finitely generated virtually free group $\mathfrak{G} \cong \pi_1(\mathfrak{G}(-), Y)$ is defined as the tuple

$$\tau(\mathfrak{G}) = \left(m_\mathfrak{G};\, \zeta_1(\mathfrak{G}), \ldots, \zeta_\kappa(\mathfrak{G}), \ldots, \zeta_{m_\mathfrak{G}}(\mathfrak{G})\right),$$

where the $\zeta_\kappa(\mathfrak{G})$ are integers indexed by the divisors of $m_\mathfrak{G}$, given by

$$\zeta_\kappa(\mathfrak{G}) = \left|\left\{e \in E(Y) : |\mathfrak{G}(e)| \,\Big|\, \kappa\right\}\right| - \left|\left\{v \in V(Y) : |\mathfrak{G}(v)| \,\Big|\, \kappa\right\}\right|.$$

It can be shown that the type $\tau(\mathfrak{G})$ is in fact an invariant of the group $\mathfrak{G}$, i.e., independent of the particular decomposition of $\mathfrak{G}$ in terms of a graph of groups $(\mathfrak{G}(-), Y)$, and that two virtually free groups $\mathfrak{G}_1$ and $\mathfrak{G}_2$ contain the same number of free subgroups of index n for every $n \in \mathbb{N}$ if and only if $\tau(\mathfrak{G}_1) = \tau(\mathfrak{G}_2)$; cf. [33, Theorem 2]. Note that, as a consequence of (10), the Euler characteristic of $\mathfrak{G}$ can be expressed in terms of the type $\tau(\mathfrak{G})$ via

$$\chi(\mathfrak{G}) = -m_\mathfrak{G}^{-1} \sum_{\kappa | m_\mathfrak{G}} \varphi(m_\mathfrak{G}/\kappa)\, \zeta_\kappa(\mathfrak{G}). \tag{55}$$

It follows in particular that if two virtually free groups have the same number of free index n subgroups for every n, then their Euler characteristics must coincide. Our approach to the function $f(\mathfrak{G}) : \mathbb{N} \to \mathbb{N}$ given by

$$f_\lambda(\mathfrak{G}) = \text{number of free subgroups of index } \lambda m_{\mathfrak{G}} \text{ in } \mathfrak{G}$$

is to relate it to another arithmetic function $g_\lambda(\mathfrak{G})$ which turns out to be easier to compute. Define a *torsion-free* $\mathfrak{G}$–action on a set Ω to be a $\mathfrak{G}$–action on Ω which is free when restricted to finite subgroups. For a finite set Ω to admit a torsion–free $\mathfrak{G}$–action it is necessary and sufficient that $|\Omega|$ be divisible by $m_{\mathfrak{G}}$. For $\lambda \in \mathbb{N}_0$ define $g_\lambda(\mathfrak{G})$ by the condition that

$$(\lambda m_{\mathfrak{G}})!\, g_\lambda(\mathfrak{G}) = \text{number of torsion–free } \mathfrak{G}\text{–actions on a set with } \lambda m_{\mathfrak{G}} \text{ elements,}$$

in particular $g_0(\mathfrak{G}) = 1$. Then the arithmetic functions $f_\lambda(\mathfrak{G})$ and $g_\lambda(\mathfrak{G})$ are related via the transformation formula[9]

$$\sum_{\mu=1}^{\lambda} f_\mu(\mathfrak{G})\, g_{\lambda-\mu}(\mathfrak{G}) = m_{\mathfrak{G}}\, \lambda\, g_\lambda(\mathfrak{G}), \quad \lambda \geq 1. \tag{56}$$

Moreover, an analysis of the universal mapping property associated with the presentation $\mathfrak{G} \cong \pi_1(\mathfrak{G}(-), Y)$ of $\mathfrak{G}$ reveals that

$$g_\lambda(\mathfrak{G}) = \frac{\prod\limits_{e \in E(Y)} \left[(\lambda m_{\mathfrak{G}}/|\mathfrak{G}(e)|)!\, |\mathfrak{G}(e)|^{\lambda m_{\mathfrak{G}}/|\mathfrak{G}(e)|} \right]}{\prod\limits_{v \in V(Y)} \left[(\lambda m_{\mathfrak{G}}/|\mathfrak{G}(v)|)!\, |\mathfrak{G}(v)|^{\lambda m_{\mathfrak{G}}/|\mathfrak{G}(v)|} \right]}, \quad \lambda \geq 0; \tag{57}$$

compare [33, Prop. 3]. From the latter formula it can be deduced that the sequence $g_\lambda(\mathfrak{G})$ is of hypergeometric type and that the generating function $\Theta_{\mathfrak{G}}(z) := \sum_{\lambda \geq 0} g_\lambda(\mathfrak{G}) z^\lambda$ satisfies a homogeneous linear differential equation

$$\vartheta_0(\mathfrak{G})\, \Theta_{\mathfrak{G}}(z) + (\vartheta_1(\mathfrak{G})\, z - m_{\mathfrak{G}})\, \Theta'_{\mathfrak{G}}(z) + \sum_{\mu=2}^{\mu(\mathfrak{G})} \vartheta_\mu(\mathfrak{G})\, z^\mu\, \Theta_{\mathfrak{G}}^{(\mu)}(z) = 0 \tag{58}$$

of order $\mu(\mathfrak{G})$ with integral coefficients given in terms of the type via

$$\vartheta_\mu(\mathfrak{G}) = \frac{1}{\mu!} \sum_{j=0}^{\mu} (-1)^{\mu-j} \binom{\mu}{j} m_{\mathfrak{G}}\, (j+1) \prod_{\kappa | m_{\mathfrak{G}}} \prod_{\substack{1 \leq k \leq m_{\mathfrak{G}} \\ (m_{\mathfrak{G}}, k) = \kappa}} (j m_{\mathfrak{G}} + k)^{\zeta_\kappa(\mathfrak{G})}, \quad 0 \leq \mu \leq \mu(\mathfrak{G}); \tag{59}$$

cf. [33, Prop. 5]. Inserting formula (57) into (56) yields a recursive description of the arithmetic function $f_\lambda(\mathfrak{G})$ attached to a finitely generated virtually free group $\mathfrak{G}$. However, formulae obtained in this way (referred to as being of Hall type) usually turn out to be quite unsatisfactory when dealing with number–theoretic aspects of the sequence $f_\lambda(\mathfrak{G})$ such as divisibility properties. Instead, one proceeds as follows. Introducing the generating function

$$\Xi_{\mathfrak{G}}(z) := \sum_{\lambda \geq 0} f_{\lambda+1}(\mathfrak{G})\, z^\lambda$$

[9] See for instance [33, Cor. 1].

one notes that, in view of equation (56), the series $\Xi_{\mathfrak{G}}(z)$ is related to $\Theta_{\mathfrak{G}}(z)$ via the identity

$$\Xi_{\mathfrak{G}}(z) = m_{\mathfrak{G}}\, \frac{d}{dz}\big(\log\, \Theta_{\mathfrak{G}}(z)\big), \tag{60}$$

and, applying Bell's formula[10]

$$\frac{d^\mu}{dz^\mu} f(g(z)) = \sum_{\pi\vdash\mu} \frac{\mu!}{\prod_{j\geq 1}\pi_j!}\left[\prod_{j\geq 1}\left(\frac{g^{(j)}(z)}{j!}\right)^{\pi_j}\right] f^{(\|\pi\|)}(g(z)) \tag{61}$$

for the derivatives of a composite function with $f(t) = \mathrm{e}^t$ and $g(z) = m_{\mathfrak{G}}^{-1}\int \Xi_{\mathfrak{G}}(z)\,dz$, one finds that

$$\Theta_{\mathfrak{G}}^{(\mu)}(z) = \mu!\;\Theta_{\mathfrak{G}}(z) \sum_{\nu=1}^{\mu} \sum_{\substack{\mu_1,\ldots,\mu_\nu>0\\ \mu_1+\cdots+\mu_\nu=\mu}} (\nu!\, m_{\mathfrak{G}}^{\nu})^{-1} \prod_{j=1}^{\nu} \frac{\Xi_{\mathfrak{G}}^{(\mu_j-1)}(z)}{\mu_j!}, \quad \mu\geq 1.$$

Combining these identities for $1 \leq \mu \leq \mu(\mathfrak{G})$ with (58), we obtain for $\Xi_{\mathfrak{G}}(z)$ the differential equation

$$\Xi_{\mathfrak{G}}(z) = \vartheta_0(\mathfrak{G}) + \sum_{\mu=1}^{\mu(\mathfrak{G})}\sum_{\nu=1}^{\mu} \sum_{\substack{\mu_1,\ldots,\mu_\nu>0\\ \mu_1+\cdots+\mu_\nu=\mu}} \binom{\mu}{\mu_1,\ldots,\mu_\nu} (\nu!\, m_{\mathfrak{G}}^{\nu})^{-1}\, \vartheta_\mu(\mathfrak{G})\, z^\mu \prod_{j=1}^{\nu} \Xi_{\mathfrak{G}}^{(\mu_j-1)}(z) \tag{62}$$

with $\vartheta_\mu(\mathfrak{G})$ as in (59). Comparing coefficients in (62) now yields the following.

Proposition 2 ([37, Prop. 1]). *Let $\mathfrak{G}$ be a finitely generated virtually free group. Then the function $f_\lambda(\mathfrak{G})$ satisfies the recursion*

$$f_{\lambda+1}(\mathfrak{G}) = \sum_{\mu=1}^{\mu(\mathfrak{G})} \sum_{\substack{\lambda_1,\ldots,\lambda_\mu>0\\ \lambda_1+\cdots+\lambda_\mu=\lambda}} (\mu!\, m_{\mathfrak{G}}^{\mu})^{-1}\, \mathcal{F}_\mu^{(\mathfrak{G})}(\lambda_1,\ldots,\lambda_\mu) \prod_{j=1}^{\mu} f_{\lambda_j}(\mathfrak{G}) \qquad (\lambda\geq 1,\ f_1(\mathfrak{G}) = \vartheta_0(\mathfrak{G})) \tag{63}$$

with coefficients

$$\mathcal{F}_\mu^{(\mathfrak{G})}(\lambda_1,\ldots,\lambda_\mu) := \sum_{\nu=\mu}^{\mu(\mathfrak{G})} \vartheta_\nu(\mathfrak{G})\, F_{\mu,\nu}(\lambda_1,\ldots,\lambda_\mu),$$

where

$$F_{\mu,\nu}(\lambda_1,\ldots,\lambda_\mu) := \nu! \sum_{\substack{\nu_1,\ldots,\nu_\mu\geq 0\\ \nu_1+\cdots+\nu_\mu=\nu-\mu}} \prod_{j=1}^{\mu}\left[\binom{\lambda_j-1}{\nu_j}\Big/(\nu_j+1)\right]$$

and $\vartheta_\nu(\mathfrak{G})$ is as in (59).

[10] By a partition π we mean any sequence $\pi = \{\pi_j\}_1^\infty$ of non-negative integers, such that $\pi_j = 0$ for all but finitely many j. $|\pi| = \sum_{j=1}^\infty j\,\pi_j$ is called the weight of π, and $\|\pi\| = \sum_{j=1}^\infty \pi_j$ is the norm or length of the partition π. If $|\pi| = 0$, π is called the empty partition, otherwise π is non-empty. As usual, we also write $\pi\vdash\mu$ for $|\pi| = \mu$, and say that π is a partition of μ.

We now restrict to the case where $\mathfrak{G} = \mathfrak{H}(\mathfrak{q})$. Then

$$m_{\mathfrak{q}} := m_{\mathfrak{H}(\mathfrak{q})} = \begin{cases} \mathfrak{q}, & \mathfrak{q} \text{ even} \\ 2\mathfrak{q}, & \mathfrak{q} \text{ odd}, \end{cases}$$

$$\zeta_\kappa^{(\mathfrak{q})} := \zeta_\kappa(\mathfrak{H}(\mathfrak{q})) = \begin{cases} 1, & 2 \nmid \kappa \text{ and } \kappa < \mathfrak{q} \\ -1, & \kappa = m_{\mathfrak{q}} \\ 0, & \text{otherwise}, \end{cases}$$

$$\chi(\mathfrak{H}(\mathfrak{q})) = -\frac{\mathfrak{q}-2}{2\mathfrak{q}}$$

and

$$\mu_{\mathfrak{q}} := \mu(\mathfrak{H}(\mathfrak{q})) = \begin{cases} \mathfrak{q}-1, & \mathfrak{q} \text{ odd} \\ \mathfrak{q}/2, & \mathfrak{q} \text{ even}. \end{cases}$$

We make use of two auxiliary results, the first one concerning the coefficients $\vartheta_\mu(\mathfrak{H}(\mathfrak{q}))$, the second a general observation concerning partitions.

Lemma 5 ([37, Lemma 1]). *For $0 \le \mu \le \mu_{\mathfrak{q}}$ the integer $\vartheta_\mu^{(\mathfrak{q})} := \vartheta_\mu(\mathfrak{H}(\mathfrak{q}))$ is divisible by $m_{\mathfrak{q}}^\mu$, and*

$$m_{\mathfrak{q}}^{-\mu}\,\vartheta_\mu^{(\mathfrak{q})} \equiv \binom{\mu_{\mathfrak{q}}}{\mu} \mod 2.$$

Lemma 6 ([37, Lemma 2]). *Let $\pi = \{\pi_j\}_{j\ge1}$ be a partition. Then we have*

$$\nu_2\Bigg(\prod_{j\ge1} j^{\pi_j}\Bigg) \le \nu_2\big((2(|\pi| - ||\pi||))!\big)$$

with equality occurring if and only if $|\pi| \le ||\pi|| + 1$.

In Proposition 2 put $\mathfrak{G} = \mathfrak{H}(\mathfrak{q})$, write $f_\lambda(\mathfrak{q})$ for $f_\lambda(\mathfrak{H}(\mathfrak{q}))$ and $\mathcal{F}_\mu^{(\mathfrak{q})}(\lambda_1, \dots, \lambda_\mu)$ for $\mathcal{F}_\mu^{(\mathfrak{H}(\mathfrak{q}))}(\lambda_1, \dots, \lambda_\mu)$, and multiply both sides of (63) by a sufficiently large odd number F depending only on $\mathfrak{q}$ to obtain

$$F\,f_{\lambda+1}(\mathfrak{q}) = \sum_{\mu=1}^{\mu_{\mathfrak{q}}} 2^{-(\nu_2(\mu!)+\mu\nu_2(m_{\mathfrak{q}}))} \sum_{\substack{\lambda_1,\dots,\lambda_\mu>0\\ \lambda_1+\dots+\lambda_\mu=\lambda}} \tilde{\mathcal{F}}_\mu^{(\mathfrak{q})}(\lambda_1, \dots, \lambda_\mu) \prod_{j=1}^{\mu} f_{\lambda_j}(\mathfrak{q}), \quad \lambda \ge 1, \tag{64}$$

with integers $\tilde{\mathcal{F}}_\mu^{(\mathfrak{q})}(\lambda_1, \dots, \lambda_\mu)$ satisfying

$$\nu_2(\tilde{\mathcal{F}}_\mu^{(\mathfrak{q})}(\lambda_1, \dots, \lambda_\mu)) = \nu_2(\mathcal{F}_\mu^{(\mathfrak{q})}(\lambda_1, \dots, \lambda_\mu)).$$

Decompose $\mathcal{F}_\mu^{(\mathfrak{q})}(\lambda_1, \dots, \lambda_\mu)$ as

$$\mathcal{F}_\mu^{(\mathfrak{q})}(\lambda_1, \dots, \lambda_\mu) = \mu!\,\vartheta_\mu^{(\mathfrak{q})} + \Delta_\mu^{(\mathfrak{q})}(\lambda) + R_\mu^{(\mathfrak{q})}(\lambda_1, \dots, \lambda_\mu),$$

where

$$\Delta_\mu^{(\mathfrak{q})}(\lambda) := \begin{cases} (\mu+1)!\,(\lambda-\mu)\,\vartheta_{\mu+1}^{(\mathfrak{q})}/2, & \mu < \mu_{\mathfrak{q}} \\ 0, & \mu = \mu_{\mathfrak{q}} \end{cases}$$

and

$$R_\mu^{(\mathfrak{q})}(\lambda_1, \dots, \lambda_\mu) := \sum_{\nu=\mu+2}^{\mu_{\mathfrak{q}}} \sum_{\substack{\nu_1,\dots,\nu_\mu\ge0\\ \nu_1+\dots+\nu_\mu=\nu-\mu}} \vartheta_\nu^{(\mathfrak{q})} \frac{\nu!}{(\nu_1+1)\dots(\nu_\mu+1)} \binom{\lambda_1-1}{\nu_1} \cdots \binom{\lambda_\mu-1}{\nu_\mu}.$$

Applying Lemmas 5 and 6 we find that, for each $\mu \in [\mu_{\mathfrak{q}}]$, the term $R_\mu^{(\mathfrak{q})}(\lambda_1, \dots, \lambda_\mu)$ is divisible by $2^{\nu_2(\mu!)+\mu\nu_2(m_{\mathfrak{q}})}$, and that

$$R_\mu^{(\mathfrak{q})}(\lambda_1, \dots, \lambda_\mu)/2^{\nu_2(\mu!)+\mu\nu_2(m_{\mathfrak{q}})}$$

is even. Moreover, Lemma 5 allows one to show that $\Delta_\mu^{(\mathfrak{q})}(\lambda)$ as well as $\mu!\vartheta_\mu^{(\mathfrak{q})}$ are also divisible by $2^{\nu_2(\mu!)+\mu\nu_2(m_{\mathfrak{q}})}$ for all $\mu \in [\mu_{\mathfrak{q}}]$, and that

$$\Delta_\mu^{(\mathfrak{q})}(\lambda)/2^{\nu_2(\mu!)+\mu\nu_2(m_{\mathfrak{q}})} \equiv \begin{cases} \binom{\mathfrak{q}/2}{\mu+1}\lambda, & \mu \equiv 0\,(2) \text{ and } \nu_2(\mathfrak{q}) = 1 \\ 0, & \text{otherwise} \end{cases} \quad \text{mod } 2,$$

respectively

$$\mu!\,\vartheta_\mu^{(\mathfrak{q})}/2^{\nu_2(\mu!)+\mu\nu_2(m_{\mathfrak{q}})} \equiv \binom{\mu_{\mathfrak{q}}}{\mu} \quad \text{mod } 2.$$

We conclude that $\tilde{\mathcal{F}}_\mu^{(\mathfrak{q})}(\lambda_1, \dots, \lambda_\mu)$ is divisible by $2^{\nu_2(\mu!)+\mu\nu_2(m_{\mathfrak{q}})}$, and that

$$\tilde{\mathcal{F}}_\mu^{(\mathfrak{q})}(\lambda_1, \dots, \lambda_\mu)/2^{\nu_2(\mu!)+\mu\nu_2(m_{\mathfrak{q}})} \equiv \begin{cases} 0, & \nu_2(\mathfrak{q}) = 1 \,\&\, \mu \equiv 0\,(2) \,\&\, \lambda \equiv 1\,(2) \\ \binom{\mu_{\mathfrak{q}}}{\mu}, & \text{otherwise} \end{cases} \quad \text{mod } 2. \tag{65}$$

Evaluating (64) modulo 2 in the light of (65), and applying the binomial law in the ring $GF(2)[[z]]$, identity (54) follows.

6.2. **The patterns $\Pi_{\mathfrak{q}}^*$.** We now pass to some of the consequences for the patterns $\Pi_{\mathfrak{q}}^*$ of the functional equation (54). Our first result collects together a miscellany of information; in particular, it resolves the classification problem, and it answers the question when the series $X_{\mathfrak{q}}^*(z)$ is rational over $GF(2)$ (namely never).

Theorem 9. (a) *Let $\mathfrak{q}_1, \mathfrak{q}_2 \geq 3$ be integers. Then we have $\Pi_{\mathfrak{q}_1}^* = \Pi_{\mathfrak{q}_2}^*$ if and only if $\mu_{\mathfrak{q}_1} = \mu_{\mathfrak{q}_2}$.*

(b) *Every entry of $\Pi_{\mathfrak{q}}^*$ is congruent to 1 modulo $2^{\nu_2(\mu_{\mathfrak{q}})}$.*

(c) *The first two entries of $\Pi_{\mathfrak{q}}^*$ are 1 and $2^{\nu_2(\mu_{\mathfrak{q}})}+1$; in particular, given an integer $\alpha^* \geq 2$, the set $\{1, \alpha^*\}$ can be extended to a free parity pattern of some Hecke group if and only if $\alpha^* - 1$ is a 2–power.*

(d) *The series $X_{\mathfrak{q}}^*(z)$ is never rational over $GF(2)$; in particular, the set $\Pi_{\mathfrak{q}}^*$ is always infinite.*

This is the contents of [37, Theorems 1 and 2]. We also obtain an explicit combinatorial description of the patterns $\Pi_{\mathfrak{q}}^*$ somewhat analogous to that afforded by Theorem 5 for the patterns $\Pi_{G,\mathfrak{q}_1,\mathfrak{q}_2}$ with $G \in \mathbf{Fin}(p)$.

Theorem 10 ([37, Theorem 3]). *Denote by $\mathfrak{s}(x)$ the sum of digits in the binary representation of the natural number x. Then, for every $\mathfrak{q} \geq 3$,*

$$\Pi_{\mathfrak{q}}^* = \Big\{\lambda \in \mathbb{N} : \ \mathfrak{s}(\lambda) + \mathfrak{s}((\mu_{\mathfrak{q}} - 1)\lambda + 1) - \mathfrak{s}(\mu_{\mathfrak{q}}\lambda) = 1\Big\}.$$

Again, it is apparent from Theorem 10 that the parity patterns $\Pi_{\mathfrak{q}}^*$ will not in general lend themselves to an explicit description in closed terms as in the case of the modular group. Instead, $\Pi_{\mathfrak{q}}^*$ generically tends to inherit the peculiar kind of fractal behaviour observed in Pascal's triangle when considered modulo 2. There is however one special case where we can describe the patterns $\Pi_{\mathfrak{q}}^*$ in a straightforward and completely explicit way, namely when $\mu_{\mathfrak{q}}$ is a 2–power. Hence, while a canonical generalization of the free parity pattern met in the modular group $\mathfrak{H}(3)$ and in $\mathfrak{H}(4)$ to all Hecke groups does not exist, the latter type of pattern precisely characterizes two infinite series of Hecke groups. For an integer $\mathfrak{q} \geq 3$ define

$$\Lambda_{\mathfrak{q}}^* := \left\{ \frac{\mu_{\mathfrak{q}}^{\sigma} - 1}{\mu_{\mathfrak{q}} - 1} : \sigma = 1, 2, \dots \right\},$$

i.e., $\Lambda_{\mathfrak{q}}^*$ is the set of partial sums of the geometric series $\sum_{\sigma=0}^{\infty} \mu_{\mathfrak{q}}^{\sigma}$ generated by the free rank $\mu_{\mathfrak{q}}$ of $\mathfrak{H}(\mathfrak{q})$.

Theorem 11 ([37, Theorem 4]). *Let $\mathfrak{q} \geq 3$ be an integer. Then the following assertions are equivalent:*

(i) $\Pi_{\mathfrak{q}}^* = \Lambda_{\mathfrak{q}}^*$.

(ii) $\mathfrak{q}$ *or* $\mathfrak{q} - 1$ *is a 2–power.*

By specializing Theorem 11 to the case where $\mathfrak{q}$ is a prime number, we obtain another characterization of Fermat primes.

Corollary 9. [37, Cor. 1] *Let $\mathfrak{q} > 2$ be a prime. Then $\mathfrak{q}$ is a Fermat prime if and only if*

$$\Pi_{\mathfrak{q}}^* = \Lambda_{\mathfrak{q}}^* = \left\{ \frac{(\mathfrak{q}-1)^{\sigma} - 1}{\mathfrak{q} - 2} : \sigma = 1, 2, \dots \right\}.$$

6.3. **An afterthought.** The results presented in the last subsection concerning parity properties of the function $f_{\lambda}(\mathfrak{G})$ in the Hecke group case, while rather striking and certainly interesting in their own right, may perhaps just be taken as promising indications that a more general theory is awaiting its discovery. Indeed, all our results on $f_{\lambda}(\mathfrak{G})$ have close analogues in Section 5. Clearly, this area is in need of further research. We would like to close this section with the following.

Problem 3. *Find results concerning the parity of the function $f_{\lambda}(\mathfrak{G})$ for a larger class of virtually free groups, containing in particular all groups of the forms $C_2 * C_{\mathfrak{q}}$ and $C_{\mathfrak{q}} * C_{\mathfrak{q}}$ with $\mathfrak{q} \geq 3$. What can be said about the divisibility of $f_{\lambda}(\mathfrak{G})$ modulo arbitrary primes?*

7. Parity patterns in one–relator groups

Most of the major developments concerning the theory of subgroup growth have so far concentrated on one of two classes of discrete groups: finitely generated nilpotent groups and finitely generated groups containing a free subgroup of finite index (i.e. fundamental groups of finite graphs of finite groups). On the other hand, almost nothing was known until recently concerning the subgroup growth of one–relator groups. If Γ is a one–relator group involving $d \geq 3$ generators, then, by a result of Baumslag and Pride [5], Γ contains a subgroup of finite index which can be mapped homomorphically onto a non–abelian free group. Hence, in this case $s_n(\Gamma)$ grows super–exponentially fast, just like the sequence of subgroup numbers of a non–abelian free group. One might feel however that, at least generically, the relationship between the subgroup growth of one–relator groups and that of free groups should be rather more intimate than the latter observation seems to imply. More specifically, one might ask, as Lubotzky does in [29], whether the limit

$$\lim{}_{n\to\infty} \frac{s_n(\Gamma)}{s_n(F_{d-1})} \tag{66}$$

is finite and positive for $d \geq 3$, and if so, what this limit is. In [41], results and methods from representation theory are employed and further developed to obtain an asymptotic estimate for the number of index n subgroups in a surface group; in particular, it is shown that Lubotzky's question as to the existence of the limit (66) has an affirmative answer for surface groups, and that the value of (66) in this case is 2. In this final section, we are concerned with a result from the more recent paper [42] describing the behaviour modulo 2 of the function $s_n(\Gamma)$ for a class of one–relator groups Γ containing in particular all surface groups.

7.1. **The result.** We shall work over the alphabet $\mathcal{A} = \{x_1, x_2, \ldots, x_1^{-1}, x_2^{-1}, \ldots\}$. Define a class of words $\mathcal{W}$ over $\mathcal{A}$ as follows.

(i) $x_i^2, [x_i, x_j] \in \mathcal{W}$ for all $i, j \in \mathbb{N}$ and $i \neq j$.

(ii) If $w_1, w_2 \in \mathcal{W}$ have no generator in common, then $w_1 w_2 \in \mathcal{W}$.

(iii) If $v \in \mathcal{W}$, and x_i is a generator not occurring in v, then $[v, x_i] \in \mathcal{W}$.

(iv) $\mathcal{W}$ is the smallest set of words over $\mathcal{A}$ satisfying (i), (ii), and (iii).

Clearly, all surface group relators

$$\prod_{i=1}^{g} [x_{2i-1}, x_{2i}] \quad \text{and} \quad \prod_{i=1}^{h} x_i^2, \quad g, h \geq 1$$

are contained in $\mathcal{W}$, as is, for instance, the word $w = [x_1^2 x_2^2, x_3]$. For a word $w = w(x_1, \ldots, x_d)$ over $\mathcal{A}$ involving the generators $x_1, \ldots, x_d$, define the one-relator group Γ_w *associated with* w via

$$\Gamma_w = \Big\langle x_1, x_2, \ldots, x_d \mid w(x_1, \ldots, x_d) = 1 \Big\rangle.$$

Our next result describes the behaviour modulo 2 of $s_n(\Gamma_w)$ for words $w \in \mathcal{W}$.

Theorem 12 ([42, Theorem 1]). *If w is in $\mathcal{W}$ and involves at least three generators, then $s_n(\Gamma_w)$ is odd if and only if $n = k^2$ or $n = 2k^2$ for some $k \geq 1$; in particular, all groups Γ_w with $w \in \mathcal{W}$ involving three or more generators share the same parity pattern, and $s_n(\Gamma_w)$ is multiplicative modulo 2.*

It appears likely that Theorem 12 is best possible in the sense that if for some word w over $\mathcal{A}$ the function $s_n(\Gamma_w)$ displays the parity pattern described in Theorem 12, then in fact $w \in \mathcal{W}$.

7.2. **Some background: A recurrence relation modulo 2, Euler's pentagonal theorem, and results concerning the parity of the partition function.** The key to Theorem 12 lies in a remarkable recurrence relation for the mod 2 behaviour of $s_n(\Gamma_w)$ with $w \in \mathcal{W}$.

Proposition 3 ([42, Theorem 3]). *Let $w \in \mathcal{W}$ be a word involving three or more generators. Then we have*

$$s_n(\Gamma_w) \equiv \sum_{\substack{k \geq 1 \\ k(k+1) < 2n}} s_{n-\frac{1}{2}k(k+1)}(\Gamma_w) + \delta(n) \mod 2, \quad n \geq 1, \tag{67}$$

where

$$\delta(n) = \begin{cases} 1, & n \text{ odd and triangular,} \\ 0, & \text{otherwise.} \end{cases}$$

The (representation theoretic) background of Proposition 3 will be discussed in section 7.3. Looking at equation (67), which descends in triangular numbers, one cannot but feel reminded of Euler's pentagonal theorem

$$\prod_{n\geq 1}(1-q^n) = 1 + \sum_{k\geq 1}(-1)^k q^{\frac{1}{2}k(3k-1)}(1+q^k) = \sum_{k=-\infty}^{+\infty}(-1)^k q^{\frac{1}{2}k(3k-1)},$$

which yields the recurrence relation

$$p(n) = \sum_{k\geq 1}(-1)^{k+1}\Big\{p\big(n-\frac{1}{2}k(3k-1)\big) + p\big(n-\frac{1}{2}k(3k+1)\big)\Big\},\quad n\geq 1$$

for the partition function $p(n)$, with the convention that $p(m) = 0$ for $m < 0$; cf. [1, Chap. 1.3]. On the other hand, while $p(n)$ is known to satisfy a number of surprising congruences (for instance modulo 5, 7, 11) when n is in certain special arithmetic progressions[11], there do not seem to be any such congruences modulo 2 or 3. In fact, the parity of $p(n)$ appears to be quite random, and, on the basis of extensive numerical evidence, it is believed that the partition function is 'about equally often' even and odd; i.e., that

$$\sum_{\substack{n\leq x\\ p(n)\equiv 0(2)}} 1 \sim \frac{x}{2} \quad (x\to\infty);$$

cf. [45]. Subbarao [51] has conjectured that, for any arithmetic progression $r \pmod t$, there are infinitely many integers $m \equiv r\ (t)$ for which $p(m)$ is odd, and also infinitely many integers $n \equiv r\ (t)$ for which $p(n)$ is even. Partial results in this direction have been obtained by Ono [44]. Recently, Nicolas, Ruzsa, and Sárközy [43] have shown that, for all r and t,

$$x^{-\frac{1}{2}}\sum_{\substack{n\leq x\\ n\equiv r(t)\\ p(n)\equiv 0(2)}} 1 \to \infty \quad \text{as} \quad x\to\infty.$$

In an appendix to the latter paper, Serre shows that the same type of result holds in fact for the coefficients of arbitrary modular forms. Against this background, Theorem 12 appears remarkable and highly surprising. Once conjectured however, it can be established by induction on n, using the recurrence relation (67) and classical results of Gauß and Legendre concerning the representation numbers of binary quadratic forms; cf. [14, § 205], [28], [8, Chap. VI.8], and [20, Satz 89].

7.3. **Some representation theory.** The proof of Proposition 3 depends on certain parity properties of character values and multiplicities for symmetric groups, which appear to be both new and of independent interest. Given a word $w = w(x_1, \ldots, x_d)$ over the alphabet $\mathcal{A}$ and an irreducible character χ of S_n, define numbers $\alpha_\chi(w) \in \mathbb{C}$ by means of the expansion

$$\begin{aligned} N_w(\pi) :&= \Big|\Big\{(x_1,\ldots,x_d)\in S_n^d:\ w(x_1,\ldots,x_d)=\pi\Big\}\Big| \\ &= (n!)^{d-1}\sum_{\chi}\alpha_\chi(w)\chi(\pi), \end{aligned}$$

[11]Cf. [25] for a comprehensive account up to the 1970's concerning divisibility properties of $p(n)$. Some exciting recent developments in this area are described in [2] and [3].

where $\pi \in S_n$ and the sum is taken over the set $\mathrm{Irr}(S_n)$ of all irreducible characters of S_n. Note that $N_w(\pi)$ is a class function, hence the coefficients $\alpha_\chi(w)$ are well defined. Applying standard results from the representation theory of symmetric groups, it is not difficult to establish the following information concerning the $\alpha_\chi(w)$, which leads to the explicit computation of these coefficients for each word $w \in \mathcal{W}$ and all $\chi \in \mathrm{Irr}(S_n)$ (cf. for instance [24] for the necessary representation theoretic background).

Lemma 7 ([42, Lemma 1]). *Let w_1, w_2, v be words over $\mathcal{A}$, and let χ be an irreducible character of S_n.*

(i) *We have $\alpha_\chi(x_i^2) = 1$ and $\alpha_\chi([x_i, x_j]) = \frac{1}{\chi(1)}$ for all $i, j \in \mathbb{N}$ with $i \neq j$.*

(ii) *If w_1 and w_2 have no generator in common, then we have*

$$\alpha_\chi(w_1 w_2) = \frac{\alpha_\chi(w_1)\,\alpha_\chi(w_2)}{\chi(1)}.$$

(iii) *If x_i does not occur among the generators of v, then*

$$\alpha_\chi([v, x_i]) = \frac{1}{\chi(1)} \sum_{\chi' \in \mathrm{Irr}(S_n)} \alpha_{\chi'}(v)\,\langle \chi^2 \mid \chi' \rangle.$$

Denote by σ the bijection between the self-conjugate partitions of n and the partitions of n into distinct odd parts, mapping a self-conjugate partition λ onto the partition given by the symmetric hooks of λ (with respect to the main diagonal). Moreover, denote by C_λ the conjugacy class of S_n whose cycle structure is given by the partition $\lambda \vdash n$, and by χ_λ the irreducible character of S_n associated with λ. Then an inductive argument based on the Murnagham–Nakayama rule enables one to prove the following result.[12]

Lemma 8 ([42, Lemma 2]). *Let λ_1, λ_2 be partitions of n with λ_1 self-conjugate. Then $\chi_{\lambda_1}(C_{\lambda_2})$ is odd if and only if $\lambda_2 = \lambda_1^\sigma$.*

Call an irreducible character χ of S_n *symmetric*, if $\chi = \varepsilon_n \chi$, where $\varepsilon_n = \chi_{(1^n)}$ is the sign character; this is equivalent to demanding that the partition associated with χ be self-conjugate. A rather subtle argument in the $GF(2)$–algebra $\mathfrak{A} = \mathrm{GF}(2)[\mathrm{Irr}(S_n)]$ generated by the irreducible characters of S_n, building heavily on Lemma 8, now establishes the following.

Lemma 9 ([42, Lemma 3]). *Let χ and χ' be irreducible characters of S_n.*

(i) *If χ is symmetric, then $\langle \chi^{2^k} \mid \chi' \rangle = \langle \chi^{2^k} \mid \varepsilon_n \chi' \rangle$ for all k.*

(ii) *If both χ and χ' are symmetric, then $\langle \chi^2 \mid \chi' \rangle$ is odd if and only if $\chi = \chi'$.*

An irreducible character χ of S_n is termed a 2-*core* character, if $\frac{n!}{\chi(1)}$ is odd. Note that, since the degree of an irreducible representation of a finite group G always divides $|G/\zeta_1(G)|$, this concept is well defined for arbitrary finite groups; cf. [23] or [21, Chap. V, Satz 17.10]. For $G = S_n$, the hook formula shows that an irreducible character χ_λ is 2-core if and only if all hook lengths of the associated partition λ are odd. The latter condition is easily seen to be equivalent to requiring that λ is of the form $\triangle = (k, k-1, \ldots, 1)$ for some $k \geq 1$. It follows that S_n has a 2-core character if and only if $n = \frac{k(k+1)}{2}$ is a triangular number, in which case $\chi_\triangle$ is the unique 2-core character; in particular, 2-core characters are symmetric. With these

[12] Note that since $\pi \sim \pi^a$ for all $\pi \in S_n$ and exponents a coprime to the order of π, characters of S_n are integer-valued.

preliminaries, Lemmas 7 and 9 allow us to derive the following result, which is the decisive tool in proving Proposition 3.

Lemma 10 ([42, Lemma 4]). *Let $w \in \mathcal{W}$ be a word involving d generators, and let χ be an irreducible character of S_n.*

(i) *If $d \geq 2$, then $(n!)^{d-2}\chi(1)\alpha_\chi(w)$ is an integer.*

(ii) *If $d \geq 3$ and χ is not 2-core, then $(n!)^{d-2}\chi(1)\alpha_\chi(w)$ is even.*

(iii) *If $d \geq 2$ and χ is 2-core, then $(n!)^{d-2}\chi(1)\alpha_\chi(w)$ is odd.*

With Lemma 10 in hand, we can now establish Proposition 3: Let $w \in \mathcal{W}$ be a word involving $d \geq 3$ generators. By the exponential principle, the subgroup numbers $s_n(\Gamma_w)$ are related to the sequence $h_n(\Gamma_w) = |\mathrm{Hom}(\Gamma_w), S_n)|/n!$ via the equation[13]

$$nh_n(\Gamma_w) = \sum_{\nu=0}^{n-1} s_{n-\nu}(\Gamma_w)h_\nu(\Gamma_w), \quad n \geq 1.$$

Also, since homomorphisms of Γ_w to S_n can be identified with solutions of the equation $w(x_1, \ldots, x_d) = 1$ in S_n, we have

$$h_n(\Gamma_w) = (n!)^{d-2} \sum_{\chi \in \mathrm{Irr}(S_n)} \alpha_\chi(w)\chi(1).$$

From Lemma 10 we know that, for $w \in \mathcal{W}$, $h_n(\Gamma_w)$ is odd if and only if n is a triangular number. Hence, for $n \geq 1$, we find that modulo 2

$$\begin{aligned} s_n(\Gamma_w) &= nh_n(\Gamma_w) \; - \sum_{\nu=1}^{n-1} s_{n-\nu}(\Gamma_w)h_\nu(\Gamma_w) \\ &\equiv \sum_{\substack{k\geq 1 \\ k(k+1)<2n}} s_{n-k(k+1)/2}(\Gamma_w) \; + \; \delta(n), \end{aligned}$$

the correction term $\delta(n)$ coming from the term $nh_n(\Gamma_w)$.

References

[1] G. Andrews, *The Theory of Partitions*, Cambridge University Press, 1984.

[2] S. Ahlgren and K. Ono, Congruence properties for the partition function, *Proc. Natl. Acad. Sciences USA* **98** (2001), 12882 – 12884.

[3] S. Ahlgren and K. Ono, Congruences and conjectures for the partition function, in: q-series with applications to combinatorics, number theory, and physics (Urbana (IL), 2000), *Contemp. Math.* **291** (2001), Amer. Math. Soc., Providence (RI), 1 – 10.

[4] R. Baer, Situation der Untergruppen und Struktur der Gruppe, *Sitzungsber. Heidelberg. Akad. Math.-Nat. Klasse* **2** (1933), 12 – 17.

[5] B. Baumslag and S. Pride, Groups with two more generators than relators, *J. London Math. Soc.* **17** (1978), 425 – 426.

[6] K. S. Brown, *Cohomology of Groups*, Springer, New York, 1982.

[7] P. J. Cameron, *Combinatorics*, Cambridge University Press, 1994.

[8] H. Davenport, *The Higher Arithmetic*, sixth edition, Cambridge University Press, 1992.

[9] R. Dedekind, Über Gruppen, deren sämtliche Teiler Normalteiler sind, *Math. Ann.* **48** (1897), 548 – 561.

[10] I. M. S. Dey, Schreier systems in free products, *Proc. Glasgow Math. Soc.* **7** (1965), 61 – 79.

[11] W. Dicks and M. J. Dunwoody, *Groups acting on Graphs*, Cambridge University Press, 1989.

[13] Cf. for instance [12, Prop. 1].

[12] A. Dress and T. Müller, Decomposable functors and the exponential principle, *Adv. in Math.* **129** (1997), 188 – 221.
[13] G. Frobenius, Verallgemeinerung des Sylow'schen Satzes, *Berl. Sitz.* (1895), 981 – 993.
[14] C. F. Gauß, *Disquisitiones Arithmeticae* (Lipsia in commissis apud Gerh. Fleischer Iun), 1801; English translation by A. Clarke (New York: Springer-Verlag), 1986.
[15] M. Grady and M. Newman, Counting subgroups of given index in Hecke groups, *Contemporary Math.* **143** (1993), 431 – 436.
[16] M. Grady and M. Newman, Some divisibility properties of the subgroup counting function for free products, *Math. Comp.* **58** (1992), 347 – 353.
[17] M. Grady and M. Newman, Residue periodicity in subgroup counting functions, *Contemporary Math.* **166** (1994), 265 – 273.
[18] M. Hall, Subgroups of finite index in free groups, *Can. J. Math.* **1** (1949), 187 – 190.
[19] P. Hall, On a theorem of Frobenius, *Proc. London Math. Soc.* **40** (1936), 468 – 501.
[20] E. Hecke, *Vorlesungen über die Theorie der Algebraischen Zahlen*, Academische Verlagsgesellschaft, Leipzig, 1923; reprinted (New York: Chelsea Publishing Company) 1970. English translation by G. Brauer and J. Goldman with R. Kotzen as *Lectures on the Theory of Algebraic Numbers* (New York: Springer-Verlag), 1981.
[21] B. Huppert, *Endliche Gruppen I*, Springer–Verlag, Berlin, 1967.
[22] I. M. Isaacs and G. R. Robinson, On a theorem of Frobenius: solutions of $x^n = 1$ in finite groups, *Am. Math. Monthly* **99** (1992), 352 – 354.
[23] N. Ito, On the degrees of irreducible representations of a finite group, *Nagoya Math. J.* **3** (1951), 5 – 6.
[24] A. Kerber, *Algebraic Combinatorics via Finite Group Actions*, BI–Wiss.–Verl., Mannheim, 1991.
[25] M. Knopp, *Modular Functions in Analytic Number Theory*, Markham, Chicago, 1970.
[26] A. Karrass, A. Pietrowski, and D. Solitar, Finite and infinite cyclic extensions of free groups, *J. Austral. Math. Soc.* **16** (1973), 458–466.
[27] H. Katsurada, Y. Takegahara, and T. Yoshida, The number of homomorphisms from a finite abelian group to a symmetric group, *Comm. in Algebra* **28** (2000), 2271–2290.
[28] A.-M. Legendre, *Essai sur la Théorie des Nombres*, Paris, 1798. Fourth edition as *Théorie des Nombres*, 1830; reprinted (Paris: Albert Blanchard) 1955.
[29] A. Lubotzky, *Subgroup Growth*, lecture notes prepared for the conference Groups '93 Galway/St Andrews, University College Galway.
[30] A. Lubotzky, Counting finite index subgroups. In: Groups '93 Galway/St Andrews, LMS Lecture Notes Series No. 212, Cambridge University Press, 1995, 368 – 404.
[31] A. Lubotzky and D. Segal, *Subgroup Growth*, Progress in Mathematics, Birkhäuser, Basel, to appear.
[32] T. Müller, Kombinatorische Aspekte endlich erzeugter virtuell freier Gruppen, Ph. D. Thesis, Universität Frankfurt am Main, 1989.
[33] T. Müller, Combinatorial aspects of finitely generated virtually free groups, *J. London Math. Soc.* (2) **44** (1991), 75 – 94.
[34] T. Müller, Counting free subgroups of finite index, *Archiv d. Math.* **59** (1992), 525 – 533.
[35] T. Müller, Remarks on the PhD thesis of A. Meyer, unpublished manuscript, 1998.
[36] T. Müller, Enumerating representations in finite wreath products, *Adv. in Math.* **153** (2000), 118 – 154.
[37] T. Müller, Parity patterns in Hecke groups and Fermat primes, *Proc. 1999 Bielefeld conference on geometric and combinatorial group theory* (H. Helling and T.W. Müller editors), to appear.
[38] T. Müller, Modular subgroup arithmetic and a theorem of Philip Hall, *Bull. London Math. Society*, in press.
[39] T. Müller, An analysis of the descent principle, preprint, 2001.
[40] T. Müller, Modular subgroup arithmetic in free products, *Forum Math.*, submitted.
[41] T. Müller and J.–C. Puchta, Character theory of symmetric groups and subgroup growth of surface groups, *J. London Math. Society*, in press.
[42] T. Müller and J.-C. Puchta, Parity patterns in one-relator groups, *J. Group Theory*, to appear.
[43] J.-L. Nicolas, I. Ruzsa, and A. Sárközy, On the parity of additive representation functions, *J. Number Theory* **73** (1998), 292 – 317.

[44] K. Ono, On the parity of the partition function in arithmetic progressions, *J. Reine u. Angew. Math.* **472** (1996), 1 – 15.
[45] T. Parkin and D. Shanks, On the distribution of parity in the partition function, *Math. Comp.* **21** (1967), 466 – 480.
[46] O. Schreier, Die Untergruppen der freien Gruppen, *Abh. Math. Sem. Univ. Hamburg* **5** (1927), 161–183.
[47] J.–P. Serre, *Trees*, Springer, Berlin, 1980.
[48] J.–P. Serre, *Cohomologie des groupes discrets*, Ann. Math. Studies 70, Princeton University Press, 1971.
[49] J. Stallings, On torsion–free groups with infinitely many ends, *Ann. of Math.* **88** (1968), 312–334.
[50] W. Stothers, The number of subgroups of given index in the modular group, *Proc. Royal Soc. Edinburgh* **78A** (1977), 105 – 112.
[51] M. Subbarao, Some remarks on the partition function, *Amer. Math. Monthly* **73** (1966), 851 – 854.
[52] H. Wielandt, Ein Beweis für die Existenz von Sylowgruppen, *Arch. Math.* **10** (1959), 401 – 402.
[53] H. Wilf, Generatingfunctionology, 2nd edition, Academic Press, San Diego, 1994.
[54] T. Yoshida, $|\mathrm{Hom}(A, G)|$, *J. Algebra* **156** (1993), 125-156 .

Counting Nets in the Monster

Simon P. Norton

Abstract

We aim to count the number of conjugacy classes of nets, i.e. triples of 6-transpositions in the Monster up to braiding. An exact answer is given subject to a conjecture.

1 Introduction

In [3] the author introduced the concept of a *net*, which was defined in terms of the concept of *quilt* introduced in [2] and developed in subsequent work by the second author of that paper. We start by summarizing these definitions, using the notation of [3].

We start with a group G and consider the set of (ordered) triples of involutions (a, b, c) where $a, b, c \in G$. We define two operations x: $(a, b, c) \mapsto (b, a^b, c)$ and y: $(a, b, c) \mapsto (a, c, b^c)$; then x and y satisfy the relation $xyx = yxy$ which defines a familiar presentation of the 3-string braid group. Indeed, we may think of x as an operation that passes a string corresponding to a under one corresponding to b, while y passes the string corresponding to b under one corresponding to c.

We may also formulate the braid group in terms of the generators s: $(a, b, c) \mapsto (b, c, a^{bc})$ and t: $(a, b, c) \mapsto (c, b^c, a^{bc})$. As $s = xy$, $t = xyx = yxy$, $x = s^{-1}t$, and $y = ts^{-1}$, it is clear that $\langle s, t\rangle = \langle x, y\rangle$. It is easily seen that the element $z = s^3 = t^2$, which is central in $\langle s, t\rangle$, corresponds to conjugation by abc, an element which is invariant under both s and t; this gives rise to another familiar presentation of the braid group as $\langle s, t | s^3 = t^2\rangle$.

We now define a *quilt* as a connected geometry associated with a subset of the orbits of z on the full set of triples. These are actually the flags of the geometry, which has rank 3. As it in fact corresponds to a type of polyhedron, we call its elements vertices (corresponding to orbits under $\langle s\rangle$), edges (corresponding to orbits under $\langle t\rangle$), and faces (corresponding to orbits under $\langle x, z\rangle$). Two elements of different types are incident if the two corresponding orbits share a flag; if they share more than one flag we regard them as being multiply incident.

As $s^3 = z$, each vertex corresponds to 3 or 1 flags. In the latter case we say that the vertex is *collapsed.* Similarly, an edge will correspond to 2 or 1 flags, and in the latter case we say that the edge is collapsed. If a face has n flags we call it an n-gon, with the usual specializations for particular values of n; n will always divide the order of ab, and if they are unequal we may describe the face as collapsed.

Allowing for multiple incidences, the number of incidences between a particular element and elements of either of the other two types will be equal to the number of flags corresponding to that particular element. In particular a collapsed vertex or edge will be incident with just 1 element of either of the other two types, and an uncollapsed vertex or edge will be incident with 3 or 2 elements of either of the other two types. If there are no collapsed vertices or edges, then the geometry corresponds to an (orientable) trivalent polyhedron, where "going round" a vertex, edge of face corresponds to applying s, t or x respectively; in other cases we may consider the geometry to have one or more degenerate vertices or edges.

If the group G is the Fischer-Griess Monster $\mathbb{M}$, and the involutions a, b, c all belong to the class called $2A$ in the ATLAS [1] (whose notation we use throughout), then we call the quilt a *net*. In this case, because of the 6-transposition property (which means that the product of any two $2A$-involutions of $\mathbb{M}$ has order at most 6), the faces of the polyhedron all have at most 6 sides. The significance of this will appear in the next section.

Some of the motivation for the study of nets is given in [3].

2 Euler Characteristics and Folded Nets

We define χ as the permutation character of the Monster on its $2A$-centralizer (which has structure $2.B$, i.e. a double cover of the Baby Monster), and S as the set of orbits of the Monster by conjugation on (ordered) triples of transpositions. It is easy to count the number of orbits in S: this is just the trace of the tensor cube of χ, which turns out to have value 1400384.

This number is too large for one to be able to contemplate a complete enumeration of such triples. To what extent is the problem reduced by passing to the set of orbits under the braid group, i.e. nets? It is the purpose of this paper to answer this question.

The *Euler Characteristic* of a net can be defined as $V - E + F$ where V, E and F are respectively the numbers of vertices, *non-collapsed* edges, and faces. A familiar argument using induction on the size of a net can be used to show that, with this definition, the value of its Euler Characteristic depends only on the topology of the surface defined by the union of closed disks corresponding to the faces, where the intersection of two such disks is determined by the edges and vertices common to the two relevant faces. In particular, if the surface has genus 0 or 1, the Euler Characteristic of the net will be 2 or 0 respectively.

Induction can also be used to show that the Euler Characteristic is one-sixth of the *defect* of the net, defined as the sum of the defects of its elements, where a non-collapsed vertex or edge has defect 0, a collapsed vertex has defect 4, a collapsed edge has defect 3, and an n-gon (whether collapsed or otherwise) has defect $6 - n$. It follows immediately from the 6-transposition property of $\mathbb{M}$ that the defect of a net will always be non-negative, so that its genus will be 0 or 1 (and its defect 12 or 0 respectively). In [3] we used the terms *netball* and *honeycomb* for nets of genus 0 and 1 respectively (in the latter case because all faces of a net with defect zero must clearly be hexagons).

We now define the concept of a *folded net*. This is the same as that of a net except that we quotient out the triples of transpositions (a, b, c) by conjugation, not by powers of abc, but by the entire Monster. In other words, a folded net can be obtained from a net by quotienting out those symmetries which correspond to conjugation by elements of the Monster.

It is clear that the number of conjugacy classes of nets is the same as the number of folded nets: both are easily seen to be equal to the number of orbits on triples of transpositions of the compositum of the braid group and the group of conjugations by elements of $\mathbb{M}$.

It is also clear that the concepts of Euler Characteristic and defect for nets pass through to folded nets unchanged, and that the folded net corresponding to a netball will have genus 0. The folded net corresponding to a honeycomb may have genus either 0 or 1, but all known honeycombs fold to surfaces of genus 0, and the arguments of [3] suggest that there may be no counterexamples. We therefore make the following conjecture:

Conjecture 1 *All folded nets have genus zero.*

3 The Calculations – Part 1

The calculations that follow were done with the aid of the character tables in the GAP library [4], together with one table at that time not in the library, namely that for the centralizer in the Monster of an element of class $2B$, which has structure $2^{1+24}.Co_1$. We call this group Z. This table was computed by the author using Fischer's "matrix" method, and, together with the table of its double cover $2^{1+24}.2.Co_1$, is in the library for GAP release 4.3.

Our procedure is to count the total defect of all the folded nets. Conjecture 1 implies that all folded nets have defect 12, on which assumption division by 12 will give the number of (folded) nets. Even if we do not assume this conjecture, we will still know the total number of folded nets which do not have genus 1.

As a first step, which will illustrate the type of argument we use, we count the number of collapsed vertices in folded nets. Let there be u uncollapsed vertices and v collapsed vertices. Then $3u + v$ is the number of orbits of the Monster on (ordered) triples of transpositions, which we have already seen to be 1400384. We write χ^3 for the tensor cube of χ, and χ_3 for the (virtual) character whose value on any element $g \in \mathbb{M}$ is $\chi(g^3)$. (Later we use similar notations with 3 replaced by 2.) Then we prove:

Theorem 1 *The total number of collapsed vertices in folded nets is the trace of* χ_3*, which has value* 683.

Proof. We start by showing that the number of vertices in folded nets is the same as the number of orbits of the Monster on triples of transpositions subject to cyclic permutation. The former is just the number of orbits of s on S, which is clearly the same as the number of orbits of $t^{-1}st$. The latter takes (a, b, c) to (c, a^c, b^c), which is conjugate to (c, a, b).

The number of orbits of the Monster on triples of transpositions subject to cyclic permutation is the trace of $(\chi^3 + 2\chi_3)/3$. It therefore follows that $u+v$ (the number of vertices in folded nets) is equal to the trace of $(\chi^3+2\chi_3)/3$. But we already know that the trace of χ^3 is $3u+v$. It follows that v is the trace of χ_3, which can be calculated to be 683. □

The vertices therefore give rise to a defect of $4.683 = 2732$. The next step is to calculate the number of collapsed edges. Let us define χ_0 as the restriction of χ to $2.B$; χ_0^+ and χ_0^- as the positive and negative parts of χ_0 with respect to the central involution; and X^{2+} and X^{2-} to be the symmetric and exterior squares of a character X.

Theorem 2 *The total number of collapsed edges in folded nets is the trace of* $(\chi_0^+ - \chi_0^-)_2$, *which has value* 5000.

Proof. Let us now write u and v for the number of uncollapsed and collapsed edges respectively. Then $2u+v$ is the number of orbits of S, i.e. the number of orbits of the Monster on triples (a,b,c). This can alternatively be seen as the number of orbits of the $2.B$ centralizing any particular transposition b on pairs (a,c), which is the trace of χ_0^2.

We consider the orbits on such pairs, up to conjugation by elements of $2.B$, of various subgroups of the group generated by $(a,c)\mapsto(c,a)$ and $(a,c)\mapsto(a,c^b)$. As $(a,c)\mapsto(a^b,c^b)$ is a conjugation by an element of $2.B$, this group is a four-group, say V_4, which has five subgroups:

1. The trivial group, say V_1. As stated above, the number of orbits is the trace of χ_0^2, which is the same as the trace of $(\chi_0^+)^2 + (\chi_0^-)^2$.

2. The group generated by $(a,c)\mapsto(c,a)$, say V_2. Here the number of orbits is the trace of χ_0^{2+}, or that of $(\chi_0^+)^{2+} + (\chi_0^-)^{2-}$.

3. The group generated by $(a,c)\mapsto(a,c^b)$, say V_3. Here the number of orbits is the trace of $(\chi_0^+)^2$.

4. The whole four-group V_4. Here the number of orbits is the trace of $(\chi_0^+)^{2+}$.

5. The group generated by $(a,c)\mapsto(c,a^b)$, say V_5. This is the one we want, as $u+v$ is the number of orbits on S of t, which takes (a,b,c) to (c,b^c,a^{bc}), or, equivalently, the element taking (a,b,c) to (c,b,a^b) (the conjugate of the previous image by c), or, equivalently, the number of orbits of pairs (a,c) up to conjugation by elements of $C_{\mathbb{M}}(b)$ of the element taking (a,c) to (c,a^b). We evaluate the number of orbits as follows.

There are five types of orbit under V_4, according to which of V_1-V_5 is the stabilizer of the pair in question. (As V_4 is abelian, all pairs in the orbit have the same stabilizer.) For each subgroup V_i we may write the number of V_i-orbits in one to the five types of V_4-orbit as a vector w_i: it is then easily seen that $w_1=(4,2,2,1,2)$, $w_2=(2,2,1,1,1)$, $w_3=(2,1,2,1,1)$, $w_4=(1,1,1,1,1)$, and $w_5=(2,1,1,1,2)$. So $w_5 = w_1 - w_2 - w_3 + 2w_4$, which means that we can express the number of V_5-orbits in terms of the numbers of V_i-orbits for $1\le i\le 4$. If we do the calculation, we find that the number of V_5-orbits is $(\chi_0^+)^{2+} + (\chi_0^-)^{2-}$, and we therefore deduce that this is the value of $u + v$. It follows that $v = (\chi_0^+ - \chi_0^-)_2$ as required, and calculation shows that this has value 5000. □

For convenience we keep a running total of the defect as we enumerate it. It is now $2732 + 3.5000 = 17732$.

4 The Calculations – Part 2

We now need to count the number of faces of various types. Recall that a face is defined as an orbit on S of $x = (a, b, c) \mapsto (b, a^b, c)$. We classify faces according to the conjugacy class of ab (or, equivalently, the conjugacy class of $\langle a, b\rangle$). The number of sides for each such face will be a divisor of the order of ab. The number of vertices on such faces will be the inner product of χ with the permutation character of $\mathbb{M}$ over $C_{\mathbb{M}}(\langle a, b\rangle)$, while the number of such faces will be the inner product of χ with the permutation character of $\mathbb{M}$ over $N_{\mathbb{M}}(\langle a, b\rangle) \cap C_{\mathbb{M}}(ab)$, a group in which $C_{\mathbb{M}}(\langle a, b\rangle)$ has index equal to the order of ab. This will enable us to compute the defect, which is 6 times the second number minus the first.

When the order of ab is not composite, this enables us to compute exactly how many such faces have a given number of sides. When ab is composite, this information can also be computed, but this requires the evaluation of further inner products of permutation characters, corresponding to the orbits of the groups generated by various powers of x. We state the relevant results in each case.

Here are the nine possible conjugacy classes for ab and the calculations for the associated defects:

1. $1A$. In this case $a = b$ and all faces are 1-gons. The number of such faces is the trace of χ^2, which is 9, giving a defect of $6.9 - 9 = 45$ and a running total of $17732 + 45 = 17777$.

2. $2A$. The inner products of χ with the permutation characters of $\mathbb{M}$ over $2^2.{}^2E_6(2).2$ and $2^2.{}^2E_6(2)$ are 72 and 111, giving rise to a defect of $6.72 - 111 = 321$, or a running total of $17777 + 321 = 18098$. In fact there are 33 1-gons and 39 2-gons.

3. $2B$. This time the subgroups for which we need the permutation characters are $2^{1+24}.Co_2$ and $2^{1+23}.Co_2$. We may calculate the permutation characters by writing down those for Z $(= C_{\mathbb{M}}(2B) = 2^{1+24}.Co_1)$ over these groups and inducing them up to $\mathbb{M}$. The relevant inner products are 115 and 183, giving rise to a defect of $6.115 - 183 = 507$, or a running total of $18098 + 507 = 18605$. In fact there are 47 1-gons and 68 2-gons.

4. $3A$. Here we use the groups $3 \times Fi_{23}$ and Fi_{23}, for which the relevant inner products are 371 and 993, giving rise to a defect of $6.371 - 993 = 1233$, or a running total of $18605 + 1233 = 19838$. In fact there are 60 1-gons and 311 triangles.

5. $3C$. This time the groups are $3 \times Th$ and Th, with inner products 682 and 2034, giving rise to a defect of $6.682 - 2034 = 2058$, or a running total of $19838 + 2058 = 21896$. In fact there are 6 1-gons and 676 triangles.

6. $4A$. The groups are $2^{1+23}.McL.2$ and $2^{1+22}.McL$, and as in Case 3 we can compute the permutation characters by inducing up to $\mathbb{M}$ the permutation characters of Z over these groups. The inner products are 7426 and 28127, giving rise to a defect of $6.7426 - 28127 = 16429$, so that the running total is $21896 + 16429 = 38325$. If we also compute the inner product of χ with the permutation character of $\mathbb{M}$ over $2^{1+23}.McL$, which is 14667, we can show that there are 185 1-gons, 511 2-gons, and 6730 squares.

7. $4B$. The groups are $4.F_4(2).2$ and $2.F_4(2)$, giving rise to inner products of 7081 and 27927, so that the defect is $6.7081 - 27927 = 14559$, with a running total of $38325 + 14559 = 52884$. Again, we can use the inner product of χ with the permutation character if $\mathbb{M}$ over $4.F_4(2)$, which is 14117, to show that there are 45 1-gons, 131 2-gons, and 6905 squares.

8. $5A$. The groups are $5 \times HN$ and HN, giving rise to inner products 88337 and 441109, and a defect of $6.88337 - 441109 = 88913$, or a running total of $52884 + 88913 = 141797$. In fact there are 144 1-gons and 88193 pentagons.

9. The final case is $6A$. The groups are $3 \times 2.Fi_{22}.2$ and $2.Fi_{22}$, giving rise to inner products 153499 and 899891, so that the defect is $6.153499 - 899891 = 21103$, or an overall total of $141797 + 21103 = 162900$. Again, we can use the inner products with χ of the permutation characters of $\mathbb{M}$ over $3 \times 2.Fi_{22}$ and $2.Fi_{22}.2$, which are 301203 and 458273 respectively, to show that there are 365 1-gons, 747 2-gons, 5430 triangles, and 146957 hexagons.

Dividing the total defect by 12 completes the proof of:

Theorem 3 *The number of folded nets with genus zero is* 13575. *If we assume Conjecture 1, this is also the total number of folded nets, or the number of conjugacy classes of nets.*

Whether this number is small enough to make a complete enumeration of nets feasible is a question that will have to await further study.

References

[1] J. H. Conway, R. T. Curtis, S. P. Norton, R. A. Parker and R. A. Wilson, *An* ATLAS *of Finite Groups*, Oxford University Press, 1985.

[2] J. H. Conway and T. M. Hsu, *Quilts and T-systems,* J. Alg. 174 (1995) pp. 856-903.

[3] S. Norton, Netting the Monster, in *The Monster and Lie Alegbras,* Vol 7 (eds. Ferrar & Harada), Walter de Gruyter & Co., 1998, pp. 111-125 (Proceedings of the 1996 Columbus conference).

[4] M. Schönert et al., *GAP 4 Manual (Groups, Algorithms and Programming)*, Lehrstuhl D für Mathematik, RWTH Aachen, 2000.

Overgroups of finite quasiprimitive permutation groups.

Cheryl E. Praeger

Abstract

A permutation group is quasiprimitive if each of its non-trivial normal subgroups is transitive. Quasiprimitive permutation groups arise naturally when studying automorphism groups of vertex-transitive graphs, and they form a family of permutation groups that properly contains all primitive permutation groups. In this chapter we describe the current state of our knowledge of the structure of finite quasiprimitive groups and in particular of the quasiprimitive permutation groups containing a given finite quasiprimitive group. To illustrate how these results can be used in graph theory we discuss their application to finite s-arc-transitive graphs and other classes of vertex-transitive graphs. Many of the results depend on the finite simple group classification.

1 Introduction

A finite transitive permutation group is *quasiprimitive* if each of its non-trivial normal subgroups is transitive. Such groups often arise as automorphism groups of combinatorial or geometric structures, and in this context a natural problem is to determine the full automorphism groups of these structures. The group theoretic equivalent of this problem is the problem of determining the overgroups of a given quasiprimitive permutation group in the symmetric group, and the heart of this problem is to find all the quasiprimitive overgroups. We shall address these problems in this chapter. As illustration and motivation we will consider the problem of understanding the structre of finite s-arc-transitive graphs.

The class of quasiprimitive permutation groups properly contains the class of primitive groups and admits a similar partition into several distinct sub-classes in an analogous fashion to that given by the O'Nan-Scott Theorem for primitive groups (see [20] or [14] for primitive groups and [18] for quasiprimitive groups). A broad-brush description of finite quasiprimitive groups will be given in §3, identifying eight 'O'Nan-Scott types' of such groups. These 'O'Nan-Scott types' are used as a means of organising investigations involving quasiprimitive groups, and they also appear in the statements of the major results.

To show how quasiprimitive groups and the O'Nan-Scott types can be used in graph theory we consider finite s-arc-transitive graphs in §4. It turns out that each non-bipartite finite s-arc-transitive graph is a normal cover of a quasiprimitive s-arc-transitive graph, and that if $s \geq 2$ then only four of the eight O'Nan-Scott types can arise. We discuss the importance of quasiprimitive groups for determining the full automorphism group of a quasiprimitive s-arc-transitive graph. In particular it is crucial to understand the quotient actions of imprimitive quasiprimitive groups and to know the quasiprimitive overgroups of a given quasiprimitive group G on a finite set Ω, that is the quasiprimitive groups H satisfying $G < H < \mathrm{Sym}(\Omega)$. We discuss what is known about such quotient actions and overgroups in §5 and §6. Then in §7 we present various applications of this theory in graph theory. Detailed information about the finite simple groups was used to prove many of the results we state.

2 Primitive and quasiprimitive groups

Let G be a transitive permutation group on a set Ω. A partition $\mathcal{B}$ of Ω is G-invariant if, for all blocks $B \in \mathcal{B}$ and elements $g \in G$, the image $B^g = \{\alpha^g \mid g \in G\}$ is also a block of $\mathcal{B}$. Since G is transitive on Ω, G permutes the parts of $\mathcal{B}$ transitively. We denote the permutation group of $\mathcal{B}$ induced by G as $G^{\mathcal{B}}$, and if $\mathcal{B}$ is *non-trivial*, that is, if $1 < |\mathcal{B}| < |\Omega|$, we refer to the action of G on $\mathcal{B}$ as a *non-trivial quotient action* of G. The group G is *primitive* on Ω if there are no nontrivial G-invariant partitions of Ω; if G is not primitive then we call it *imprimitive.*

If $\mathcal{B}$ consists of the orbits in Ω of a normal subgroup N of G, then $\mathcal{B}$ is called a *G-normal partition* relative to N. Each G-normal partition $\mathcal{B}$ is G-invariant and the action of G on $\mathcal{B}$ is called a *normal quotient action* of G. Thus G is quasiprimitive if and only if the only G-normal partitions are the trivial ones, and hence every primitive group is quasiprimitive. The converse does not hold since, for example, every transitive action of a non-abelian simple group is quasiprimitive.

Suppose that G is a quasiprimitive group on Ω and $\mathcal{B}$ is a G-invariant partition of Ω. If a normal subgroup of G is intransitive on $\mathcal{B}$ then it is also intransitive on Ω, and since G is quasiprimitive on Ω, the only normal subgroup intransitive on $\mathcal{B}$ is the identity subgroup. Hence $G^{\mathcal{B}} \cong G$ and $G^{\mathcal{B}}$ is quasiprimitive. Thus every non-trivial quotient action of a quasiprimitive group is quasiprimitive, but the O'Nan-Scott types of these actions, as described in §3, may differ. Suppose in addition that $\mathcal{B}$ is maximal in the sense that if $\mathcal{B}'$ is a G-invariant partition of Ω properly refined by $\mathcal{B}$, then $\mathcal{B}' = \{\Omega\}$. In this case $G^{\mathcal{B}}$ is primitive. Thus each finite quasiprimitive permutation group has at least one primitive quotient action. However in various combinatorial applications it is not appropriate to 'pass to a primitive quotient action' since this action may not have the required combinatorial properties, see §4.

3 Types of quasiprimitive groups

The main theorem in [18] gives a subdivision of the family of finite quasiprimitive permutation groups analogous to that given by the O'Nan-Scott Theorem for primitive groups.

Theorem 1 [18] *Each finite quasiprimitive permutation group belongs to exactly one of eight pairwise disjoint types, described in Table* 1.

Type	Name	soc(G)	Description
HA	Affine	$N = Z_p^d$	$G = NG_0$, $G_0 \leq \mathrm{GL}(N)$, G_0 irreducible
HS	Holomorph of a simple group	$N \times C_G(N)$	$N \cong C_G(N) \cong T$, $G \leq \mathrm{Hol}(N)$
HC	Holomorph of a compound group	$N \times C_G(N)$	$N \cong C_G(N) \cong T^k$, $G \leq \mathrm{Hol}(N)$, $k \geq 2$
AS	Almost simple	$N = T$	$G \leq \mathrm{Aut}(T)$
SD	Simple diagonal	$N = T^k$	$G \leq N \cdot (\mathrm{Out}(T) \times S_k)$, $k \geq 2$
CD	Compound diagonal	$N = T^{k\ell}$	$G \leq G_0 \,\mathrm{wr}\, S_\ell$ in product action G_0 of type SD, $k \geq 2, \ell \geq 2$
TW	Twisted wreath	$N = T^\ell$	N regular, $\ell \geq 2$
PA	Product action	$N = T^\ell$	$G \leq G_0 \,\mathrm{wr}\, S_\ell$, G_0 almost simple, $\ell \geq 2$

Table 1: O'Nan-Scott types for finite quasiprimitive groups

All quasiprimitive groups of the first three types are primitive, whereas each of the remaining five types contains imprimitive quasiprimitive groups as well as primitive groups. Each quasiprimitive group G in the first three types is permutationally isomorphic to a primitive subgroup of the *holomorph* $\mathrm{Hol}(M)$ of a certain group M, where $\mathrm{Hol}(M)$ is the semidirect product $M{\cdot}\mathrm{Aut}M$ formed with respect to the natural action of $\mathrm{Aut}M$ on M, and G contains the base group M of $\mathrm{Hol}(M)$. Moreover M is *regular* on Ω, that is, M is transitive and only the identity element fixes a point of Ω. Table 1 lists the possible O'Nan-Scott types of a finite quasiprimitive group $G \leq \mathrm{Sym}(\Omega)$, together with information about a minimal normal subgroup N of G and the *socle* soc(G), that is, the product of the minimal normal subgroups of G. In the table T denotes a non-abelian simple group. In type CD, $G_0 \leq \mathrm{Sym}(\Delta)$ is quasiprimitive of type SD, and $G \leq G_0 \,\mathrm{wr}\, S_\ell$ acts on $\Omega = \Delta^\ell$ in product action. In type PA, $G_0 \leq \mathrm{Sym}(\Delta)$ is quasiprimitive of type AS, and there is a G-invariant partition of Ω which may be identified with Δ^ℓ on which G acts in product action.

4 Finite s-arc-transitive graphs

A graph $\Gamma = (\Omega, E)$ consists of a vertex set Ω and a subset E of unordered pairs from Ω, called edges. An automorphism of Γ is an element of $\mathrm{Sym}(\Omega)$ that leaves E invariant, and the full automorphism group of Γ is the subgroup of $\mathrm{Sym}(\Omega)$ consisting of all the automorphisms and is denoted $\mathrm{Aut}(\Gamma)$. An *s-arc* is a vertex sequence $(\alpha_0, \alpha_1, \ldots, \alpha_s)$ such that $\{\alpha_{i-1}, \alpha_i\} \in E$ for $1 \leq i \leq s$, and $\alpha_{i-1} \neq \alpha_{i+1}$ for $1 \leq i < s$; and a 1-arc is often called simply an arc. For $G \leq \mathrm{Aut}(\Gamma)$, we say that Γ is (G, s)-arc transitive if G acts transitively on the s-arcs of Γ.

Let $\Gamma = (\Omega, E)$ be a graph and $G \leq \mathrm{Aut}(\Gamma)$. For any partition $\mathcal{B}$ of Ω the *quotient graph* $\Gamma_\mathcal{B}$ is the graph $(\mathcal{B}, E_\mathcal{B})$ where $\{B, C\} \in E_\mathcal{B}$ if there exist $\alpha \in B, \beta \in C$ such that $\{\alpha, \beta\} \in E$. If Γ is connected then also $\Gamma_\mathcal{B}$ is connected. If $\mathcal{B}$ is G-invariant then the quotient action of G on $\mathcal{B}$ leaves $E_\mathcal{B}$ invariant and hence G induces a group of automorphisms of $\Gamma_\mathcal{B}$. In this case if G is vertex-transitive or arc-transitive on Γ then $G^\mathcal{B}$ will also be vertex-transitive or arc-transitive respectively on $\Gamma_\mathcal{B}$. However if Γ is (G, s)-arc transitive then $G^\mathcal{B}$ will not in general be transitive on the s-arcs of $\Gamma_\mathcal{B}$ if $s \geq 2$. This means that in general an s-arc-transitive graph Γ will not have a vertex-primitive s-arc-transitive quotient. Thus there is no hope that the problem of classifying finite s-arc-transitive graphs, or even giving a useful description of their structure, can be reduced to the case of vertex-primitive s-arc-transitive graphs.

Happily when $\mathcal{B}$ is a G-normal partition then G does act s-arc transitively on $\Gamma_{\mathcal{B}}$ and moreover, if $|\mathcal{B}| > 2$, then Γ is a cover of $\Gamma_{\mathcal{B}}$ in the sense that, for $\{B, C\} \in E_{\mathcal{B}}$, each vertex in B is adjacent in Γ to exactly one vertex in C and vice versa. In this case we say that Γ is a *normal cover* of $\Gamma_{\mathcal{B}}$. If $\mathcal{B}$ is the set of orbits of a maximal intransitive normal subgroup N of G and if $|\mathcal{B}| > 2$ (possibly $N = 1$), then G is both vertex-quasiprimitive and s-arc-transitive on $\Gamma_{\mathcal{B}}$, see [18]. Such a normal subgroup exists provided that Γ is not bipartite. It was the wish to understand vertex-quasiprimitive s-arc transitive graphs that led to the development of the theory for finite quasiprimitive permutation groups described in §3. It turns out that only four of the eight O'Nan-Scott types of quasiprimitive permutation groups can occur as s-arc-transitive automorphism groups of graphs (for $s \geq 2$).

Theorem 2 [18] *If G is vertex-quasiprimitive and s-arc-transitive on a finite graph Γ with $s \geq 2$, then G is of type* HA, AS, TW, *or* PA. *Moreover there are examples for each of these types.*

Those with G of type HA were classified in [9], and the almost simple examples with $\mathrm{soc}(G) = \mathrm{PSL}_2(q), \mathrm{Sz}(q)$ or $\mathrm{Ree}(q)$ have also been classified, see [4, 5, 8]. The first classification depends on the finite simple group classification. However even in the latter classifications where the groups G were well-known, it was necessary to determine the full automorphism groups of the graphs constructed in order to decide whether certain pairs of graphs were isomorphic. Finding the full automorphism group of a graph Γ, given a vertex-quasiprimitive s-arc transitive subgroup G of $\mathrm{Aut}(\Gamma)$ $(s \geq 2)$, can be quite difficult. Several questions arise.

Question 1 *Can* $\mathrm{Aut}(\Gamma)$ *be much larger than G? Is* $\mathrm{Aut}(\Gamma)$ *quasiprimitive? If* $\mathrm{Aut}(\Gamma)$ *is quasiprimitive, do G and* $\mathrm{Aut}(\Gamma)$ *have the same O'Nan-Scott types, or the same socles?*

There are examples known of $(G, 2)$-arc-transitive graphs Γ with G vertex-quasiprimitive of type TW or AS, for which $\mathrm{Aut}(\Gamma)$ is not vertex-quasiprimitive, see [1, Section 6] and [10]. However, even if $\mathrm{Aut}(\Gamma)$ is not quasiprimitive, quasiprimitive groups play a major role in determining $\mathrm{Aut}(\Gamma)$. If N is a maximal intransitive normal subgroup of $\mathrm{Aut}(\Gamma)$, then both G and $\mathrm{Aut}(\Gamma)$ induce vertex-quasiprimitive 2-arc-transitive actions on the normal quotient $\Gamma_{\mathcal{B}}$ where $\mathcal{B}$ is the set of N-orbits, and $G^{\mathcal{B}} \cong G$. Thus to analyse the possibilities successfully we need to know (at least) the following.

1. The possible O'Nan-Scott types for the (quasiprimitive) quotient actions of finite quasiprimitive groups for each O'Nan-Scott type.

2. The quasiprimitive overgroups of a given finite quasiprimitive group.

We discuss the current state of our knowledge of quotient actions and overgroups of quasiprimitive groups in the next two sections. The analysis described here for finite s-arc-transitive graphs is effective also for other classes of finite arc-transitive graphs such as locally primitive graphs or locally quasiprimitive graphs (see [13]). Bipartite locally s-arc transitive graphs require a specialised analysis that parallels the one described here, and again quasiprimitive groups play a major role (see [7]).

5 Quotients actions of quasiprimitive graphs

In §2 we defined the quotient action of a transitive permutation group $G \leq \mathrm{Sym}(\Omega)$ on a G-invariant partition $\mathcal{B}$ of Ω, and we saw that if G is quasiprimitive on Ω, then the group $G^{\mathcal{B}}$ induced on $\mathcal{B}$ is isomorphic to G and is also quasiprimitive. We saw in §4 above that it is important when computing the full automorphism group of a vertex-transitive graph with vertex set Ω to understand the possibilities for the O'Nan-Scott types of G^{Ω} and $G^{\mathcal{B}}$. Suppose that $\mathcal{B}$ is nontrivial, that is $1 < |\mathcal{B}| < |\Omega|$, so that the quasiprimitive group G is imprimitive of type $X \in$ {AS, SD, CD, TW, PA} (see §3). Also, since a minimal normal subgroup N of G must be transitive on Ω, it follows that $|N| \geq |\Omega| > |\mathcal{B}|$ and hence N is not regular on $\mathcal{B}$. Thus $G^{\mathcal{B}}$ is quasiprimitive of type $X^{\mathcal{B}}$, where $X^{\mathcal{B}} \in$ {AS, SD, CD, PA}. In [19] some further restrictions on $X, X^{\mathcal{B}}$ were found as follows.

Theorem 3 [19, Theorem 1] *Let G be a quasiprimitive permutation group of O'Nan-Scott type X on a finite set Ω, and let $\mathcal{B}$ be a non-trivial G-invariant partition of Ω. Then $G^{\mathcal{B}}$ is quasiprimitive of type $X^{\mathcal{B}}$ for some X, $X^{\mathcal{B}}$ such that the $(X, X^{\mathcal{B}})$-entry in Figure 3 is the symbol ✓, and all such pairs $(X, X^{\mathcal{B}})$ can occur.*

	AS	SD	CD	PA
AS	✓	–	–	–
SD	–	–	✓	–
CD	–	–	✓	–
TW	–	✓	✓	✓
PA	–	✓	✓	✓

Figure 1: Quasiprimitive Quotient Action Matrix

Again this result depends on the finite simple group classification.

6 Overgroups of finite quasiprimitive groups

Let $G < H < \mathrm{Sym}(\Omega)$ where Ω is finite and G, H are quasiprimitive with O'Nan-Scott types X, Y respectively. If both G and H are primitive then the possibilities for (X, Y) and a description of the possible inclusions $G < H$ are given in [17]. In particular the examples in the special case where $X = Y =$ AS but $\mathrm{soc}(G) \neq \mathrm{soc}(H)$ may be read off from the tables in [15].

Information about the possible pairs X, Y in the case where H is primitive but G is imprimitive is given in [2]. This paper also describes the possible inclusions $G < H$ in all cases except for $(X, Y) =$ (AS, AS), (TW,PA) and (PA,PA). For the latter two cases, the results of [3] yield a great deal of information about the corresponding inclusions. The case $(X, Y) = (\mathrm{AS}, \mathrm{AS})$ deserves some comments. Suppose that $X = Y = \mathrm{AS}$, that $N = \mathrm{soc}(G), M = \mathrm{soc}(H)$, and $N \neq M$. Let $\alpha \in \Omega$. Then since H is primitive, the stabiliser H_α is maximal in H, and since G is quasiprimitive, N is transitive so that $H = H_\alpha N$. Thus to classify the possible inclusions $G < H$ in this case we need a solution to the following problem, work on which is in progress by Liebeck, Saxl and the author.

Problem 1 *Classify all factorisations $H = AB$ where H is almost simple, A is maximal but does not contain $\mathrm{soc}(H)$, and B is a non-abelian simple group.*

This leaves the case where both G and H are imprimitive. In this case, if $\mathcal{B}$ is a maximal H-invariant partition of Ω, then it follows from the discussion in §2 that $G^{\mathcal{B}} \cong G, H^{\mathcal{B}} \cong H$, and $G^{\mathcal{B}}$ is a quasiprimitive subgroup of the primitive group $H^{\mathcal{B}}$. Thus information about the possible O'Nan-Scott types $X^{\mathcal{B}}$, $Y^{\mathcal{B}}$ of $G^{\mathcal{B}}$, $H^{\mathcal{B}}$ respectively is given by [2, 17]. Using this information as a starting point, it was shown in [19] that either (i) $X = Y$ and $\mathrm{soc}(G) = \mathrm{soc}(H)$, or (ii) $Y = Y^{\mathcal{B}}$ and either $X = X^{\mathcal{B}}$ or $X = \mathrm{TW}$, $X^{\mathcal{B}} = \mathrm{PA}$. Moreover in case (ii) either $(X, Y) = (\mathrm{AS}, \mathrm{AS})$, or $Y = \mathrm{PA}$ and $X \in \{\mathrm{AS}, \mathrm{TW}, \mathrm{PA}\}$, and there are examples for each of these possibilities.

A summary statement of these results could be expressed as follows. Putting the results from [17] into this format is non-trivial and the details for doing this are given in [19, Section 3]. We note that proofs of the results discussed in this section depend on detailed information about finite simple groups.

Theorem 4 [2, 17, 19] *Let $G < H \leq \mathrm{Sym}(\Omega)$ where $|\Omega| = n$, G, H are quasiprimitive of type X, Y respectively such that $H \neq A_n$ or S_n. Then either $X = Y$ and* $\mathrm{soc}(G) = \mathrm{soc}(H)$, *or $X \neq \mathrm{SD}$, $Y \neq \mathrm{TW}$, and the (X, Y)-entry of Figure* 2 *is the symbol* ✓. *Moreover all such pairs (X, Y) can occur.*

	HA	HS	HC	AS	SD	CD	PA
HA	–	–	–	✓	–	–	✓
HS	–	–	–	–	✓	–	–
HC	–	–	–	–	–	✓	✓
AS	✓	✓	–	✓	✓	–	✓
CD	–	–	–	–	–	–	✓
TW	–	–	✓	–	✓	✓	✓
PA	✓	–	–	–	✓	✓	✓

Figure 2: Quasiprimitive Inclusions Matrix

7 Further applications in graph theory

We complete this chapter on overgroups of finite quasiprimitive groups with a brief look at several applications of this theory in addition to those mentioned in §4.

All of these applications depend on the finite simple group classification, either directly in their proofs or indirectly through their use of the results of §6. The first application is to finite 4-arc-transitive graphs, where this theory has led to some interesting restrictions on the number of vertices.

Theorem 5 [11, 12] *Suppose that Γ is a finite s-arc-transitive graph with $s \geq 4$. Then the number of vertices is even and not a power of* 2.

Next we consider arc-transitive graphs. In §4 we saw that, if $\Gamma = (\Omega, E)$ is a finite connected graph, $G \leq \mathrm{Aut}(\Gamma)$ with G arc-transitive on Γ, and $\mathcal{B}$ is a G-invariant partition of Ω, then the quotient graph $\Gamma_{\mathcal{B}}$ is connected and admits $G^{\mathcal{B}}$ as an arc-transitive group of automorphisms. If in addition $\mathcal{B}$ is maximal then $G^{\mathcal{B}}$ is vertex-primitive. The questions in Question 1 can equally well be asked about the subgroup $G^{\mathcal{B}}$ of $\mathrm{Aut}(\Gamma_{\mathcal{B}})$. These are essentially questions about the full automorphism group of a finite connected graph with a given vertex-primitive, arc-transitive subgroup of automorphisms. A satisfactory answer is given in [16].

Theorem 6 [16] *Let G be a vertex-primitive arc- or edge-transitive group of automorphisms of a finite connected graph Γ. Then either G and* $\mathrm{Aut}(\Gamma)$ *have the same socle, or $G < H \leq$* $\mathrm{Aut}(\Gamma)$ *where* $\mathrm{soc}(G) \neq \mathrm{soc}(H)$ *and G, H are explicitly listed.*

Finally we consider graphs Γ that are vertex-transitive and edge-transitive but not arc-transitive. Such graphs have been called *half-transitive* in the literature, and also a subgroup of $\mathrm{Aut}(\Gamma)$ with these properties is said to be half-transitive on Γ. Suppose that Γ is given with a half-transitive subgroup G of automorphisms. In order to decide whether Γ is half-transitive it is often necessary to determine $\mathrm{Aut}(\Gamma)$ and then to check whether or not $\mathrm{Aut}(\Gamma)$ contains an element interchanging the two vertices of an edge.

The problem of recognising half-transitive graphs was studied in [6] for the family of Cayley graphs of simple groups. For a group G and a subset S of G such that $1_G \notin S$ and $S^{-1} = S$, the *Cayley graph* $\mathrm{Cay}(G, S)$ of G relative to S is defined as the graph (G, E) where $\{x, y\} \in E$ if and only if $yx^{-1} \in S$; $\mathrm{Cay}(G, S)$ is connected if and only if S is a generating set for G. Each Cayley graph $\Gamma = \mathrm{Cay}(G, S)$ admits the group G acting by right multiplication as a subgroup of automorphisms regular on vertices, and also admits the subgroup $A(G, S) := \{x \in \mathrm{Aut}(G) \mid S^x = S\}$ in its natural action on G. Thus $\mathrm{Aut}(\Gamma)$ contains the semidirect product $G \cdot A(G, S)$ and in particular is vertex-transitive. Now $G \cdot A(G, S) = N_{\mathrm{Aut}(\Gamma)}(G)$ and Γ is called a *normal Cayley graph* if $\mathrm{Aut}(\Gamma) = G \cdot A(G, S)$.

In [6, Theorem 1.1] non-normal Cayley graphs of non-abelian simple groups were analysed and several distinct possibilities were identified for overgroups of $G \cdot A(G, S)$ in $\mathrm{Aut}(\Gamma)$. Next a set of technical conditions was developed such that if these conditions were satisfied then none of the possibilities identified in [6, Theorem 1.1] was allowable and consequently the Cayley graph was normal (see [6, Theorem 1.3]). These conditions were designed in such a way that, if they were satisfied, then $\mathrm{Aut}(\Gamma) = G \cdot A(G, S)$ would be half-transitive. Several constructions of new half-transitive graphs were obtained by finding non-abelian simple groups G and subsets S satisfying the conditions.

G	p
A_{p+1}	$p \equiv 3 \pmod 4$, $p+1 \neq 2^a$
Ree(q)	p divides $q^2 - q + 1$
BM	47
J_1	19
J_4	43
Ly	37 and 67

Table 2: Half-transitive Cayley graphs

Theorem 7 [6, Constructions 4.1, 4.2, 4.4] *Let G be a simple group as in Table* 2. *Then there exists a half-transitive, connected normal Cayley graph* Cay(G, S) *of valency $|S| = 2p$, where p is a prime as in Table* 2.

This overview of finite quasiprimitive permutation groups focussed on the possible structures of such groups using a similar framework to that of the O'Nan-Scott Theorem for finite primitive groups. Special attention was given to the quasiprimitive overgroups of a given quasiprimitive group in terms of the O'Nan-Scott types of these groups. Various applications of this theory were presented for vertex-transitive graphs. The effectiveness of the theory in these and other applications is largely due to its use of the classification of the finite simple groups. The theory seems particularly well-suited for applications to classes of objects which are closed under some natural quotient operation.

References

[1] Robert W. Baddeley, Two-arc transitive graphs and twisted wreath products, *J. Alg. Combin* **2** (1993), 215–237.

[2] Robert W. Baddeley and Cheryl E. Praeger, On primitive overgroups of quasiprimitive permutation groups, Research Report No. 2002/03, University of Western Australia, 2002.

[3] Robert W. Baddeley, Cheryl E. Praeger, and Csaba Schneider, Quasiprimitive permutation groups preserving Cartesian decompositions, (in preparation), 2002.

[4] Xin Gui Fang and Cheryl E. Praeger, Finite two-arc transitive graphs admitting a Suzuki simple group, *Comm. Algebra* **27** (1999), 3727–3754.

[5] Xin Gui Fang and Cheryl E. Praeger, Finite two-arc transitive graphs admitting a Ree simple group, *Comm. Algebra* **27** (1999), 3755–3769.

[6] Xin Gui Fang, Cheryl E. Praeger and Jie Wang, On the automorphism groups of Cayley graphs of finite simple groups, *J. London Math. Soc.*, to appear.

[7] Michael Giudici, Cai Heng Li and Cheryl E. Praeger, Analysing finite locally s-arc transitive graphs, preprint, 2002.

[8] Akbar Hassani, Luz R. Nochefranca and Cheryl E. Praeger, Two-arc transitive graphs admitting a two-dimensional projective linear group, *J. Group Theory* **2** (1999), 335–353.

[9] A. A. Ivanov and Cheryl E. Praeger, On finite affine 2-arc transitive graphs, *European J. Combin.* **14** (1993), 421–444.

[10] Cai Heng Li, A family of quasiprimitive 2-arc transitive graphs which have non-quasiprimitive full automorphism groups, *European J. Combin.* **19** (1998), 499-502.

[11] C. H. Li, Finite s-arc transitive graphs of prime-power order, *Bull. London Math. Soc.* **33** (2001), 129-137.

[12] C. H. Li, On finite s-arc transitive graphs of odd order, *J. Combin. Theory Ser. B* **81** (2001), 307-317.

[13] Cai Heng Li, Cheryl E. Praeger, Akshay Venkatesh and Sanming Zhou, Finite locally-quasiprimitive graphs, *Disc. Math.* **246** (2002), 197-218.

[14] M. W. Liebeck, Cheryl E. Praeger and Jan Saxl, On the O'Nan-Scott Theorem for finite primitive permutation groups, *J. Austral. Math. Soc.* (A) **44** (1988), 389–396.

[15] M. W. Liebeck, Cheryl E. Praeger and Jan Saxl, *The maximal factorisations of the finite simple groups and their automorphism groups*, *Memoirs Amer. Math. Soc.* **86** (1990), No. 432.

[16] M. W. Liebeck, C. E. Praeger and J. Saxl, Primitive permutation groups with a common suborbit, and edge-transitive graphs, *Proc. London Math. Soc.* (3) **84** (2002), 405–438.

[17] Cheryl E. Praeger, The inclusion problem for finite primitive permutation groups, *Proc. London Math. Soc.* (3) **60** (1990), 68–88.

[18] Cheryl E. Praeger, An O'Nan-Scott Theorem for finite quasiprimitive permutation groups and an application to 2-arc transitive graphs, *J. London Math. Soc.* (2), **47** (1993), 227–239.

[19] Cheryl E. Praeger, Quotients and inclusions of finite permutation groups, Research Report No. 2002/05, Univ. of Western Australia, 2002.

[20] L. L. Scott, Representations in characteristic p, in *Santa Cruz conference on finite groups*, *Proc. Sympos. Pure Math.* **37** (1980), 318–331.

Old groups can learn new tricks

László Pyber

Abstract

In 1937 B. H. Neumann gave a simple construction of continuously many non-isomorphic finitely generated groups, using families of finite alternating groups. We describe a generalisation of this construction, designed to settle various open problems in the area of subgroup growth.

As an unexpected byproduct our groups yield new examples related to a question of Grothendieck concerning isomorphism of groups with isomorphic profinite completions.

We also give a quick survey of some areas of infinite group theory in which these and related constructions based on finite simple groups play a role.

1 Introduction

Let Γ be a finitely generated group and $\mathcal{N}$ the set of normal subgroups of finite index in Γ. Supposing we are given all the finite quotients Γ/N where $N \in \mathcal{N}$, what can we say about Γ itself?

The finite quotients Γ/N together with the natural epimorphisms $\varphi_{N,M} : \Gamma/M \to \Gamma/N$ whenever $M \leq N$ form an inverse system. The inverse limit of this system is the *profinite completion* $\widehat{\Gamma}$ of Γ. Using this language the above question asks for the properties of Γ determined by the profinite group $\widehat{\Gamma}$.

Another closely related question is the following: Which profinite groups are profinite completions of finitely generated groups Γ?

This question seems to be too general to admit a useful answer. As we will see in specific instances, however, one can obtain reasonable answers with nice applications.

Denote by $s_n(\Gamma)$ the number of subgroups of index at most n in Γ. If Γ is finitely generated then $s_n(\Gamma)$ is always finite. It is clear that $s_n(\Gamma) = s_n(\widehat{\Gamma})$ for all n, that is, the subgroup growth function encodes information about the profinite completion of Γ.

The above questions concerning the relationship between Γ and $\widehat{\Gamma}$ lead to the following

Question 1 *Supposing we are given the function $s_n(\Gamma)$ what can we say about Γ?*

Question 2 *What are the possible subgroup growth functions for finitely generated groups Γ?*

Research supported by grants OTKA T034878 and T038059.

There is a well-developed theory around these questions [Lu2], [Lu3], [MS]. Indeed the behaviour of $s_n(\Gamma)$ is the main topic of the forthcoming book of Lubotzky and Segal [LS].

Our main contribution [Py] is a construction (described in Section 5) which shows that all "reasonable" functions between $n^{\log n}$ (small growth) and n^n (maximal growth) can be realised as subgroup growth functions of 4-generator groups. This essentially completes the investigation of the "spectrum" of possible subgroup growth types and settles several questions posed by Lubotzky, Mann and Segal (see Section 3).

Why should one consider the problem of finding groups of given subgroup growth? Apart from its intrinsic interest such an investigation may lead to the discovery of groups with unusual properties.

One of the highlights in the parallel theory of word growth of finitely generated groups was the construction by Grigorchuk [Gri2] of groups of intermediate word growth. Grigorchuk's construction has led to the solution of many other problems such as the construction of finitely presented amenable but not elementary amenable groups [Gri6].

We will briefly review some of these applications related to our main theme (in Section 2).

Similarly the family of groups we construct has some additional interesting properties.

Extending a classical result of B. H. Neumann [Ne] we prove that there exist continuously many non-isomorphic 4-generator residually finite groups with isomorphic profinite completions.

This (and other examples) indicate that for the profinite completion $\widehat{\Gamma}$ to determine Γ some rather stringent conditions must hold.

In another direction Grothendieck [Gro2] suggested the following problem (in a slightly different formulation).

Problem 3 *Let Γ be a finitely generated residually finite group. Let Γ_0 be a finitely generated subgroup which is dense in the profinite topology of Γ. Suppose that $\widehat{\Gamma}_0 \cong \widehat{\Gamma}$. Under what conditions does this imply that $\Gamma_0 = \Gamma$?*

Platonov and Tavgen [PT1] gave the first example of a pair of groups $\Gamma_0 < \Gamma$ satisfying the above hypothesis for which $\Gamma_0 \neq \Gamma$. In Section 6 we describe rather different examples using our main construction.

2 Word growth

Let Γ be a group generated by a finite set S. Denote by $w_n^S(\Gamma)$ the number of elements of Γ of length at most n with respect to $S \cup S^{-1}$. The *word growth* of Γ (with respect to S) is the growth of the sequence $w_n^S(\Gamma)$.

A word growth function $\gamma(n)$ is dominated by $\delta(n)$ written $\gamma \precsim \delta$ if there is a constant c such that $\gamma(n) \leq \delta(cn)$ for all n. Two functions are *equivalent* if $\gamma \precsim \delta$ and $\delta \precsim \gamma$. If S_1 and S_2 are two generating sets of Γ, the corresponding word growth functions are in the same equivalence class of functions called the *word growth degree* of Γ.

The word growth of groups has received considerable attention following the observation [Sch], [Mi] that if $\widetilde{M}$ is the universal cover of the compact Riemannian manifold M then the (geometric) growth of $\widetilde{M}$ is equivalent to the word growth of the fundamental group $\pi_1(M)$ of the manifold M.

If $\Gamma = \mathbb{Z}^d$ is the free abelian group of rank d then its word growth is polynomial, more precisely it is equivalent to n^d. By a major result of Gromov [Gro1] a group Γ has polynomial word growth exactly if Γ is virtually nilpotent.

If $\Gamma = \mathbb{F}_d$ is a free group of rank $d \geq 2$ then its word growth is exponential and clearly this is the largest possible word growth degree.

In [Gri1] Grigorchuk constructed an example of an infinite finitely generated 2-group Γ as a group acting on the binary rooted tree. This group was designed to provide a simple solution to the general Burnside problem. In [Gr2] Grigorchuk proved that the group Γ has intermediate word growth (strictly between polynomial and exponential) thereby answering a question of Milnor.

Recently Bartholdi [Bar1], [Bar2] (see also [MP]) has shown that for the word growth degree $\gamma(n)$ of the above Γ we have

$$e^{n^\alpha} \leq \gamma(n) \leq e^{n^\beta}$$

where

$$\alpha = 0.5157 \quad \text{and} \quad \beta = 0.767.$$

A remaining open problem is the question of the existence of groups with word growth degree exactly $e^{\sqrt{n}}$. Such groups would have interesting additional properties (see [BG]).

The above mentioned construction has been generalised in various ways. Using these generalisations Grigorchuk [Gri3] has shown that the set of word growth degrees of finitely generated groups contains both chains and antichains of continuously large size.

It is also an open problem however, whether there exist groups Γ with growth degrees strictly between polynomial and $e^{\sqrt{n}}$. It is expected that no such group Γ can be residually finite. This has been confirmed for residually p-groups [Gri4] and afterwards for residually nilpotent groups [LM].

As we will see later the analogous questions for subgroup growth have been answered in a rather satisfactory way.

In particular, using generalised Grigorchuk groups Segal [Se] showed that there exist finitely generated groups of arbitrarily small non-polynomial subgroup growth. This was proved by realising certain profinite groups as profinite completions.

Theorem 2.1 *Let (p_n) be any sequence of primes exceeding* 3. *Then there exists a* 4*-generator group* Γ *such that*

$$\widehat{\Gamma} \cong W = \overset{\lim W_n}{\longleftarrow} \quad \textit{(the inverse limit of the } W_n\textit{)}$$

where $W_n = PSL(2, p_n)$ wr $PSL(2, p_{n-1}) \ldots$ wr $PSL(2, p_0)$ *the (iterated) permutational wreath product using the natural permutation representation of each* $PSL(2,p)$ *on the points of the projective line over* $\mathbb{F}_p$.

In the same paper [Se] Segal proves a general result concerning realisations of profinite groups as profinite completions of finitely generated groups.

Theorem 2.2 *Let S be any non-empty collection of non-abelian finite simple groups. Then there exists a 63-generator just infinite group whose upper composition factors comprise exactly the set S.*

3 Subgroup growth

Denote by $R(\Gamma)$ the intersection of all finite index subgroups of a group Γ. Obviously $s_n(\Gamma/R(\Gamma)) = s_n(\Gamma)$. So when investigating subgroup growth there is no harm in assuming $R(\Gamma) = 1$ i.e. that Γ is a residually finite group. In this respect the study of subgroup growth is more restricted than the study of word growth.

On the other hand, there is some interest in investigating the subgroup growth of certain infinitely generated groups. Moreover, $s_n(\Gamma)$ (and not only its growth) is determined by Γ itself. Accordingly $s_n(\Gamma)$ and the related function $a_n(\Gamma)$ (= the number of subgroups of index exactly n) has been the subject of investigations of a number-theoretic flavour (see [Lu2], [LS] and the references therein).

Here we consider the asymptotic behaviour of $s_n(\Gamma)$. Given a function f we say that Γ has *subgroup growth type* f, if there exist positive constants a and b such that

(1) $s_n(\Gamma) \leq f(n)^a$ for all n;

(2) $s_n(\Gamma) \geq f(n)^b$ for infinitely many n.

If moreover (2) holds for all sufficiently large n we say that Γ has *strict growth type* f. Note that having strict growth type f is an equivalence relation.

A classical result of M. Hall [Ha] implies that $\mathbb{F}_d$ $(d \geq 2)$ has subgroup growth type n^n. Clearly this is the upper limit for finitely generated groups.

Let us mention that Hurwitz had already studied a question which is essentially counting finite index subgroups of surface groups. Most surface groups turn out to have subgroup growth type n^n (see [LS, 14.4]).

At the other extreme, the smallest possible growth type for an infinite finitely generated residually finite group Γ is achieved by the infinite cyclic group $\mathbb{Z}$. For a relatively elementary proof see [Sh2]. This observation also follows from an (equally deep) analogue of Gromov's theorem; as proved by Lubotzky, Mann and Segal [LMS] a finitely generated residually finite group Γ has polynomial subgroup growth exactly if Γ is virtually soluble of finite rank.

In his 1994 ICM talk [Lu3] Lubotzky stated that "it is widely open as to what the possible types of subgroup growth are for finitely generated groups". Similar remarks have been made in [MS].

As proved by Lubotzky [Lu1] arithmetic groups in characteristic zero with the congruence subgroup property (e.g. $\Gamma = SL(d,\mathbb{Z})$, $d \geq 3$) have subgroup growth type $n^{\log n/\log\log n}$.

Recently Abért, Nikolov and Szegedy [ANSz] proved that many arithmetic groups in characteristic p (e.g. $\Gamma = SL(d, F_p[t])$, $d \geq 3$) have subgroup growth type $n^{\log n}$.

Some other groups with "small growth", that is of subgroup growth type at most $n^{\log n}$ have been constructed in [LPSh]. This construction provided a starting point for the investigations in [Py].

Segal [Se] proved using certain generalised Grigorchuk groups that all functions $f : \mathbb{N} \to \mathbb{N}$ of the form $n^{g(n)}$ such that $g(n)$ is non-decreasing and $g(n) \leq \log\log n$ can be realised as subgroup growth types. In particular (as noted before) there is no gap between polynomial and non-polynomial subgroup growth. This construction can also be used to realise many other small subgroup growth types (see [LS]).

The subgroup growth type of finitely generated free soluble groups (of derived length ≥ 2) and various related groups is exponential (see [PSh1]). In [PSh1] we asked whether there exists a finitely generated group Γ of subgroup growth type strictly between 2^n and n^n.

Metabelian groups with fractionally exponential subgroup growth, that is of growth type e^{n^γ} have been constructed by Segal and Shalev [SSh] in the case when $\gamma = \frac{1}{d}$ for some positive integer d. In [MS] the following question is raised: "Can a finitely generated group have growth type e^{n^γ} where γ is irrational?"

The following somewhat unexpected result [Py] answers the questions mentioned above and essentially completes the picture.

Theorem 3.1 *Let $f : \mathbb{N} \to \mathbb{N}$ be a function such that $f(n) = n^{g(n)}$, where $g(n)$ is non-decreasing, $\log n \leq g(n)$ and $g(n) = o(n)$. Then there exists a 4-generator group Γ having strict growth type f.*

Lubotzky [Lu2] asked whether a finitely generated amenable group has at most exponential subgroup growth. The above groups Γ turn out to be elementary amenable which implies a negative answer to this question.

Still we wonder whether the growth type of a finitely generated amenable group is strictly less than n^n.

More provocatively one can ask whether a finitely generated group Γ of subgroup growth type n^n always has a virtually free quotient. This would be an interesting counterpart of the Lubotzky–Mann–Segal theorem.

4 Cartesian products of simple groups

Let G be a the Cartesian product of an infinite family of non-abelian finite simple groups. G is naturally a profinite group with the product topology.

Such groups occur naturally as quotients of various profinite groups. For example if Γ is a non-soluble finitely generated linear group, then Γ has an open subgroup H and a (closed) subgroup N, $N \triangleleft H$ such that $G = H/N$ is of the above type [DPSSh].

On the other hand, Cartesian products of certain families of simple groups have been used to realise various small subgroup growth types [Sh3]. Moreover, in [Ma] the following result is proved using families of finite alternating groups.

Theorem 4.1 *Let f be a non-decreasing function with $n \le f(n) \le n^n$ for all n. Then there exists a 2-generator profinite group whose maximal subgroup growth is of type f.*

It is suggested in [MS] that perhaps all intermediate subgroup growth types between $n^{\log n/(\log\log n)^2}$ and 2^n are achieved by such 2-generator profinite groups. This is essentially confirmed by the next result [Py] for growth types $\ge n^{\log n}$.

Theorem 4.2 *Let $f : \mathbb{N} \to \mathbb{N}$ be a function such that $\frac{\log f(n)}{\log n}$ is non-decreasing. Let G be the profinite group $\prod_{n=5}^{\infty} \mathrm{Alt}(n)^{f(n)}$. Then*

$$f(n) \le s_n(G) \le f(n) n^{23+12\log n}.$$

Since $\mathrm{Alt}(n)$ has a subgroup $\mathrm{Alt}(n-1)$ of index n, it is clear that G has at least $f(n)$ open subgroups of index n for every n. The key to proving the upper bound is the following amusing elementary result on finite permutation groups.

Proposition 4.3 *Let Ω be a finite set and H a subgroup of index n in $\mathrm{Alt}(\Omega)$. Then the set Ω has a partition $\Omega = \Omega_1 \cup \cdots \cup \Omega_t$ such that the group $A = \mathrm{Alt}(\Omega_1) \times \cdots \times \mathrm{Alt}(\Omega_t)$ is contained in H and $|H : A| \le n^3$.*

As noted in [KL] the finite group $\mathrm{Alt}(n)^{n!/8}$ is generated by two elements for n sufficiently large, say $n \ge N$. This implies that the profinite group $\prod_{n=N}^{\infty} \mathrm{Alt}(n)^{f(n)}$ is generated by two elements if $f(n) \le \frac{n!}{8}$ for all $n \ge N$. Therefore 2-generator profinite groups of "alternating type" realise all subgroup growth types f between $n^{\log n}$ and n^n such that $\frac{\log f(n)}{\log n}$ is non-decreasing.

Using not necessarily finitely generated groups one can easily realise arbitrarily fast growing functions as subgroup growth functions. However, it may be worth to point out the following consequence of Theorem 4.2.

Corollary 4.4 *Let $f : \mathbb{N} \to \mathbb{N}$ be a function such that $f(n) = n^{g(n)}$ where $g(n)$ is non-decreasing and $\log n \le g(n)$. There is a group G such that G has strict growth type f.*

Proof. It is clear that our statement holds for $G = \prod_{n=5}^{\infty} \mathrm{Alt}(n)^{f(n)}$ as a profinite group.

It was proved independently by Saxl and Wilson [SW] and Martinez and Zelmanov [MZ] that in a profinite group G which is a Cartesian product of nonabelian finite simple groups such that each finite simple constituent occurs only finitely many times, every finite index subgroup is open (see [LSh] for a more general result). Hence considering G as an abstract group we obtain the corollary. ∎

It would be interesting to decide whether the condition $\log n \le g(n)$ can be eliminated in the above result.

In [Py] we also obtain some non-trivial restrictions on the subgroup growth functions. Suppose for example that $f(n) = s_n(G)$ for some group G and say that $f(n) \geq n^{2\log n}$. Then [Py, Proposition 4.3] implies that we have

$$f(n^{\log n}) \geq f(n)^{\frac{1}{2}\log n}.$$

This shows that at least a weak version of the condition that $\frac{\log f(n)}{\log n}$ should be non-decreasing holds for any subgroup growth type f.

We end this section by considering yet another growth function.

For a finitely generated profinite group G set $b_n(G) = |G/\overline{G^n}|$ where $\overline{G^n}$ is the closed subgroup generated by the n-th powers in G. The finiteness of $b_n(G)$ follows from the solution by Zelmanov to the restricted Burnside problem. For a finitely generated discrete group Γ set $b_n(\Gamma) = b_n(\widehat{\Gamma})$.

The behaviour of the function $b_n(\Gamma)$, the *index growth* of Γ has been considered for example in [VZ], [Lu1], [Sh1]. Finitely generated groups of polynomial index growth do not seem to admit a simple characterisation (see [BMP]). However, a finitely generated pro-p group G has polynomial index growth exactly if G is p-adic analytic. Moreover, by a result of Lazard (see [DDMS, Chapter 11] if G is a non-p-adic analytic pro-p-group, then its index growth is at least exponential.

Here we point out that for arbitrary finitely generated profinite groups there is no such gap.

Proposition 4.5 *Let $f : \mathbb{N} \to \mathbb{N}$ be a non-decreasing function with $n \leq f(n) \leq 2^n$. Then there exists a 2-generator profinite group whose index growth is of type f.*

Proof. Consider profinite groups G of the form $G = \prod_i PSL(n_i, 2)^{f_i}$. Choose the sequence f_i in such a way that the order l_i of $PSL(n_i, 2)$ is divisible by the product of the orders of the preceding factors $PSL(n_j, 2)^{f_j}$ if $f_i \geq 1$.

If e_i denotes the exponent of $PSL(n_i, 2)$, then we have

$$l_i^{f_i} \leq |G : \overline{G^{e_i}}| \leq l_i^{f_i+1} \text{ if } f_i \geq 1.$$

By [BMP, Proposition 2.4] $l_i \leq e_i^R$ holds for some absolute constant R.

Assume that the f_i are chosen in such a way that $l_i^{f_i} \leq f(e_i)$ or $f_i \leq 1$. Then we have $|G : \overline{G^{e_i}}| \leq f(e_i)^{2R}$ if $f_i \geq 1$ (we use $f(e_i) \geq e_i$ if $f_i = 1$). It follows that $|G : \overline{G^n}| \leq f(n)^{2R}$ holds for all n.

Furthermore $l_i^{f_i} \leq f(e_i) \leq 2^{e_i} \leq 2^{l_i}$ implies that $f_i \leq \frac{l_i}{\log l_i}$. By the results in [KL] a group of the form $PSL(n, q)^f$ can be generated by two elements if $f \leq \frac{|PSL(n,q)|}{|Out(PSL(n,q))|}$ and n is sufficiently large. It follows that if $f_i = 0$ for small i, then the groups $PSL(n, 2)^{f_i}$ are 2-generator groups and so is G.

It is easy to see that one can choose the f_i to further satisfy $l_i^{f_i+2} \geq f(e_i)$ and $f_i \geq 1$ for infinitely many i. For such i we have $|G : G^{e_i}|^3 \geq f(e_i)$. This completes the proof of the proposition. ■

5 A general construction

Let $\Omega_1, \Omega_2, \ldots$ be finite sets of odd size $5 \leq n_1 < n_2 < \ldots$. Denote the elements of Ω_i by $w_i^1, w_i^2, \ldots$ and set $\Omega = \cup\Omega_i$. The profinite group $G = \prod_{i=1}^{\infty} \mathrm{Alt}(\Omega_i)$ acts as a permutation group on the set Ω. Define two permutations π and τ of Ω; π acts on Ω_i as the 3-cycle $\pi_i = (w_i^1, w_i^2, w_i^3)$, τ acts on Ω_i as the n_i-cycle $\tau_i = (w_i^1, w_i^2, \ldots, w_i^{n_i})$ for all i.

The groups $\Gamma = \langle \pi, \tau \rangle$ were first considered by B. H. Neumann [Ne] who proved that two such groups Γ are not isomorphic if the corresponding sequences $\{n_i\}$ are different (thereby proving the existence of continuously many non-isomorphic 2-generator groups).

At the end of his paper Neumann observes that such a Γ has a chain of normal subgroups $1 \triangleleft D \triangleleft N \triangleleft \Gamma$ such that D is the restricted direct product of the corresponding finite alternating groups, N/D is isomorphic to $\mathrm{Alt}(\mathbb{Z})$ (the group of even permutations of $\mathbb{Z}$) and Γ/N is an infinite cyclic group. The condition that the finite alternating groups should have different degrees is used in an essential way.

Essentially the same groups were rediscovered in [LW] where it is shown that Γ is amenable and that Γ is dense in the profinite group G. Later in [LPSh] it was shown that in fact $\widehat{\Gamma} \cong G \times \widehat{\mathbb{Z}}$ holds. The starting point to finding useful generalisations of the groups Γ was the observation that this follows easily from the remarks in [Ne] on the structure of Γ.

In [Py] we give a general construction in which some (but only finitely many) of the alternating groups used may have the same degrees. What is surprising is that one can give such a construction where the groups constructed satisfy a similar strong structure theorem.

We proceed by describing the construction.

Let $f : \mathbb{N} \to \mathbb{N}$ be a function. Let $\Omega_1, \Omega_2, \ldots$ be finite sets of size $7 \leq n_1 \leq n_2 \leq \ldots$ such that for every $n \geq 7$ exactly $f(n)$ sets have size n. We define Ω, G, D, τ, π as above.

Let l_i be a non-decreasing unbounded sequence of positive integers with $2l_i \leq n_i$ and set $v_i = \left[\frac{n_i}{l_i} - 1\right] l_i$. Denote the subset $w_i^{l_i}, u_i^{2l_i}, \ldots, w_i^{v_i}$ of Ω_i by L_i and set $L = \bigcup_{i=1}^{\infty} L_i$.

Let Q be a subgroup of G and Q_i the permutation group induced by Q on the set of upper indices $\{1, \ldots, n_i\}$ of the elements $n_i^1, n_i^2 \ldots$ of Ω_i. Suppose that the group Q satisfies the following hypothesis supp(q) $\subseteq L$ for every $q \in Q$ (where supp Q is the set of elements which are not fixed by q),

$$Q \cap D = 1 \tag{1}$$

if $n_i = n_j$ and $i \neq j$ then the permutation groups Q_i, Q_j are different (we allow them to be equivalent as permutation groups).

The main tool in proving the results in [Py] is the family of groups of the form $\Gamma = \langle \pi, \tau, Q \rangle$.

We describe the structure of such a group Γ in terms of its subgroups $D, N = \langle \pi^\Gamma \rangle$ and $H = D\langle \tau, Q \rangle$.

Theorem 5.1 (Structure Theorem) *Let Γ be as above. The following hold:*
$\Gamma = NH$ and $N \cap H = D$.
$N/D = \mathrm{Alt}(\mathbb{Z})$.
$H/D \cong Q$ wr $\mathbb{Z}$ (i.e. the wreath product of Q and $\mathbb{Z}$).

If Q is a perfect group for example if Q is a nonabelian finite simple group, then the profinite completion of Γ turns out to be not much larger than G.

Corollary 5.2 *If Q is a perfect group then $\widehat{\Gamma}$ has a subgroup of index ≤ 2 isomorphic to $G \times \widehat{\mathbb{Z}}$. Moreover, if for all even numbers i we have $f(i) = 0$ than $\widehat{\Gamma} \cong G \times \widehat{\mathbb{Z}}$.*

In particular, choosing Q to be isomorphic to an appropriate Alt(5) subgroup of G we obtain the following

Theorem 5.3 *Let $f : \mathbb{N} \to \mathbb{N}$ be a function such that $f(n) = n^{g(n)}$ where $g(n) = o(n)$. There exists an integer $N \geq 7$ such that for some Γ the profinite completion $\widehat{\Gamma}$ has a subgroup of index at most 2 isomorphic to $G \times \widehat{\mathbb{Z}}$, where $G = \prod_{n \geq N}^{\infty} \mathrm{Alt}(n)^{f(n)}$. Moreover for $n \geq N$ we have $s_n(\Gamma) \geq f(n)$.*

Together with Theorem 4.2 this easily implies Theorem 3.1.

Combining Theorem 5.3 and Theorem 4.1 we obtain the following

Theorem 5.4 *Let $f : \mathbb{N} \to \mathbb{N}$ be a function such that $f(n) = n^{g(n)}$ where $1 \leq g(n) = o(n)$. There exists a 4-generator profinite group whose maximal subgroup growth is of type f.*

We remark that our groups Γ give the first examples of finitely generated groups with intermediate maximal subgroup growth. The results in [PSh2] indicate these are indeed the natural examples.

A similar construction based on sequences of groups of the form $PSL(n,2)$ (see [LPSh] for a special case) can be used to show that functions as in Proposition 4.5 can be realised as index growth functions of finitely generated groups.

We end this section with an amusing open problem. Does the group $G = \prod_{\substack{i \text{ odd} \\ i\geq 5}} \mathrm{Alt}(i)$ have a finitely generated subgroup such that $\widehat{\Gamma} \cong G$?

We remark that a random pair of elements of G generates a dense subgroup isomorphic to $\mathbb{F}_2$ [DPSSh].

6 Groups with given profinite completions

Let Γ and Γ_0 be finitely generated groups. It is known [DFPR] that $\widehat{\Gamma} \cong \widehat{\Gamma}_0$ exactly if the set of isomorphism classes of finite quotients of Γ is equal to the set of isomorphism classes of finite quotients of Γ_0.

Many examples have been given of nonisomorphic pairs of polycyclic groups Γ, Γ_0 with $\widehat{\Gamma} \cong \widehat{\Gamma}_0$ (see [Pi], [Bau]). On the other hand, by a deep result of Grünewald, Pickel and Segal [GPS] the polycyclic groups with a given profinite completion lie in finitely many isomorphism classes.

Pickel [Pi] has constructed infinitely many non-isomorphic finitely presented metabelian groups with isomorphic profinite completions. Of course there can only be a countable number of such groups. Using Corollary 5.2 in [Py] we show the following

Theorem 6.1 *There are continuously many non-isomorphic* 4*-generator residually finite groups with profinite completions isomorphic to*

$$G = \prod_{\substack{j \text{ odd} \\ j\geq 7}} \mathrm{Alt}(j) \times \widehat{\mathbb{Z}}.$$

In fact, the same is true for many other profinite groups of the form $G = \prod_{\substack{j \text{ odd} \\ j\geq 7}} \mathrm{Alt}(j)^{f(j)} \times \widehat{\mathbb{Z}}$.

It would be interesting to decide whether there exist continuously many finitely generated residually finite, soluble groups with isomorphic profinite completions.

What if besides $\widehat{\Gamma} \cong \widehat{\Gamma}_0$ we assume that Γ_0 is a profinitely dense subgroup of Γ? Grothendieck arrived at this problem when he discovered a remarkable close connection between profinite completions and representation theory [Gro2].

For a group Γ and a commutative ring A denote by $\mathrm{Rep}_A(\Gamma)$ the category of finitely presented A-modules on which the group Γ operates.

Theorem 6.2 *Let $u : \Gamma_0 \to \Gamma$ be a homomorphism of finitely generated groups. The following are equivalent:*

a) *The continuous homomorphism $\widehat{u} : \widehat{\Gamma}_0 \to \widehat{\Gamma}$ induced by u is an isomorphism.*

b) *The "restriction functor"*

$$u_A^* : \mathrm{Rep}_A(\Gamma) \to \mathrm{Rep}_A(\Gamma_0)$$

is an equivalence of categories for all commutative rings A.

Grothendieck [Gro2] investigated conditions under which one could conclude that a homomorphism u as in the above theorem is actually an isomorphism. In particular he asked whether it is sufficient to assume that Γ and Γ_0 are finitely presented.

To our knowledge this question remains open. Relaxing the condition "finitely presented" to "finitely generated" Platonov and Tavgen [PT1] gave the first construction of negative examples. Their construction is based on a construction of Higman [Hi] of an infinite finitely presented group with no nontrivial finite quotients. Using a similar approach Bass and Lubotzky [BL] gave new, interesting examples based on certain hyperbolic superrigid lattices. Soluble examples of derived length 3 were found by Tavgen [Ta].

Some rather different examples can be given using our main construction [Py].

Theorem 6.3 *Let I be a set of odd integers (≥ 7) and set $G = \prod_{j \in I} \mathrm{Alt}(j)$. There exists a pair of groups $\Gamma_0 < \Gamma$ such that Γ_0 is dense in the profinite topology of Γ, $\widehat{\Gamma}_0 \cong G \times \widehat{\mathbb{Z}} \cong \widehat{\Gamma}$ but $\Gamma \not\cong \Gamma_0$.*

One can construct in a similar way an abundance of other "alternating type" examples. It would be interesting to decide whether one can use our construction to obtain finitely presented ones.

We note that the answer to Grothendieck's problem is positive in many interesting cases for example for finitely generated soluble linear groups. Moreover, Platonov and Tavgen [PT2] obtained a positive solution when Γ is a subgroup of $SL(2, K)$ where K is either the field of real or rational numbers.

It is open however in the case $\Gamma = SL(d, \mathbb{Z})$, $d \geq 3$.

References

[ANSZ] M. Abért, N. Nikolov, B. Szegedy, Congruence subgroup growth of arithmetic groups in positive characteristic, *Duke Math. J.*, to appear.

[Bar1] L. Bartholdi, The growth of Grigorchuk's torsion group, *Int. Math. Res. Notices* **20** (1998), 1349–1356.

[Bar2] L. Bartholdi, Lower bounds on the growth of a group acting on the binary rooted tree, *Int. J. Alg. Comp.* **11** (2001), 73–88.

[Bau] G. Baumslag, Residually finite groups with the same finite images, *Compositio Math.* **29** (1974), 249–252.

[BG] L. Bartholdi, R. I. Grigorchuk, Lie methods in growth of groups and groups of finite width, Computational and Geometric aspects of Modern Algebra (M. Atkinson et al. ed.) LMS Lecture Notes Ser. 275, Cambridge Univ. Press, Cambridge, 2000, pp. 1–27.

[BL] H. Bass, A. Lubotzky, Nonarithmetic superrigid groups: Counterexamples to Platonov's conjecture, *Ann. Math.* **151** (2000), 1151–1173.

[BMP] A. Balog, L. Pyber, A. Mann, Polynomial index growth groups, *Int. J. Alg. Comp.* **10** (2000), 773–782.

[DDMS] J. D. Dixon, M. P. F. du Sautoy, A. Mann, D. Segal, *Analytic pro-p groups*, 2nd edition, Cambridge University Press, Cambridge, 1999.

[DFPR] J. D. Dixon, E. W. Formanek, J. L. Poland, L. Ribes, Profinite completions and isomorphic finite quotients, *J. Pure Appl. Algebra* **23** (1982), 227–231.

[DPSSh] J. D. Dixon, L. Pyber, Á. Seress, A. Shalev, Residual properties of free groups and probabilistic methods, J. Reine Angew. Math., to appear.

[GPS] F. J. Grünewald, P. F. Pickel, D. Segal, Polycyclic groups with isomorphic finite quotients, *Ann. Math.* **111** (1980), 155–195.

[Gri1] R. I. Grigorchuk, On Burnside's problem on periodic groups, *Funktsional Anal. i Prilozhen.* **14** (1980), 53–54.

[Gri2] R. I. Grigorchuk, On Milnor's problem of group growth, *Dokl. Akad. Nauk SSSR* **271** (1983), 31–33.

[Gri3] R. I. Grigorchuk, On the growth degrees of finitely generated groups and the theory of invariant means, *Izv. Akad. Nauk SSSR, Ser. Math.* **48** (1984), 939–985.

[Gri4] R. I. Grigorchuk, On the Hilbert–Poincaré series of graded algebras that are associated with groups, *Mat. Sb.* **182** (1989), 207–225.

[Gri5] R. I. Grigorchuk, On growth in group theory, Proceedings of the International Congress of Mathematicians, Kyoto, 1990 (Math. Soc. of Japan, 1991), 325–338.

[Gri6] R. I. Grigorchuk, An example of a finitely presented amenable group which does not belong to the class EG, *Matem. Sbornik* **189** (1998), 75–98.

[Gro1] M. Gromov, Groups of polynomial growth and expanding maps, *Publ. Math. I.H.E.S.* **53** (1981), 53–73.

[Gro2] A. Grothendieck, Représentations linéaires et compactification profinie des groupes discrets, *Manuscripta Math.* **2** (1970), 375–396.

[Ha] M. Hall, Subgroups of finite index in free groups, *Canad. J. Math.* **1** (1949), 187–190.

[Hi] G. Higman, A finitely generated infinite simple group, *J. London Math. Soc.* **26** (1951), 61–64.

[KL1] W. M. Kantor, A. Lubotzky, The probability of generating a finite classical group, *Geom. Dedicata* **36** (1990), 67–87.

[LM] A. Lubotzky, A. Mann, On groups of polynomial subgroup growth, *Invent. Math.* **104** (1991), 521–533.

[LMS] A. Lubotzky, A. Mann, D. Segal, Finitely generated groups of polynomial subgroup growth, *Israel J. Math.* **82** (1993), 363–371.

[LPSh] A. Lubotzky, L. Pyber, A. Shalev, Discrete groups of slow subgroup growth, *Israel J. Math.* **96** (1996), 399–418.

[LS] A. Lubotzky, D. Segal, Subgroup growth, in preparation.

[LSh] M. W. Liebeck, A. Shalev, Diameters of finite simple groups: sharp bounds and applications, *Ann. Math.* **154** (2001), 383–406.

[Lu1] A. Lubotzky, Subgroup growth and congruence subgroups, *Invent. Math.* **119** (1995), 267–295.

[Lu2] A. Lubotzky, Counting finite index subgroups, in: *Groups '93, Galway/St. Andrews*, LMS Lecture Notes Ser. 211 (Cambridge Univ. Press, Cambridge 1995), 368–404.

[Lu3] A. Lubotzky, Subgroup growth, Proceedings of the International Congress of Mathematicians, Zürich, 1994 (Birkhäuser, Basel, 1995), 309–317.

[LW] A. Lubotzky, B. Weiss, Groups and Expanders, *DIMACS Series in Discrete Mathematics and Theoretical Computer Science* **10** (1993), 95–109.

[Ma] A. Mann, Positively finitely generated groups, *Forum Math.* **8** (1996), 424–459.

[Mi] J. Milnor, A note on curvature and fundamental group, *J. Diff. Geom.* **2** (1968), 1–7.

[MP] R. Muchnik, I. Pak, On growth of Grigorchuk groups, *Int. J. Alg. Comp.* **11** (2001), 1–17.

[MS] A. Mann, D. Segal, Subgroup growth: some current developments, in: *Infinite Groups 94* (Ravello) (de Gruyter, Berlin, 1995), 179–197.

[MZ] C. Martinez, E. I. Zelmanov, Products of powers in finite simple groups, *Israel J. Math.* **96** (1996), 469–479.

[Ne] B. H. Neumann, Some remarks on infinite groups, *J. London Math. Soc.* **12** (1937), 120–127.

[Pi] P. F. Pickel, Metabelian groups with the same finite quotients, *Bull. Austral. Math. Soc.* **11** (1974), 115–120.

[PT1] V. Platonov, O. I. Tavgen, Grothendieck's problem on profinite completions of groups, *Soviet Math. Dokl.* **33** (1986), 822–825.

[PT2] V. Platonov, O. I. Tavgen, Grothendieck's problem on profinite completions and representations of groups, *K-Theory* **4** (1990), 89–101.

[PSh1] L. Pyber, A. Shalev, Groups with super-exponential subgroup growth, *Combinatorica* 16 (1996), 527–533.

[PSh2] L. Pyber, A. Shalev, Asymptotic results for primitive permutation groups, *J. Algebra* **188** (1997), 103–124.

[Py] L. Pyber, Groups of intermediate subgroup growth and a problem of Grothendieck, in preparation.

[Sch] A. S. Schwarzc, A volume invariant of coverings, *Dokl. Akad. Nauk USSR* **105** (1955), 32–34.

[Se] D. Segal, The finite images of finitely generated groups, *Proc. London Math. Soc.* (3) **82** (2001), 597–613.

[Sh1] A. Shalev, Growth functions, *p*-adic analytic groups and groups of finite coclass, *J. London Math. Soc.* **46** (1992), 111–122.

[Sh2] A. Shalev, Groups whose subgroup growth is less than linear, *Int. J. Alg. Comp.* 7 (1997), 77–91.

[Sh3] A. Shalev, Subgroup growth and sieve methods, *Proc. London Math. Soc.* **74** (1997), 335–359.

[SSh] D. Segal, A. Shalev, Groups with fractionally exponential subgroup growth, *J. Pure Appl. Algebra* **88** (1993), 205–223.

[SW] J. Saxl, J. S. Wilson, A note on powers in simple groups, *Math.Proc. Cambridge Philos. Soc.* **122** (1997), 91–94.

[Ta] O. I. Tavgen, The problem of Grothendieck in the class of soluble groups, *Dokl. Ak. Nauk, BSSR* **31** (1987), 873–876.

[VZ] M. R. Vaughan-Lee, E. I. Zelmanov, Bounds in the restricted Burnside problem, *J. Austral. Math. Soc. (Ser. A)* **67** (1999), 261–271.

SHADOWS OF ELEMENTS, SOLVABILITY OF FINITE QUOTIENTS AND THE MARGULIS-PLATONOV CONJECTURE

YOAV SEGEV*

1. Introduction

In this note we wish to keep the presentation as accessible as possible and we have two goals. The first is to give a brief description of the recent proof obtained by Rapinchuk, Seitz and myself [12] of the theorem which states that finite quotients of the multiplicative group $D^{\times}$ of a finite dimensional division algebra D are solvable. In particular, we will briefly discuss the interplay between the "shadows" of elements of $D^{\times}$ on a finite index normal subgroup $N \leq D^{\times}$, and properties of the commuting graph of $D^{\times}/N$.

The solvability of finite quotients of $D^{\times}$ is closely related to the Margulis–Platonov conjecture ([12, section 6]). Since we will not need the complete formulation of the Margulis-Platonov conjecture ((MP) for short), we refere the reader to [11, appendix A], or [8, chapter 9] for that and for more information about (MP). Here we briefly mention that (MP) describes the normal subgroup structure of the K-rational points of absolutely simple simply connected algebraic K-groups, where K is a global field (i.e. a finite extension of $\mathbb{Q}$ or of the field of rational functions $F(t)$, where F is a finite field). One can thus sense that (MP) is a fundumental conjecture.

What we will do in this note, and this is our second goal, is describle the two formidable challenges left open in conjecture (MP). These are the anisotropic unitary groups $\mathrm{SU}_2(V, f)$ and $\mathrm{SU}_1(D, \circ)$, the first described in §5 and the second in §6. We will also include a proof for the well known reduction from the anisotropic $\mathrm{SU}_n(V, f)$ case of (MP), $n \geq 2$, to the anisotropic $\mathrm{SU}_2(V, f)$ case and some well known information about reflections. The two open cases above are basically the only remaining open cases of (MP), we formulate them precisely: Conjecture 5.1 covers the $\mathrm{SU}_2(V, f)$ case and Conjecture 6.1 covers the $\mathrm{SU}_1(D, \circ)$ case.

I would like to thank G. Prasad, L. Rowen and G. Seitz for their helpful remarks. The version of the reduction from the anisotropic $\mathrm{SU}_n(V, f)$ case, $n \geq 2$, to the $\mathrm{SU}_2(V, f)$ case, brought in this note, was observed by M. Aschbacher, G. Prasad, G. Seitz and myself in Caltech at the summer of 1999. Thanks are due also to J. P. Tignol for pointing out to me [7, Prop. 6.1] (see also [5, exercise 12, pg. 202]) which lead to Lemma 5.3 and consequently Proposition 5.2.

* Partially supported by BSF grant no. 2000-171.

2. Finite quotients of $D^\times$ are solvable

The purpose of this section is to give some insight into the proof of the following theorem.

Theorem 2.1 ([12]). *Finite quotients of the multiplicative group $D^\times$ of a finite dimensional division algebra D are solvable.*

Here D is a division algebra, i.e., D is a field, except that multiplication in D may not be commutative (D is sometimes called a skew-field). The center of D, denoted in this section by K, is a (commutative) field and D is finite dimensional as a vector space over K. Let us recall that the dimension $\dim_K D$ is a square, say n^2, and n is called the *degree* of D. Before we continue, two remarks concerning Theorem 2.1 are in order. The first is:

Remark 2.2. We note that by [4], for any noncomutative division algebra D (finite dimensional or infinite dimensional), $D^\times$ is never solvable (see also [1, Thm. 19 and Thm. 20], [14, Thm. 14.4.1, pg. 439] and [18]). Also, using the Zariski density of $\mathcal{G}(K)$ in $\mathcal{G}$, for any connected nonabelian reductive algebraic group $\mathcal{G}$ defined over (the arbitrary infinite field) K, it is proved in [17, 12.2(3), pg. 219], that $\mathcal{G}(K)$ is not solvable in this case. It follows immediately ([17, 12.3, pg. 220]) that $D^\times$ is not solvable in the finite dimensional case. Here is a well known elementary proof of this fact.

Lemma 2.3. *Let D be a finite dimensional division algebra with center K and degree n. Suppose D is not commutative. Then,*

(1) *if $x \in D^\times \smallsetminus K^\times$, then the number of conjugates of x in $D^\times$ is infinite;*

(2) *if αx is conjugate in $D^\times$ to x, where $\alpha \in K^\times$, then $\alpha^n = 1$;*

(3) *let $H \leq D^\times$ be a normal subgroup such that $[H, H] \leq K^\times$. Then $H \leq K^\times$;*

(4) *$D^\times$ is not solvable.*

Proof. (1): Notice that the centralizer $C_D(x)$ is a division subalgebra of D and if the number of conjugates of x in $D^\times$ is finite, then D is a finite union of subsets of the form $yC_D(x)$. However, viewing D as a (right) vector space over $C_D(x)$ it can be easily proved that D is not a finite union of proper subspaces.

(2): Let $m[X]$ be the minimal polynomial of x over K. If αx is a conjugate of x, then αx is also a root of $m[X]$. But then x is also a root of the polynomial $m[X] - m[\alpha X]$ which is a polynomial whose free coefficient is zero. Thus x is a root of the polynomial $X^{-1}(m[X] - m[\alpha X])$, which is possible only if $m[X] - m[\alpha X] = 0$. It follows that $\alpha^k = 1$, where k is the degree of $m[X]$. But k divides n, so $\alpha^n = 1$.

(3): Let $H \leq D^\times$ be a normal subgroup such that $[H, H] \leq K^\times$. Suppose first that H is abelian and consider the K-subspace E of D spanned by H. This is clearly a subalgebra of D, hence a division subalgebra of D. Of course E is normalized by $D^\times$. By the Cartan-Brauer-Hua Theorem (see [3, Theorem 3.9.2, pg. 144] for a very easy proof) either $E = D$, or $E \subseteq K$. Since E is abelian, we see that $E \subseteq K$.

The above argument shows that the center of H, $\mathrm{Z}(H)$, is contained in $K^\times$. Let $x \in H \smallsetminus \mathrm{Z}(H)$. By induction on n we see that $C_H(x)$ is contained in the center of $C_D(x)$ so, in particular, $C_H(x)$ is abelian. Further for each $y \in H$, $y^{-1}xy = \alpha x$, for some $\alpha \in K^\times$, so by (2), $[x, y]^n = 1$ (where $[x, y] = x^{-1}y^{-1}xy$). It is easy to check that since $[H, H] \subseteq K^\times$, the map $y \to [x, y]$ is a homomorphism $H \to K^\times$ whose image is finite (it is contained in the subgroup of n-th roots of 1). It follows that the kernel of this map, $C_H(x)$, has finite index in H, and as we saw it is abelian. As this holds for all $x \in H \setminus \mathrm{Z}(H)$ one easily checks that $\mathrm{Z}(H)$ has finite index in H. Replacing H by $HK^\times$ if necessary we may assume that $\mathrm{Z}(H) = K^\times$. But now we have that $K^\times$ has finite index in H, and also by (1), any $x \in H \setminus K^\times$ has an infinite number of conjugates in $D^\times$ which are all contained in H. Thus x has an infinite number of conjugates of the form αx, $\alpha \in K^\times$. This contradicts (2).

(4): This is an immediate consequence of (3). □

Our second remark is,

Remark 2.4. Theorem 2.1 is false when D is not finite dimensional over K. Indeed the following example was communicated to me by A. Lichtman and is due to him. Consider the following Malcev-Neimann power series skew-field:

Let G be an ordered group, i.e. G is a group with a total ordering such that $s \leq s'$ and $t \leq t'$ implies $st \leq s't'$, for all $s, t, s', t' \in G$. Let K be a commutative field. The Malcev-Neimann power series skew-field $K((G))$ is the skew-field whose elements are formal series $x = \sum sa_s$, where the sum runs over all $s \in G$, $a_s \in K$ for all $s \in G$, and the support of x (i.e. the set $\mathrm{supp}(x) := \{s \in G \mid a_s \neq 0\}$) is well ordered (via the ordering induced from G). It is easy to check that the natural definitions of addition and multiplication are valid in $K((G))$. By [3, Thm. 2.4.5, pg. 75], $K((G))$ is a skew-field. Let e denote the identity element of G and let $N := \{x \in K((G)) \mid \min \mathrm{supp}(x) = e\}$. We note that given $x \in K((G))$, if $\min \mathrm{supp}(x) = e$ and the coefficient a_e, of e in x is 1, then we can write $x = e - y$, such that $y = \sum sb_s$ $(s > e)$. The inverse of x in $K((G))$ is then given by $e + y + y^2 + \ldots$ (it can be shown that this element belongs to $K((G))$). This shows how to compute the inverse of x and as an easy consequence the inverse of any element in $K((G))$. Using this one shows that N is a normal subgroup of $K((G))^\times$ and it is easy to check that $K((G))^\times / N \cong G$. Taking G to be a free group, we see that G, and hence any finite group, is a quotient of $K((G))^\times$, this shows that the hypothesis in Theorem 2.1 that D be finite dimensional is essential. For more information about the Malcev-Neimann construction see [3, Section 2.4].

It seems to us that it is rather accurate to say that the structure of $D^\times$ had been quite mysterious until recently. In fact, even now many questions about $D^\times$ are left open. For example:

Question 2.5 (Rapinchuk, Prasad). Are finite quotients of $\mathrm{SL}_1(D)$ solvable?

Even this question, which is of course closely related to Theorem 2.1, remains open (see Question 2 in the introduction of [12] for a generalization of the above question due to Rapinchuk and Prasad). Let us recall that $\mathrm{SL}_1(D) = \{x \in D^\times \mid \mathrm{Nrd}_{D/K}(x) = 1\}$, here $\mathrm{Nrd}_{D/K}$ is the *reduced norm* defined in the first paragraph of §3 in a slightly more general context. Of course questions regarding normal subgroups of infinite index in $D^\times$ had not been answered or even addressed at all.

Going back to what is known about the structure of $D^\times$, in the paper [15] new techniques were introduced to obtain information about finite quotients of $D^\times$. These techniques were further developed and sharpened in [11], where valuation theory entered the arguments and the results in an explicit and much more significant way. Finally, the most general result, Theorem 2.1, was proved in [12]. Since we want to keep this exposition as simple as possible, we will now only give a hint at the new techniques developed to investigate finite quotients of $D^\times$ referring the interested reader to the actual papers for much more detail.

Let us consider a normal subgroup of finite index $N \leq D^\times$. The two most important notions for us here are: *The commuting graph of* $D^\times/N$ and the set $N(a)$, for $a \in D^\times$, which we call the *shadow* of a (on the normal subgroup N). We will define both terms in Definition 2.6 below. It is the interplay between the commuting graph and the shadows which enables us to prove our Theorem 2.1.

Definitions 2.6. (1) Let $a \in D^\times$. The *shadow* of a on N is denoted $N(a)$ and defined by $N(a) = \{n \in N \mid a + n \in N\}$.

(2) Suppose $K^\times \subseteq N$ and let $a \in D^\times$. The *K-shadow* of a is denoted $\dot{N}(a)$ and defined by $\dot{N}(a) = N(a) \cap K^\times$.

(3) The *commuting graph* of a group H is the graph whose vertex set is $H \setminus \{1\}$ and whose edges are pairs of distinct commuting elements. We denote it by $\Delta(H)$. This graph has a natural distance function $d_H(\ ,\ )$ and we let $\operatorname{diam}(\Delta(H))$ be the diameter of $\Delta(H)$ (the largest distance between two vertices, being infinity if the graph is disconnected – thus the diameter of the complete graph is one).

The reasons that one is led to consider the shadows $N(a)$ of elements $a \in D^\times$ are revealed in the following considerations. First we have,

Theorem 2.7 ([2], [19]). *Let D be a division algebra (not necessarily finite dimensional), and let $N \subseteq D^\times$ be a normal subgroup of finite index. Then $D = N - N = \{n - m \mid n, m \in N\}$.*

Note that a main feature of Theorem 2.7 is that multiplicative properties of N (being of finite index in $D^\times$), yield additive properties. Indeed many of the results presented in this section may be viewed as an interplay between additive and multiplicative properties of D. Here is another basic example. Let

$$*\colon D^\times \to D^\times/N,$$

be the canonical homomorphism.

Lemma 2.8. *Let $a \in D^\times \setminus N$ and $n \in N$. Then a^* commutes with $(a+n)^*$.*

Proof. Indeed, $(a+n)^* = (n^{-1}a+1)^*$ and $(n^{-1}a+1)^*$ commutes with $(n^{-1}a)^* = a^*$. □

Using Lemma 2.8 one is tempted to consider the following false argument: take two arbitrary non-identity elements $a^*, b^* \in D^\times/N$ and let $d(\ ,\ )$ be the distance function on $\Delta(D^\times/N)$. We want to restrict $d(a^*, b^*)$. By Theorem 2.7, $a-b = n-m$, for some $m, n \in N$, and hence, by Lemma 2.8, $a^*, (a+m)^* = (b+n)^*, b^*$ is a path in Δ, proving hereby that $\operatorname{diam}(\Delta(D^\times/N)) \leq 2$. The reason this argument fails is that it may happen that $a + m = b + n \in N$, i.e. that m is in the shadow of a and n is in the shadow of b. So we are forced to consider the shadows.

Notice that by Theorem 2.7, $N(a)$ is never empty. Further, given $a \in D^\times \setminus N$, if $n \in N(a^{-1})$, then since $a + n^{-1} = a(a^{-1} + n)n^{-1}$, it follows that $n^{-1} \notin N(a)$. Thus the shadow of a is a proper nonempty subset of N. Going back to the elements a^*, b^* above, it turns out that the failure of the argument showing that $d(a^*, b^*) \leq 2$ caused by the fact that $m \in N(a)$ (and $n \in N(b)$) has some gains as well: assuming that $d(a^*, b^*) > 2$ and working, not only with the shadows of a and b, but with the shadows of all elements in the coset of a, aN and the coset of b, bN, one sees that a miraculous thing happens: these shadows tend to align!

Proposition 2.9 (Prop. 6.11 and Lemma 6.12(1) in [11]). *Let $a, b \in D^\times \setminus N$ such that $d(a^*, b^*) \geq 4$. Then given any $a' \in aN$ and $b' \in bN$, either $N(a') \subseteq N(b')$ of $N(b') \subseteq N(a')$; furthermore, after perhaps interchanging a and b, we have that given any $a', a'' \in aN$, either $N(a') \subseteq N(a'')$ or $N(a'') \subseteq N(a')$.*

Once we obtain an element $a \in D^\times \setminus N$ such that given any $am, an \in aN$, either $N(am) \subseteq N(an)$ or $N(an) \subseteq N(am)$, as in Proposition 2.9, we can define a preorder relation on N by letting $m \leq n$ if and only if $N(am) \subseteq N(an)$. This is just the beginning of a series of arguments leading to a valuation on D, which can be employed to restrict the structure of $D^\times/N$. Thus we either have that $\text{diam}(\Delta(D^\times/N))$ is "small" or the shadows of elements can be used to restrict the structure of $D^\times/N$, in either case, the structure of $D^\times/N$ is restricted.

The situation when $\text{diam}(\Delta(D^\times/N)) \geq 4$ is handled in [11]. However, for the proof of Theorem 2.1, this situation is too restrictive. The reason is as follows. Suppose that $D^\times/N$ is not solvable. We wish to employ our commuting graph techniques, combined with our "shadows techniques" to show that this is impossible. To do that we have to be able to say something about $\Delta(D^\times/N)$, when $D^\times/N$ is not solvable. For that we first pass to a minimal nonsolvable quotient, i.e., we replace N (if necessary) by a larger normal subgroup, so that we obtain that $D^\times/N$ is a *minimal nonsolvable group* (MNS-group for short), i.e., a nonsolvable group all of whose proper quotients are solvable. Can we say something about $\Delta(D^\times/N)$ now? If we could show that $\text{diam}(\Delta(D^\times/N)) \geq 4$, then the machinery of [11] would apply perfectly to obtain a contradiction. However, by [16], $\text{diam}(\Delta(D^\times/N))$ may be 3 (and is always ≥ 3). Luckily the following (somewhat strange) property is satisfied by $H = D^\times/N$.

Property $(3\frac{1}{2})$: The group H possess the property $(3\frac{1}{2})$ if there are two elements $x, y \in \Delta(H)$ such that $d_H(x, y) \geq 3$ and such that if x, a, b, y is a path in $\Delta(H)$, then there exists $h \in H$ such that $d(x^h, y) \geq 3$ and x^h, a^h, b, y is *not* a path in $\Delta(H)$.

We call this property Property $(3\frac{1}{2})$ because it is immediate that it is satisfied when the elements $x, y \in H$ satisfy $d_H(x, y) \geq 4$ and it is (of course) stronger than $\text{diam}(\Delta(H)) \geq 3$.

Theorem 2.10 ([12]). *Let H be a nonsolvable finite group such that every proper quotient of H is solvable. Then H has the Property $(3\frac{1}{2})$.*

Relating Theorem 2.10 to the shadows, a crucial outcome of Property $(3\frac{1}{2})$ is the following proposition. Though we can not obtain the full strength of Proposition 2.9, it is very useful,

Proposition 2.11 (Prop. 5.7 and Lemma 5.6(1) in [12]). *Let $x, y \in D^\times \setminus N$ such that x^*, y^* satisfy property $(3\frac{1}{2})$ (where $H = D^\times/N$). Then given any $x' \in xN$ and $y' \in yN$, either $\dot{N}(x') \subseteq \dot{N}(y')$ of $\dot{N}(y') \subseteq \dot{N}(x')$. Furthermore, after perhaps interchanging x and y, we have that given any $x', x'' \in xN$, either $\dot{N}(x') \subseteq \dot{N}(x'')$ or $\dot{N}(x'') \subseteq \dot{N}(x')$.*

Notice that Proposition 2.11 says something about the K-shadows $\dot{N}(x)$ and not about the shadows $N(x)$. Note further that unlike $N(x)$, $\dot{N}(x)$ may be empty. However Proposition 2.11, together with a series of arguments eventually lead to a valuation of D and show that $D^\times/N$ must be solvable. To conclude this section we mention that the proof of Theorem 2.10 (and hence, of Theorem 2.1) involves a heavy use of the classification of finite simple groups, see [12, section 7].

3. Hermitian spaces and unitary groups, notation and definitions

The purpose of this section is to define what we mean by a hermitian space (V, f) and to define the groups $\mathrm{U}(V, f)$ and $\mathrm{SU}(V, f)$. Let us start by recalling that a simple algebra A is just a ring with 1 whose only ideals are $\{0\}$ and A. Since for every element $z \neq 0$ in the center of A, Az is an ideal, we see that z is invertible in A so the center of A is a field which we will denote by L. We say that A is finite dimensional if the dimension of A as a vector space over L is finite. It is well known that when A is finite dimensional, $A \cong \mathrm{End}_D(V) \cong \mathrm{M}_n(D)$, where V is some (right) vector space over a division algebra D with center L. Let us recall also that the *reduced norm* $\mathrm{Nrd}_{A/L}\colon A \to L$ is given as follows. Let $\bar{L}$ be the algebraic closure of L, then $A \otimes_L \bar{L} \cong \mathrm{M}_n(\bar{L})$ (for some n). Thus we have the maps $a \to a \otimes 1 \xrightarrow{\varphi} \mathrm{M}_n(\bar{L})$ and given $a \in A$, the reduced norm of a is defined by $\mathrm{Nrd}_{A/L}(a) \stackrel{\mathrm{def}}{=} \det(\varphi(a \otimes 1))$. It turns out that $\mathrm{Nrd}_{A/L}(a) \in L$, for all $a \in A$.

Now let us fix D to denote a finite dimensional division algebra with center L. Let us fix V to denote a finite dimensional (right) vector space over D. Let $A := \mathrm{End}_D(V)$. We view L as the center of A (via $\alpha \to \alpha \cdot id$, for $\alpha \in L$, where id is the identity map on V).

An *involution of the second kind* on A is an antiautomorphism $\tau\colon A \to A$ of order 2 whose restriction to L is nontrivial. So for all $a, b \in A$ we have, $\tau(a + b) = \tau(a) + \tau(b)$, $\tau(ab) = \tau(b)\tau(a)$ and $\tau(\tau(a)) = a$. Further, the restriction $\tau\colon L \to L$ is nontrivial. We will denote by $K \subset L$ the subfield of elements fixed by τ. When $A = D$ (i.e. $\dim(V) = 1$) we sometimes write a° in place of $\tau(a)$.

Definition 3.1. Let $\circ$ be an involution of the second kind on D. A *hermitian form* on V with respect to $\circ$ is a map $f\colon V \times V \to D$ such that for all $v, w \in V$ and $d \in D$,

(1) f is biadditive;

(2) $f(vd, w) = d^\circ f(v, w)$ and $f(v, wd) = f(v, w)d$;

(3) $f(w, v) = (f(v, w))^\circ$.

The form f is called *anisotropic* if in addition,

(4) $f(v, v) = 0$ iff $v = 0$.

Let us note that,

Lemma 3.2. *Let $\circ$ be an involution of the second kind on D and let $s \in D^\times$ be a symmetric element ($s^\circ = s$). Define $*\colon D \to D$ by $a^* = s^{-1}d^\circ s$. Then $*$ is an involution of the second kind on D. Let f be a hermitian form on V, with respect to $\circ$. Then the form f' on D defined by $f'(v, w) = s^{-1}f(v, w)$ is a hermitian form on V with respect to the involution $*$.*

Proof. This is an easy calculation. □

Definition 3.3. If f is a hermitian form on V (with respect to an involution of the second kind on D which is clear from the context) then we say that (V, f) is a *hermitian space.* If the form f is anisotropic, we say that (V, f) is anisotropic and if the form f is nondegenerate ($f(v, w) = 0, \forall w \in V \Rightarrow v = 0$), we say that (V, f) is nondegenerate. Whenever we say that (V, f) is a hermitian space we denote the involution on D by $\circ$.

Definition 3.4. Let (V, f) be a hermitian space. We denote by $\mathrm{U}(V, f)$ the group of all linear transformations g of V such that $f(g(v), g(w)) = f(v, w)$, for all $v, w \in V$. If $\dim(V) = n$ and we wish to emphasize that, we write $\mathrm{U}_n(V, f)$ in place of $\mathrm{U}(V, f)$. $\mathrm{SU}(V, f)$ denotes the subgroup of $\mathrm{U}(V, f)$ of all transformations having reduced norm 1. We write $\mathrm{SU}_n(V, f)$ when we wish to emphasize the dimension.

4. Generalized reflections

We continue with the notation of §3. Let (V, f) be an anisotropic hermitian space over the division algebra D with the involution of the second kind $\circ: D \to D$ (see Definition 3.3). For $v, w \in V$ we denote $f(v, w) = < v, w >$. Let $A = \mathrm{End}_D(V)$ and for an element $a \in A$, let $\mathrm{Fix}(a) := \{v \in V \mid a(v) = v\}$.

Definition 4.1. A *generalized reflection* is an element $r \in \mathrm{U}(V, f)$ such that $\mathrm{Fix}(r)$ is a hyperplane (i.e. a subspace of codimension 1) of V. Given a vector $0 \neq w \in V$ and $\alpha \in D^{\times}$ such that $\alpha^{\circ} < w, w > \alpha =< w, w >$, we denote by $r_{w,\alpha}$ the reflection

$$r_{w,\alpha}(v) = v + w(\alpha - 1) < w, w >^{-1}< w, v > .$$

$r_{w,\alpha}$ is the (unique) reflection such that $r_{w,\alpha}(w) = w\alpha$ and $\mathrm{Fix}(r_{w,\alpha}) = w^{\perp} = \{v \in V \mid < v, w > = 0\}$.

Lemma 4.2. (1) $r_{w,\alpha}^{-1} = r_{w,\alpha^{-1}}$;

(2) *for all* $g \in \mathrm{U}(V, f)$, *we have* $g r_{w,\alpha} g^{-1} = r_{g(w),\alpha}$;

(3) $r_{w\alpha,\beta} = r_{w,\alpha\beta\alpha^{-1}}$;

(4) *if* $r = r_{w,\alpha}$ *and* $s = r_{z,\beta}$ *are two reflections such that* $r(y) = s(y) \neq y$, *for some* $y \in V$, *then* $r = s$.

Proof. (1): We have $w = r_{w,\alpha}^{-1}(w\alpha) = r_{w,\alpha}^{-1}(w)\alpha$, so $r_{w,\alpha}^{-1}(w) = w\alpha^{-1}$. Of course $\mathrm{Fix}(r_{w,\alpha}^{-1}) = w^{\perp}$, so $r_{w,\alpha}^{-1} = r_{w,\alpha^{-1}}$.

(2): Let $r = r_{w,\alpha}$. Then

$$\begin{aligned} grg^{-1}(u) &- g(g^{-1}(u) + w(\alpha - 1) < w, w >^{-1}< w, g^{-1}(u) >) \\ &= g(g^{-1}(u) + w(\alpha - 1) < g(w), g(w) >^{-1}< g(w), u >) \\ &= u + g(w)(\alpha - 1) < g(w), g(w) >^{-1}< g(w), u > \\ &= r_{g(w),\alpha}(u). \end{aligned}$$

(3): Both $r_{w\alpha,\beta}$ and $r_{w,\alpha\beta\alpha^{-1}}$ take $w \to w\alpha\beta\alpha^{-1}$ and centralize $w^{\perp}$.

(4): Write $y = w\gamma + w' = z\delta + z'$, where w' (resp. z') are some vectors in $w^{\perp}$ (resp. $z^{\perp}$). Then $r(y) = w\alpha\gamma + w' = z\beta\delta + z' = s(y)$. Subtracting we get $w(1 - \alpha)\gamma = z(1 - \beta)\delta$. Hence we can take $z = w\mu$. Then, $y = w\gamma + w' = w\mu\delta + z'$, so $\gamma = \mu\delta$. Also, $w\alpha\gamma + w' = w\mu\beta\delta + z'$, so $\alpha\gamma = \mu\beta\delta$. It follows that $\alpha\mu\delta = \mu\beta\delta$, so $\alpha\mu = \mu\beta$, or $\alpha = \mu\beta\mu^{-1}$, so by (3), $r_{z,\beta} = r_{w\mu,\beta} = r_{w,\mu\beta\mu^{-1}} = r_{w,\alpha}$. □

Lemma 4.3. *Let $v, w \in V$ be distinct vectors such that $< v, v > = < w, w >$. Set $\gamma = < v - w, v >$ and $\beta = -\gamma^{-1}\gamma^{\circ}$. Then,*

(1) $< v - w, v - w > = \gamma + \gamma^{\circ}$ *and hence* $\beta^{\circ} < v - w, v - w > \beta = < v - w, v - w >$. *In particular $r_{v-w,\beta}$ is a generalized reflection.*

(2) *For all $z \in V$ we have $r_{v-w,\beta}(z) = z - (v - w)\gamma^{-1} < v - w, z >$, in particular, $r_{v-w,\beta}$ is the unique reflection such that $r_{v-w,\beta}(v) = w$.*

Proof. Let $s = < v, v >$ and notice that $\gamma = < v - w, v > = s - < w, v >$ and $< v - w, w > = < v, w > - s = -\gamma^{\circ}$. Hence $< v - w, v - w > = < v - w, v > - < v - w, w > = \gamma + \gamma^{\circ}$ (in particular $\gamma \neq 0$). Also

$$\begin{aligned}\beta^{\circ} < v - w, v - w > \beta &= \beta^{\circ}(\gamma + \gamma^{\circ})\beta = \gamma(\gamma^{\circ})^{-1}(\gamma + \gamma^{\circ})\gamma^{-1}\gamma^{\circ} \\ &= \gamma(\gamma^{-1} + (\gamma^{\circ})^{-1})\gamma^{\circ} = \gamma + \gamma^{\circ} = < v - w, v - w > .\end{aligned}$$

This shows (1).

Set $r = r_{v-w,\beta}$. Given $z \in V$ we have

$$\begin{aligned} r(z) &= z + (v - w)(\beta - 1) < v - w, v - w >^{-1} < v - w, z > \\ &= z + (v - w)(-\gamma^{-1}\gamma^{\circ} - 1)(\gamma + \gamma^{\circ})^{-1} < v - w, z > \\ &= z - (v - w)\gamma^{-1}(\gamma + \gamma^{\circ})(\gamma + \gamma^{\circ})^{-1} < v - w, z > \\ &= z - (v - w)\gamma^{-1} < v - w, z > .\end{aligned}$$

We have $r(v) = v - (v - w)\gamma^{-1} < v - w, v > = v - (v - w)\gamma^{-1}\gamma = w$. The uniqueness of r follows from Lemma 4.2(4). □

Corollary 4.4. *Suppose* $\dim(V) = 2$, *then every element of* $\mathrm{U}(V, f)$ *is either a reflection or a product of two reflections.*

Proof. Let $g \in \mathrm{U}(V, f)$ and let $v \in V$ such that $v \neq g(v)$. Then $g(v) \in V$ and of course $< g(v), g(v) > = < v, v >$. By Lemma 4.3(2), there exists a reflection r such that $r(v) = g(v)$. Hence $r^{-1}g$ centralizes v, so $r^{-1}g$ is a reflection. □

Remark 4.5. Notice that by the definition of the reduced norm, given a generalized reflection $r = r_{w,\alpha}$, we have $\mathrm{Nrd}_{A/L}(r) = \mathrm{Nrd}_{D/L}(\alpha)$. Note further that $\alpha \in \{d \in D \mid d\tau(d) = 1\}$, where $\tau\colon D \to D$ is the involution of the second kind on D given by $\tau(d) = s^{-1}d^{\circ}s$, with $s = \langle w, w\rangle$ (see Lemma 3.2).

Remark 4.6. By Lemma 4.3(2) it follows immediately that if $\dim(V) \geq 2$, then $\mathrm{U}(V, f)$ is generated by generalized reflection. The following question seems elementary and a positive answer should be useful.

Question 4.7. Is $\mathrm{SU}(V, f)$ generated by the generalized reflections contained in it? Does this hold when $\dim(V) = 2$ and the center of D is a global field?

5. The reduction from $\dim(V) \geq 2$ to $\dim(V) = 2$

We continue with the notation of §3. Let D be a finite dimensional division algebra with center L, let $\circ\colon D \to D$ be an involution of the second kind and let K be the fixed field of $\circ$. Let (V, f) be an anisotropic hermitian space over D (with respect to $\circ$). As usual let $A = \mathrm{End}_D(V)$. In this section we wish to reduce the anisotropic unitary case of conjecture (MP), when $\dim(V) \geq 2$ to the following conjecture:

Conjecture 5.1 ((MP) for anisotropic $\mathrm{SU}(V, f)$, $\dim(V) \geq 2$). *Let (V, f) be a two dimensional anisotropic hermitian space over a finite dimensional division algebra D. Suppose that the center of D is a global field. If $N \leq \mathrm{SU}_2(V, f)$ is a normal subgroup of finite index, then $N = \mathrm{SU}_2(V, f)$.*

In fact, (MP) in this case says that $\mathrm{SU}(V, f)$ must be projectively simple, i.e., $\mathrm{SU}(V, f)/Z$ is simple (as an abstract group), where Z is the center of $\mathrm{SU}(V, f)$, and when $\dim(V) \geq 2$ is arbitrary. But, if K is a global field, then by [6] and [9] any noncentral normal subgroup of $\mathrm{SU}(V, f)$ has finite index. Further, we will show in this section that the case $\dim(V) \geq 2$, follows from the case $\dim(V) = 2$, i.e., we will prove the following well known result.

Proposition 5.2. *Suppose that* $\dim(V) \geq 2$. *Assume that for any subspace $V' \subseteq V$ of dimension two, $\mathrm{SU}(V', f')$ is projectively simple (i.e. proper normal subgroups are central), where f' is the restriction of f to $V' \times V'$. Then $\mathrm{SU}(V, f)$ is projectively simple.*

Proposition 5.2 follows easily from the following two lemmas.

Lemma 5.3. *Suppose that B is a finite dimensional simple algebra with center L. Let τ and σ be two involutions of the second kind on B such that K is the subfield of elements of L fixed by both τ and σ. Let $\mathrm{U}(B, \tau) := \{b \in B \mid b\tau(b) = 1\}$ and define $\mathrm{U}(B, \sigma)$ similarly. Then $\{\mathrm{Nrd}_{B/L}(b) \mid b \in \mathrm{U}(B, \tau)\} = \{\mathrm{Nrd}_{B/L}(b') \mid b' \in \mathrm{U}(B, \sigma)\}$.*

Proof. This is an immediate consequence of [7, Proposition 6.1, pg. 261] (see also [5, exercise 12, pg. 202]) which says that

$$\mathrm{Nrd}_{B/L}(\mathrm{U}(B, \tau)) = \{z\iota(z)^{-1} \mid z \in \mathrm{Nrd}_{B/L}(B^{\times})\},$$

where ι is the unique nontrivial element in $\mathrm{Gal}(L/K)$. So we see that the group of norms $\mathrm{Nrd}_{B/L}(\mathrm{U}(B, \tau))$ is independent of the involution τ. □

Lemma 5.4. *Suppose* $\dim(V) \geq 2$ *and let* $G = \mathrm{SU}(V, f)$. *For a subset $S \subseteq V$, let $G_S := \{g \in G \mid g(w) = w, \forall w \in S\}$. Let $\mathcal{H} := \{H \leq G \mid H = G_U$, for some subspace $U \subseteq V$ with* $\dim(U) = \dim(V) - 2\}$. *Then G is generated by $\mathcal{H}$.*

Proof. We must show that $G = \langle \mathcal{H} \rangle$. If $\dim(V) = 2$, there is nothing to prove, so suppose $\dim(V) > 2$. Fix a vector $v \in V$, $v \neq 0$. Of course, $G_v \cong \mathrm{SU}(U, f')$, where $U = v^{\perp} = \{u \in V \mid f(u, v) = 0\}$ and f' is the restriction of f to $U \times U$. We will show that $G = \langle G_v, \mathcal{H} \rangle$, so the lemma will follow by induction on $\dim(V)$. Let $g \in G$. If $g \in G_v$, then $g \in \langle G_v, \mathcal{H} \rangle$, so suppose $g(v) \neq v$ and let $r = r_{w,\alpha} \in \mathrm{U}(V, f)$ be the unique reflection so that $r(g(v)) = v$ (see Lemma 4.3(2)). By Remark 4.5, $\mathrm{Nrd}_{A/L}(r) = \mathrm{Nrd}_{D/L}(\alpha)$, and, in the notation of Proposition 5.3, $\alpha \in \mathrm{U}(D, \tau)$, where τ is as in Remark 4.5. Let $u \in U$, $u \neq 0$, set $t = \langle u, u \rangle$ and let $\sigma\colon D \to D$

be the involution $\sigma(d) = t^{-1}d^{\circ}t$. By Proposition 5.3 there exists $\beta \in \mathrm{U}(D,\sigma)$ such that $\mathrm{Nrd}_{D/L}(\beta) = \mathrm{Nrd}_{D/L}(\alpha)$. Let $s = r_{u,\beta}$. Then, by Remark 4.5 $\mathrm{Nrd}_{A/L}(s) = \mathrm{Nrd}_{A/L}(r)$. Let $X = w^{\perp} \cap u^{\perp}$. Then $s^{-1}r \in G_X \leq \langle \mathcal{H} \rangle$ and $s^{-1}rg \in G_v$. It follows that $g \in \langle G_v, G_X \rangle \leq \langle G_v, \mathcal{H} \rangle$. □

PROOF OF PROPOSITION 5.2: Let $N \leq G = \mathrm{SU}(V,f)$ be a noncentral normal subgroup. By [6] and [9], N has finite index in G. Let $\mathcal{H}$ be as in Lemma 5.4. Then, by the definition of $\mathcal{H}$ and by the hypothesis of Proposition 5.2, each $H \in \mathcal{H}$ is projectively simple. Since $N \cap H$ is a normal subgroup of finite index in H, for all $H \in \mathcal{H}$, it follows that $N \cap H$ is noncentral in H, and hence $H \leq N$. We conclude that $G = \langle \mathcal{H} \rangle \leq N$. □

6. THE CASE WHEN $\dim(V) = 1$

The purpose of this section is to state (MP) in the anisotropic unitary case when $\dim(V) = 1$ in a simple way. We continue with the notation of §5 except that here $\dim(V) = 1$. Let $\{v\}$ be a basis of V and set $s = f(v,v)$. Then it is clear that $\mathrm{U}(V,f) \cong \{d \in D^{\times} \mid d^{\circ}sd = s\}$, and $\mathrm{SU}(V,f) \cong \{d \in \mathrm{U}(V,f) \mid \mathrm{Nrd}_{D/L}(d) = 1\}$. Let $*\colon D \to D$ be the map $a \to s^{-1}a^{\circ}s$. Then by Lemma 3.2, $*$ is an involution of the second kind on D and of course, $\mathrm{U}(V,f) \cong \{d \in D^{\times} \mid d^{*}d = 1\}$. Thus replacing $\circ$ by $*$ if necessary, we may always assume that

$$\mathrm{U}(V,f) \cong U_1(D,\circ) = \{d \in D^{\times} \mid d^{\circ}d = 1\} \text{ and}$$
$$\mathrm{SU}(V,f) \cong \mathrm{SU}_1(D,\circ) = \{d \in \mathrm{U}(D,\circ) \mid \mathrm{Nrd}_{D/L}(d) = 1\}.$$

To state conjecture (MP) for $\mathrm{SU}_1(D,\circ)$ we must recall that a *valuation* of D is a homomorphism $v\colon D^{\times} \to \Gamma$ from $D^{\times}$ onto a totally ordered group Γ such that $v(a+b) \geq \min\{v(a), v(b)\}$, whenever $a + b \neq 0$. Given a valuation v of D and an element $\alpha \in \Gamma$ with $\alpha \geq 0$, let

$$\mathfrak{m}_{D,v}(\alpha) = \{x \in D^{\times} \mid v(x) > \alpha\} \cup \{0\}$$

be the two sided ideal of the valuation ring $O_{D,v} = \{x \in D^{\times} \mid v(x) \geq 0\} \cup \{0\}$. The set of all elements in $O_{D,v}$ which are congruent to 1 modulo this ideal,

$$1 + \mathfrak{m}_{D,v}(\alpha),$$

is a normal subgroup of $D^{\times}$, let us denote it by $N_{v,\alpha}$. We thus have the normal subgroups of $D^{\times}$,

$$N_{\vec{v},\vec{\alpha}} := \bigcap_{i=1}^{m} N_{v_i,\alpha_i} \quad \vec{v} = \{v_1,\ldots,v_m\} \text{ and } \vec{\alpha} = \{\alpha_1,\ldots,\alpha_m\}.$$

Conjecture (MP) says that when K is a global field, the normal subgroups $\mathrm{SU}(D,\circ) \cap N_{\vec{v},\vec{\alpha}}$ are basically *all* the proper noncentral normal subgroups of $\mathrm{SU}(D,\circ)$.

Conjecture 6.1 ([(MP) for anisotropic $\mathrm{SU}(V,f)$, $\dim(V) = 1$). *Let D be a finite dimensional division algebra with center the global field L. Let $\circ\colon D \to D$ be an involution of the second kind. Let $N < \mathrm{SU}(D,\circ)$ be a proper noncentral normal subgroup (necessarily of finite index). Then there are valuations $v_i\colon D^{\times} \to \Gamma_i$ and nonnegative elements $\alpha_i \in \Gamma_i$, $1 \leq i \leq m$, such that $N \supseteq \mathrm{SU}(D,\circ) \cap N_{\vec{v},\vec{\alpha}}$.*

References

[1] A. S. Amitsur, *Rational identities and applications to algebra and geometry,* J. Algebra **3**(1966), 304–359.

[2] V. Bergelson, D. B. Shapiro, *Multiplicative subgroups of finite index in a ring,* Proc. AMS **116**(1992), 885–896.

[3] P.M. Cohn, *Skew fields*, Encyclopedia of Mathematics and its applications Vol. 57, Cambridge Univ. press, 1995.

[4] L. Hua, *On the multiplicative group of a field,* Acad. Sinica Sience, **3**(1950), 1–6.

[5] M-A. Knus, A. Merkurjev, M. Rost, J-P. Tignol, *The book of involutions*, AMS Coll. Pub. Vol. 44(1998).

[6] G. A. Margulis, *Finiteness of quotients of discrete groups,* Funct. Analysis and Appl. **13**(1979), 178–187.

[7] A. S. Merkurjev, *The norm principle for algebraic groups,* Algebra i Analiz **7**(1995), no. 2, 77–105, English transl.: St. Petersburg Math. J. **7**(1996), no. 2, 243-264.

[8] V. P. Platonov, A.S. Rapinchuk, *Algebraic Groups and Number Theory,* "Pure and Applied Mathematics" series, N 139, Academic Press, 1993.

[9] G. Prasad, *Strong approximation for semi-simple groups over function fields,* Ann. Math. **105**(1977), 553–572.

[10] M. S. Raghunathan, *On the group of norm 1 elements in a division algebra,* Math. Ann. **279**(1988), 457–484.

[11] A. S. Rapinchuk, Y. Segev, *Valuation-like maps and the congruence subgroup property,* Invent. Math. **144**(2001), 571–607.

[12] A. Rapinchuk, Y. Segev, G. Seitz, *Finite quotients of the multiplicative group of a finite dimensional division algebra are solvable*, Journal of the AMS, to appear.

[13] W. Scharlau, *Quadratic and hermitian forms*, Springer-Verlag 1985.

[14] W. R. Scott, *Group theory,* Prentice-Hall, Inc., Englewood Cliffs, N.J. 1964.

[15] Y. Segev, *On finite homomorphic images of the multiplicative group of a division algebra,* Ann. Math. **149**(1999), 219-251.

[16] ——— *The commuting graph of minimal nonsolvable groups*, Geom. Ded. **88**(2001), 55–66.

[17] Y. Segev, G. M. Seitz, *Anisotropic groups of type A_n and the commuting graph of finite simple groups,* Pacific J. Math. **202**(2002), 125-226.

[18] C. J. Stuth, *A generalization of the Cartan-Brauer-Hua theorem,* Proc. AMS **15**(1964), 211–217.

[19] G. Turnwald, *Multiplicative subgroups of finite index in rings,* Proc. AMS **120**(1994), 377-381.

Applications of random generation to residual properties of some infinite groups

Aner Shalev

1 Results

The purpose of this paper is to survey recent results on random generation of finite simple groups, and to indicate their use in the study of residual properties of some infinite groups. These applications form yet another contribution of the probabilistic approach to some problems in abstract groups. While the first application we outline provides a new and shorter proof of an already proven result (Magnus conjecture), most of the applications we describe constitute new results in residual properties, and it is unclear whether they can also be established without a probabilistic approach.

By a finite simple group we mean a nonabelian finite simple group. We assume below the classification of finite simple groups. Thus, to prove an asymptotic statement for finite simple groups one needs to check it for alternating groups and for the finite groups of Lie type. Recall that a group G is said to be residually C (where C is some collection of groups) if the intersection of the normal subgroups $N \lhd G$ such that $G/N \in C$ is trivial. The infinite groups whose residual properties we shall investigate are free groups, the modular group, as well as general free products of finite groups. It remains to be seen whether our methods can be applied for more general groups appearing in combinatorial group theory, such as one relator groups, free products with amalgamation, hyperbolic groups, etc.

In this section we present three 'pairs' of results: the first result in each pair deals with random generation, and the second result is a corresponding application to residual properties. These applications are by no means straightforward corollaries. Hints of their proof appear in the second section.

The starting point for our discussion is the following theorem, proved by Dixon [D] for alternating groups, by Kantor and Lubotzky [KL] for classical groups and some low rank exceptional groups, and by Liebeck and myself [LiSh1] for the remaining exceptional groups.

Theorem 1 *Let S be a finite simple group, and let $x, y \in S$ be two randomly chosen elements. Then the probability that x, y generate S tends to 1 as $|S| \to \infty$.*

This verifies a conjecture of Dixon [D] from 1969. We note that this statement for alternating groups was already conjectured by Netto in the 19th century. This result has various interesting applications; the one which is relevant for us is the following.

Theorem 2 *Let C be an infinite set of finite simple groups. Then the free group F_2 is residually C.*

Here F_k denotes the free group on k generators. The result for F_2 extends for F_k for all $k > 1$, since F_k is known to be residually $\{F_2\}$. Theorem 2, which was conjectured by Magnus, has a long history, and partial proofs were given by Katz and Magnus [KM], Gorchakov and Levchuk [GL], Lubotzky [Lu], Wiegold [Wi], Wilson [W] and others. Theorem 2 was finally proved by Weigel in a series of papers [We1, We2, We3]. The new proof, by Dixon, Pyber Seress and myself [DPSSh], applies Theorem 1 to obtain a much shorter proof of a somewhat stronger result (see Section 2 for more details).

Our next results on random generation deal with generators of specific orders.

Theorem 3 *Let S be a finite simple group, and suppose $S \neq PSp_4(q), Sz(q)$. Let $x, y \in S$ be randomly chosen elements of orders $2, 3$ respectively. Then the probability that x, y generate S tends to 1 as $|S| \to \infty$.*

This result for classical and alternating groups is proved in [LiSh2] (we note that the alternating case follows from a more general yet unpublished result of Müller and Pyber). More recently Guralnick and myself proved it for exceptional groups of Lie type [GSh], based on the fact that these groups are $(2,3)$-generated [LM]. We note that if $S = PSp_4(q)$ where q is not a power of 2 or 3, then the probability that randomly chosen elements of orders $2, 3$ generate S tends to $1/2$ [LiSh2], and this result is also useful for us here.

The main application of Theorem 3 is the determination of the finite simple quotients of the modular group $PSL_2(\mathbb{Z})$ up to finitely many exceptions, a project which started already a hundred years ago. In a yet unpublished paper [LiSh4] Liebeck and myself apply this theorem in the study of residual properties of the modular group.

Theorem 4 *Let C be an infinite set of finite simple groups not containing $PSp_4(q)$ (q a power of 2 or 3) or $Sz(q)$. Then $PSL_2(\mathbb{Z})$ is residually C.*

The exceptions in the theorem are genuine, since $PSp_4(2^k), PSp_4(3^k)$ and $Sz(q)$ are not images of the modular group. The case of alternating groups already follows from a more general result of Tamburini and Wilson [TW] (see below), so the main novelty of Theorem 4 is when C consists of groups of Lie type.

Our next result deals with random (r, s)-generation, namely random generation by elements of arbitrary prime orders r, s. Of course the case $r = s = 2$ has to be excluded, since the group generated is dihedral. The fact that alternating groups are randomly (r, s)-generated is proved in [LiSh2]. The case of classical groups has just been established in [LiSh3].

Theorem 5 *Let r, s be prime numbers, not both 2. Then there exists a number $f(r, s)$ such that if S is a finite simple classical group in dimension at least $f(r, s)$, and $x, y \in S$ are randomly chosen elements of orders r, s respectively, then the probability that x, y generate S tends to 1 as $|S| \to \infty$.*

Note that some assumption on S is needed, since S might not contain elements of orders r, s. It would be interesting to find out whether the conclusion of Theorem 5 holds assuming only that S has such elements (instead of the assumption of large rank). Here the case of exceptional groups is also of interest.

Theorem 5 can be used to derive some random (A, B)-generation results, where A, B are finite groups. This means generation by a random copy of A and a random copy of B in S. For alternating groups a result of this type was conjectured by Lubotzky and proved by Müller and Pyber (yet unpublished). Some versions for classical groups are obtained in [LiSh4] using Theorem 5. The idea is that, if A, B are non-trivial and not both 2-groups, then we may pick elements $a \in A$ and $b \in B$ of prime orders r, s which are not both 2, and then deduce some kind of random (A, B)-generation of S using the random (r, s)-generation of S. This approach does not allow us to deal with the case where A, B are both 2-groups (not both of order 2). The core of the problem is where $A = C_2, B = C_2 \times C_2$. It would be interesting to find out which families of finite simple groups of Lie type are randomly $(C_2, C_2 \times C_2)$ generated.

The results mentioned above, and additional tools, enable us to establish new residual properties of free products $A * B$ of finite groups A, B. Tamburini and Wilson showed that, if A, B are non-trivial finite groups, not both of order 2, and C is an infinite collection of alternating groups, then the free product $A * B$ is residually C [TW]. The case where C consists of finite simple classical groups can be solved using a probabilistic argument, with some extra-assumption on the finite groups A, B.

Theorem 6 *Let A, B be non-trivial finite groups, not both 2-groups. Then there exists a number $f(|A|, |B|)$ depending only on $|A|$ and $|B|$ such that, for every infinite set C of finite simple classical groups of rank at least $f(|A|, |B|)$, the free product $A * B$ is residually C.*

This theorem is proved in [LiSh4] when the ranks of the groups in C are unbounded, and in [LiSh5] for the case of bounded (large) rank. A random $(C_2, C_2 \times C_2)$-generation result for classical group of large rank would enable us to allow A, B to be both 2-groups (not both of order 2).

2 Hints of proofs

The probabilistic method, which was applied in many branches of mathematics, enables one to prove existence theorems in non-constructive ways. Instead of constructing an object with the desired properties one shows, using counting arguments or more general probability measures, that most objects (in some relevant space) have these properties, and therefore such an object exists.

The possible relevance of this approach to residual properties stems from the observation that a residual property is an existence statement. Indeed, to say that G is residually C amounts to saying that, given a non-identity element $g \in G$, there exists an epimorphism ϕ from G to some group $S \in C$ such that $\phi(g) \neq 1$. In the traditional approach to residual properties one tries to construct explicitly such a homomorphism ϕ, which in many cases proves very difficult. The hope is that, in some of these cases, the existence of ϕ could be established probabilistically.

Suppose G is finitely generated, and that the groups in the collection C are all finite. Fix an element $g \in G$ with $g \neq 1$. For a group $S \in C$ consider the space $X = Hom(G, S)$ of all homomorphisms from G to S. Then X is finite, and can be viewed as a probability space, equipped with the uniform distribution. Our aim is to show that, for a randomly chosen $\phi \in X$ we have

1. $\mathrm{Prob}(\phi \text{ is onto and } \phi(g) \neq 1) \to 1$ as $|S| \to \infty$.

This would imply that, for a large enough $S \in C$, the required homomorphism $\phi : G \to S$ exists, and so G is residually C.

Now, the task above can be naturally devided into two parts.

2. Show that $\mathrm{Prob}(\phi \text{ is onto}) \to 1$ as $|S| \to \infty$.

3. Show that $\mathrm{Prob}(\phi(g) \neq 1) \to 1$ as $|S| \to \infty$.

As we shall see Task 2 is related to random generation, and using existing random generation results (formulated in the previous section) we shall sometimes get it for free. In these cases carrying out Task 3 would suffice in order to establish that G is residually C.

Let us now demonstrate this method in specific situations. The simplest case is that of $G = F_2$. Fix free generators a, b for G. Then $X = Hom(G, S)$ can be identified with $S \times S$, by attaching to $\phi \in X$ the pair $(x, y) \in S \times S$, where $x = \phi(a)$ and $y = \phi(b)$.

Then ϕ is onto S if and only if x, y generate S, and the probability that this happens tends to 1 by Theorem 1. Therefore Task 2 is carried out.

For Task 3 we may write $g = w(a, b)$, a non-identity word in the free generators. Then $\phi(g) = w(x, y)$, and our aim is to show that, when $x, y \in S$ are chosen at random,

4. $\mathrm{Prob}(w(x, y) \neq 1) \to 1$ as $|S| \to \infty$.

For alternating groups and classical groups of unbounded rank this can be shown using some combinatorial arguments. For simple groups of Lie type of bounded rank we show this using algebraic geometry (counting q-rational points). See [DPSSh] for more details.

We note that statement 4 has some additional applications. For example, it immediately implies a result of Jones [J] that an infinite collection of finite simple groups generates the variety of all groups. Applying it for the power word $w = a^n$ yields some Burnside-type applications, as shown by Mann and Martinez [MM].

Let us now turn to the modular group $G = PSL_2(\mathbb{Z})$. It is well known that $G \cong C_2 * C_3$, a free product of groups of orders 2 and 3. Let a, b be canonical generators for G of orders 2 and 3 respectively. For a positive integer k let $I_k(S)$ be the set of elements $s \in S$ satisfying $s^k = 1$.

Then $X = Hom(G, S)$ can be identified with $I_2(S) \times I_3(S)$, attaching to $\phi \in X$ the pair $(x, y) \in I_2(S) \times I_3(S)$ such that $x = \phi(a)$ and $y = \phi(b)$. It is clear that ϕ is onto if and only if x, y generate S. The probability that this happens is essentially the $(2,3)$-generation probability, which in most cases tends to 1 by Theorem 3 (I say essentially, since here we also allow x or y to be 1, but this does not really matter). In these cases Task 2 is carried out, and we can focus on Task 3.

Again we may write $g = w(a,b)$, a non-identity canonical word in the free product. Then $\phi(g) = w(x,y)$, and we have to show that

5. $\mathrm{Prob}(w(x,y) \neq 1 \mid x^2 = y^3 = 1) \to 1$ as $|S| \to \infty$.

For technical reasons this is shown in a slight variation: instead of letting (x,y) range over $I_2(S) \times I_3(S)$, we let them range over $D \times E$, where D, E are certain large conjugacy classes in S of elements of orders 2 and 3 respectively. Random generation results where x and y are chosen from such classes are available, and by applying them essentially the same argument works.

Note that a priori it is not even clear that the probability in 5 is non-zero when $S \in C$ is large enough. Indeed, perhaps for some infinite series of finite simple groups S, any elements $x, y \in S$ of order 2 and 3 respectively also satisfy some extra relation $w(x,y) = 1$ which does not follow from the relations $x^2 = y^3 = 1$. We show that (for $S \neq Sz(q)$) this is not the case, by passing to algebraic groups and constructing a subgroup of type $PSL_2(\mathbb{Z})$ there, using Bass-Serre theory of groups acting on trees. The general proof of statement 5 is rather long, and applies combinatorial arguments, algebraic groups, and algebraic geometry.

Let us now examine the case of a general free product $G = A * B$, where A, B are non-trivial finite groups, not both 2-groups. To sketch the proof of Theorem 6 let S be a classical group of large rank (given the groups A, B). The assumption on the rank ensures that S contains copies of A, B, and we choose such copies which satisfy some technical extra assumption. This defines injections $f : A \to S$, $g : B \to S$, and therefore a homomorphism $\phi = f * g : G = A * B \to S$. Consider a twist ϕ_t of that homomorphism, where $t \in S$, $\phi_t = f * (g^t)$, and $g^t(b) = t^{-1}g(b)t$. The proof then proceeds by showing that, for a fixed non-identity element g in the free product $A * B$, and for a randomly chosen element $t \in S$, we have

6. $\mathrm{Prob}(\phi_t \text{ is onto}) \to 1$ as $|S| \to \infty$.

7. $\mathrm{Prob}(\phi_t(g) \neq 1) \to 1$ as $|S| \to \infty$.

Statement 6 amounts to saying that, viewing A and B in their embeddings in S, the probability that A and B^t generate S tends to 1. This is indeed the random (A,B)-generation result proved in [LiSh4] using Theorem 5 as a main tool. The proof of statement 7 is rather long and technical, and will not be described here.

Finally, let us note that some of the ideas and methods outlined in this paper can be applied in the context of profinite groups. Recall that a profinite group G (and its Cartesian powers G^k) can be viewed as a probability space with respect to the normalized Haar measure. Here too results on random generation serve as a useful tool. A profinite group G is said to be positively finitely generated (PFG) if for some k the measure $P(G,k)$ of the set of k-tuples generating G is positive in G^k. Various groups were shown to be PFG by Kantor and Lubotzky [KL], Mann [Ma] and others. The most general result of this type, which appears in [BPSh], is the following.

Theorem 7 *Let G be a finitely generated profinite group, d a positive integer, and suppose the alternating group A_d does not occur as a section H/K, where H, K are open subgroups of G and $K \lhd H$. Then G is positively finitely generated.*

This enables us to prove the following [DPSSh].

Corollary 8 *Let G be the profinite completion of $SL_d(\mathbb{Z})$, $d \geq 3$. Then G has a dense free subgroup of finite rank.*

Let us sketch the proof. The group G has arbitrarily large nonabelian simple quotients S modulo open subgroups. Now, if $w(a_1, \ldots, a_k)$ is a non-identity element of the free group F_k on $a_1, \ldots, a_k$, then for randomly chosen elements $x_1, \ldots, x_k \in S$, the probability that $w(x_1, \ldots, x_k) \neq 1$ tends to 1. Indeed, for $k = 2$ this is statement 4 above, and our method actually establishes it for any k. Using this it easily follows that a random k-tuple of elements of G generates a discrete subgroup isomorphic to F_k.

On the other hand, by Theorem 7, G is PFG (this special case is already obtained in [Ma]). Choose a number k (depending on d) such that $P(G, k) > 0$. It follows that the measure of k-tuples in G^k generating a dense F_k subgroup is positive, and so at least one such k-tuple exists.

It turns out that, combining Theorem 7 and other tools, one can obtain the following analogue of the well known Tits alternative for linear groups [DPSSh].

Theorem 9 *Let Γ be a finitely generated group which is linear over some field, and G its profinite completion. Then either Γ is virtually soluble, or G has an open subgroup G_0 having a dense free subgroup of finite rank.*

Pyber conjectures that we may actually take $G_0 = G$.

References

[BPSh] A.V. Borovik, L. Pyber, A. Shalev, Maximal subgroups in finite and profinite groups, *Trans. Amer. Math. Soc.* **348** (1996), 3745–3761.

[D] J.D. Dixon, The probability of generating the symmetric group, *Math. Z.* **110** (1969), 199–205.

[DPSSh] J.D. Dixon, L. Pyber, Á. Seress, A. Shalev, Residual properties of free groups and probabilistic methods, to appear in *J. reine angew. Math.*

[GL] Yu.M. Gorchakov and V.M. Levchuk, On approximation of free groups, *Algebra i Logika* **9** (1970), 415–421.

[GSh] R. Guralnick and A. Shalev, Zero-one laws for finite Chevalley groups, in preparation.

[J] G.A. Jones, Varieties and simple groups, *J. Austr. Math. Soc.* **17** (1974), 163–173.

[KL] W.M. Kantor, A. Lubotzky, The probability of generating a finite classical group, *Geom. Ded.* **36** (1990), 67–87.

[KM] R. Katz and W. Magnus, Residual properties of free groups, *Comm. Pure Appl. Math.* **22** (1969), 1–13.

[LiSh1] M.W. Liebeck, A. Shalev, The probability of generating a finite simple group, *Geom. Ded.* **56** (1995), 103–113.

[LiSh2] M.W. Liebeck, A. Shalev, Classical groups, probabilistic methods, and the $(2,3)$-generation problem, *Annals of Math.* **144** (1996), 77–125.

[LiSh3] M.W. Liebeck and A. Shalev, Random (r,s)-generation of finite classical groups, *Bull. London Math. Soc.* **34** (2002), 185–188.

[LiSh4] M.W. Liebeck and A. Shalev, Residual properties of the modular group and other free products, submitted.

[LiSh5] M.W. Liebeck and A. Shalev, Residual properties of free products of finite groups, submitted.

[LM] F. Lübeck and G. Malle, $(2,3)$-generation of exceptional groups, *J. London Math. Soc. (2)* **59** (1999), 109–122.

[Lu] A. Lubotzky, On a problem of Magnus, *Proc. Amer. Math. Soc.* **98** (1986), 583–585.

[Ma] A. Mann, Positively finitely generated groups, *Forum Math.* **8** (1996), 429–459.

[MM] A. Mann, C. Martinez, The exponent of finite groups, *Arch. Math.* **67** (1996), 8–10.

[TW] M.C. Tamburini, J.S. Wilson, A residual property of free products, *Math. Z.* **186** (1984), 525–530.

[We1] T.S. Weigel, Residual properties of free groups, *J. Algebra* **160** (1993), 16–41.

[We2] T.S. Weigel, Residual properties of free groups II, *Comm. Algebra* **20** (1992), 1395–1425.

[We3] T.S. Weigel, Residual properties of free groups III, *Israel J. Math.* **77** (1992), 65–81.

[Wi] J. Wiegold, Free groups are residually alternating of even degree, *Arch. Math. (Basel)* **28** (1977), 337–339.

[W] J.S. Wilson, A residual property of free groups, *J. Algebra* **138** (1991), 36–47.

LOW DIMENSIONAL REPRESENTATIONS OF FINITE QUASISIMPLE GROUPS

PHAM HUU TIEP

1. Introduction

Let G be a finite quasisimple group and $\mathbb{F}$ an algebraically closed field of characteristic ℓ. In many applications it is very important to know the *smallest degree* $\mathfrak{d}_\ell^1(G)$ of nontrivial irreducible $\mathbb{F}G$-representations. In this article, we will survey recent results on the following problem:

Problem 1.1. *Given a finite quasisimple group G and ℓ, determine $\mathfrak{d}_\ell^1(G)$ and all nontrivial irreducible $\mathbb{F}G$-representations of degree $\mathfrak{d}_\ell^1(G)$.*

It is usually the case that G has a few irreducible representations of degree $\mathfrak{d}_\ell^1(G)$, $\mathfrak{d}_\ell^1(G)+1$, and maybe $\mathfrak{d}_\ell^1(G)+2$, and then there is a relatively big *gap* between these degrees and the next degree. In various applications it is important to know this gap. Let us denote by $\mathfrak{d}_\ell^2(G)$ the next degree of nontrivial irreducible $\mathbb{F}G$-representations after $2\mathfrak{d}_\ell^1(G)$.

Problem 1.2. *Given a finite quasisimple group G and ℓ, determine $\mathfrak{d}_\ell^2(G)$.*

It turns out that Problem 1.2 is closely related to

Problem 1.3. *Given a finite quasisimple group G, ℓ, and $\varepsilon > 0$, classify all irreducible $\mathbb{F}G$-modules of dimension less than $(\mathfrak{d}_\ell^1(G))^{2-\varepsilon}$.*

We note that a solution for the case $\varepsilon = 1/2$ of Problem 1.3 would be good enough for a number of applications that we have in mind.

2. Bound and gap results for symmetric groups and alternating groups

First we briefly mention the results on Problems 1.1 and 1.2 for covering groups of the alternating groups $\mathbb{A}_n$. It is convenient to consider representations of covering groups G of $\mathbb{A}_n$ together with those of the symmetric groups $\mathbb{S}_n$.

In the case of linear representations of symmetric and alternating groups, Problems 1.1 and 1.2 have been solved by Wagner, resp. by James. It is well known that $\mathfrak{d}_\mathbb{C}^1(\mathbb{S}_n) = \mathfrak{d}_\mathbb{C}^1(\mathbb{A}_n) = n-1$, and $\mathbb{A}_n$, resp. $\mathbb{S}_n$, has exactly 1, resp. 2, irreducible complex representation of degree $n-1$, provided that $n \geq 5$. Let $\kappa_{n,\ell}$ equal 1 if $\ell | n$ and 0 otherwise.

Theorem 2.1. [Wag1, Wag2] *Assume $n \geq 9$. Then $\mathfrak{d}_\ell^1(\mathbb{S}_n) = \mathfrak{d}_\ell^1(\mathbb{A}_n) = n-1-\kappa_{n,\ell}$. Moreover, the representations of smallest degree can be obtained by reducing the smallest complex representations modulo ℓ.*

The author gratefully acknowledges the support of the NSF (grant DMS-0070647).

Theorem 2.2. [J] *If* $n \geq 15$ *then* $\mathfrak{d}_\ell^2(\mathbb{S}_n) = \mathfrak{d}_\ell^2(\mathbb{A}_n) \geq n(n-5)/2$.

Actually, [J] gives an exact formula for $\mathfrak{d}_\ell^2(\mathbb{S}_n)$ which depends on ℓ. If $n \leq 14$ one can also determine $\mathfrak{d}_\ell^2(\mathbb{S}_n)$ and $\mathfrak{d}_\ell^2(\mathbb{A}_n)$ using [JLPW] and various results on decomposition matrices for $\mathbb{S}_n$ and $\mathbb{A}_n$.

Next, we consider the *spin representations*, i.e. faithful representations of G, where G is either a double cover $2\mathbb{S}_n = 2^+\mathbb{S}_n$ or $2^-\mathbb{S}_n$ of $\mathbb{S}_n$, or the double cover $2\mathbb{A}_n$ of $\mathbb{A}_n$. Certainly, $\ell \neq 2$ in this case. From Schur's classification of complex spin representations of $2\mathbb{S}_n$ and $2\mathbb{A}_n$, it follows that $\mathfrak{d}_{\mathbb{C}}^1(2\mathbb{S}_n) = 2^{[(n-1)/2]}$ and $\mathfrak{d}_{\mathbb{C}}^1(2\mathbb{A}_n) = 2^{[(n-2)/2]}$. Moreover, $2\mathbb{S}_n$, resp. $2\mathbb{A}_n$, has one or two faithful complex representations of degree equal to $\mathfrak{d}_{\mathbb{C}}^1(2\mathbb{S}_n)$, resp. $\mathfrak{d}_{\mathbb{C}}^1(2\mathbb{A}_n)$ – those are the so-called *basic spin representations* and correspond to the partition (n) in Schur's classification. By definition, a *basic spin representation* in characteristic ℓ is any irreducible constituent of the reduction modulo ℓ of a complex basic spin representation. In the modular case, a lower bound for $\mathfrak{d}_\ell^2(2\mathbb{S}_n)$ and $\mathfrak{d}_\ell^2(2\mathbb{A}_n)$ was obtained by Wagner in [Wag3], where he showed that $\mathfrak{d}_\ell^1(2\mathbb{S}_n) \geq 2^{[(n-s)/2]}$ and $\mathfrak{d}_\ell^1(2\mathbb{A}_n) \geq 2^{[(n-s-1)/2]}$, if s is the number of nonzero terms in the 2-adic decomposition of n and $n \geq 9$. A precise formula for $\mathfrak{d}_\ell^1(2\mathbb{S}_n)$ and $\mathfrak{d}_\ell^1(2\mathbb{A}_n)$ has recently been found by Kleshchev and the author [KT]. They have also established a lower bound for $\mathfrak{d}_\ell^2(2\mathbb{S}_n)$ and $\mathfrak{d}_\ell^2(2\mathbb{A}_n)$.

Theorem 2.3. [KT] *Assume* $n \geq 8$.

(i) $\mathfrak{d}_\ell^1(2\mathbb{S}_n) = 2^{[(n-1-\kappa_{n,\ell})/2]}$ *and* $\mathfrak{d}_\ell^1(2\mathbb{A}_n) = 2^{[(n-2-\kappa_{n,\ell})/2]}$.

(ii) *Let* $H = 2\mathbb{S}_n$ *or* $2\mathbb{A}_n$, *and let* V *be an irreducible faithful* $\mathbb{F}H$*-module of dimension less than* $2\mathfrak{d}_\ell^1(H)$. *Then* $\dim V = \mathfrak{d}_\ell^1(H)$, *and* V *is a basic spin module.*

The proof relies particularly on the following characterization of (modular) basic spin representations of $2\mathbb{S}_n$ and $2\mathbb{A}_n$, obtained by Meierfrankenfeld [Me] and Wales [Wa]. See also section §5.

Proposition 2.4. *Let* $n \geq 5$ *and let* V *be an irreducible* $2\mathbb{A}_n$*-module on which (an inverse image) of a 3-cycle has a quadratic minimal polynomial. Then* V *is a basic spin module.*

It is likely that $\mathfrak{d}_\ell^2(H) = \mathfrak{d}_\ell^1(H) \cdot O(n)$ for $H = 2\mathbb{S}_n$ or $2\mathbb{A}_n$.

Formulae for $\mathfrak{d}_\ell^1(G)$ and $\mathfrak{d}_\ell^2(G)$ with G being a covering group of most of the 26 sporadic finite simple groups are available in [JLPW] or GAP. The remaining cases will probably be settled soon, once enough information about decomposition matrices becomes available. In fact $\mathfrak{d}_\ell^1(G)$ have been determined for all covers G of sporadic groups, cf. [?]. We also mention the paper [HM2] of Hiss and Malle, where representations of finite quasisimple groups of degree up to 250 were classified. Keeping this in mind, we will focus on the case of finite quasisimple groups of Lie type.

3. Bound results for finite groups of Lie type

From now on to the end of the paper, we assume that G is a finite (quasisimple) group of Lie type, of simply connected type, in characteristic p. Representations of G in the defining characteristic p of small degree are investigated in [KL] and more recently in [Lu2]. We will concentrate on the cross characteristic case and assume that $\mathbb{F}$ is an algebraically closed field of characteristic $\ell \neq p$. We will also assume that G is none of the following groups: $SL_2(q)$, $SL_3(q)$ with $q = 2, 4$, $SL_4(q)$ with $q = 2, 3$, $Sp_4(2)'$, $SU_4(q)$ with $q = 2, 3$, $\Omega_8^+(2)$, $\Omega_7(3)$, $G_2(q)$ with $q = 3, 4$, ${}^2B_2(8)$, $F_4(2)$, ${}^2F_4(2)'$, ${}^2E_6(2)$. Information about irreducible representations of the groups in this list can be found in [Atlas, JLPW]. Our assumptions imply in particular that G is the universal cover of the simple group $G/Z(G)$.

Lower bounds for the degree of irreducible representations of finite groups of Lie type in cross characteristic were found by Landazuri and Seitz [LS] and improved later by Seitz and Zalesskii [SZ]. These bounds, $\mathfrak{b}_{LSZ}(G)$, have proved to be very useful in a vast number of applications. For the reader's convenience, we reproduce $\mathfrak{b}_{LSZ}(G)$ in Table I, where ϕ_n stands for the nth cyclotomic polynomial in q.

TABLE I. The Landazuri-Seitz-Zalesskii bounds for $\mathfrak{d}_\ell^1(G)$, and $\mathfrak{d}_\mathbb{C}^1(G)$

G	$\mathfrak{b}_{LSZ}(G)$	$\mathfrak{d}_\mathbb{C}^1(G)$
$SL_n(q)$	$\frac{q^n-q}{q-1}\mathbf{-n+1}$	$\frac{q^n-q}{q-1}$
$Sp_{2n}(q),\ 2 \nmid q$	$(q^n-1)/2$	$(q^n-1)/2$
$Sp_{2n}(q),\ 2 \mid q$	$\frac{(q^n-1)(q^n-q)}{2(q+1)}$	$\frac{(q^n-1)(q^n-q)}{2(q+1)}$
$SU_n(q)$	$\left[\frac{q^n-1}{q+1}\right]$	$\left[\frac{q^n-1}{q+1}\right]$
$Spin_{2n}^+(q),\ q>3$	$\frac{(q^n-1)(q^{n-1}+q)}{q^2-1}\mathbf{-n}$	$\frac{(q^n-1)(q^{n-1}+q)}{q^2-1}$
$Spin_{2n}^+(q),\ q=3$	$\frac{(q^n-1)(q^{n-1}-1)}{q^2-1}$	$\frac{(q^n-1)(q^{n-1}-1)}{q^2-1}$
$Spin_{2n}^+(q),\ q=2$	$\frac{(q^n-1)(q^{n-1}-1)}{q^2-1}\mathbf{-7}$	$\frac{(q^n-1)(q^{n-1}-1)}{q^2-1}$
$Spin_{2n}^-(q)$	$\frac{(q^n+1)(q^{n-1}-q)}{q^2-1}\mathbf{-n+2}$	$\frac{(q^n+1)(q^{n-1}-q)}{q^2-1}$
$Spin_{2n+1}(q),\ q>3$	$\frac{q^{2n}-1}{q^2-1}\mathbf{-n}$	$\frac{q^{2n}-1}{q^2-1}$
$Spin_{2n+1}(q),\ q=3$	$\frac{(q^n-1)(q^n-q)}{q^2-1}$	$\frac{(q^n-1)(q^n-q)}{q^2-1}$
${}^2B_2(q)$	$(q-1)\sqrt{q/2}$	$(q-1)\sqrt{q/2}$
${}^2G_2(q)$	$q(q-1)$	q^2-q+1
$G_2(q),\ q \equiv \epsilon = \pm1 (\bmod 3)$	$q^3\mathbf{-q}$	$q^3+\epsilon$
$G_2(q),\ q \equiv 0 (\bmod 3)$	$\mathbf{q^3-q}$	q^4+q^2+1
${}^3D_4(q)$	$q^5-q^3\mathbf{+0}$	q^5-q^3+q
${}^2F_4(q)$	$(q^5\mathbf{-q^4})\sqrt{q/2}$	$(q^3+1)(q^2-1)\sqrt{q/2}$
$F_4(q),\ 2 \nmid q$	$q^8\mathbf{-q^6}$	q^8-q^4+1
$F_4(q),\ 2 \mid q$	$(q^3-1)(\mathbf{q^8-q^7})/2$	$(q^3-1)^2(q^5+q)/2$
${}^2E_6(q)$	$q^{11}\mathbf{-q^9}$	$(q^5+q)(q^6-q^3+1)$
$E_6(q)$	$q^{11}\mathbf{-q^9}$	$(q^5+q)(q^6+q^3+1)$
$E_7(q)$	$q^{17}\mathbf{-q^{15}}$	$q\phi_7\phi_{12}\phi_{14}$
$E_8(q)$	$q^{29}\mathbf{-q^{27}}$	$q\phi_4^2\phi_8\phi_{12}\phi_{20}\phi_{24}$

In the particular case of complex representations, that is where $\ell = 0$, one can find the precise values of $\mathfrak{d}^1_{\mathbb{C}}(G)$ (and $\mathfrak{d}^2_{\mathbb{C}}(G)$) using the Deligne-Lusztig theory. This has been done by Zalesskii and the author [TZ1] for classical groups, and by Lübeck [Lu1] for exceptional groups, and the values of $\mathfrak{d}^1_{\mathbb{C}}(G)$ are displayed in Table I.

Certainly, $\mathfrak{b}_{LSZ}(G) \leq \mathfrak{d}^1_\ell(G) \leq \mathfrak{d}^1_{\mathbb{C}}(G)$, and $\mathfrak{d}^1_{\mathbb{C}}(G)$ should be a good approximation for $\mathfrak{d}^1_\ell(G)$. Moreover, empirical data seem to imply that the inequality $\mathfrak{d}^1_{\mathbb{C}}(G) - 2 \leq \mathfrak{d}^1_\ell(G) \leq \mathfrak{d}^1_{\mathbb{C}}(G)$ holds in all cases.

Comparing $\mathfrak{b}_{LSZ}(G)$ with $\mathfrak{d}^1_{\mathbb{C}}(G)$, one sees that the Landazuri-Seitz-Zalesskii bounds are very good, and in fact they are the best possible ones in many cases. In several cases (the lines in Table I with a part of $\mathfrak{b}_{LSZ}(G)$ printed boldface), the difference between $\mathfrak{d}^1_{\mathbb{C}}(G)$ and $\mathfrak{b}_{LSZ}(G)$ is still a polynomial of q or of the rank of G. It turns out that one can still improve the bound in these cases. We record the improvements on the Landazuri-Seitz-Zalesskii bounds in Table II.

TABLE II. Improvements on the Landazuri-Seitz-Zalesskii bounds

G	Improved bound for $\mathfrak{d}^1_\ell(G)$	$\mathfrak{d}^1_{\mathbb{C}}(G)$	cf.
$SL_n(q)$	$\frac{q^n - q}{q-1} - 1$	$\frac{q^n - q}{q-1}$	[GPPS]
$Spin^+_{2n}(q),\ q > 3$	$\frac{(q^n-1)(q^{n-1}+q)}{q^2-1} - 2$	$\frac{(q^n-1)(q^{n-1}+q)}{q^2-1}$	[Hof1]
$Spin^+_{2n}(q),\ q \leq 3$	$\frac{(q^n-1)(q^{n-1}-1)}{q^2-1}$	$\frac{(q^n-1)(q^{n-1}-1)}{q^2-1}$	[Hof1]
$Spin^-_{2n}(q)$	$\frac{(q^n+1)(q^{n-1}-q)}{q^2-1} - 1$	$\frac{(q^n+1)(q^{n-1}-q)}{q^2-1}$	[Hof1]
$Spin_{2n+1}(q),\ q > 3$	$\frac{q^{2n}-1}{q^2-1} - 2$	$\frac{q^{2n}-1}{q^2-1}$	[Hof1]
$Spin_{2n+1}(q),\ q = 3$	$\frac{(q^n-1)(q^n-q)}{q^2-1}$	$\frac{(q^n-1)(q^n-q)}{q^2-1}$	[Hof1]
$G_2(q),\ q \equiv 1 (\mathrm{mod}\ 3)$	q^3	q^3+1	[Hiss2]
$G_2(q),\ q \equiv 2 (\mathrm{mod}\ 3)$	q^3-1	q^3-1	[Hiss2]
$G_2(q),\ q \equiv 0 (\mathrm{mod}\ 3)$	q^4+q^2	q^4+q^2+1	[Hiss2]
${}^3D_4(q)$	q^5-q^3+q-1	q^5-q^3+q	[Lu1, MMT]
$F_4(q),\ 2 \nmid q$	q^8+q^4-2	q^8+q^4+1	[MT2]
${}^2E_6(q)$	$(q^5+q)(q^6-q^3+1)-2$	$(q^5+q)(q^6-q^3+1)$	[MMT]
$E_6(q)$	$(q^5+q)(q^6+q^3+1)-1$	$(q^5+q)(q^6+q^3+1)$	[Hof2]
$E_7(q)$	$q\phi_7\phi_{12}\phi_{14} - 2$	$q\phi_7\phi_{12}\phi_{14}$	[Hof2]
$E_8(q)$	$q\phi_4^2\phi_8\phi_{12}\phi_{20}\phi_{24} - 3$	$q\phi_4^2\phi_8\phi_{12}\phi_{20}\phi_{24}$	[Hof2]

Notice that there still remain two cases where the difference between $\mathfrak{d}^1_{\mathbb{C}}(G)$ and $\mathfrak{b}_{LSZ}(G)$ is a polynomial of $\sqrt{q}$, namely the cases of ${}^2F_4(q)$ and $F_4(q)$ with q even.

Question 3.1. *Improve the Landazuri-Seitz-Zalesskii bounds for ${}^2F_4(q)$ and $F_4(q)$ with q even.*

A partial answer to Question 3.1 in the case of ${}^2F_4(q)$ and

$$2, 3 \neq \ell | (q^2 - q + 1)(q^4 - q^2 + 1)$$

was given in [Lu1].

4. Gap results for classical groups

Now we proceed to describe further results on Problem 1.1, namely on determining smallest cross characteristic representations of finite groups of Lie type, and also on Problems 1.2 and 1.3. At present, these results are in best shape for the groups $SL_n(q)$, $SU_n(q)$, and $Sp_{2n}(q)$ with q odd; in particular, $\mathfrak{d}^2_\ell(G) \approx (\mathfrak{d}^1_\ell(G))^{2-\varepsilon}$ in these cases. These groups share the common property that $\mathfrak{d}^1_{\mathbb{C}}(G)$ is fairly small, and complex representations of degree $\mathfrak{d}^1_{\mathbb{C}}(G)$ (and $\mathfrak{d}^1_{\mathbb{C}}(G)+1$) are the so-called *Weil representations*. This kind of representations was first constructed by Weil in [W] for classical groups defined over local rings. A key ingredient of Weil's construction is the action of certain classical groups on *Heisenberg groups*. It turns out that Weil's construction can also be carried over to the case of classical groups over finite fields. This has been done in [Ge, Hw1, Is, S, Ward] and gives rise to the class of *complex Weil representations*. By definition, *ℓ-modular Weil representations* are the nontrivial irreducible constituents of the reduction modulo ℓ of complex Weil representations. Since the construction relies on the splitting of certain extensions of extraspecial p-groups by classical groups, this class of representations exists only for the three aforementioned types of classical groups. But see the subsection 4.4 for an extension of this construction to other finite groups of Lie type. Weil representations constitute one of the most interesting classes of (complex and modular) representations of finite groups of Lie type, with a lot of remarkable features (cf. for instance [Go, Gr, T1, TZ2]), and they give answers to many questions concerning the representation theory of finite groups of Lie type.

4.1. **Special linear groups.** Let $G = SL_n(q)$ with $n \geq 3$, and let $\kappa_{n,q,\ell} = 1$ if $\ell | (q^n-1)/(q-1)$ and 0 otherwise. Over $\mathbb{C}$, G has $q-1$ irreducible Weil representations, one of degree $(q^n-q)/(q-1)$, and $q-2$ of degree $(q^n-1)/(q-1)$. Reduced modulo ℓ, these complex representations yield $(q-1)_{\ell'}$ (inequivalent) irreducible *ℓ-modular Weil representations*, one of degree $(q^n-q)/(q-1)-\kappa_{n,q,\ell}$ and $(q-1)_{\ell'}-1$ of degree $(q^n-1)/(q-1)$, cf. for instance [GT1]. Here and below, $N_{\ell'}$ denotes the ℓ'-share of the integer N.

Theorem 4.1. [GT1] *Let $G = SL_n(q)$ with $n \geq 3$, and $(n,q) \neq (3,2)$, $(3,4)$, $(4,2)$, $(4,3)$, $(6,2)$, $(6,3)$.*

(i) *Then $\mathfrak{d}^1_\ell(G) = (q^n-q)/(q-1) - \kappa_{n,q,\ell}$.*

(ii) *Let Φ be a nontrivial irreducible $\mathbb{F}G$-representation of degree less than*

$$\mathbf{d}^2_\ell(G) := \begin{cases} (q-1)(q^2-1)/(3,q-1), & \text{if } n=3,\\ (q-1)(q^3-1)/(2,q-1), & \text{if } n=4,\\ (q^{n-1}-1)\left(\frac{q^{n-2}-q}{q-1} - \kappa_{n-2,q,\ell}\right), & \text{if } n \geq 5. \end{cases}$$

Then Φ is one of $(q-1)_{\ell'}$ irreducible ℓ-modular Weil representations.

We also mention that $\mathfrak{d}^2_\ell(G) = 217$ if $G = SL_6(2)$ and 6292 if $G = SL_6(3)$, see [GT1].

The main idea of the proof of Theorem 4.1 is as follows. Suppose Φ is an irreducible $\mathbb{F}G$-representation of degree less than $\mathbf{d}^2_\ell(G)$. By restricting Φ to the first parabolic subgroup P of G (which is the stabilizer in G of an 1-space in the natural module $\mathbb{F}_q^n$), we show that Φ is a constituent of the Harish-Chandra induction $R^G_P(\Psi)$, where Ψ is a "small" representation of the Levi subgroup L of P. At this stage, the Dipper-James theory of cross characteristic representations of $GL_n(q)$ can be used to decompose $R_P^G(\Psi)$ and determine which irreducible constituents of it may have degree less than $\mathbf{d}_\ell^2(G)$.

The precise value of $\mathfrak{d}_\ell^2(G)$ has been determined by Brundan and Kleshchev in [BrK], where the authors follow the method of [J] closely and invoke the representation theory of $GL_n(q)$ as developed in the work of Brundan, Dipper, and Kleshchev [BrDK]. Generically,

$$\mathfrak{d}_\ell^2(G) = \frac{(q^n-1)(q^{n-1}-q^2)}{(q^2-1)(q-1)} - \frac{q^n-q}{q-1} \cdot \kappa_{n-2,q,\ell} - \epsilon$$

where $\epsilon = 0, \pm 1$.

4.2. **Special unitary groups.** It was already shown in [LS, S] that $\mathfrak{d}_\ell^1(SU_n(q)) = [(q^n-1)/(q+1)]$ if $n \geq 3$. Over $\mathbb{C}$, G has $q+1$ irreducible Weil representations, one of degree $(q^n+q(-1)^n)/(q+1)$, and q of degree $(q^n-(-1)^n)/(q+1)$. Reduced modulo ℓ, these complex representations yield $(q+1)_{\ell'}$ (inequivalent) irreducible *ℓ-modular Weil representations*, one of degree $(q^n+q(-1)^n)/(q+1)$ and $(q+1)_{\ell'}-1$ of degree $(q^n-(-1)^n)/(q+1)$, cf. for instance [HM1]. The following gap result was obtained by Hiss and Malle:

Theorem 4.2. [HM1] *Let $G = SU_n(q)$ with $n \geq 4$ and $(n,q) \neq (4,2)$, $(4,3)$. Suppose Φ is a nontrivial irreducible $\mathbb{F}G$-representation of degree less than*

$$\begin{cases} (q^2+1)(q^2-q+1)/(2,q-1)-1, & \text{if } n=4, \\ (q^{n-2}-1)(q-1)[(q^{n-2}-1)/(q+1)], & \text{if } n \geq 5. \end{cases}$$

Then Φ is one of $(q+1)_{\ell'}$ Weil representations.

As in the case of Theorem 4.1, part of the proof of Theorem 4.2 is to show that any $\mathbb{F}G$-representation of "small" degree has to occur in the Harish-Chandra induction $R_P^G(\Psi)$ of a "small" representation Ψ of the Levi subgroup L of P, where P is the first parabolic subgroup. The other key ingredients of the proof are the results of Broué and Michel [BM] on unions of ℓ-blocks, and Geck's theorem [G2] about unitriangular shape of the decomposition matrix for $GU_n(q)$.

The lower bound for $\mathfrak{d}_\ell^2(G)$ given in Theorem 4.2 has been improved further in [GMST]. To state the result, let

$$\kappa'_{n,q,\ell} = \begin{cases} 1, & \ell \Big| \dfrac{q^{2[n/2]}-1}{q^2-1}, \\ 0, & \text{otherwise.} \end{cases}$$

Theorem 4.3. [GMST] *Let $n \geq 5$ and $G = SU_n(q)$. Suppose that Φ is a nontrivial irreducible $\mathbb{F}G$-representation of degree less than*

$$\mathbf{d}_\ell^2(G) := \begin{cases} (q^n-1)(q^{n-1}-q)/(q^2-1)(q+1), & \text{if } 2|n \text{ and } q=2, \\ (q^n-1)(q^{n-1}+1)/(q^2-1)(q+1)-1-\kappa'_{n,q,\ell}, & \text{if } 2|n \text{ and } q>2, \\ (q^n+1)(q^{n-1}-q^2)/(q^2-1)(q+1)-\kappa'_{n,q,\ell}, & \text{if } n \geq 7 \text{ is odd}, \\ (q^n+1)(q^{n-1}-q^2)/(q^2-1)(q+1)-1, & \text{if } n=5. \end{cases}$$

Then Φ is one of $(q+1)_{\ell'}$ Weil representations.

Observe that $SU_n(q)$ has an irreducible complex representation of degree

$$\begin{cases} (q^n-1)(q^{n-1}-q)/(q^2-1)(q+1), & \text{if } n \geq 6 \text{ is even and } q=2, \\ (q^n-1)(q^{n-1}+1)/(q^2-1)(q+1), & \text{if } n \geq 6 \text{ is even and } q>2, \\ (q^n+1)(q^{n-1}-q^2)/(q^2-1)(q+1), & \text{if } n \geq 5 \text{ is odd,} \end{cases}$$

cf. [GMST]. Thus we have determined $\mathfrak{d}_\ell^2(SU_n(q))$ (up to a constant ≤ 2). Moreover, it is shown in [GMST] that the lower bound $\mathbf{d}_\ell^2(SU_n(q))$ for $\mathfrak{d}_\ell^2(SU_n(q)$ established in Theorem 4.3 is best possible

- if $\ell|(q+1)$ and $n \geq 5$, or
- if $q=2$ and n is even.

The main ideas of the proof of Theorem 4.3 will be explained in the next subsection.

4.3. **Symplectic groups in odd characteristic.** Let $G = Sp_{2n}(q)$ with $n \geq 2$ and q odd. It is well known that $\mathfrak{d}_\ell^1(Sp_{2n}(q)) = (q^n-1)/2$, see [LS, S]. Over $\mathbb{C}$, G has 4 irreducible Weil representations, two of degree $(q^n-1)/2$, and two of degree $(q^n+1)/2$. Reduced modulo ℓ, these complex representations yield 4 (inequivalent) irreducible *ℓ-modular Weil representations*, if $\ell \neq 2$. If $\ell = 2$, we get 2 irreducible *2-modular Weil representations* of degree $(q^n-1)/2$, cf. for instance [GMST]. It was shown by Guralnick, Penttila, Praeger, and Saxl in [GPPS] that the degree of any nontrivial $\mathbb{F}G$-representation is either $\mathfrak{d}_\ell^1(G)$, or $\mathfrak{d}_\ell^1(G)+1$, or at least $2\mathfrak{d}_\ell^1(G)$. Guralnick, Magaard, and Saxl also proved that any irreducible $\mathbb{F}G$-module of dimension $(q^n \pm 1)/2$ is a Weil module.

A complete solution to Problems 1.2 and 1.3 is given by the following theorem:

Theorem 4.4. [GMST] *Let $G = Sp_{2n}(q)$ with $n \geq 2$ and q odd, and let V be a nontrivial irreducible $\mathbb{F}G$-module of dimension less than $(q^n-1)(q^n-q)/2(q+1)$. Then V is a Weil module of dimension $(q^n \pm 1)/2$. In particular, if $(n,q) \neq (2,3)$ then*

$$\mathfrak{d}_\ell^2(G) = (q^n-1)(q^n-q)/2(q+1).$$

The case $\ell = 0$ of Theorem 4.4 was proved in [TZ2]. Also, $\mathfrak{d}_\ell^2(Sp_4(3))$ equals 10 if $\ell \neq 2,3$ and 14 if $\ell = 2$, cf. [Atlas, JLPW].

Sketch of Proof of Theorem 4.4.

Unlike the case of unitary groups, unitriangular shape for the decomposition matrix of $G = Sp_{2n}(q)$ (or of the conformal symplectic groups $CSp_{2n}(q)$) has not been established yet. One of the main novelties of [GMST] is the study of *local properties* of low dimensional representations. Let V be a nontrivial irreducible $\mathbb{F}G$-module with $\dim(V) < (q^n-1)(q^n-q)/2(q+1)$. Then we show that

(1) every long root subgroup of G affords only $(q-1)/2$ distinct nontrivial linear characters on V.

The second main novelty of [GMST] is the *gluing method*, which is to "glue" V from its restrictions to a collection $\mathcal{C}$ of natural subgroups of G. Here, $\mathcal{C}$ consists of maximal parabolic subgroups, semisimple subgroups $Sp_{2k}(q) \times Sp_{2n-2k}(q)$ with $1 \leq k \leq n-1$, and $SL_2(q^n)$ (naturally embedded in G). The main feature of $\mathcal{C}$ is that every element $g \in G$ is conjugate to an element in a member of $\mathcal{C}$. Now suppose V has property (1) and let φ be the Brauer character of V. Then we show that there is a formal sum ψ of the trivial character and the ℓ-modular Weil characters of G such that $\varphi|_C = \psi|_C$ for all $C \in \mathcal{C}$. It follows by irreducibility of V that φ is a Weil character, and so V is a Weil module.

This method works well provided that $n \geq 3$. When $n = 2$, one may use the results of White [Wh1, Wh2, Wh3] on the decomposition matrices of $Sp_4(q)$ in cross characteristics.

Sketch of Proof of Theorem 4.3. Let $G = SU_n(q)$ and let V be a nontrivial irreducible $\mathbb{F}G$-module with $\dim(V) < \mathbf{d}^2_\ell(G)$. As in the proof of Theorem 4.4, we also show that V has certain local properties. By a standard subgroup $SU_3(q)$ in G we mean the pointwise stabilizer in G of a non-degenerate $(n-3)$-dimensional subspace of the natural module $\mathbb{F}^n_{q^2}$. Then one of these local properties is the following, which was first introduced in [TZ2] for complex representations:

(2) The restriction of V to a standard subgroup $SU_3(q)$ involves only irreducible Weil modules and maybe the trivial module.

Assume $n \geq 6$. Then we aim to "push down" $\dim(V)$ to below the Hiss-Malle bound; namely, we show that any $\mathbb{F}G$-module V with property (2) has dimension less than the bound given in Theorem 4.2. This will imply that V is either the trivial module or a Weil module. This step also involves using GAP to do some computation with a parabolic subgroup of $SU_5(2)$.

As a rule, the case of groups of low rank ($n = 4, 5$ in this proof) requires fairly delicate consideration. Some of the ingredients of this consideration are

- the aforementioned results of Broué and Michel [BM],
- the results of Fong and Srinivasan [FS], and of Geck and Hiss [GH] on basic sets of Brauer characters in an ℓ-block, and
- the study of certain linear relations between Green functions in the ℓ-block containing V (which follow from property (2)).

4.4. **Symplectic groups in even characteristic.** Let $G = Sp_{2n}(q)$ with $n \geq 2$ and q even. Until very recently, the only available information about low dimensional cross characteristic representations of G with $n \geq 3$ (and $(n,q) \neq (3,2)$, $(4,2)$) is that $\mathfrak{d}^1_\ell(Sp_{2n}(q)) = (q^n-1)(q^n-q)/2(q+1)$, see [LS]. When $n = 2$, the decomposition matrices of G were determined by White [Wh4].

Another principal difficulty of this case is that the classical construction of Weil representations does not work here, so a priori it is unclear what is the prototype of low dimensional cross characteristic representations of G, and what plays the role of Weil representations here.

Clearly, G may be naturally embedded in $SL_{2n}(q)$ and in $SU_{2n}(q)$. By restricting the Weil modules of $SL_{2n}(q)$ and $SU_{2n}(q)$ to G, we have defined in [GT2] a collection $\mathcal{W}$ consisting of

- four (uniquely determined) irreducible $\mathbb{F}G$-modules of dimension

$$\frac{(q^n-1)(q^n-q)}{2(q+1)}, \quad \frac{(q^n+1)(q^n+q)}{2(q+1)} - \begin{cases}0\\1\end{cases},$$
$$\frac{(q^n+1)(q^n-q)}{2(q-1)} - \begin{cases}0\\1\end{cases}, \quad \frac{(q^n-1)(q^n+q)}{2(q-1)} - \begin{cases}0\\1\end{cases},$$

(whether 0 or 1 is chosen in the formulae depends on n, q, ℓ);

• $((q-1)_{\ell'}-1)/2$ (uniquely determined) irreducible $\mathbb{F}G$-modules of dimension $(q^{2n}-1)/(q-1)$;

• $((q+1)_{\ell'}-1)/2$ (uniquely determined) irreducible $\mathbb{F}G$-modules of dimension $(q^{2n}-1)/(q+1)$.

In many instances, the representations in $\mathcal{W}$ behave themselves very similarly to the Weil representations of $SL_{2n}(q)$ and $SU_{2n}(q)$. This justifies why we have called the representations belonging to $\mathcal{W}$ *Weil representations* of $Sp_{2n}(q)$ with q even. The Brauer characters of these representations can be worked out using the concept of *dual pairs*, invented by Howe [Hw2] for odd characteristic, and developed in [T2] for characteristic 2.

Let $\alpha = 19/15$ if $(n,q) = (5,2)$, $\alpha = 2$ if $(n,q) = (5,4)$ or $(6,2)$, and $\alpha = 0$ otherwise.

Theorem 4.5. [GT2] *Let $G = Sp_{2n}(q)$ with $n \geq 2$ even and q even. Let V be a nontrivial irreducible $\mathbb{F}G$-module of dimension less than*

$$\mathbf{d}_\ell^2(G) := \begin{cases} q^2(q-1), & n=2, \\ 21, & (n,q)=(3,2), \\ q^2(q^3-1), & n=3,\ q>2, \\ 203, & (n,q)=(4,2), \\ (q^4-1)(q^3-1)q^2, & n=4,\ q>2, \\ \left(\dfrac{(q^{n-1}+1)(q^{n-2}-q)}{q^2-1}-1-\alpha\right)\dfrac{q^{n-1}(q^{n-1}-1)(q-1)}{2}, & n \geq 5. \end{cases}$$

Then V belongs to the collection $\mathcal{W}$ defined above.

Notice that

$$\mathfrak{d}_{\mathbb{C}}^2(G) = \frac{(q^{2n}-1)(q^{n-1}-1)(q^{n-1}-q^2)}{2(q^4-1)} = \frac{q^{4n-6}}{2}\left(1+q^{-4}+O(q^{-5})\right);$$

$$\mathbf{d}_\ell^2(G) = \frac{q^{4n-6}}{2}\left(1-q^{-1}+O(q^{-5})\right),$$

provided $n \geq 5$, so $\mathbf{d}_\ell^2(G)$ is the asymptotically correct bound for $\mathfrak{d}_\ell^2(G)$. Furthermore, $\mathfrak{d}_\ell^2(G) \approx \mathfrak{d}_\ell^1(G)^{2-\varepsilon}$, similarly to the case of $SL_n(q)$, $SU_n(q)$, and $Sp_{2n}(q)$ with q odd.

As in the case of Theorem 4.4, the proof of Theorem 4.5 [GT2] also involves the study of local properties of low dimensional representations, and the gluing method. But both ingredients needed much refinements. We also relied more heavily on the Deligne-Lusztig theory, and on the results of Broué and Michel [BM].

The methods we used in the proof of Theorem 4.5 should be successful as well in the case of orthogonal groups, which is being handled now.

5. Minimal Polynomial Problem

In the present and the next sections, we will highlight considerable progress on a number of problems that has been achieved by using the results on low dimensional representations described in previous sections.

Let G be a finite quasisimple group and V be an irreducible $\mathbb{F}G$-module. For $g \in G \setminus Z(G)$, let $o(g)$ denote the order of g modulo $Z(G)$, and $d_V(g)$ denote the degree of the minimal polynomial of g on V. Clearly, $d_V(g) \leq o(g)$; moreover, one may expect that $d_V(g) = o(g)$ in "generic" position. So the general *minimal polynomial problem* may be stated as follows:

Problem 5.1. *Under the above notation, classify all triples (G, V, g) such that $d_V(g) < o(g)$.*

This problem dates back at least to the classical works of Blichfeldt, Hall and Higman, and Thompson, and it is far from being solved. We will concentrate on the case where $o(g)$ is an r-power, where r is a given prime.

5.1. $\mathrm{char}(\mathbb{F}) = \ell \neq r$ **case.** Referring to ongoing works of Zalesskii, and of Kleshchev and Zalesskii concerning the cases where $G/Z(G)$ is a sporadic group, respectively an alternating group, we will assume that G is a finite quasisimple group of Lie type in characteristic p.

5.1.1. *Unipotent subcase:* $r = p$. The following basic result has been established by Zalesskii:

Theorem 5.2. [Z] *Let G be a finite quasisimple group of Lie type in characteristic p, of simply connected type. Suppose that V is an irreducible $\mathbb{F}G$-module with* $\mathrm{char}(\mathbb{F}) = \ell \neq p$, *and $g \in G$ is an element of order p, such that $1 < d_V(g) < p$. Then one of the following holds.*

(i) $G = Sp_{2n}(p)$ *or* $SU_3(p)$, *g a transvection.*

(ii) $G = SL_2(p^2)$ *or* $Sp_4(p)$.

It remains to determine the possible modules V in Theorem 5.2. The case 5.2(ii) can easily be done directly. Assume we are in the case 5.2(i). It was shown by Zalesskii and the author in [TZ2] that V is a Weil module, if $\ell = 0$. It turns out that the same conclusion is true in any cross characteristic.

Theorem 5.3. [GMST] *In case* (i) *of Theorem 5.2, V is a Weil module.*

5.1.2. *Semisimple subcase:* $r \neq p$. This subcase turns out to be more subtle than the unipotent subcase. Nevertheless, DiMartino and Zalesskii have proved the following theorem:

Theorem 5.4. [DZ] *Let G be a finite classical group in characteristic p and $W = \mathbb{F}_q^n$ the natural module for G with $n \geq 4$. Let r be a prime other than p, and $g \in G$ be an r-element that fixes a nonzero totally singular subspace of W. Let V be an irreducible $\mathbb{F}G$-module in characteristic ℓ coprime to p such that $1 < d_V(g) < |g|$. Then one of the following holds.*

(i) $G = Sp_{2n}(q)$, *q odd*, $|g| = q + 1$, $\dim(g - 1)W = 2$.

(ii) $G = GU_n(q)$, $|g| = q + 1$, $\dim(g - 1)W = 1$.

(iii) $G = GU_n(q)$, $q = 2$, $|g| = 9$, $n > 4$, $\dim(g - 1)W = 3$.

Again, we are interested in determining the possible modules V in Theorem 5.4. If $\ell = 0$ and $G \neq Sp_4(3)$, then it was shown by Zalesskii and the author in [TZ2] that V is a Weil module. The same conclusion is true in any cross characteristic.

Theorem 5.5. [GMST] *In all cases of Theorem 5.4, if* $G \neq Sp_4(3)$ *then* V *is a Weil module.*

If $G = Sp_4(3)$, there is one more possibility for V, namely the unipotent representation of degree 6, cf. [GMST].

To identify the modules V in Theorems 5.2 and 5.4, different methods have been employed in [TZ2] and [GMST]. In [TZ2] one shows that $\dim(V)$ is less than $\mathfrak{d}^2_{\mathbb{C}}(G)$, therefore V is a Weil module by Theorem 4.4. In [GMST], one shows that V possesses a local property (property (1) in the case of $Sp_{2n}(q)$), whence V is a Weil module by the local characterization of Weil modules obtained in [GMST].

5.2. $\mathrm{char}(\mathbb{F}) = \ell = r$ **case.** The most interesting subcase is the case of *quadratic modules* in characteristic ℓ, that is, G is generated by the set of all elements $g \in G$ for which $[g, g, V] = 0$. Quadratic pairs (G, V) with $F^*(G)$ being quasisimple were investigated by Thompson [Th] and Ho [H1, H2] in the seventies. The interest in this problem has recently been renewed by a possible application in the *third generation proof* of the classification of finite simple groups theorem (CFSG), which is being developed by Meierfrankenfeld, Stellmacher, Stroth, and others. Motivated by this, one would like to classify all quadratic modules for *known* quasisimple groups. Using CFSG, Meierfrankenfeld and Chermak have identified possible groups that can possess quadratic modules:

Theorem 5.6. [Ch] *Let* G *be a finite group with* $F^*(G)$ *quasisimple,* $\ell > 2$ *a prime, and let* V *be a faithful irreducible* $\mathbb{F}_\ell G$*-module. Suppose that there is an elementary abelian* ℓ*-subgroup* A *such that* $G = \langle A^G \rangle$ *and* $[A, A, V] = 0$*. Then one of the following holds.*

(a) $F^*(G)/Z(F^*(G)) \in Lie(\ell)$.

(b) $\ell = 3$, $|A| = 3$, *and either*

(i) $G = PGU_n(2)$, $n \geq 5$;

(ii) $G = 2\mathbb{A}_n$, $n \geq 5$, $n \neq 6$; *or*

(iii) $Z(G)$ *is a nontrivial* 2*-group and* $G/Z(G)$ *is* $Sp_6(2)$, $\Omega^+_8(2)$, $G_2(4)$, Co_1, Sz, J_2.

It remains to classify quadratic modules for the above groups G. Case (a) (under the assumption that G is perfect) was done by Premet and Suprunenko in [PS]. Case (b)(ii) was handled by Wales in [Wa] and by Meierfrankenfeld in [Me], where they showed that V is a basic spin module (Meierfrankenfeld has actually found all *indecomposable* quadratic modules, not just the irreducible ones). The following two results of [GMST] settle the remaining cases (b)(i) and (b)(iii):

Theorem 5.7. [GMST] *In case* (b)(i) *of Theorem 5.6,* V *is a Weil module.*

As in the proof of Theorem 5.5, we show that the modules V in 5.6(b)(i) possess some local property, which implies that V is a Weil module, by means of the local characterization of Weil modules in [GMST].

Theorem 5.8. [GMST] *Each of the groups* $2Sp_6(2)$, $2\Omega^+_8(2)$, $2J_2$, $2G_2(4)$, $2Sz$, *and* $2Co_1$, *has a unique irreducible quadratic* $\mathbb{F}_3$*-module* V. *In the first two cases* V *can be obtained by reducing the root lattice of type* E_8 *modulo* 3, *and in the last four cases* V *can be obtained by reducing the Leech lattice modulo* 3.

One of the main ideas of the proof of Theorem 5.8 is the following. Given a quadratic module V of a group G in Theorem 5.8, first we show that the quadratic element a (a generator of A in the notation of Theorem 5.6) is contained in a small and "easy to handle" subgroup H of G. Next we classify quadratic modules for H, and then try to identify V using the now available information about the restriction $V|_H$. For a given group G, this idea may need to be applied repeatedly.

6. Rank 3 permutation modules

In this section we display some new results concerning the submodule structure of rank 3 permutation modules.

Let G be a rank 3 permutation group on a finite set Ω, and $\mathbb{F}$ an algebraically closed field of characteristic ℓ. Many questions of combinatorial nature can be answered once we know the submodule structure of the corresponding permutation module $\mathbb{F}\Omega$.

We will focus on the following situation: G is a finite classical group in characteristic p, V is the natural module for G, and Ω is the set of singular 1-spaces in V. The action of G on Ω has permutation rank ≤ 3, and yields the main examples among all rank 3 actions of finite classical groups, whose classification is due to Kantor and Liebler. When $\ell = 0$, the theory is well known and dates back to work of D. G. Higman in the sixties. In contrast to this, hardly anything at all is known about the permutation modules in the natural characteristic, that is $\ell = p$, except for the case of the groups $Sp_{2m}(q)$, where the composition factors of $\mathbb{F}\Omega$ were given by Zalesskii and Suprunenko [ZS], and the case of the groups $Sp_{2m}(p)$, for which the submodule lattice of $\mathbb{F}\Omega$ was determined by Sin [Sin].

The study of the cross characteristic case, i.e. $\ell \neq p$, was first taken up by Liebeck [Li1, Li2]. To describe his results, we fix some notation. Let $(\cdot,\cdot)$ denote the G-invariant, hermitian or bilinear, form on V. For $x \in \Omega$, let

$$\Delta(x) = \{y \in \Omega \mid (x,y) \neq 0\}, \ a := |\Delta(x)|,$$

and define the endomorphism δ of the G-module $\mathbb{F}\Omega$ by

$$\delta(x) = \sum_{y \in \Delta(x)} y .$$

Then δ has three eigenvalues: a, c, d, for some integers c, d. Liebeck defined the *subgraph submodules* U_c, U_d, where $U_e := \langle e(x-y) + \delta(x) - \delta(y) \rangle_{\mathbb{F}}$ for $e = c, d$, and proved the following basic result:

Theorem 6.1. [Li1, Li2] *Let U be any submodule of $\mathbb{F}\Omega$. Then either*

(i) *U is contained in the (unique) trivial submodule $\boldsymbol{I}$ of $\mathbb{F}\Omega$; or*

(ii) *U contains U_c or U_d.*

6.1. **Case I : $c \neq d$ in $\mathbb{F}$.** In this case, δ has three different eigenvalues. Using Theorem 6.1, Liebeck has determined the submodule lattice of $\mathbb{F}\Omega$. It turns out that $\mathbb{F}\Omega$ has composition length 3 or 4, and the submodule lattice can be pictured as in Figure III.

FIGURE III.

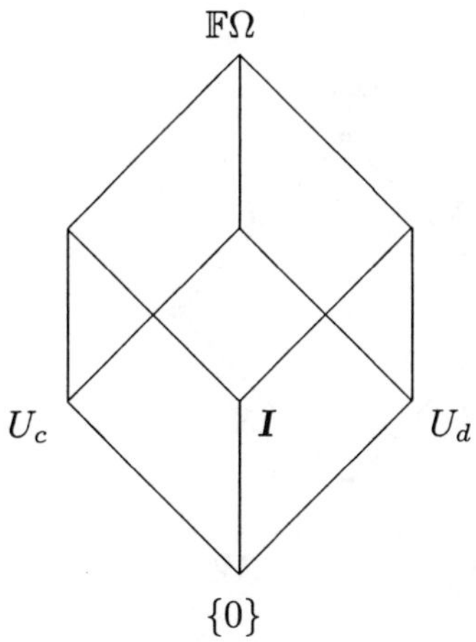

Length 3

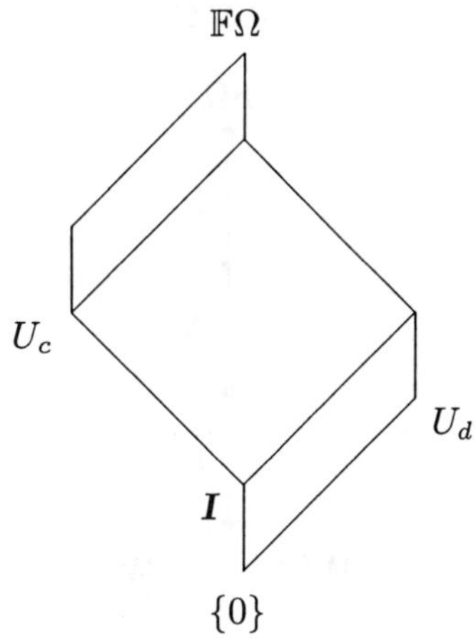

Length 4

(Depending on (ℓ, n, q), one may have to switch U_c and U_d in the picture)

6.2. **Case II : $c = d$ (in $\mathbb{F}$).** In this case, the structure of $\mathbb{F}\Omega$ is more complicated, and the composition length may be 8 or higher. Using the results on low dimensional cross characteristic representations, we have been able to determine the submodule lattice of $\mathbb{F}\Omega$ for

(i) $G = Sp_{2n}(q)$ [LST] (here $c = d$ means q is odd and $\ell = 2$), and

(ii) $G = GU_{2n}(q)$, $SO^{\pm}_{2n}(q)$, $SO_{2n+1}(q)$ [ST] (here $c = d$ means $\ell|(q+1)$, $\ell|(q+1)$, and $(\ell, q) = (2, \text{odd})$, respectively).

In case (ii), we can also handle the perfect groups $SU_{2n}(q)$ and $\Omega^{\pm}_{n}(q)$, except for an ambiguity in the case of $SU_4(q)$ with $\ell = 2$.

The remaining case is $G = SU_{2n+1}(q)$ (with $n \geq 1$). In this case, we have determined all composition factors of $\mathbb{F}\Omega$, but we do not know the multiplicity say α of a Weil module as a composition factor of $\mathbb{F}\Omega$, cf. [ST]. What we do know is that $2 \leq \alpha \leq q+1$. It is sensible to conjecture that $\alpha = 2$, which is *Geck's conjecture* [G1] when $n = 1$. A proof of Geck's conjecture was announced by Okuyama. Also, a computer calculation performed by Lux has confirmed our conjecture in the case of $SU_5(3)$.

In future papers we will consider other rank 3 permutation actions of finite groups of Lie type.

To illustrate our results, we display in Figure IV the submodule lattice of $\mathbb{F}\Omega$ in the case $G = Sp_{2n}(q)$, q odd, and $\ell = 2$, cf. [LST].

FIGURE IV.

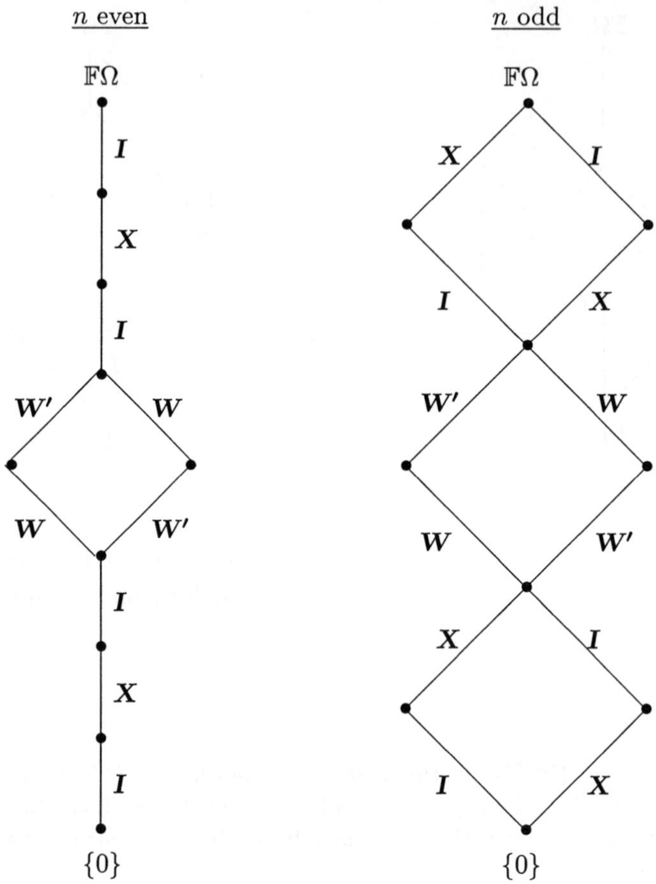

Here, $\boldsymbol{W}$ and $\boldsymbol{W'}$ are Weil modules of dimension $(q^n-1)/2$, $\boldsymbol{I}$ is the trivial module, and

$$\dim(\boldsymbol{X}) = \frac{(q^n+1)(q^n-q)}{2(q-1)} - \begin{cases} 1, & n \text{ even} \\ 0, & n \text{ odd} \end{cases}$$

7. More applications

In this section we mention some more applications in which the results on low dimensional representations have helped achieve considerable progress.

A typical scheme of applying the results on low dimensional representations is as follows. Suppose we want to prove some statement $\mathcal{P}$ involving representations Φ of finite groups G. First one tries to reduce to the case G is quasisimple (or almost quasisimple). Next, with G being (almost) quasisimple, one shows that either $\mathcal{P}$ holds for Φ, or Φ has degree less than some bound d. At this stage, results on low dimensional representations should allow one to identify the representations Φ with $\deg(\Phi) < d$, for which some brute force arguments may be needed to establish $\mathcal{P}$.

1. Classification of maximal subgroups of finite classical groups.

Let G be a finite classical group. According to the fundamental theorem of Aschbacher [A], any maximal subgroup M of G is a member of either one of the eight collections $\mathcal{C}_i$, $1 \leq i \leq 8$, of naturally defined subgroups of G, or of the collection $\mathcal{S}$ of certain quasisimple subgroups of G. Conversely, if $M \in \cup_{i=1}^{8}\mathcal{C}_i$, then the maximality of M has been determined by Kleidman and Liebeck in [KL]. It remains to determine, which $M \in \mathcal{S}$ are indeed maximal subgroups of G. This in turn leads to a number of questions concerning modular representations of finite quasisimple groups, the resolution of which requires a lot of information about low dimensional representations of finite quasisimple groups, in particular, a solution of Problem 1.3. We refer the reader to the paper of Magaard [M] in this volume for a detailed account of this topic.

2. Recognition of finite linear groups.

Results on low dimensional representations are obviously useful for recognizing finite linear groups, in both theoretical and (computer) computation settings, once the degree of the representation is given or is bounded. For an illustration, we refer the reader to the aforementioned paper [HM2] of Hiss and Malle. See also [GPPS], where results on low dimensional representations have been used to classify the maximal subgroups of classical groups containing an element of prime order acting irreducibly on a subspace of large dimension.

3. Results on low dimensional representations have made it possible to achieve significant progress in a number of more specific problems. We refer the reader

• to [Gu], where low dimensional modules in characteristic p for groups with no normal p-subgroup are shown to be semisimple,

• to [GT3], where we explore a new approach to the $k(GV)$-problem, and

• to [GT4], in which derangements of finite primitive permutation groups are studied.

We expect more applications to evolve in near future.

References

[A] M. Aschbacher, On the maximal subgroups of the finite classical groups, *Invent. Math.* **76** (1984), 469 – 514.

[BM] M. Broué and J. Michel, Blocs et séries de Lusztig dans un groupe réductif fini, *J. reine angew. Math.* **395** (1989), 56 – 67.

[BrK] J. Brundan and A. S. Kleshchev, Lower bounds for the degrees of irreducible Brauer characters of finite general linear groups, *J. Algebra* **223** (2000), 615 – 629.

[BrDK] J. Brundan, R. Dipper, and A. S. Kleshchev, Quantum linear groups and representations of $GL_n(F_q)$, *Mem. Amer. Math. Soc.* **149** (2001), no. 706.

[Ch] A. Chermak, Quadratic groups in odd characteristic, (to appear).

[Atlas] J. H. Conway, R. T. Curtis, S. P. Norton, R. A. Parker, and R. A. Wilson, '*An ATLAS of Finite Groups*', Clarendon Press, Oxford, 1985.

[DZ] L. DiMartino and A. E. Zalesskii, Minimal polynomials and lower bounds for eigenvalue multiplicities of prime power order elements in representations of classical groups, *J. Algebra* **243** (2001), 228 – 263.

[FS] P. Fong and B. Srinivasan, The blocks of finite general linear and unitary groups, *Invent. Math.* **69** (1982), 109 – 153.

[G1] M. Geck, Irreducible Brauer characters of the 3-dimensional special unitary groups in non-describing characteristic, *Comm. Algebra* **18** (1990), 563 – 584.

[G2] M. Geck, On the decomposition numbers of the finite unitary groups in nondefining characteristic, *Math. Z.* **207** (1991), 83 – 89.

[GH] M. Geck and G. Hiss, Basic sets of Brauer characters of finite groups of Lie type, *J. reine angew. Math.* **418** (1991), 173 – 188.

[Ge] P. Gérardin, Weil representations associated to finite fields, *J. Algebra* **46** (1977), 54–101.

[Go] R. Gow, Even unimodular lattices associated with the Weil representations of the finite symplectic group, *J. Algebra* **122** (1989), 510 – 519.

[Gr] B. H. Gross, Group representations and lattices, *J. Amer. Math. Soc.* **3** (1990), 929–960.

[Gu] R. M. Guralnick, Small representations are completely reducible, *J. Algebra* **220** (1999), 531 – 541.

[GMST] R. M. Guralnick, K. Magaard, J. Saxl, and Pham Huu Tiep, Cross characteristic representations of symplectic and unitary groups, *J. Algebra* (to appear).

[GPPS] R. M. Guralnick, T. Penttila, C. Praeger, and J. Saxl, Linear groups with orders having certain large prime divisors, *Proc. London Math. Soc.* **78** (1999), 167 – 214.

[GT1] R. M. Guralnick and Pham Huu Tiep, Low-dimensional representations of special linear groups in cross characteristic, *Proc. London Math. Soc.* **78** (1999), 116 – 138.

[GT2] R. M. Guralnick and Pham Huu Tiep, Cross characteristic representations of even characteristic symplectic groups, (submitted).

[GT3] R. M. Guralnick and Pham Huu Tiep, The $k(GV)$-problem, (in preparation).

[GT4] R. M. Guralnick and Pham Huu Tiep, Derangements in finite primitive groups, (in preparation).

[Hiss1] G. Hiss, Regular and semisimple blocks of finite reductive groups, *J. London Math. Soc.* **41** (1990), 63 – 68.

[Hiss2] G. Hiss, Zerlegungszahlen endlicher Gruppen vom Lie-Typ in nicht-definierender Characteristik, Habilitationsschrift, Lehrstuhl D für Mathematik, RWTH Aachen, Germany, 1993.

[HM1] G. Hiss and G. Malle, Low dimensional representations of special unitary groups, *J. Algebra* **236** (2001), 745 – 767.

[HM2] G. Hiss and G. Malle, Low-dimensional representations of quasi-simple groups, *LMS J. Comput. Math.* **4** (2001), 22 – 63.

[H1] C. Y. Ho, Chevalley groups of odd characteristic as quadratic pairs, *J. Algebra* **41** (1976), 202 – 211.

[H2] C. Y. Ho, On the quadratic pairs, *J. Algebra* **43** (1976), 338 – 358.

[Hof1] C. Hoffman, Cross characteristic projective representations for some classical groups, *J. Algebra* **229** (2000), 666 – 677.

[Hof2] C. Hoffman, Projective representations for some exceptional finite groups of Lie type, in: '*Modular Representation Theory of Finite Groups*', M. J. Collins, B. J. Parshall, L. L. Scott, eds., Walter de Gruyter, Berlin et al, 2001, 223 – 230.

[Hw1] R. Howe, On the characters of Weil's representations, *Trans. Amer. Math. Soc.* **177** (1973), 287 – 298.

[Hw2] R. Howe, θ-series and invariant theory, *Proc. Symp. Pure Math.* **33** (1979), part 1, pp. 257 – 285.

[Is] I. M. Isaacs, Characters of solvable and symplectic groups, *Amer. J. Math.* **95** (1973), 594 – 635.

[J] G. D. James, On the minimal dimensions of irreducible representations of symmetric groups, *Math. Proc. Cam. Phil. Soc.* **94** (1983), 417 – 424.

[Jan] C. Jansen, Minimal degrees of faithful representations for sporadic simple groups and their covering groups, (in preparation).

[JLPW] C. Jansen, K. Lux, R. A. Parker, and R. A. Wilson, '*An ATLAS of Brauer Characters*', Oxford University Press, Oxford, 1995.

[KL] P. B. Kleidman and M. W. Liebeck, '*The Subgroup Structure of the Finite Classical Groups*', London Math. Soc. Lecture Note Ser. no. 129, Cambridge University Press, 1990.

[KT] A. S. Kleshchev and Pham Huu Tiep, On restrictions of modular spin representations of symmetric and alternating groups, (submitted).

[LS] V. Landazuri and G. Seitz, On the minimal degrees of projective representations of the finite Chevalley groups, *J. Algebra* **32** (1974), 418 – 443.

[LST] J. M. Lataille, P. Sin and Pham Huu Tiep, The modulo 2 structure of rank 3 permutation modules for odd characteristic symplectic groups, *J. Algebra* (to appear).

[Li1] M. W. Liebeck, Permutation modules for rank 3 unitary groups, *J. Algebra* **88** (1984), 317 – 329.

[Li2] M. W. Liebeck, Permutation modules for rank 3 symplectic and orthogonal groups, *J. Algebra* **92** (1985), 9 – 15.

[Lu1] F. Lübeck, Smallest degrees of representations of exceptional groups of Lie type, *Comm. Algebra* **29** (2001), 2147 – 2169.

[Lu2] F. Lübeck, Small degree representations of finite Chevalley groups in defining characteristic, *LMS J. Comput. Math.* **4** (2001), 135 – 169.

[M] K. Magaard, these Proceedings.

[MT] K. Magaard and Pham Huu Tiep, Irreducible tensor products of representations of quasi-simple finite groups of Lie type, in: '*Modular Representation Theory of Finite Groups*', M. J. Collins, B. J. Parshall, L. L. Scott, eds., Walter de Gruyter, Berlin et al, 2001, pp. 239 – 262.

[MT2] K. Magaard and Pham Huu Tiep, The classes $\mathcal{C}_6$ and $\mathcal{C}_7$ of maximal subgroups of finite classical groups, (in preparation).

[MMT] K. Magaard, G. Malle, and Pham Huu Tiep, Irreducibility of tensor squares, symmetric squares, and alternating squares, *Pacific J. Math.* **202** (2002), 379 – 427.

[Me] U. Meierfrankenfeld, A characterization of the spin module for $2 \cdot A_n$, *Arch. Math.* **57** (1991), 238 – 246.

[PS] A. A. Premet and I. D. Suprunenko, Quadratic modules for Chevalley groups over fields of odd characteristics, *Math. Nachr.* **110** (1983), 65 – 96.

[S] G. Seitz, Some representations of classical groups, *J. London Math. Soc.* **10** (1975), 115 – 120.

[SZ] G. Seitz and A. E. Zalesskii, On the minimal degrees of projective representations of the finite Chevalley groups, II, *J. Algebra* **158** (1993), 233 – 243.

[Sin] P. Sin, The permutation representation of $\mathrm{Sp}(2m, \mathbb{F}_p)$ acting on the vectors of its standard module, *J. Algebra* **241** (2001), 578 – 591.

[ST] P. Sin and Pham Huu Tiep, Rank 3 permutation modules of finite classical groups, (in preparation).

[Th] J. G. Thompson, Quadratic pairs, in: Actes du Congrès International des Mathématiciens (Nice 1970), Gauthier-Villars, Paris (1971), Tome 1, pp. 375 – 376.

[T1] Pham Huu Tiep, Globally irreducible representations of finite groups and integral lattices, *Geometriae Dedicata* **64** (1997), 85 – 123.

[T2] Pham Huu Tiep, Dual pairs and low-dimensional representations of finite classical groups, (in preparation).

[TZ1] Pham Huu Tiep and A. E. Zalesskii, Minimal characters of the finite classical groups, *Comm. Algebra* **24** (1996), 2093 – 2167.

[TZ2] Pham Huu Tiep and A. E. Zalesskii, Some characterizations of the Weil representations of the symplectic and unitary groups, *J. Algebra* **192** (1997), 130 – 165.

[Ward] H. N. Ward, Representations of symplectic groups, *J. Algebra* **20** (1972), 182 – 195.

[Wag1] A. Wagner, The faithful linear representations of least degree of S_n and A_n over a field of characteristic 2, *Math. Z.* **151** (1976), 127 – 137.

[Wag2] A. Wagner, The faithful linear representations of least degree of S_n and A_n over a field of odd characteristic, *Math. Z.* **154** (1977), 103 – 114.

[Wag3] A. Wagner, An observation on the degrees of projective representations of the symmetric and alternating groups over an arbitrary field, *Arch. Math.* **29** (1977), 583 – 589.

[Wa] D. B. Wales, Some projective representations of S_n, *J. Algebra* **61** (1979), 37–57.

[W] A. Weil, Sur certaines groupes d'opérateurs unitaires, *Acta Math.* **111** (1964), 143 – 211.

[Wh1] D. White, The 2-decomposition numbers of $Sp(4,q)$, q odd, *J. Algebra* **131** (1990), 703 – 725.

[Wh2] D. White, Decomposition numbers of $Sp(4,q)$ for primes dividing $q \pm 1$, *J. Algebra* **132** (1990), 488 – 500.

[Wh3] D. White, Brauer trees of $Sp(4,q)$, *Comm. Algebra* **20** (1992), 645 – 653.

[Wh4] D. White, Decomposition numbers of $Sp_4(2^a)$ in odd characteristics, *J. Algebra* **177** (1995), 264 – 276.

[Z] A. E. Zalesskii, Eigenvalues of matrices of complex representations of finite groups of Lie type, Lecture Note Math. no. **1352**, Springer, Berlin et al, 1988, 206 – 218.

[ZS] A. E. Zalesskii and I. D. Suprunenko, Permutation representations and a fragment of the decomposition matrix of symplectic and special linear groups over a finite field, *Siberian Math. J.* **31** (1990), 744 – 755.

STRUCTURE AND PRESENTATIONS OF LIE-TYPE GROUPS

F. G. TIMMESFELD

§ 1 Introduction.

Let $\mathcal{B}$ be an irreducible spherical building over the type set I of rank $\ell \geq 2$ (i.e. $|I| = \ell$), considered as a chamber system (for definition see [Ro, chapter 3] or [Ti1, I(4.4)]) and $\mathcal{A}$ an apartment of $\mathcal{B}$. Denote by Φ the set of roots (half-apartments) of $\mathcal{A}$ (See [Ro, p. 14] or [Ti1, I(4.6)]). For each chamber $c \in \mathcal{B}$ and each $i \in I$ denote by $\Delta_i(c)$ the set of chambers of $\mathcal{B}$, which are i-adjacent to c. If $\Delta_i(c) \cap \mathcal{A} \neq \emptyset$ we call $\Delta_i(c) \cap \mathcal{A}$ an *i-panel* (or simply *panel*, if the type is of no importance) of $\mathcal{A}$. Notice that, since $\mathcal{A}$ is a Coxeter (chamber) complex, such an i-panel of $\mathcal{A}$ just consists of a pair of i-adjacent chambers of $\mathcal{A}$.

Denote by $\mathrm{Aut}(\mathcal{B})$ the group of type preserving automorphisms of $\mathcal{B}$, i.e. $\mathrm{Aut}(\mathcal{B})$ is the set of bijections $\sigma : \mathcal{B} \to \mathcal{B}$ with $c \overset{i}{\sim} d$, if and only if $c^\sigma \overset{i}{\sim} d^\sigma$ for $c, d \in \mathcal{B}$ and $i \in I$. For $r \in \Phi$ let

$$A_r := \{\sigma \in \mathrm{Aut}(\mathcal{B}) \mid c^\sigma = c \text{ for each } c \in \mathcal{B} \text{ such that } \Delta_i(c) \cap \mathcal{A} \text{ is a panel of } \mathcal{A} \text{ contained in } r \text{ for some } i \in I\}.$$

A_r is called the *root subgroup* of $\mathrm{Aut}(\mathcal{B})$ corresponding to the root r of $\mathcal{A}$.

Since it is fairly complicated to see through this definition lets keep the following (easiest possible) example in mind (building of type A_ℓ).

(1.1) **Example.** Let V be an $(\ell + 1)$-dimensional vectorspace with basis $B = (v_1, \cdots, v_{\ell+1})$ and let $\mathcal{B}$ be the set of all maximal flags (chambers) of the projective space of V, i.e. a chamber $c \in \mathcal{B}$ is of the form $c = (V_1, \cdots, V_\ell)$ with subspaces V_i of V satisfying $\dim V_i = i$ and $V_i \subseteq V_{i+1}$ for $i = 1, \cdots, \ell\,(V_{\ell+1} = V)$. Call two such chambers $c = (V_1, \cdots, V_\ell)$ and $d = (W_1, \cdots, W_\ell)$ i-adjacent, $i \in I = \{1, \cdots, \ell\}$, if they just differ on the i-th component, i.e. $V_j = W_j$ for all $j \neq i$. Let $\mathcal{A}$ be the set of such chambers of $\mathcal{B}$ which are spanned by vectors in B. Then $\mathcal{B}$ is a building (of type A_ℓ), with apartment set consisting of all possible $\mathcal{A}'s$, when B runs over all possible bases of V. Moreover $\mathrm{Aut}(\mathcal{B}) = P\Gamma L(V)$.

Now consider the natural action of $W = \Sigma_{\ell+1}$ on B and $\mathcal{A}$. If $c = (W_1, \cdots, W_\ell) \in \mathcal{A}$ use the convention $W_{\ell+1} = V$. Then it follows from the definition of $\mathcal{A}$, that if $\{v_i, v_j\} \subseteq W_k, 2 \le k \le \ell+1$, then $v_i \in W_{k-1}$ or $v_j \in W_{k-1}$. Thus, if we set for the 'reflection' $w_r = (ij) \in W$:

$$r := \{c = (W_1, \cdots, W_\ell) \in \mathcal{A} \mid \text{If } k \text{ is the smallest index with } \{v_i, v_j\} \subseteq W_k, \text{ then } v_i \in W_{k-1}\}$$

and similarly

$$-r = \{d = (U_1, \cdots, U_\ell) \in \mathcal{A} \mid \text{If } k \text{ is the smallest index with } \{v_i, v_j\} \subseteq U_k, \text{ then } v_j \in U_{k-1}\}$$

it follows that $\mathcal{A} = r \cup (-r), r \cap (-r) = \emptyset$ and $r^{w_r} = -r$. In this situation r and $-r$ are the two 'opposite' roots corresponding to the reflection w_r. Then it can be shown that the root group A_r is the image under the homomorphism $P : \Gamma L(V) \to P\Gamma L(V)$ of the transvection group corresponding to the point $\langle v_i \rangle$ and hyperplane $H = \langle v_k \mid k = 1, \cdots, \ell+1, k \ne j \rangle$. Moreover it can be shown with elementary matrix manipulations that

$$(*) \qquad PSL(V) = \langle A_r \mid r \in \Phi \rangle \text{ (i.e. } r \text{ a root corresponding to some reflection } (ij) \in W!).$$

Now for each root r of $\mathcal{A}$ let

$\mathcal{W}(r)$ be the set of apartments of $\mathcal{B}$ containing r (as subset).

Then it can be shown, see [Ti1, I(4.9)], that A_r acts fixed-point-freely on $\mathcal{W}(r)$. If A_r acts also transitively, whence regularly, on $\mathcal{W}(r)$ for each $r \in \Phi$, then the building $\mathcal{B}$ is called a *Moufang-building.* An important theorem of J. Tits [Tits1] shows, that if $\ell \ge 3$, $\mathcal{B}$ is always a Moufang-building. It is well known that for $\ell = 2$ this is no longer true.

(1.2) **Definition.** If $\mathcal{B}$ is an irreducible, spherical Moufang building of rank $\ell \ge 2$, $\mathcal{A}$ an apartment of $\mathcal{B}$ and Φ the set of roots of $\mathcal{A}$, then we call

$G := \langle A_r \mid r \in \Phi \rangle \le \mathrm{Aut}(\mathcal{B})$ the group of *Lie-type* $\mathcal{B}$, where A_r is the root-subgroup of $\mathrm{Aut}(\mathcal{B})$ corresponding to r.

This definition seems to depend on the choice of the apartment $\mathcal{A}$ of $\mathcal{B}$. But, as will be seen in §3, this is as for $\mathrm{Aut}(\mathcal{B}) = P\Gamma L(V)$ and $G = PSL(V)$, not the case. In fact one obtains the same group, starting with any apartment of $\mathcal{B}$. This definition of a group of Lie-type $\mathcal{B}$ generalizes the usual, somewhat vague definition of a group of Lie-type. It includes

- the simple classical groups of Witt-index ℓ, $2 \le \ell < \infty$.
- the simple algebraic groups of relative rank $\ell \ge 2$.
- the finite groups of Lie-type of rank $\ell \ge 2$.

The development of a general structure theory (including simplicity) of such groups of Lie-type $\mathcal{B}$ has been started in [Ti2], see also [Ti1, I §4 and II §5]. One important ingredient of this structure theory is the theory of rank one groups, see [Ti1, I]. Here a *rank one group* $X = \langle A, B \rangle$ is a group generated by two different nilpotent subgroups A and B satisfying:

For each $1 \neq a \in A$ there exists a $b \in B$ such that $A^b = B^a$ and vice versa.

Rank one groups come into the theory of groups of Lie-type $\mathcal{B}$ through the following observation, see [Til, I(4.12)]:

If r and $-r$ are opposite roots of some apartment $\mathcal{A}$ of $\mathcal{B}$, then $X_r = \langle A_r, A_{-r} \rangle$ is a rank one group.

One should think of such a rank one group as a generalization of $SL_2(K)$. In fact in example (1.1) we always have $X_r \simeq SL_2(K)$, K the division ring over which V is defined.

§ 2 Rank one groups

In this section we state, without proof, some properties of rank one groups, since, as mentioned already in the introduction, the rank one groups are of central importance for the Lie-type groups of higher rank. Proofs of all these properties can be found in [Til, I].

(2.1) **Definition.** A group X generated by two different nilpotent subgroups A and B satisfying:

(*) For each $a \in A^{\#}$ there exists a $b \in B$ satisfying $A^b = B^a$ and vice versa,

will be called a *rank one group.* The conjugates of A (and B) are called the *unipotent subgroups* of X and the conjugates of $H = N_X(A) \cap N_X(B)$ are called *diagonal subgroups.*

If A is abelian, X is a rank one group with *abelian unipotent subgroups*, abbreviated AUS. Moreover, if for each $a \in A^{\#}$ and $b \in B$ satisfying (*), also

$$(**) \qquad a^b = b^{-a}$$

holds, X is called a *special rank one group.* It follows from [Til, I(1.2)(2)] that this definition of special is symmetric in A and B. Finally, if for some $N_X(A)$ invariant subgroup $1 \neq A_0 \leq A$ we have $a^b = b^{-a}$ for each $a \in A_0^{\#}$, then X is called *relatively special* with respect to A_0. By [Til, I(1.12)] the following are equivalent:

(1) X is relatively special with respect to A_0.
(2) $X_0 = \langle A_0, B_0 \rangle$ is a special rank one group, where $B_0 = A_0^x$ for $x \in X$ with $A^x = B$.

(Since A_0 is $N_X(A)$-invariant we have $A_0^x = A_0^y$ for all $x, y \in X$ with $A^x = B = A^y$.)

(2.2) **Example.** Let k be a field. Then $St_2(k)$ is the group generated by the symbols $a(t), b(t), t \in k$ subject to the relations:

(a) $a(t)a(\tau) = a(t+\tau), b(t)b(\tau) = b(t+\tau); t, \tau \in k.$
(b) $a(u)^{n(t)} = b(-t^{-2}u); u \in k$ and $t \in k^*$, where $n(t) = a(-t)b(t^{-1})a(-t)$.

Then by [Til, I(5.1)] $St_2(k)$ is a special rank one group with unipotent subgroups $A = \{a(t) \mid t \in k\}$ and $B = \{b(t) \mid t \in k\}$. Moreover, if $|k| > 4$ and $|k| \neq 9$, then by Theorem 10 of [St] $St_2(k)$ is the universal central extension of $PSL_2(k)$.

Let in the following $X = \langle A, B \rangle$ be a rank one group and let $\Omega = A^X$. Then we have

(2.3) **Lemma.**

(1) The element $b = b(a)$ satisfying (2.1)(*) is uniquely determined. Moreover the map $\chi : A^\# \to B^\#, B^\# \to A^\#$ with $\chi(a) = b(a), \chi(b) = a(b)$ is a bijection with $\chi^2 = \mathrm{id}$.
(2) $X = \langle C, D \rangle = \langle C, d \rangle$ for all $C \neq D \in \Omega$ and $d \in D^\#$.
(3) X is doubly transitive on Ω and A is a nilpotent normal subgroup of $N_X(A)$ acting regularly on $\Omega - \{A\}$.

See [Til, I] (1.2) - (1.4).

From (2.3)(3) one obtains, see [Til, I(1.3)], that the concept of a rank one group and of a group with a split BN-pair of rank one are equivalent. Namely, if Y has a split BN-pair of rank one and $B = U \cdot H$, $H = B \cap N$ and $U \trianglelefteq B$ nilpotent with $U \cap H = 1$, then clearly $X = \langle U^Y \rangle$ is a rank one group.

(2.4) **Lemma.** Suppose $N \trianglelefteq X$. Then either $N \leq Z(X)$ or $X = NA$. In particular we obtain:

(1) $Z(X) = Z_2(X)$.
(2) X is quasisimple if it is perfect.
(3) $\langle a^X \rangle$ is not nilpotent for each $a \in A^\#$.

(2.5) **Theorem.** Suppose X is special with AUS. Then one of the following holds:

(1) A is an elementary abelian p-group. Moreover for each $a \in A^\#$ we have $X(a) = \langle a, b(a) \rangle \simeq (P)SL_2(p)$.
(2) A is torsion free and divisible. For each $a \in A^\#$ let $A(a) = \{a^{\frac{m}{n}} \mid 0 \neq n, m \in \mathbb{Z}\}$ and $B(a) = \{b(a)^{\frac{m}{n}} \mid 0 \neq n, m \in \mathbb{Z}\}$. Then $X(a) = \langle A(a), B(a) \rangle$ is an epimorphic image of $St_2(\mathbb{Q})$.

This is (5.2) and (5.6) of [Til, I]. The proof occupies most of section 5 of chapter I of [Til]. From (2.5) we obtain the following simplicity result for special rank one groups.

(2.6) **Corollary.** Suppose X is special with AUS. Then one of the following holds:

(1) X is quasisimple.
(2) A is an elementary abelian 2 or 3-group. Moreover X' is quasisimple and $|A \cap X'| > 3$.
(3) $X \simeq (P)SL_2(3)$ or $SL_2(2)$.

Proof. In case (2.5)(2) or if $p \geq 5$ in (2.5)(1) it follows from (2.4)(2) that (1) holds. That we obtain either (2) or (3) if $p \leq 3$ follows from [Ti1, I(2.10)].

I actually do not know any example in which neither (1) nor (3) holds. If one could show this, this would simplify the simplicity proofs for Lie-type groups, which are not defined over $GF(2)$ or $GF(3)$.

Although there is no complete classification, (2.5) and (2.6) show that special rank one groups with AUS are reasonably well understood. Hence it is of great importance to find criteria under which arbitrary rank one groups are special respectively relatively special. Here the condition is quadratic action.

(2.7) **Proposition.** Suppose $1 \neq A_0 \leq Z(A)$ is H-invariant and let $X_0 = \langle A_0, B_0 \rangle$ where $B_0 = A_0^x$ for $x \in X$ with $A^x = B$. Suppose X acts on some abelian group N such that

(a) $[N, X, X_0] \neq 0$.
(b) $[N, A, A_0] = 0$.

Then X_0 is special (i.e. X is relatively special with respect to A_0). In particular, if $[N, X, X] \neq 0 = [N, A, A]$, then X is special with AUS.

Proof. This is [Ti1, I(2.4)]. Notice that, if K is the kernel of the action of X on N, then by (2.4) $A \cap K = 1$. Hence if $[N, A, A] = 0$, then A is abelian by the 3-subgroup lemma. Thus, setting $A = A_0$, we obtain the second statement.

(2.8) **Example.**

(1) Let G be a unitary group of Witt-index 1 over some division ring with natural module V. Let P be an isotropic point of V and

$$U_P := \{\sigma \in G \mid P^{\sigma - \mathrm{id}} = 0 \text{ and } (P^\perp)^{\sigma - \mathrm{id}} \subseteq P\}$$
$$T_P := \{\sigma \in G \mid V^{\sigma - \mathrm{id}} \subseteq P, (P^\perp)^{\sigma - \mathrm{id}} = 0\}$$

Then $R = \langle U_P^G \rangle$ is a relatively special rank one group with respect to T_P.

That R is a rank one group has been shown in [Ti1, I(1.9)]. Now $[V, T_P, U_P] = 0 = [V, U_P, T_P]$ and (2.7) show that R is relatively special with respect to T_P.

(2) Let G be an orthogonal group of Witt-index 1 and U_P as in (1). Suppose that there exist more than two singular points. Then $R = \langle U_P^G \rangle$ is a special rank one group with AUS.

Namely if W is the so called spin-module for G, then it is well known that $[W, U_P, U_P] = 0$.

Finally a result which will be of great importance for us in §4 and which somehow explains, why the characteristic two case is much harder for Steinberg-presentation type results.

(2.9) **Lemma.** Suppose V is a faithful $\mathbb{Z}X$-module, satisfying:

(i) $V = [V, X]$ and $C_V(X) = 0$.
(ii) $[V, A, A] = 0$.

Then there exists an element $\tau \in Z(X)$ with $\tau = -\mathrm{id}$ on V.

Proof. By (2.7) X is special with AUS. Pick $a \in A^{\#}$. Then $a \in A(a) \leq X(a)$ and $X(a)$ is by (2.5) an image of $St_2(k)$, k a prime field. Hence we may set $a = a(1)$ in the notation of (2.2). Hence $\tau = n(1)^2$ acts by [Ti1, I(3.5)(2)] as $-\mathrm{id}$ on V. In particular $\tau \in Z(X)$. □

If now A is not an elementary abelian 2-group, then V is not an elementary abelian 2-group. Hence $\tau \neq 1 = \tau^2$.

§ 3 On the structure of Lie-type groups.

Let in this section $\mathcal{B}$ be an irreducible spherical Moufang building over I of rank $\ell \geq 2$, $\mathcal{A}$ an apartment of $\mathcal{B}$ and Φ the set of roots of $\mathcal{A}$. Let $G = \langle A_r \mid r \in \Phi \rangle$ be the group of Lie-type $\mathcal{B}$, as defined in (1.2). We start to investigate the structure of G.

(3.1) **Definitions.** Fix a chamber c of $\mathcal{A}$ and set $\Phi^+ := \{r \in \Phi \mid c \in r\}$. Using the geometric realization of the Weyl-group of $\mathcal{A}$ ($\mathcal{A}$ is a Coxeter complex!) we can identify Φ with a root-system in the original sense such that Φ^+ becomes the set of positive roots in this root system. (I.e. Φ is of one of the following types $A_\ell, B_\ell, C_\ell, D_\ell, E_\ell, F_4, G_2$ or $I_2(m)$. By [Tits3] resp. by the Theorem of Tits and Weiss, see [VM, (5.3.4)] Moufang buildings of type H_3 and H_4 do not exist, and if $\mathcal{B}$ is of type $I_2(m)$ then $m = 3, 4, 6$ or 8. In the last case, i.e. Φ of type $I_2(8)$ we call Φ of type ${}^2F_4$, see [VM,(5.4)].) For $r \in \Phi$ we fix the following notation:

$$\begin{aligned} A_r &:= \text{the root subgroup of } \mathrm{Aut}(\mathcal{B}) \text{ corresponding to } r \\ X_r &:= \langle A_r, A_{-r} \rangle. \end{aligned}$$

Then, by [Ti1, I(4.12)] X_r is a rank one group with unipotent subgroups A_r and A_{-r}.

$$\begin{aligned} H_r &:= N_{X_r}(A_r) \cap N_{X_r}(A_{-r}) \\ G &:= \langle A_r \mid r \in \Phi \rangle \\ U &:= \langle A_r \mid r \in \Phi^+ \rangle \\ H &:= \langle H_r \mid r \in \Phi \rangle. \end{aligned}$$

By [Ti1, II(5.1)] G acts chambertransitively on $\mathcal{B}$ and c is the only chamber of $\mathcal{B}$ fixed by U. (c is obviously fixed by U by definition of the A_r in §1!) Let Π be the fundamental system contained in Φ^+. Then each root in Φ^+ can be written as a linear combination of roots in Π with nonnegative coefficients.

Our first aim is to enlarge Φ to a possibly nonreduced root system $\widetilde{\Phi}$, such that for the A_r, $r \in \widetilde{\Phi}$ some sort of "Chevalley-commutator-relations" hold. For this we may assume that Φ is not of type $I_2(8)$, since in this case by a theorem of [Tits3], see also [VM, 5.4.5] these commutator relations hold for $\widetilde{\Phi} = {}^2F_4$. Now we will have $\Phi = \widetilde{\Phi}$ except possibly if $\mathcal{B}$ is of type C_ℓ. In the latter case let

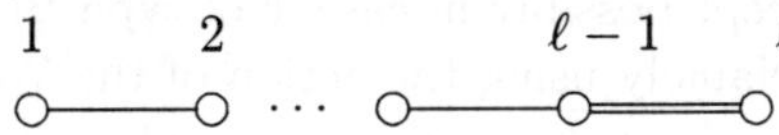

be the Dynkindiagram of Π and let W be the Weyl-group of Φ acting naturally on Φ. Now by [Tits2, 7.4] a building of type C_ℓ can be regarded as a polar space $\mathcal{P}$ of rank ℓ. Let r be a root of Φ conjugate to r_ℓ under W. Then r corresponds to some point P of $\mathcal{P}$. (Up to duality in case $\ell = 2$. If $\mathcal{P}$ is classical with natural G-module V, then

$$A_r \leq U_P = \{\sigma \in G \mid P^{\sigma - \mathrm{id}} = 0, (P^\perp)^{\sigma - \mathrm{id}} \subseteq P\}!)$$

Let

$$\begin{aligned} T_P &:= \text{the set of central elations on } \mathcal{P} \text{ corresponding to } P \\ &:= \{\sigma \in \mathrm{Aut}(\mathcal{B}) \mid \sigma \text{ fixes each point on each line through } P\}. \end{aligned}$$

Then we have three cases to distinguish:

(a) $T_P = 1$. In this case choose $\Phi = \widetilde{\Phi}$ to be of type B_ℓ. (Since originally Φ is just the set of half apartments of $\mathcal{A}$, we still have the freedom to choose the length of the roots appropriately!)
(b) $A_r = T_P$. In this case choose $\Phi = \widetilde{\Phi}$ to be of type C_ℓ.
(c) $1 < T_P < A_r$. In this case set $A_{2r} = T_P$. Moreover, choose Φ to be of type B_ℓ and set

$$\widetilde{\Phi} = \Phi \cup \{2r \mid r \in \Phi \text{ short }\}.$$

Then $\widetilde{\Phi}$ is a root system of type BC_ℓ. Moreover $\Phi_0 = \{s \in \Phi \mid s \text{ long }\} \cup \{2r \mid r \in \Phi \text{ short }\}$ is a root-subsystem of type C_ℓ.

With all this notation we have:

(3.2) Lemma. Suppose $\widetilde{\Phi}$ is of type BC_ℓ and $r \in \Phi$ is short. Then the following hold:

(i) $A_r' \leq A_{2r} \leq Z(A_r)$.
(ii) A_{2r} is H_r-invariant.
(iii) $X_{2r} = \langle A_{2r}, A_{-2r} \rangle$ is also a rank one group.

Proof. (i) If $\mathcal{P}$ is a classical polar space, (i) follows immediately by an application of the 3-subgroup-lemma to the action on the natural module. If $\mathcal{P}$ is not classical (i) follows from the commutator relations of [Tits3] for the root groups on Moufang-quadrangles, see [Ti1, II(5.4)]. (ii) follows directly from the definition of $A_{2r} = T_P$, since H_r fixes P. For the proof of (iii), which is not so easy, see [Ti1, II(5.19)]. (The idea of the proof is to apply (2.7) for the action of X_r on some section of a unipotent subgroup. This X_r-invariant unipotent subgroup is constructed using again the commutator relations of [Tits3] on Moufang quadrangles!) □

If $\mathcal{B}$ is not of type C_ℓ (nor $I_2(8)$), then we simply set $\widetilde{\Phi} := \Phi$. Notice that, if $s \in \widetilde{\Phi}$ then A_s is abelian except possibly in case $\widetilde{\Phi}$ of type BC_ℓ and s is short. The proof for this fact is easy. Namely using the action of the Weyl group on $\widetilde{\Phi}$ it is easy to see that there exists, except $\widetilde{\Phi}$ of type BC_ℓ and s short, a root $r \in \widetilde{\Phi}$ such that $\langle r, s\rangle$ is of type A_2. Now it follows from the commutator-relations of root-groups on a Moufang-plane that A_r and A_s are abelian. The next theorem is of central importance to the structure of Lie-type groups.

(3.3) **Theorem.** Suppose $r, s \in \widetilde{\Phi}$ with $s \neq -r$ or $-2r$. Then $[A_r, A_s] \leq \langle A_{\lambda r + \mu s} \mid \lambda, \mu \in \mathbb{N} \text{ and } \lambda r + \mu s \in \widetilde{\Phi}\rangle$.

In (3.3) we use the convention $\langle \emptyset \rangle = \{1\}$. Hence the fact that $A_s' = 1$ if $2s \notin \widetilde{\Phi}$ is a part of (3.3). For the proof see [Ti1, II(5.7)(2)]. Notice that by the commutator-relations of [Tits3] on Moufang-polygons, see also [VM, (5.4.6)], (3.3) holds if $\ell = 2$. Now the proof of (3.3) consists of showing that $\langle A_r, A_s\rangle$ acts faithfully on some rank 2 residue of $\mathcal{B}$, such that A_r and A_s induce root-groups on this residue. From (3.3) we obtain as an immediate Corollary.

(3.4) **Corollary.** Let h be the greatest "height" of a root in $\widetilde{\Phi}^+$. Then

(1) U is nilpotent of class at most h.
(2) $A_h \leq Z(U)$, where h also denotes a root of height h in $\widetilde{\Phi}^+$.

In (3.4) is, as usual, $\widetilde{\Phi}^+$ the set of all roots of $\widetilde{\Phi}$ which are linear combinations of roots in Π with nonnegative coefficients.

(3.5) **Notation.** Notice that by (2.3)(2) we have in the situation of (3.2) $H_{2r} = N_{X_{2r}}(A_r) \cap N_{X_{2r}}(A_{-r}) \leq H_r$. Moreover, if $w \in X_{2r}$ interchanges (by conjugation) A_{2r} and A_{-2r}, it also interchanges A_r and A_{-r}. Hence for $r \in \Phi$ we may choose an $n_r \in X_r$ interchanging A_r and A_{-r}, with the convention that, if $2r \in \widetilde{\Phi}$, then $n_r \in X_{2r}$ and also interchanges A_{2r} and A_{-2r}. Now set $N = \langle H, n_r \mid r \in \Phi\rangle$. We have

(3.6) **Lemma.** Let $r, s \in \Phi$. Then the following hold:

(1) $[H_r, H_s] \leq H_r \cap H_s$.
(2) $[H_r, n_s] \leq H_s$.
(3) $H \trianglelefteq N$ and $W = N/H$ is generated by the involutions $w_\alpha = n_\alpha H$ for $\alpha \in \Phi$.
(4) H fixes all chambers of $\mathcal{A}$ and W acts as the Weyl-group (of $\mathcal{A}$) on $\mathcal{A}$. (Possibly with kernel $\neq 1$!)

Proof. By the proof of (1.3) in [Ti1, I(4.12)] H_r fixes the roots r and $-r$ globally, whence it also fixes $\mathcal{A} = r \dot{\cup} -r$. Moreover, again by [Ti1, I(4.12)] H_r fixes some chamber in $\mathcal{A}$ and thus all chambers in $\mathcal{A}$, since $\mathcal{A}$ is a Coxeter complex. In particular H_r fixes s and $-s$, whence A_s, A_{-s}, X_s and H_s. This proves (1) and the first part of (4). Now

$$H_s n_s = \{n \in X_s \mid n \text{ interchanges } A_s \text{ and } A_{-s}\}.$$

Hence H_r normalizes $H_s n_s$ and $\langle H_s n_s \rangle = H_s \langle n_s \rangle$. Since $|H_s \langle n_s \rangle : H_s| = 2$ this implies (2). (3) is a consequence of (2). Finally, the second part of (4) follows from the fact that by [Ti1, I(4.12)] n_r acts as the "reflection" corresponding to r on $\mathcal{A}$.
□

By (3.6)(4) H normalizes all A_s, $s \in \widetilde{\Phi}^+$ and whence normalizes U. Thus we may set $B = U \cdot H$. With this notation we have

(3.7) **Theorem.** The following hold:

(1) B, N is a BN-pair of G.
(2) $U \cap H = 1$ and $H = B \cap N$.
(3) H is the kernel of the action of N on $\mathcal{A}$.

For a proof of (3.7) see [Ti1, II(5.12)(4) and (5.13)]. Notice that by (3.6)(4) we have

(*) $\quad A_s^{n_r} = A_{s^{w_r}}$ for all $r, s \in \widetilde{\Phi}$; where w_r also denotes the reflection induced by w_r on $\mathcal{A}$ and $\widetilde{\Phi}$.

This equation will be of importance for us in §4. Finally, if we set

$$\widehat{H} := \{\sigma \in \mathrm{Aut}(\mathcal{B}) \mid c^\sigma = c \text{ for each chamber } c \text{ of } \mathcal{A}\},$$

then by (3.6)(4) $H \leq \widehat{H}$ and we have

(3.8) **Corollary.** $G \trianglelefteq \mathrm{Aut}(\mathcal{B})$ and $\mathrm{Aut}(\mathcal{B}) = G\widehat{H}$.

For a proof see [Ti1, II(5.18)]. (3.8) follows from (3.7) using the fact, see [Ti1, II(5.16)], that one can show using the BN-pair decomposition of G that G acts transitively on the pairs $(c, \mathcal{A})$, where c is a chamber of $\mathcal{B}$ and $\mathcal{A}$ an apartment of $\mathcal{B}$ containing c. (So called strong transitivity in the theory of buildings.)

(3.9) **Notation.** A conjugacy class Σ of abelian subgroups of some group R is called a class of *abstract root subgroups* of R, if $R = \langle \Sigma \rangle$ and for all $A, B \in \Sigma$ one of the following holds:

(1) $\langle A, B \rangle$ is a rank one group with unipotent subgroups A and B.
(2) $[A, B] = 1$.
(3) $\langle A, B \rangle$ is nilpotent of class at most two and $[a, B] = [A, b] = [A, B] \in \Sigma$ for all $a \in A^\#$, $b \in B^\#$.

Nearly simple groups generated by a class of abstract root subgroups have been classified in [Ti3]. In this classification theory many details about so called Σ-subgroups, i.e. subgroups generated by elements of Σ, are provided. Hence, as the next theorem shows, this classification theory also gives a lot of information about subgroups of Lie-type groups.

(3.10) **Theorem.** Suppose $\mathcal{B}$ is not a Moufang octagon. Let h be a highest root in $\widetilde{\Phi}^+$. Then the following hold:

(1) $\Sigma = A_h^G$ is a class of abstract root subgroups of the normal subgroup $G_0 = \langle \Sigma \rangle$ of G.
(2) G_0' is simple.

For the proof of (1) see [Ti1, II(5.20)]. (2) follows from (1) using simplicity criteria for groups generated by abstract root subgroups, see [Ti1, II(5.21)]. In (5.21), (5.22) of [Ti1, II] it is also shown that G/G_0' is "small". Notice that in (3.10)(2) we indeed must take G_0' as the following examples show:

$$\begin{aligned} G &= G_2(2), G_0' = G_0 = G' \simeq U_3(3) \\ G &= G_0 = \mathrm{Sp}(4,2) \simeq \Sigma_6, G_0' \simeq A_6. \end{aligned}$$

§ 4 The Steinberg-presentation.

Let G be a Chevalley-group over the field k with root-system Φ of rank $\ell \geq 2$ (i.e. G is not $A_1(k)$!). Then, if $|k| > 4$, by Theorem 10 of [St] the set of symbols $a_r(t), r \in \Phi, t \in k$ satisfying the relations

(4.1)

(1) $a_r(t)a_r(\tau) = a_r(t+\tau)$, $r \in \Phi$; $t, \tau \in k$.
(2) If $\alpha, \beta \in \Phi$ with $\beta \neq \pm\alpha$, then

$$[a_\alpha(t), a_\beta(\tau)] = \Pi a_{i\alpha+j\beta}(c_{ij}t^i\tau^j);$$

where i, j runs over all positive integers such that $i\alpha + j\beta \in \Phi$ and the $c_{ij} \in k$ are the so called structure constants. Moreover the product is taken in order of increasing $i + j$.

is a presentation for the universal central extension $\widehat{G}$ of G. Notice that in general (i.e. over infinite fields) $\widehat{G}$ is different from the universal Chevalley-group.

Now the proof of this theorem works in principal as follows. Let $\widehat{G}$ be the group given by the above presentation and $\pi : \widehat{G} \to G$ be the natural homomorphism mapping the $a_r(t)$ onto the root-elements $x_r(t)$ of G. Then one has to show:

(i) π is a central extension with $\ker \pi \leq \widehat{G}'$.
(ii) $\widehat{G}$ is centrally closed, i.e. each central extension of $\widehat{G}$ splits.

To prove (i) let $A_r = \{a_r(t) \mid t \in k\}$ for $r \in \Phi$ and $\widehat{U} = \langle A_r \mid r \in \Phi^+\rangle$ and $U = \langle x_r(t) \mid r \in \Phi^+, t \in k\rangle$. Then it follows from the fact that each element of U has a unique expression as a product of the root-elements $x_r(t), r \in \Phi^+, t \in k$ that $\pi\mid_{\widehat{U}}$ is an isomorphism (and similarly for $\widehat{U}^- = \langle A_r \mid r \in \Phi^-\rangle$!). From this it follows already, see [St, L 37], that the relations (2.2)(b) hold for $a(t) = a_r(t)$ and $b(t) = a_{-r}(t)$, i.e. that $X_r = \langle A_r, A_{-r}\rangle$ is a rank one group. Now $\ker\pi \leq H = \langle H_r \mid r \in \Phi\rangle$, where H_r is defined as in §3. Hence $[\widehat{U}, \ker\pi] \leq \widehat{U} \cap \ker\pi = 1$ and similarly $[\widehat{U}^-, \ker\pi] = 1$ and thus $\ker\pi \leq Z(\widehat{G})$, since $\widehat{G} = \langle \widehat{U}, \widehat{U}^-\rangle$. Now, as $|k| > 4$ and $\ell \geq 2$, it is easy to see that $\widehat{G} = \widehat{G}'$. This shows (i). The proof of (ii) is more difficult and occupies the central part of §7 of [St].

Now one would like to have a similar theorem for the groups of Lie-type $\mathcal{B}$ as defined in §3. But unfortunately we obtained in (3.3) just "global" commutator-relations for the root-groups. In fact it seems to be very difficult even to know how elementwise commutator-relations ought to look like in the general case. Hence we consider in the rest of this section the following hypothesis:

(4.2) **Hypothesis.** (St) Let Φ be an irreducible, spherical, possibly nonreduced root-system satisfying the cristallographic condition of rank $\ell \geq 2$ (i.e. Φ is of type $A_\ell, B_\ell, C_\ell, BC_\ell, \ell \geq 2, D_\ell, \ell \geq 4, E_\ell, 6 \leq \ell \leq 8, F_4, G_2$ or ${}^2F_4$.) and G is a group generated by subgroups $A_r \neq 1, r \in \Phi$ satisfying:

(1) For $r, s \in \Phi$ with $s \neq -r$ or $-2r$ we have

$$[A_r, A_s] \leq \langle A_{\lambda r + \mu s} \mid \lambda, \mu \in \mathbb{N} \text{ and } \lambda r + \mu s \in \Phi\rangle$$

(Here we use again the convention $\langle\emptyset\rangle = \{1\}$!)

(2) $X_r = \langle A_r, A_{-r}\rangle$ is a rank one group with unipotent subgroups A_r and A_{-r} for each $r \in \Phi$.

(Notice that by (1) A_r is nilpotent of class at most two!)

Clearly by (4.1)(2) and the remarks after (4.1) the universal central extension of a Chevalley-group satisfies hypothesis (St). Moreover, by §3, all groups of Lie-type $\mathcal{B}$ satisfy (St). Hence this hypothesis seems to be a good substitute to the Chevalley-commutator-relations (4.1)(1) and (2) in our more general situation. We have:

(4.3) **Main-Theorem.** Suppose G is a group satisfying (St) with Φ not of type G_2 or ${}^2F_4$. Then one of the following holds:

(i) G is perfect and there exists a surjective homomorphism $\pi : G \to \overline{G}$, where $\overline{G}$ is a group of Lie-type $\mathcal{B}$, mapping the A_r with $r \neq 2s$; $r, s \in \Phi$, onto the root-subgroups of $\overline{G}$ corresponding to the roots of some apartment $\mathcal{A}$ of $\mathcal{B}$ (in the sense of §3). Moreover $\ker\pi \leq Z(G) \cap H$, where

$$H = \langle H_r \mid r \in \Phi\rangle \text{ and } H_r = N_{X_r}(A_r) \cap N_{X_r}(A_{-r}).$$

(ii) $\Phi = J \dot\cup K$ with $J \neq \emptyset \neq K$ and either $J = \{\pm r\}$ or $J = \{\pm r, \pm 2r\}$ or J carries the structure of an irreducible root system Ψ of rank ≥ 2. Moreover $G = G(J) * G(K)$, where $G(J) = \langle X_r \mid r \in J\rangle$ (and similarly $G(K)$) and either $G(J)$ is a rank one group or $G(J)$ satisfies the conclusion of (i) with respect to Ψ.

(iii) $J' = \{r \in \Phi \mid A_r$ is an elementary abelian 2-group $\} \neq \emptyset$. Let $J = J' \cup \{s \in \Phi \mid 2s \in J'\}$ and $K = \Phi - J$. Then $G = G(J) * G(K)$.

In particular our Main-theorem tells us, that if Φ can not be decomposed into the disjoint union of two nonempty subsets and G is the central product of subgroups corresponding to these subsets and if no A_r, $r \in \Phi$ is an elementary abelian 2-group, G is a perfect central extension of a group of Lie-type $\mathcal{B}$. In case (ii) the root-system Ψ is not necessarily a root subsystem of Φ. For example, if $\Phi = B_2$ the group $G = SL_3(K) * SL_2(K)$ satisfies (St) with respect to Φ and with Ψ of type A_2. It is clear that case (ii) has to occur, since the commutator relations of (4.2)(1) might degenerate. (For example all commutators in (4.2)(1) might be 1, in which case obviously $G = *_{r \in \Phi} X_r$.)

Unfortunately the Main-theorem only gives us real information if the characteristic is different from two. This is due to the fact, that central involutions in some X_r obtained from (2.7) and (2.9) play a central role in the proof. Namely using these central involutions one can show that one either is in case (ii) or that one always has equality in the commutator relations (4.2)(1). Now in the second case one is able to construct a BN-pair and then show that the A_r with $r \neq 2s$ act as root-groups in the sense of §3 on the corresponding building.

Now, since the proof of this Main-theorem is spread over several papers, and since the auxiliary results obtained in these papers might also be useful, we will discuss the proof of (4.3) in more detail in the rest of this section.

The starting point is [Ti4] in which the following two theorems were proven.

(4.4) **Theorem.** Suppose G satisfies (St) and

(*) For all $r, s \in \Phi$ and all $n_r \in X_r$ interchanging A_r and A_{-r} we have $A_s^{n_r} = A_{s^{w_r}}$, where w_r is the reflection on Φ corresponding to r.

Then there exists a group $\overline{G}$ of Lie-type $\mathcal{B}$, $\mathcal{B}$ an irreducible, spherical Moufang-building and a surjective homomorphism $\pi : G \to \overline{G}$, mapping the A_r with $r \neq 2\alpha$ onto the root-subgroups of $\overline{G}$ corresponding to the roots of some apartment $\mathcal{A}$ of $\mathcal{B}$. Moreover $\ker \pi \leq Z(G) \cap H$, H as in (4.3).

This is theorem 1 of [Ti4]. Contrary to (4.3) it also holds for root-systems of type G_2, ${}^2F_4$ and also of type H_3, H_4 and $I_2(m)$, in the latter case showing that for Φ of type H_3, H_4 or $I_2(m)$ with $m > 8$ no such group exists. The proof is constructive in so far, that from the conditions we construct the building $\mathcal{B}$ and then show that the A_r act as root-groups in the sense of §3 on $\mathcal{B}$.

The second theorem of [Ti4] is the starting point for all the later development.

(4.5) **Theorem.** Suppose G satisfies (St) with Φ not of type ${}^2F_4$ and

(+) always equality holds in (4.2)(1).

Then G satisfies the hypothesis $(*)$ of (4.4) and whence the conclusion.

Notice that for Chevalley-groups of type C_ℓ and F_4 in characteristic two condition $(+)$ is not satisfied. This is one of the reasons why characteristic two is an exception in (4.3). The proof of (4.5) consists of extensive commutator calculations and applications of the theory of rank one groups, in particular (2.7), (2.5) and [Ti1, I(3.7)], which shows that a quadratic module V for $X = (P)SL_2(k)$ is a direct sum of natural modules, if $V = [V, X]$ and $C_V(X) = 0$.

Next in [Ti5] the case when Φ is of type A_ℓ, D_ℓ or E_ℓ is treated, which is particularly easy:

(4.6) **Proposition.** Suppose G satisfies (St) with Φ of type A_ℓ, D_ℓ or E_ℓ. Then $\Phi = \Phi_1 \cup \cdots \cup \Phi_k$ with root-subsystems Φ_i of Φ and $G = G(\Phi_1) * \cdots * G(\Phi_k)$ where $G(\Phi_i) = \langle X_r \mid r \in \Phi_i \rangle$. Moreover either $G(\Phi_i) = X_r$ (if $\Phi_i = \{\pm r\}$) or $G(\Phi_i)$ satisfies the conclusion of (4.4) with respect to the root-system Φ_i.

The proof of (4.6) consists of two steps. First a lemma, which solves (4.6) in case $\Phi = A_2$, which is used as induction basis:

(4.7) **Lemma.** Suppose G satisfies (St) with $\Phi = \{\pm r, \pm s, \pm(r+s)\}$ of type A_2. Then one of the following holds:

(i) $G = X_r * X_s * X_{r+s}$.
(ii) Always equality holds in the commutator-relations (4.2)(1).

This is [Ti1, II(1.1)]. The proof is very elementary, but is the basis for similar proofs in more complicated cases. (I.e. Φ of type B_2 or BC_2!) And secondly a 3-transposition argument applied to the Weyl-group of G.

Now, as the example after (4.3) shows, (4.6) is no longer true in case Φ is of type B_ℓ, C_ℓ, BC_ℓ or F_4. In fact the possible decompositions of Φ and G are quite complicated and in case $\Phi = F_4$ it is possible that one central factor of G is of type A_5, i.e. is of higher rank than Φ. But still one needs to describe the situation in case $\Phi = B_2 = C_2$ first. This is done in the main-result of my student C. Müller [Mü]:

(4.8) **Proposition.** Suppose G satisfies (St) with Φ of type C_2. Then one of the following holds:

(i) $G = X_\alpha * C(X_\alpha)$ for some long root $\alpha \in \Phi$. Moreover $X_\beta \leq C(X_\alpha)$ for all $\beta \in \Phi - \{\pm\alpha\}$.
(ii) A_α is an elementary abelian 2-group for all $\alpha \in \Phi$.
(iii) Always equality holds in (4.2)(1).

For the proof of (4.8) central involutions in $X_\alpha, \alpha \in \Phi$ long, obtained from (2.9), play a central role. Using (4.8), in the next step the case when Φ is of type D_ℓ, C_ℓ or F_4 was treated in [Ti6]. We can state the main result of this paper as follows.

(4.9) **Theorem.** Suppose G satisfies (St) with Φ of type $B_\ell, C_\ell, \ell \geq 2$ or F_4. Then one of the cases of (4.3) holds.

For the proof of (4.9) one needs first to sharpen (4.8) a little bit. That is one shows that in case (i) either G is a central product of rank one groups or $\Psi = \Phi - \{\pm\alpha\}$ carries the structure of a root-system of type A_2, $G(\Psi)$ satisfies case (i) of the Main-theorem with $\mathcal{B}$ a Moufang-plane and $G = X_\alpha * G(\Psi)$. Let now $H = \langle H_r \mid r \in \Phi\rangle$, $N = \langle H, n_r \mid r \in \Phi\rangle$, where $n_r \in X_r$ interchanges A_r and A_{-r}, and $\overline{N} = N/H$. Then it follows essentially from (4.7) and the slightly strengthened version of (4.8), that $\{\overline{n}_r \mid r \in \Phi\}$ is a normal set of $\{3,4\}$-transpositions of $\overline{N}$. (For example if Φ is of type B_2 but $G = X_\alpha * G(\Psi)$, then $W(\Phi) \simeq D_8$ passes into $\Sigma_3 \times \mathbb{Z}_2$. Hence if Φ is of type B_ℓ, C_ℓ or F_4 and $\alpha, \beta \in \Phi$ and $\langle\alpha,\beta\rangle$ is the root-subsystem spanned by α and β, we know the structure of $G(\langle\alpha,\beta\rangle)$ and whence of $\langle\overline{n}_\gamma \mid \gamma \in \langle\alpha,\beta\rangle\rangle$. From this one obtains that $o(\overline{n}_\alpha\overline{n}_\beta) \leq 4$ and $\overline{n}_\alpha^{\overline{n}_\beta} = \overline{n}_\delta$ for some $\delta \in \langle\alpha,\beta\rangle$, which shows that $\{\overline{n}_r \mid r \in \Phi\}$ is a set of $\{3,4\}$-transpositions of $\overline{N}$.) Now the proof of (4.9) proceeds by discussing the possibilities for $\overline{N}$, which in turn give us the corresponding possibilities for G. Since very many possibilities arise, which are simply put together as case (ii) in (4.3), the proof is quite complicated in detail.

Finally the case Φ of type BC_ℓ remains to be treated, which is done in [Ti7]. Here the result is the same.

(4.10) **Theorem.** Suppose G satisfies (St) with Φ of type $BC_\ell, \ell \geq 2$. Then one of the cases of (4.3) hold.

For the proof of (4.10) one first has to prove a similar result as (4.8) in case Φ is of type BC_2. Now one considers the root-subsystem

$$\Phi_0 = \{2r \mid r, 2r \in \Phi\} \cup \{s \in \Phi \mid 2s \notin \Phi\} \text{ of type } C_\ell.$$

One finally gets the possibilities for G from the possibilities for $G(\Phi_0)$ with the help of the description of the rank two subgroups. (I.e. structure of $G(\langle\alpha,\beta\rangle)$; $\alpha,\beta \in \Phi$!) Notice that in case (i) of the Main-theorem with Φ of type BC_ℓ, G is not necessarily of "type BC_ℓ". For example it may happen that $A_r = A_{2r}$ for all $r \in \Phi$ with $2r \in \Phi$. In that case $G = G(\Phi_0)$ is of "type C_ℓ".

It should be mentioned that C. Müller is working on the case Φ of type G_2 of the Main-theorem. It is well known that in this case characteristic three is an exceptional case, i.e. if G is a Chevalley-group of type G_2 in characteristic three one does not have equality in the commutator-relations (4.2)(1), whence one can not use (4.5) to identify G. Hence the main problem in case Φ of type G_2 is to discover, on a purely group theoretic level, why characteristic three is different from the others.

§ 5 The Curtis-Tits-presentation.

Let $\mathcal{B}$ be an irreducible, spherical Moufang-building of rank $\ell \geq 2$, $\mathcal{A}$ an apartment of $\mathcal{B}$ and Φ the set of roots of $\mathcal{A}$. Choose a fundamental system $\Pi = \{r_1, \cdots, r_\ell\}$ in Φ and let $G = \langle A_r \mid r \in \Phi\rangle$ be the group of Lie-type $\mathcal{B}$ in the notation of §3. Then it follows from (3.6)(4) that $G = \langle X_r \mid r \in \Pi\rangle$, since each root $r \in \Phi$ is by (3.6)(4) conjugate under W to some fundamental root and since $X_r^n = X_{r^n}$ for $r \in \Phi, n \in N$

and $N = H\langle n_\alpha \mid \alpha \in \Pi\rangle$. (See [Ti1, II(5.12)]). Let now R be the amalgamated product of the $X_{r,s} = \langle X_r, X_s\rangle$, $r, s \in \Pi$ amalgamated over the $X_r, r \in \Pi$. Then the following "theorem" is known as the Curtis-Tits-presentation for G.

"Theorem." Let $\pi : R \to G$ be the natural homomorphism. Then $\ker \pi \leq Z(R)$.

Clearly, by definition of R as amalgamated product, such a natural surjective homomorphism $\pi : R \to G$ exists. Hence the whole problem is to show $\ker \pi \leq Z(R)$.

Now I do not know what the exact status of this "Theorem" is, since Tits in [Tits2, 13.32] has a stronger hypothesis. Namely instead of the groups $X_r, r \in \Pi$ he takes the subgroups $Y_r = X_r H$, $r \in \Pi$ and $Y_{r,s} = X_{r,s}H$, $r, s \in \Pi$. (Notice that by (3.6)(4) H normalizes all $A_r, r \in \Phi$ and whence all $X_r, r \in \Phi$!) and then forms the amalgamated product R of the $Y_{r,s}$ over the Y_r, $r, s \in \Pi$. Under this stronger hypothesis his conclusion is also stronger, namely he shows that R is isomorphic to G. Notice that such a conclusion is of course false in the more general situation of the "theorem", since in the special situation when G is an adjoint Chevalley-group the universal Chevalley-group might be an image of R. Also Curtis in [Cu] only treated a special case of our "Theorem". But it should be mentioned that Gorenstein, Lyons and Solomon in [GLS] proved the above theorem in case of finite groups of Lie-type.

Now for applications one would like to have a theorem without the hypothesis that R is the amalgamated product of the $X_{r,s}$, $r, s \in \Pi$. Just taking for R a group generated by subgroups Y_r isomorphic to X_r such that $\langle Y_r, Y_s\rangle$ is also isomorphic to $X_{r,s}$ and the diagram obtained naturally in this situation is the diagram Δ of $\mathcal{B}$. Such a theorem has been proved as Theorem 3 of [Ti2]:

(5.1) **Theorem.** Let Δ be a spherical Dynkin-diagram of rank ℓ and let R be a group generated by rank one groups Y_i, $i \in I = \{1, \cdots, \ell\}, \ell \geq 2$, with unipotent subgroups A_i and A_{-i} and diagonal subgroups $H_i = N_{Y_i}(A_i) \cap N_{Y_i}(A_{-i})$ satisfying:

(1) $H_i \leq N(Y_j)$ for $1 \leq i, j \leq \ell$.
(2) $Y_{ij} = \langle Y_i, Y_j\rangle = Y_i * Y_j$ if and only if i and j are not connected in Δ (and $i \neq j$).
(3) If i and j are connected in Δ, then there exist a Moufang-plane or classical Moufang-quadrangle $\mathcal{B}_{ij}$ with corresponding Lie-type group G_{ij} and a surjective homomorphism $\pi : Y_{ij} \to G_{ij}$ mapping the unipotent subgroups of Y_i and Y_j onto root-subgroups of G_{ij} (in the sense of §3) with $\ker \pi \leq Z(Y_{ij})$. Moreover, if $\mathcal{B}_{ij}$ is a Moufang quadrangle, then root-subgroups corresponding to short and long roots occur as images.
(4) If i and j are connected in Δ, then G_{ij} is not defined over $GF(2)$ and $GF(3)$. Moreover, if $G_{ij} \simeq PSL_3(4)$, then $2 \not| |Z(Y_{ij})|$.

Then there exist a spherical Moufang-building $\mathcal{B}$ of rank ℓ with corresponding Lie-type group G and a surjective homomorphism $\sigma : R \to G$ with $\ker \sigma \leq Z(R)$ mapping each $Y_i, i = 1, \cdots, \ell$ onto X_{r_i}, where $\Pi = \{r_1, \cdots, r_\ell\}$ is a fundamental root-system with $\Delta = \Delta(\Pi)$. Moreover $\ker \sigma \leq H = \langle H_i \mid i = 1, \cdots, \ell\rangle$.

In the above theorem classical Moufang-quadrangles are just the Moufang quadrangles obtained from classical groups of Witt-index 2. Notice that by the classification of spherical buildings of rank $\ell \geq 3$ only classical Moufang-quadrangles occur as residues if rank $\mathcal{B} \geq 3$. Hence all groups of Lie-type $\mathcal{B}$ with rank $\mathcal{B} \geq 3$ satisfy the hypothesis of (5.1), so long they are not defined over $GF(2)$ and $GF(3)$. Using the classification of Moufang polygons by Tits and Weiss, it might be possible to get rid of the condition that the Moufang-quadrangles have to be classical. But for this one would need to have a good knowledge of the exceptional Moufang-quadrangles and corresponding Lie-type groups, which I don't have.

The proof of (5.1) works in principal as follows:

(a) First, using the nontrivial action of the H_i on A_i, one shows that one can arrange in (3) the "local" homomorphisms $\pi : Y_{ij} \to G_{ij}$ by applying an automorphism of G_{ij}, such that π maps A_i and A_j onto "fundamental" root-subgroups of G_{ij}.
(b) Next one constructs a Weyl-group of R and, using the action of this Weyl-group, one extends the commutator relations between the fundamental root groups $A_i, A_j, 1 \leq i \neq j \leq \ell$ to arbitrary root-groups.
(c) Now, using (b) one constructs a BN-pair in R, similarly as in the proof of (4.4).
(d) Now, since unfortunately arbitrary groups with a spherical BN-pair are not known, I proved a theorem similar to the theorem of Seitz [Se] determining chamber transitive subgroups of finite Lie-type groups, to identify G.

Now, using (4.4) and the arguments in part (b) of the proof of (5.1) one can show:

(5.2) **Theorem.** Let Φ be an irreducible spherical root-system of rank $\ell \geq 2$ with fundamental system Π (BC_ℓ and ${}^2F_4$ are allowed). For each $J \subseteq \Pi$ with $|J| = 2$ let $\Phi_J = \{r \in \Phi \mid r \text{ is a linear combination of the roots in } J\}$ and let $\Phi' = \bigcup \Phi_J$, $J \subseteq \Pi$ with $|J| = 2$. Let G be a group generated by nonidentity subgroups $A_r, r \in \Phi'$ satisfying:

(1) $X_\alpha = \langle A_\alpha, A_{-\alpha} \rangle$ is a rank one group for all $\alpha \in \Phi'$.
(2) If $\alpha, \beta \in \Phi_J$ with $\beta \neq -\alpha$ or -2α, then

$$[A_\alpha, A_\beta] \leq \langle A_{\lambda\alpha+\mu\beta} \mid \lambda, \mu \in \mathbb{N} \text{ and } \lambda\alpha + \mu\beta \in \Phi_J \rangle.$$

(3) If $\alpha, \beta \in \Phi_J$ and $n_\beta \in X_\beta$ interchanging A_β and $A_{-\beta}$, then $A_\alpha^{n_\beta} = A_{\alpha^{w_\beta}}$; where w_β is the reflection on Φ_J corresponding to β.

Then (1) - (3) hold for all $\alpha, \beta \in \Phi$. In particular there exists by (4.4) a group $\overline{G}$ of Lie-type $\mathcal{B}$, $\mathcal{B}$ an irreducible spherical Moufang building and a surjective homomorphism $\pi : G \to \overline{G}$ with $\ker \pi \leq Z(G) \cap H$ (H as in (4.3)) mapping the $A_r, r \in \Phi$ onto the root-subgroups corresponding to the roots of some apartment of $\mathcal{B}$.

The proof of this theorem is quite easy. Namely after having constructed a Weyl-group one extends (1) - (3) by conjugation of the Weyl-group. Now the theorem follows from (4.4).

It is clear that the original Curtis-Tits presentation as stated is a consequence of (5.2), since (1) - (3) of (5.2) can be checked inside the group $X_{rs}; r, s \in \Pi$. In fact, using (4.3) instead of (4.4) one can prove a version of (5.2) without assuming (3), if the characteristic is different from 2. From (5.2) one can obtain a version of (5.1), which holds for all fields. (I.e. also for G_{ij} defined over $GF(2)$ and $GF(3)$.) Moreover the proof of this generalization of (5.1) is much easier than the original proof.

References

[Cu] Curtis, Ch. W., Central extensions of groups of Lie-type, J. Reine Angew. Math. 220 (1965) pp. 174-185.

[GLS] Gorenstein, D. Lyons, R., Solomon, R., The classification of the finite simple groups. Number 3, Part I, Chapter A: Almost simple K-groups. Vol 40.3 of Math. Surveys and Mon. AMS, Providence 1998.

[Mü] Müller, C., On the Steinberg-presentation for Lie-type-groups of type C_2, J. of Alg. 252 (2002) 150–160

[Ro] Ronan, M., Lectures on buildings, Vol. 7, Perspectives in Math., Academic Press, Boston 1989.

[Se] Seitz, G., Flag-transitive subgroups of Chevalley-groups. Ann. of Math. 97 (1973) 27-56.

[St] Steinberg, R., Lectures on Chevalley Groups. Lecture Notes, Yale University 1967.

[Ti1] Timmesfeld, F. G., Abstract Root Subgroups and simple groups of Monographs in Math. 95, Birkhäuser Verlag, Basel 2001.

[Ti2] Timmesfeld, F. G., Structure and Presentations of Lie-type groups, Proceedings of the LMS 81, (2000), 428 - 484 .

[Ti3] Timmesfeld, F. G., Abstract root subgroups and quadratic action, Adv. Math 142 (1999) 1-150.

[Ti4] Timmesfeld, F. G., On the Steinberg-presentation for Lie-type groups, to appear in Forum Math. (2002).

[Ti5] Timmesfeld, F. G., A remark on presentations of certain Chevalley groups, to appear in Archiv der Mathematik (2002).

[Ti6] Timmesfeld, F. G., Groups with a central factor of Lie-type. to appear in J. of Algebra.

[Ti7] Timmesfeld, F. G., Groups with Root-system of type BC_ℓ, to appear in Beiträge zur Algebra und Geometrie.

[Tits1] Tits, J., Endliche Spiegelungsgruppen die als Weylgruppen auftreten. Invent. Math. 43 (1977) 283-295.

[Tits2] Tits, J., Buildings of spherical type and finite BN-pairs. Lecture Notes in Math. Vol 386, Springer Verlag 1974.

[Tits3] Tits, J., Moufang polygons I: Root data, Bull. Belg. Math. Soc. 1 (1994) 455-468.

[VM] Van Maldeghem, H., Generalized polygons, Monographs in Math. 95, Birkhäuser Verlag, Basel, 1998.

Vertex stabilizers of locally projective groups of automorphisms of graphs. A summary

V.I. TROFIMOV

1. Introduction

For a graph Γ (in this paper only undirected graphs without loops or multiply edges are considered), let $V(\Gamma)$, $E(\Gamma)$ and $Aut(\Gamma)$ denote the vertex set, the edge set and the automorphism group, respectively. For $x \in V(\Gamma)$, let $\Gamma(x) = \{y \in V(\Gamma) | \{x, y\} \in E(\Gamma)\}$. For $G \leq Aut(\Gamma)$, let G_x be the stabilizer in G of x, and $G_x^{\Gamma(x)}$ the restriction of G_x on $\Gamma(x)$.

Assume that the following conditions hold:

(*) Γ is a connected graph, G is a vertex-transitive group of automorphisms of Γ, $x \in V(\Gamma)$, G_x is finite, and the group $G_x^{\Gamma(x)}$ has a normal subgroup which is isomorphic as a permutation group to $PSL_n(q)$, where $n \geq 2$ and q is a power of a prime p, acting in the natural way on the set of points of the projective space $PG_{n-1}(q)$.

Under this assumption, what is the possible structure of G_x?

This problem is important in different contexts. In particular, investigations on the pushing up problem (see, for example, [12]), diagram geometries (see, for example, [9], [10]), and 2-transitive graphs (see, for example, [28]) can be indicated.

The starting point of consideration of this problem was the case $n = 2$, $q = 2$ treated in [23], [24]. The case $n = 2$, q an arbitrary, was considered in [5]-[7] and in [27]. In this case the description of G_x can be also derived from [14] and from [4] (for $p = 2$, also from [2]).

Turn to the case $n > 2$. After a period when some basic observations were made and some rather restricted subcases were handled (see below), this case was considered in [17] (the proof was published in the series [19]-[21]).

In the present paper we describe this result (for $n > 2$) with some details and examples. Note that in the case $n = 3$ the result can be also derived from [13]. In addition, note that the classification of finite simple groups did not use in the proof of the result.

2. Preliminaries

This section contains notation and background results used throughout the paper.

Let Γ be a graph, and $G \leq Aut(\Gamma)$. For $x \in V(\Gamma)$ and a non-negative integer i, $G_x^{[i]}$ denotes the pointwise stabilizer in G of the set of vertices of Γ which are at distance at most i from x. Note that $G_x^{[i]} = G_x^{[i+1]}$ implies $G_x^{[i]} = 1$ in the case Γ

is connected and G is vertex-transitive. For $y, y', ... \in V(\Gamma)$, put $G^{[i]}_{y,y',...} = G^{[i]}_y \cap G^{[i]}_{y'} \cap ...$ (in [17] and [19]-[21], following [26], this group is denoted by $G_i(y, y', ...)$). As usually, we write $G_{y,y',...}$ for $G^{[0]}_{y,y',...}$. For a non-negative integer l, an l-arc of Γ is a path $(x_0, ..., x_l)$ of Γ such that $x_{k-1} \neq x_{k+1}$ for all $0 < k < l$. If G is vertex-transitive, then $s(G)$ is the maximum of l such that G acts transitively on the set of l-arcs of Γ, in the case the maximum exists, or ∞, otherwise.

From now on, we will assume that Γ and G satisfy (*) with $n > 2$, unless otherwise stipulated.

For any $z \in V(\Gamma)$, there is a unique structure of projective space $PG_{n-1}(q)$ on the set $\Gamma(z)$ (as the set of points) for which the group $G_z^{\Gamma(z)}$ is a group of collineations. Throughout the paper, considering $\Gamma(z)$ as a projective space we mean this structure.

For $\{z, z'\} \in E(\Gamma)$, the set $\Gamma(z)/\langle z' \rangle$ of all lines of $\Gamma(z)$ containing z' has an obvious structure of projective space $PG_{n-2}(q)$, and the group $G_{z,z'}$ induces on $\Gamma(z)/\langle z' \rangle$ a group of collineations.

For $\{z, z'\} \in E(\Gamma)$ and $Z \subseteq \Gamma(z')$, we denote by $T_{z'}(z)$ the subgroup of $G_{z'}^{\Gamma(z')}$ generated by all (projective) transvections with the center z, and by $T_{z'}(z, Z)$ the pointwise stabilizer in $T_{z'}(z)$ of Z. It is easy to see that the kernel of the action of $G_{z,z'}^{\Gamma(z')}$ on $\Gamma(z')/\langle z \rangle$ is an extension of $T_{z'}(z) = O_p(G_{z,z'}^{\Gamma(z')})$ by a cyclic group of order dividing $q-1$, and either $G_z^{[1]} = 1$ or $T_{z'}(z) = O_p((G_z^{[1]})^{\Gamma(z')})$.

It follows from [5], (2.3) (which is an analog of the Thompson–Wielandt theorem), that, for $\{z, z'\} \in E(\Gamma)$, $G^{[1]}_{z,z'}$ is a p-group (see [29]). Hence either $G_z^{[1]} = 1$, or $O_p(G_z)$ is the preimage of the subgroup $T_{z'}(z)$ of $(G_z^{[1]})^{\Gamma(z')}$ in $G_z^{[1]}$. As a result, $O_p(G_x)/G_x^{[2]}$ and $G_x^{[i]}/G_x^{[i+1]}$ for $i \geq 2$ are elementary abelian p-groups. Another consequence is that $s(G)$ is equal to 2 or 3 (cf. [25], [29]).

If $s(G) = 3$, then for each 2-arc (z_0, z_1, z_2) of Γ there is a (unique determined) collineation φ_{z_0,z_1,z_2} of the projective space $\Gamma(z_0)/\langle z_1 \rangle$ onto the projective space $\Gamma(z_2)/\langle z_1 \rangle$ which commutes with the natural action of the group G_{z_0,z_1,z_2}. If $s(G) = 2$, then for each 1-arc (z_0, z_1) of Γ there is a (unique determined) bijection φ_{z_0,z_1} of the set of subspaces of the projective space $\Gamma(z_0)/\langle z_1 \rangle$ with the set of subspaces of the projective space $\Gamma(z_2)/\langle z_1 \rangle$ which commutes with the natural action of the group G_{z_0,z_1}. Excluding the case $n = 3$ (when φ_{z_0,z_1} is as a correlation as a collineation), the mapping φ_{z_0,z_1} is either a correlation (for each 1-arc (z_0, z_1)) or a collineation (for each 1-arc (z_0, z_1)). (See [26], [29].) Accordingly, we distinguish for Γ and G the **case** $s(G) = 3$, the **case** $s(G) = 2$, $n = 3$, the **correlation case** (i.e. the case where $s(G) = 2$, $n > 3$ and φ_{z_0,z_1} is a correlation for each 1-arc (z_0, z_1) of Γ), and the **collineation case** (i.e. the case where $s(G) = 2$, $n > 3$ and φ_{z_0,z_1} is a collineation for each 1-arc (z_0, z_1) of Γ).

Assume $s(G) = 3$. For $y \in \Gamma(x)$, the group $G_x^{[1]}$ induces on the projective space $\Gamma(y)/\langle x \rangle$ a group of collineations containing the projective special linear group (see [29]). Since the kernel of this action of $G_x^{[1]}$ is an extension of $O_p(G_x)$ by a cyclic group of order dividing $q-1$, it follows that $G_x/O_p(G_x)$ has a (unique) subgroup $H_1 \times H_2$ where H_1 is isomorphic to the quotient group of $SL_n(q)$ by a subgroup of

$Z(SL_n(q))$, and H_2 is isomorphic to the quotient group of $SL_{n-1}(q)$ by a subgroup of $Z(SL_{n-1}(q))$. The group H_1 acts naturally on $\Gamma(x)$ inducing the projective special linear group of collineations. In the case $H_1 \cong SL_n(q)$, this action lifts to the natural action of H_1 on the n-dimensional vector space V_1 over $\mathbf{F}_q$ associated with $\Gamma(x)$. As a module of H_1 over $\mathbf{F}_q$ (respectively $\mathbf{F}_p$), V_1 is the *natural* $\mathbf{F}_q H_1$-module (respectively the *natural* $\mathbf{F}_p H_1$-module). Similarly, the group $H_2 \leq G_x^{[1]}/O_p(G_x)$ acts naturally on $\Gamma(y)/\langle x\rangle$ inducing the projective special linear group. In the case $H_2 \cong SL_{n-1}(q)$, this action lifts to the natural action of H_2 on the $(n-1)$-dimensional vector space V_2 over $\mathbf{F}_q$ associated with $\Gamma(y)/\langle x\rangle$. As a module of H_2 over $\mathbf{F}_q$ (respectively $\mathbf{F}_p$), V_2 is the *natural* $\mathbf{F}_q H_2$-module (respectively the *natural* $\mathbf{F}_p H_2$-module).

Assume now $s(G) = 2$. The group $G_x^{[1]}$ is an extension of $O_p(G_x)$ by a cyclic group of order dividing $q-1$. Therefore the group $G_x/O_p(G_x)$ has a (unique) subgroup H_1 which is isomorphic to the quotient group of $\hat{H}_1 \cong SL_n(q)$ by a subgroup K of $Z(\hat{H}_1)$ (we identify $\hat{H}_1$ with H_1 in the case $K = 1$). The group H_1 acts naturally on $\Gamma(x)$ inducing the projective special linear group of collineations. This action lifts to the natural action of $\hat{H}_1$ on the n-dimensional vector space V_1 over $\mathbf{F}_q$ associated with $\Gamma(x)$. As a module of $\hat{H}_1$ over $\mathbf{F}_q$ (respectively $\mathbf{F}_p$), V_1 is the *natural* $\mathbf{F}_q\hat{H}_1$-module (respectively the *natural* $\mathbf{F}_p\hat{H}_1$-module). An $\mathbf{F}_p\hat{H}_1$-module centralized by K is also regarded as an $\mathbf{F}_p H_1$-module.

3. The case $G_x^{[2]} = 1$

In the case $G_x^{[2]} = 1$, the group G_x can be easily reconstructed. We outline the corresponding arguments, since the result was only announced in [18].

Suppose first that $G_x^{[1]} = 1$. Then $s(G) = 2$ and either $n = 3$ or the collineation case holds. In fact, for $y \in \Gamma(x)$ and $a \in O_p(G_{x,y})^\#$, the stabilizer in $G_{x,y}$ of the axis of $a^{\Gamma(x)} \in T_x(y)^\#$ must coincide with the stabilizer in $G_{x,y}$ of the axis of $a^{\Gamma(y)} \in T_y(x)^\#$, and the assertion follows.

EXAMPLE 3.1. Let V be an elementary abelian group of order 2^m generated by elements $v_1, ..., v_m$, where $m = 1 + q + ... + q^{n-1}$, q a power of a prime p, $n > 2$. Let S be the stabilizer in $Aut(V)$ of the set $\{v_1, ..., v_m\}$. Then S acts faithfully on $\{v_1, ..., v_m\}$, inducing the symmetric group Sym_m. Let X be a subgroup of S such that the group induced by X on $\{v_1, ..., v_m\}$ has a normal subgroup which is isomorphic as a permutation group to $PSL_n(q)$ acting naturally on the set of points of $PG_{n-1}(q)$. Denote by G the split extension of V by X. Put $Y = v_1 X v_1$. Let Γ be the graph whose vertices are all subgroups $g^{-1}Xg$, $g \in G$, and whose edges are all pairs $\{g^{-1}Xg, g^{-1}Yg\}$, $g \in G$. (The graph Γ is the m-dimensional cube.) Then the group G, acting on $V(\Gamma)$ by conjugation, can be regarded as a group of automorphisms of Γ. If $n = 3$, for the graph Γ and the group G the case $s(G) = 2$, $n = 3$ with $G_x^{[1]} = 1$ holds. If $n > 3$, for the graph Γ and the group G the collineation case with $G_x^{[1]} = 1$ holds.

Suppose now that $G_{x,y}^{[1]} = 1 \neq G_x^{[1]}$, where $\{x, y\} \in E(\Gamma)$. Then $s(G) = 3$, $H_2 \cong SL_{n-1}(q)$ and the group $O_p(G_x)$ (acting faithfully on $\Gamma(y)$ as $T_y(x)$) is the natural $\mathbf{F}_p H_2$-module and is centralized by H_1. In fact, $[G_x^{[1]}, G_y^{[1]}] \leq G_{x,y}^{[1]} = 1$

implies that the group $C_{G_x}(G_x^{[1]})^{\Gamma(x)}$ contains $PSL_n(q)$. It is not the case if $G_x^{[1]} \neq 1$ and $s(G) = 2$.

EXAMPLE 3.2. Let M_1 be a maximal parabolic subgroup of the finite simple group $G_1 = PSL_n(q) = A_{n-1}(q)$, $n > 2$, correlated to the node 1 of the Dynkin diagram of type A_{n-1}. Let $G_2 \cong G_1$. Put $G = (G_1 \times G_2)\langle h \rangle \leq Aut(G_1 \times G_2)$ where $h^2 = 1$ and $hG_1h = G_2$ (we identify $G_1 \times G_2$ with $Inn(G_1 \times G_2)$), $X = M_1 \times G_2 \leq G$, $Y = G_1 \times hM_1h \leq G$. Let Γ be the graph whose vertices are all subgroups $g^{-1}Xg$, $g \in G$, and whose edges are all pairs $\{g^{-1}Xg, g^{-1}Yg\}$, $g \in G$. (The graph Γ is the complete bipartite graph $K_{m,m}$, $m = 1 + q + ... + q^{n-1}$.) Then the group G, acting on $V(\Gamma)$ by conjugation, can be regarded as a group of automorphisms of Γ. For the graph Γ and the group G the case $s(G) = 3$ with $G_{x,y}^{[1]} = 1 \neq G_x^{[1]}$, $\{x, y\} \in E(\Gamma)$, holds.

Suppose finally that $G_x^{[2]} = 1 \neq G_{x,y}^{[1]}$, where $\{x, y\} \in E(\Gamma)$. Observe that now $O_p(G_x)$ is a non-trivial $\mathbf{F}_pH_1$-module and, in the case $s(G) = 3$, also a non-trivial $\mathbf{F}_pH_2$-module. Let $\{y_1, ..., y_n\}$ be a frame in $\Gamma(x)$. Since $O_p(G_{y_i})^{\Gamma(x)} = T_x(y_i)$ and $[O_p(G_{y_i}), G_{x,y_i}^{[1]}] \leq G_{y_i}^{[2]} = 1$ for $1 \leq i \leq n$, the group $O_p(G_x)$ acts faithfully on $\Gamma(y_1) \cup ... \cup \Gamma(y_n)$, inducing on $\Gamma(y_i)$ the group $T_{y_i}(x)$, $1 \leq i \leq n$. If $s(G) = 3$, it follows that $H_1 \cong SL_n(q)$, $H_2 \cong SL_{n-1}(q)$ and $O_p(G_x)$ is the tensor product of the dual of the natural $\mathbf{F}_qH_1$-module by the natural $\mathbf{F}_qH_2$-module, regarded in a natural way as $\mathbf{F}_p(H_1 \times H_2)$-module. Assume $s(G) = 2$. For $1 \leq i \neq j \leq n$, each hyperplane of $\Gamma(y_j)$ fixed by the stabilizer in $O_p(G_{y_i})$ of y_j contains $\varphi_{x,y_j}(\langle y_i, y_j \rangle)$. Since $(G_{x,y_i}^{[1]})^{\Gamma(y_j)} \leq T_{y_j}(x)$ and $O_p(G_{y_i}) \leq C_G(G_{x,y_i}^{[1]})$, it follows $(G_{x,y_i}^{[1]})^{\Gamma(y_j)} \leq T_{y_j}(x, \varphi_{x,y_j}(\langle y_i, y_j \rangle))$. (Moreover, consideration of the action of G_{x,y_i,y_j} on $\Gamma(y_j)$ gives $(G_{x,y_i}^{[1]})^{\Gamma(y_j)} = T_{y_j}(x, \varphi_{x,y_j}(\langle y_i, y_j \rangle))$.) We conclude that the group $O_p(G_x)$ is of order not greater than q^n if either $n = 3$ or the correlation case holds, and of order not greater than $q^{n(n-1)/2}$ if the collineation case holds. Now it is easy to see, taking in attention that the group $O_p(G_x)$ contains the $G_{x,y}$-invariant subgroup $G_{x,y}^{[1]}$ of index q^{n-1}, that in the case $G_x^{[2]} = 1 \neq G_{x,y}^{[1]}$, $\{x, y\} \in E(\Gamma)$, with $s(G) = 2$ the following assertions hold: if $n = 3$ or the correlation case holds, then $O_p(G_x)$ is the natural $\mathbf{F}_pH_1$-module; if the collineation case holds, then $O_p(G_x)$ is the exterior square of the dual of the natural $\mathbf{F}_q\hat{H}_1$-module, regarded in a natural way as $\mathbf{F}_pH_1$-module.

EXAMPLE 3.3. Let M_{n-1} and M_n be maximal parabolic subgroups of the finite simple group $PSL_{2n-1}(q) = A_{2n-2}(q)$, $n > 2$, correlated respectively to nodes $n-1$ and n of the Dynkin diagram of type A_{2n-2}, such that $M_{n-1} \cap M_n$ contains a Borel subgroup of $PSL_{2n-1}(q)$. Thus M_{n-1} and M_n are the stabilizers in $PSL_{2n-1}(q)$, acting naturally on the projective space $PG_{2n-2}(q)$, of an $(n-2)$-dimensional subspace and an $(n-1)$-dimensional subspace which are incident. Put $G = Aut(PSL_{2n-1}(q))$, $X = N_G(M_{n-1})$, $Y = N_G(M_n)$ (we identify $PSL_{2n-1}(q)$ with $Inn(PSL_{2n-1}(q))$). Let Γ be the graph whose vertices are all subgroups $g^{-1}Xg$, $g \in G$, and whose edges are all pairs $\{g^{-1}Xg, g^{-1}Yg\}$, $g \in G$. Then the group G, acting on $V(\Gamma)$ by conjugation, can be regarded as a group of automorphisms of Γ. For the graph Γ and the group G the case $s(G) = 3$ with $G_x^{[2]} = 1 \neq G_{x,y}^{[1]}$, $\{x, y\} \in E(\Gamma)$, holds.

EXAMPLE 3.4. Let M_1 and M_n be maximal parabolic subgroups of the finite simple group $PSL_{n+1}(q) = A_n(q)$, $n > 2$, correlated respectively to nodes 1 and n of the Dynkin diagram of type A_n, such that $M_1 \cap M_n$ contains a Borel subgroup of $PSL_{n+1}(q)$. Thus M_1 and M_n are the stabilizers in $PSL_{n+1}(q)$, acting naturally on the projective space $PG_n(q)$, of a point and a hyperplane which are incident. Put $G = Aut(PSL_{n+1}(q))$, $X = N_G(M_1)$, $Y = N_G(M_n)$ (we identify $PSL_{n+1}(q)$ with $Inn(PSL_{n+1}(q))$). Let Γ be the graph whose vertices are all subgroups $g^{-1}Xg$, $g \in G$, and whose edges are all pairs $\{g^{-1}Xg, g^{-1}Yg\}$, $g \in G$. Then the group G, acting on $V(\Gamma)$ by conjugation, can be regarded as a group of automorphisms of Γ. If $n = 3$, for the graph Γ and the group G the case $s(G) = 2$, $n = 3$ with $G_x^{[2]} = 1 \neq G_{x,y}^{[1]}$, $\{x, y\} \in E(\Gamma)$, holds. If $n > 3$, for the graph Γ and the group G the correlation case with $G_x^{[2]} = 1 \neq G_{x,y}^{[1]}$, $\{x, y\} \in E(\Gamma)$, holds.

EXAMPLE 3.5. Let M_{n-1} and M_n be maximal parabolic subgroups of the finite simple group $P\Omega_{2n}^+(q) = D_n(q)$, $n > 3$, correlated respectively to nodes $n-1$ and n of the Dynkin diagram of type D_n, such that $M_{n-1} \cap M_n$ contains a Borel subgroup of $P\Omega_{2n}^+(q)$. Thus M_{n-1} and M_n are the stabilizers in $P\Omega_{2n}^+(q)$, acting naturally on the associated polar space, of $(n-1)$-dimensional subspaces whose intersection is $(n-2)$-dimensional. Put $G = P\Omega_{2n}^+(q)\langle h \rangle \leq Aut(P\Omega_{2n}^+(q))$ where $h^{-1}M_{n-1}h = M_n$ and $h^{-1}M_n h = M_{n-1}$ (we identify $P\Omega_{2n}^+(q)$ with $Inn(P\Omega_{2n}^+(q))$), $X = N_G(M_{n-1})$, $Y = N_G(M_n)$. Let Γ be the graph whose vertices are all subgroups $g^{-1}Xg$, $g \in G$, and whose edges are all pairs $\{g^{-1}Xg, g^{-1}Yg\}$, $g \in G$. Then the group G, acting on $V(\Gamma)$ by conjugation, can be regarded as a group of automorphisms of Γ. For the graph Γ and the group G the collineation case with $G_x^{[2]} = 1 \neq G_{x,y}^{[1]}$, $\{x, y\} \in E(\Gamma)$, holds.

REMARK 3.1. A remark in [21, I], p. 1222, concerning the structure of G_x, for an arbitrary group G acting transitively on the set of 2-arcs of a connected graph Γ of finite valency, in the case $G_{x,y}^{[1]} = 1$ where $\{x, y\} \in E(\Gamma)$, is somewhat ambiguous. What was meant is that there exists a monomorphism χ of G_x into the stabilizer of a vertex of $K_{|\Gamma(x)|,|\Gamma(x)|}$ in $Aut(K_{|\Gamma(x)|,|\Gamma(x)|})$ and a mapping ψ of the set of vertices of Γ at distance at most 2 from x into $V(K_{|\Gamma(x)|,|\Gamma(x)|})$ such that $\psi(\Gamma(x)) = K_{|\Gamma(x)|,|\Gamma(x)|}(\psi(x))$ and, for all $g \in G_x$, $\chi(g)\psi = \psi g$ on the set of vertices of Γ at distance at most 2 from x. It is clear (see, for example, the case $G_{x,y}^{[1]} = 1 \neq G_x^{[1]}$ of this Section for $n = 3$, $q = 2, 3$) that there are normal subgroups of point stabilizers of finite doubly transitive groups, which are permutationally isomorphic to no $(G_x^{[1]})^{\Gamma(y)}$ with Γ, G, x, y as above in this Remark.

4. The case $s(G) = 3$

In [26], Weiss showed that $G_x^{[2]} = 1$ in the case $s(G) = 3$ with $p > 3$.

In [20, II], it was proved that $G_{x,y}^{[3]} = 1$ in the case $s(G) = 3$ with $p = 3$, where $\{x, y\} \in E(\Gamma)$. As it was mentioned in [20, II], Introduction, the latter result implies that $G_x^{[2]} = 1$ in the case $s(G) = 3$ with $p = 3$. The corresponding proof was given in [21, I], Appendix.

Turn to the case $s(G) = 3$ with $p = 2$, $n = 3$. It was proved in [20, II], Propositions 3.2 and 3.3, that in this case $G_{x,y}^{[5]} = 1$, where $\{x, y\} \in E(\Gamma)$. Using this result, a detail description of G_x was obtained in [21, I], Proposition 1.1. It implies that in the case $s(G) = 3$ with $p = 2$, $n = 3$ either $G_x^{[2]} = 1$ or the following assertion (a) holds:

(a) $H_1 \cong SL_3(q)$ and $H_2 \cong SL_2(q)$, $G_x^{[6]} = 1$, $G_x^{[5]}$ is the natural $\mathbf{F}_2 H_2$-module and is centralized by H_1, $G_x^{[4]}/G_x^{[5]}$ is the dual of the natural $\mathbf{F}_2 H_1$-module and is centralized by H_2, $G_x^{[3]}/G_x^{[4]}$ is the natural $\mathbf{F}_2 H_1$-module and is centralized by H_2, $G_x^{[2]}/G_x^{[3]}$ is the tensor product of the natural $\mathbf{F}_q H_1$-module by the natural $\mathbf{F}_q H_2$-module, regarded in a natural way as $\mathbf{F}_2(H_1 \times H_2)$-module, $O_2(G_x)/G_x^{[2]}$ is the tensor product of the dual of the natural $\mathbf{F}_q H_1$-module by the natural $\mathbf{F}_q H_2$-module, regarded in a natural way as $\mathbf{F}_2(H_1 \times H_2)$-module.

REMARK 4.1. It can be derived from the proof of Proposition 1.1 of [21, I] that in the case $s(G) = 3$ with $p = 2$, $n = 3$ and (a) the following holds: $\Phi(O_2(G_x)) = [O_2(G_x), O_2(G_x)] = G_x^{[3]}$, $C_{G_x}(G_x^{[3]}) = G_x^{[2]}$, $[O_2(G_x), G_x^{[3]}] = G_x^{[5]}$, $Z(O_2(G_x)) = G_x^{[4]}$.

EXAMPLE 4.1 ([26]). Let M_2 and M_3 be maximal parabolic subgroups of the finite simple group $F_4(q)$, q even, correlated respectively to nodes 2 and 3 of the Dynkin diagram of type F_4, such that $M_2 \cap M_3$ contains a Borel subgroup of $F_4(q)$. Put $G = Aut(F_4(q))$, $X = N_G(M_2)$, $Y = N_G(M_3)$ (we identify $F_4(q)$ with $Inn(F_4(q))$). Let Γ be the graph whose vertices are all subgroups $g^{-1}Xg$, $g \in G$, and whose edges are all pairs $\{g^{-1}Xg, g^{-1}Yg\}$, $g \in G$. Then the group G, acting on $V(\Gamma)$ by conjugation, can be regarded as a group of automorphisms of Γ. For the graph Γ and the group G the case $s(G) = 3$ with $p = 2$, $n = 3$ and (a) holds.

It was proved in [21, I] that $G_x^{[2]} = 1$ in the case $s(G) = 3$ with $p = 2$, $n > 3$.

5. The case $s(G) = 2$, $n = 3$

In [19], it was proved that $G_x^{[2]} = 1$ in the case $s(G) = 2$, $n = 3$ with $p > 3$.

It was proved in [20, I], Theorem 1, that in the case $s(G) = 2$, $n = 3$ with $p = 3$ either $G_x^{[2]} = 1$ or the following assertion (b) holds:

(b) $q = 3$ and $G_x^{[3]} = 1$, G_x is a split extension of the group $O_3(G_x) = G_x^{[1]}$ by the group $G_x/G_x^{[1]} = H_1 \cong SL_3(3)$, $G_x^{[2]}$ is the dual of the natural $\mathbf{F}_2 H_1$-module, $G_x^{[1]}/G_x^{[2]}$ is the natural $\mathbf{F}_2 H_1$-module, $[G_x^{[1]}, G_x^{[1]}] = Z(G_x^{[1]}) = G_x^{[2]}$.

EXAMPLE 5.1. The group $G = Aut(Fi_{22})$ has a subgroup X (which is maximal in a maximal subgroup of $Inn(Fi_{22})$ isomorphic to $\Omega_7(3)$) such that $|O_3(X)| = 3^6$ and $X/O_3(X) \cong SL_3(3)$. There exists Y conjugated to X in G such that, for the graph Γ whose vertices are all subgroups $g^{-1}Xg$, $g \in G$, and whose edges are all pairs $\{g^{-1}Xg, g^{-1}Yg\}$, $g \in G$, and for the group G, acting on $V(\Gamma)$ by conjugation and regarded as a group of automorphisms of Γ, the case $s(G) = 2$, $n = 3$ with $p = 3$ and (b) holds.

In [16], Timmesfeld treated the case $s(G) = 2$, $n = 3$ with $p = 2$ in a broader context of amalgams with rank 2 groups of Lie type in characteristic 2. It follows from his result that in the case $s(G) = 2$, $n = 3$ with $p = 2$ either $G_x^{[2]} = 1$ or the following assertion (c) holds:

(c) $q = 2$ and $G_x^{[3]} = 1$, G_x is a split extension of the elementary abelian group $G_x^{[1]}$ of order 2^4 by the group $G_x/G_x^{[1]} = H_1 \cong SL_3(2)$, such that the group $G_x^{[1]}$ (in additive notation) is the direct sum of the natural $\mathbf{F}_2H_1$-module $[G_x^{[1]}, G_x]$ and the trivial $\mathbf{F}_2H_1$-module $G_x^{[2]}$ of order 2.

REMARK 5.1. Let Γ and G satisfy the case $s(G) = 2$, $n = 3$ with $p = 2$ and (c). For any $z \in V(\Gamma)$, denote by c_z the involution of $G_z^{[2]}$. For any $y' \in \Gamma(x)$, $y'' \in \Gamma(x) \setminus \{y'\}$, denote by $y' + y''$ the vertex from $\langle y', y'' \rangle \setminus \{y', y''\}$. Then $c_x c_{y'} c_x c_{y''} = c_x c_{y'+y''}$ for all $y' \in \Gamma(x)$, $y'' \in \Gamma(x) \setminus \{y'\}$, and $[G_x^{[1]}, G_x]^{\#} = \{c_x c_y | y \in \Gamma(x)\}$. There are two conjugacy classes of complements to $G_x^{[1]}$ in G_x. A complement L to $G_x^{[1]}$ in G_x can be chosen such that, for an arbitrary $y \in \Gamma(x)$, there exists a collineation ω_y of $\Gamma(x)$ onto $\Gamma(y)$ which commutes with the natural action of the group L_y (on $\Gamma(x) \cup \Gamma(y)$). For such L, if $y \in \Gamma(x)$ and $z \in \Gamma(x) \setminus \{y\}$, then $c_z c_{\omega_y(z)} \in L$ and $(c_z c_{\omega_y(z)})^{\Gamma(x)} \in T_x(y, \langle y, z \rangle)^{\#}$. If L is a complement from another conjugacy class, then for L we have a situation which is analogous to one from Remark 6.1 below.

EXAMPLE 5.2. The group $G = Aut(M_{22})$ has a subgroup X (which is maximal in $Inn(M_{22})$) such that $|O_2(X)| = 2^4$ and $X/O_2(X) \cong SL_3(2)$. There exists Y conjugated to X in G such that, for the graph Γ whose vertices are all subgroups $g^{-1}Xg$, $g \in G$, and whose edges are all pairs $\{g^{-1}Xg, g^{-1}Yg\}$, $g \in G$, and for the group G, acting on $V(\Gamma)$ by conjugation and regarded as a group of automorphisms of Γ, the case $s(G) = 2$, $n = 3$ with $p = 2$ and (c) holds (see [9], [10]).

REMARK 5.2. Another example of Γ and G for which the case $s(G) = 2$, $n = 3$ with $p = 2$ and (c) holds is given in Remark 6.2 below.

6. The correlation case

It was proved in [19] that $G_x^{[2]} = 1$ in the correlation case with $p > 3$.

In [20, I], Theorem 3, it was shown that $G_{x,y}^{[3]} = 1$ in the correlation case with $p = 3$, and $G_{x,y}^{[2]} = 1$ in the correlation case with $p = 2$, where $\{x, y\} \in E(\Gamma)$. In [21, II], Appendix, it was derived from this result that $G_x^{[2]} = 1$ as in the correlation case with $p = 3$, as in the correlation case with $q > p = 2$. As for the correlation case with $q = 2$, in [21, II], Appendix, it was shown (using the equation $G_{x,y}^{[2]} = 1$, $\{x, y\} \in E(\Gamma)$, stated in [20, I], Theorem 3) that either $G_x^{[2]} = 1$ or the following assertion (d) holds:

(d) $G_x^{[3]} = 1$, G_x is a split extension of the elementary abelian group $G_x^{[1]}$ of order 2^{n+1} by the group $G_x/G_x^{[1]} = H_1 \cong SL_n(2)$, such that the group $G_x^{[1]}$ (in additive notation) is the direct sum of the natural $\mathbf{F}_2H_1$-module $[G_x^{[1]}, G_x]$ and the trivial $\mathbf{F}_2H_1$-module $G_x^{[2]}$ of order 2.

REMARK 6.1. Let Γ and G satisfy the correlation case with $q = 2$ and (d). For any $z \in V(\Gamma)$, denote by c_z the involution of $G_z^{[2]}$. For any $y' \in \Gamma(x)$, $y'' \in \Gamma(x) \setminus \{y'\}$, denote by $y' + y''$ the vertex from $\langle y', y''\rangle \setminus \{y', y''\}$. Then (see [21, II], Appendix) $c_x c_{y'} c_x c_{y''} = c_x c_{y'+y''}$ for all $y' \in \Gamma(x)$, $y'' \in \Gamma(x) \setminus \{y'\}$, and $[G_x^{[1]}, G_x]^\# = \{c_x c_y | y \in \Gamma(x)\}$ (cf. Remark 5.1). In addition, if $L \cong SL_n(2)$ is a complement to $G_x^{[1]}$ in G_x (all such complements are conjugate in G_x), then, for an arbitrary $y \in \Gamma(x)$, the group L_y stabilizes a unique hyperplane $Z_{y,L}$ of $\Gamma(y)$ ($x \notin Z_{y,L}$). For $z \in Z_{y,L}$, we have $c_y c_z \in L$ and $(c_y c_z)^{\Gamma(x)} \in T_x(y, \varphi_{y,x}(\langle x, z\rangle))^\#$.

EXAMPLE 6.1 ([21, II], Appendix). For each positive integer $n > 3$, we give an example of a graph Γ and a group G for which the correlation case with $q = 2$ and (d) holds. Let H be the group $SL_{n+1}(2) \times SL_{n+1}(2)$ realized in a natural way by ordered pairs of matrices. Define $\sigma_1, \sigma_2 \in Aut(H)$ by putting $\sigma_1(\mathrm{A}, \mathrm{B}) = (\mathrm{B}, \mathrm{A})$ and $\sigma_2(\mathrm{A}, \mathrm{B}) = ((\mathrm{A}^t)^{-1}, (\mathrm{B}^t)^{-1})$ for any element (A, B) of H, where A^t and B^t are the transposes of A and B. Let G be the split extension of H by the elementary abelian group $\langle \sigma_1, \sigma_2\rangle$. Let X be the subgroup of G generated by σ_1 and all elements of H of the form (A, A), where the matrix $\mathrm{A} = (\alpha_{ij})$ is such that $\alpha_{i,n+1} = 0$ for $i = 1, ..., n$. Let h be the element of G such that $\sigma_2 h = (\mathrm{M}', \mathrm{M}'') \in H$, where

$$\mathrm{M}' = \begin{pmatrix} 0 & 0 & 1 \\ 0 & \mathrm{E} & 0 \\ 1 & 0 & 0 \end{pmatrix}, \quad \mathrm{M}'' = \begin{pmatrix} 1 & 0 & 1 \\ 0 & \mathrm{E} & 0 \\ 1 & 0 & 0 \end{pmatrix}$$

(E is the $n-1$ by $n-1$ identity matrix). Put $Y = h^{-1}Xh$. Let Γ be the graph whose vertices are all subgroups $g^{-1}Xg$, $g \in G$, and whose edges are all pairs $\{g^{-1}Xg, g^{-1}Yg\}$, $g \in G$. Then the group G, acting on $V(\Gamma)$ by conjugation, can be regarded as a group of automorphisms of Γ. For the graph Γ and the group G the correlation case with $q = 2$ and (d) holds. The group $G_X^{[2]}$ coincides with $\langle \sigma_1\rangle$.

REMARK 6.2. For $n = 3$, the construction of Example 6.1 gives an example of a graph Γ and a group G for which the case $s(G) = 2$, $n = 3$ with $p = 2$ and (c) holds.

7. The collineation case

In [19], it was proved that $G_x^{[2]} = 1$ in the collineation case with $p > 3$.

In [20, I], Theorem 2, it was shown that $G_x^{[2]} = 1$ in the collineation case with $p = 3$. In addition, according to [20, I], §4, Remark 2, $G_x^{[2]} = 1$ in the collineation case with $q > p = 2$.

Turn to the collineation case with $q = 2$. Observe that now $O_2(G_x) = G_x^{[1]}$ and $G_x/G_x^{[1]} = H_1 \cong SL_n(2)$. In [21, II], it was proved that in the collineation case with $q = 2$ and $n = 4$ either $G_x^{[2]} = 1$ or one of the following assertions (e), (f) holds:

(e) $|G_x^{[1]}| = 2^{11}$, $G_x^{[4]} = 1$, $G_x^{[3]}$ is the trivial $\mathbf{F}_2H_1$-module of order 2, $G_x^{[2]}/G_x^{[3]}$ is the natural $\mathbf{F}_2H_1$-module, $G_x^{[1]}/G_x^{[2]}$ is the exterior square of the natural $\mathbf{F}_2H_1$-module;

(f) $|G_x^{[1]}| = 2^{14}$, $G_x^{[4]} = 1$, $G_x^{[3]}$ is the dual of the natural $\mathbf{F}_2H_1$-module, $G_x^{[2]}/G_x^{[3]}$ is the natural $\mathbf{F}_2H_1$-module, $G_x^{[1]}/G_x^{[2]}$ is the exterior square of the natural $\mathbf{F}_2H_1$-module.

REMARK 7.1. The structure of the vertex stabilizer G_x satisfying (e) (of the collineation case with $q = 2$ and $n = 4$) is as follows (see [21, IV]). $G_x^{[1]} = E_1 \times E_2$, where E_1 is an extraspecial group 2_+^{1+6} and E_2 is an elementary abelian group of order 2^4 which is normal in G_x, $G_x^{[3]} = Z(E_1)$, $G_x^{[2]} = Z(E_1) \times E_2$, and the group $G_x/Z(E_1)$ is the split extension of the elementary abelian subgroup $G_x^{[1]}/Z(E_1)$ (of order 2^{10}) by a subgroup $L \cong SL_4(2)$. Moreover, the $\mathbf{F}_2L$-module $G_x^{[1]}/Z(E_1)$ is the indecomposable extension of the natural $\mathbf{F}_2L$-module $E_2Z(E_1)/Z(E_1)$ by its exterior square $G_x^{[1]}/E_2Z(E_1)$, and the preimage of L in G_x is isomorphic to the universal covering group of L. For any $z \in V(\Gamma)$, denote by c_z the involution of $G_z^{[3]}$. For any $y' \in \Gamma(x)$, $y'' \in \Gamma(x) \setminus \{y'\}$, denote by $y' + y''$ the vertex from $\langle y', y''\rangle \setminus \{y', y''\}$. Then $c_{y'}c_{y''} = c_{y'+y''}$ for all $y' \in \Gamma(x)$, $y'' \in \Gamma(x) \setminus \{y'\}$, and $E_2^{\#} = \{c_y | y \in \Gamma(x)\}$. For an arbitrary $y \in \Gamma(x)$, there exists a collineation ω_y of $\Gamma(x)$ onto $\Gamma(y)$ which commutes with the natural action of the group L_y (on $\Gamma(x) \cup \Gamma(y)$).

REMARK 7.2. Let $H \cong SL_5(2)$ act naturally on an elementary abelian group W of order 2^5. Let R be the split extension of an elementary abelian group E of order 2^{10} by H acting on E as on the exterior square of W regarded as $\mathbf{F}_2H$-module. Let w be an arbitrary element of $W^{\#}$, and H_w the centralizer of w in H. It can be shown that the vertex stabilizer G_x satisfying (f) (of the collineation case with $q = 2$ and $n = 4$) is isomorphic to the subgroup WH_w of R.

EXAMPLE 7.1. The group $G = Co_2$ has a maximal subgroup X (the centralizer of an involution of class $2B$) such that $|O_2(X)| = 2^{11}$ and $X/O_2(X) \cong SL_4(2)$. There exists Y conjugated to X in G such that, for the graph Γ whose vertices are all subgroups $g^{-1}Xg$, $g \in G$, and whose edges are all pairs $\{g^{-1}Xg, g^{-1}Yg\}$, $g \in G$, and for the group G, acting on $V(\Gamma)$ by conjugation and regarded as a group of automorphisms of Γ, the collineation case with $q = 2$ and $n = 4$ holds (see [9], [10]). Thus for the graph Γ and the group G the collineation case with $q = 2$, $n = 4$ and (e) holds.

EXAMPLE 7.2. The group $G = J_4$ has a subgroup X (which is maximal in a maximal subgroup of G isomorphic to $2^{10} : L_5(2)$) such that $|O_2(X)| = 2^{14}$ and $X/O_2(X) \cong SL_4(2)$. There exists Y conjugated to X in G such that, for the graph Γ whose vertices are all subgroups $g^{-1}Xg$, $g \in G$, and whose edges are all pairs $\{g^{-1}Xg, g^{-1}Yg\}$, $g \in G$, and for the group G, acting on $V(\Gamma)$ by conjugation and regarded as a group of automorphisms of Γ, the collineation case with $q = 2$ and $n = 4$ holds (see [9], [10]). Thus for the graph Γ and the group G the collineation case with $q = 2$, $n = 4$ and (f) holds.

In [21, III], it was proved that in the collineation case with $q = 2$ and $n = 5$ either $G_x^{[2]} = 1$ or the following assertion (g) holds:

(g) $|G_x^{[1]}| = 2^{30}$, $G_x^{[5]} = 1$, $G_x^{[4]}$ is the dual of the natural $\mathbf{F}_2H_1$-module, $G_x^{[3]}/G_x^{[4]}$

is the natural $\mathbf{F}_2H_1$-module, $G_x^{[2]}/G_x^{[3]}$ is the exterior square of the natural $\mathbf{F}_2H_1$-module, $G_x^{[1]}/G_x^{[2]}$ is the exterior square of the dual of the natural $\mathbf{F}_2H_1$-module.

REMARK 7.3. In detail the structure of the vertex stabilizer G_x satisfying (g) (of the collineation case with $q = 2$ and $n = 5$) was investigated by A. Ivanov and the author for the visit of the author in Imperial College in May, 2001. In particular, it was shown that $\Phi(G_x^{[1]}) = [G_x^{[1]}, G_x^{[1]}] = G_x^{[3]}$, $C_{G_x}(G_x^{[3]}) = G_x^{[2]}$, $[G_x^{[1]}, G_x^{[2]}] = [G_x^{[1]}, G_x^{[3]}] = [G_x^{[2]}, G_x^{[2]}] = G_x^{[4]} = Z(G_x^{[1]})$.

EXAMPLE 7.3. The group $G = B$ has a maximal subgroup X such that $|O_2(X)| = 2^{30}$ and $X/O_2(X) \cong SL_5(2)$. There exists Y conjugated to X in G such that, for the graph Γ whose vertices are all subgroups $g^{-1}Xg$, $g \in G$, and whose edges are all pairs $\{g^{-1}Xg, g^{-1}Yg\}$, $g \in G$, and for the group G, acting on $V(\Gamma)$ by conjugation and regarded as a group of automorphisms of Γ, the collineation case with $q = 2$ and $n = 5$ holds (see [9], [10]). Thus for the graph Γ and the group G the collineation case with $q = 2$, $n = 5$ and (g) holds.

In [21, IV], it was proved that $G_x^{[2]} = 1$ in the collineation case with $q = 2$ and $n > 5$.

8. Some additional remarks

REMARK 8.1. (On the problem.) It is not difficult to see that the considered problem of description of the possible structure of G_x under the assumption that (*) holds is equivalent (for any fixed n and q) to the problem of description of the possible structure of a group P_1 under the following assumption: P_1, P_2 are finite groups with one and the same identity and coinciding group operations on $P_1 \cap P_2$, such that the following conditions (i)-(iii) hold:

(i) $PSL_n(q) \leq P_1/\bigcap_{g \in P_1}(P_1 \cap P_2)^g \leq P\Gamma L_n(q)$, and $(P_1 \cap P_2)/\bigcap_{g \in P_1}(P_1 \cap P_2)^g$ contains a maximal parabolic subgroup of the group $PSL_n(q) = A_{n-1}(q)$, correlated to the node 1 (or $n - 1$) of the Dynkin diagram of type A_{n-1};

(ii) no non-trivial subgroup of $P_1 \cap P_2$ is normalized by $P_1 \cup P_2$;

(iii) there exists a permutation h on $P_1 \cup P_2$ such that the restriction of h on P_i is an isomorphism of P_i onto P_{3-i} for $i = 1, 2$.

REMARK 8.2. (More about the problem.) It is natural to consider the problem of description of G_x satisfying (*) as a case of a general problem of reconstruction of the stabilizer of a vertex of a connected graph in a vertex-transitive group of automorphisms by the restriction of this stabilizer on the neighborhood of the vertex. The latter problem can be refined in the following way. Let R be a permutation group of finite degree. Denote by $\mathcal{P}(R)$ the set of pairs (Γ, G) where Γ is a connected graph and G is a vertex-transitive group of automorphisms of Γ such that, for $x \in V(\Gamma)$, the group G_x is finite and the group $G_x^{\Gamma(x)}$ is isomorphic as a permutation group to R. Denote by $\mathcal{S}(R)$ the set of isomorphic types of G_x for all $(\Gamma, G) \in \mathcal{P}(R)$ and $x \in V(\Gamma)$. Then the Vertex Stabilizer Reconstruction Problem for R (briefly, VSRP for R) is formulated as follows:

Is the set $\mathcal{S}(R)$ finite? (And in the case, what is the structure of elements of $\mathcal{S}(R)$?)

This problem is of interest only for groups R of some special types. There are many (transitive) groups R for which VSRP is solved negatively. In [28], Weiss conjectured that VSRP is solved affirmatively for any primitive group R. If R is primitive, $(\Gamma, G) \in \mathcal{P}(R)$, and $G^{[1]}_{x,y} \neq 1$ for $\{x,y\} \in E(\Gamma)$, then $G^{[1]}_{x,y}$ is a p-group for a prime p, and $F^*(G_x) = O_p(G_x)$ (see [11]). It follows that, in the case R is primitive and $G^{[1]}_{x,y} \neq 1$ for some $(\Gamma, G) \in \mathcal{P}(R)$ and $\{x,y\} \in E(\Gamma)$, the stabilizer in R of a point is a local subgroup.

The Weiss Conjecture was verified for many primitive groups R. Here we only remark on the case $Soc(R)$ is a primitive ($\mathcal{K}$-)group. The amalgam method and certain results on representations of nearly simple groups made it possible to confirm the conjecture in this case when $Soc(R)$ is distinct from $PSL_n(q)$, $n > 2$, acting on a class of maximal parabolics. (We give some details only in the case $G^{[1]}_{x,y}$ is a 2-group for an edge $\{x,y\}$, since for odd primes the appropriate results were not published. If $Soc(R)$ is a primitive ($\mathcal{K}$-)group, $(\Gamma, G) \in \mathcal{P}(R)$ and $G^{[1]}_{x,y}$ is a non-trivial 2-group for $\{x,y\} \in E(\Gamma)$, then [1] implies that one of the following holds: (1) $Soc(R)$ is a group of Lie type in characteristic 2 (or the derived group of a non-solvable group of Lie type in characteristic 2) acting on a class of maximal parabolics, (2) R is Sym_n, $n > 2$ and $n \neq 4, 6$, acting on the class of transpositions, (3) R is $SO^+_{2n}(2)$, $n > 3$, or $SO^-_{2n}(2)$, $n > 2$, acting on the class of transvections. If (1) with $Soc(R)$ distinct from $PSL_n(q)$, $n > 2$, holds, then it follows from [12] and [3] that $G^{[3]}_{x,y} = 1$. If (2) or (3) holds, then the amalgam method easily gives $G^{[3]}_{x,y} = 1$. Note that the amalgam method can be also applied in the case $Soc(R)$ is $PSL_3(q)$ acting on a class of maximal parabolics, see [16], [13].) If $Soc(R)$ is $PSL_n(q) = A_{n-1}(q)$ acting on a class of maximal parabolics, say on the class correlated to a node i of the Dynkin diagram of type A_{n-1}, then either (*) or $1 < i < n-1$ holds. The case when (*) holds is the subject of the present paper. For the case when $1 < i < n-1$ holds see [30], [22]. In conclusion, note that, taking into account the classification of finite doubly transitive groups, the result described in the present paper is the final step for affirmative solution of VSRP for doubly transitive groups R (see a survey in [28]).

REMARK 8.3. (On the Examples.) Let G be a vertex-transitive group of automorphisms of a connected graph Γ. Let $\tilde{\Gamma}$ be the regular tree of the same valency as Γ. There exists a mapping (called universal covering or fibering of Γ) $\pi : V(\tilde{\Gamma}) \to V(\Gamma)$ such that, for any $\tilde{x} \in V(\tilde{\Gamma})$, the restriction of π on $\tilde{\Gamma}(\tilde{x})$ is a bijection with $\Gamma(\pi(\tilde{x}))$. Let $\tilde{G} = \{\tilde{g} \in Aut(\tilde{\Gamma}) |$ there exists $g \in G$ such that $\pi\tilde{g} = g\pi\}$ be the covering of G with respect to π. It is well known (and can be easily seen) that $s(\tilde{G}) = s(G)$ and, for any $\tilde{x} \in V(\tilde{\Gamma})$ and each non-negative integer i, the group $\tilde{G}^{[i]}_{\tilde{x}}$ is isomorphic as an abstract group to $G^{[i]}_{\pi(\tilde{x})}$. In addition, the restriction of π on $\tilde{\Gamma}(\tilde{x})$ realizes a permutation isomorphism of $\tilde{G}^{\tilde{\Gamma}(\tilde{x})}_{\tilde{x}}$ with $G^{\Gamma(\pi(\tilde{x}))}_{\pi(\tilde{x})}$. (As a result, in VSRP for a given R, formulated in Remark 8.2, without loss we can assume that Γ is a tree.) Suppose that, in addition, $G^{\Gamma(x)}_x$ is transitive, where $x \in V(\Gamma)$. For $\{\tilde{x}, \tilde{y}\} \in E(\tilde{\Gamma})$, the group $\langle G_{\tilde{x}}, G_{\tilde{y}} \rangle$ coincides with the edge-transitive (not vertex-transitive) subgroup $\tilde{G}^+$ of index 2 in $\tilde{G}$, generated by all vertex stabilizers in $\tilde{G}$. The group $\tilde{G}^+$ is isomorphic to the amalgamated product $\tilde{G}_{\tilde{x}} *_{\tilde{G}_{\tilde{x}} \cap \tilde{G}_{\tilde{y}}} \tilde{G}_{\tilde{y}}$ (see [15]). It follows that, in the case G_x is finite and $|\Gamma(x)| > 1$, there are infinitely many normal subgroups N of finite index in $\tilde{G}$ such that the restriction of the natural mapping $\tilde{G} \to \tilde{G}/N$ on $\tilde{G}_{\tilde{x}} \cup \tilde{G}_{\tilde{y}}$ is an injection. For any such N, the quotient group $\hat{G} = \tilde{G}/N$ can be regarded, in a natural way, as a vertex-transitive group of automorphism of the quotient graph $\hat{\Gamma} = N \setminus \tilde{\Gamma}$ under the action of N. Moreover, it is easy to see that, for any $\hat{x} \in V(\hat{\Gamma})$ and each non-negative integer i, the group $\hat{G}_{\hat{x}}^{[i]}$ is isomorphic as an abstract group to $\tilde{G}_{\tilde{x}}^{[i]}$ and hence to $G_x^{[i]}$, the group $\hat{G}_{\hat{x}}^{\hat{\Gamma}(\hat{x})}$ is permutationally isomorphic to $\tilde{G}_{\tilde{x}}^{\tilde{\Gamma}(\tilde{x})}$ and hence to $G_x^{\Gamma(x)}$, and $s(\hat{G}) = s(\tilde{G}) = s(G)$. Varying N, we get infinitely many finite graphs $\hat{\Gamma}$ and groups $\hat{G}$ with these properties. (Note also that under the assumption that (*) holds, for Γ, G and for $\hat{\Gamma}$, $\hat{G}$ one and the same case from ones marked in Section 2 holds.) In this connection, the examples of Γ and G with $G_x^{\Gamma(x)}$ permutationally isomorphic to a given group and $G_x^{[i]}$ isomorphic to a given group, for $x \in V(\Gamma)$ and $i \geq 0$, are of special interest in the case the group G has no non-trivial normal subgroup K such that the restriction of the natural mapping $G \to G/K$ on $G_x \cup G_y$, $\{x, y\} \in E(\Gamma)$, is an injection. All Examples from Sections 3-7 have this property.

REMARK 8.4. (On the proof of the result.) Very roughly, the method of the proof in [19]-[21] can be described as an analog of amalgam method, realized along a track. In the proof, without loss we may assume that Γ is a tree (see Remark 8.3). For $g \in G$, a g-track is a sequence $(..., x_{-1}, x_0, x_1, ...)$ of distinct vertices of the tree Γ such that $x_i = g^i(x_0)$ and $\{x_i, x_{i+1}\} \in E(\Gamma)$ for all $i \in \mathbf{Z}$. The most difficult part of the proof (for $n > 3$) is to prove that there is $g \in G$ for which a g-track $(..., x_{-1}, x_0, x_1, ...)$ with some special properties exists. Among these properties we emphasize the property $G^{[1]}_{...,x_{-1},x_0,x_1,...} = 1$. This property plays a crucial role in the approach since it allows to model an amalgam method machinery along the track (for example, we are certain that $Z(O_p(G_{x_0,x_1})) \not\leq G^{[1]}_{...,x_{-1},x_0,x_1,...})$. It should be mentioned that similar concepts of tracks were introduced earlier (see [15], [8]), but the main problem of the existence of the above mentioned special tracks is original. The method can be also applied to VSRP (formulated in Remark 8.2) for many other groups R.

Another ingredient of the proof (excluding the proof in the correlation case) is the following construction of subgraphs allowing to use induction to investigate an action of G_x around x.

Suppose that for a tree Γ and G the case $s(G) = 3$ holds. Let X be a subspace of dimension $n' \geq 1$ of the projective space $\Gamma(x)$, $y \in X$, and Y a subspace of dimension n' of the projective space $\Gamma(y)$ such that $x \in Y$. Denote by $\Gamma_{X,Y}$ the regular subtree of Γ such that $x \in V(\Gamma_{X,Y})$, $\Gamma_{X,Y}(x) = X$, $\Gamma_{X,Y}(y) = Y$, and $\varphi_{z_0,z_1,z_2}(\Gamma_{X,Y}(z_0)) = \Gamma_{X,Y}(z_2)$ for each 2-arc (z_0, z_1, z_2) of $\Gamma_{X,Y}$. Observe that the stabilizer in G of the set $X \cup Y$ is contained in $G_{\{V(\Gamma_{X,Y})\}}$ the stabilizer in G of the set $V(\Gamma_{X,Y})$. Let $G^{\Gamma_{X,Y}}$ denote the restriction of $G_{\{V(\Gamma_{X,Y})\}}$ on $V(\Gamma_{X,Y})$. Then $G^{\Gamma_{X,Y}} \leq Aut(\Gamma_{X,Y})$, and $\Gamma_{X,Y}$ and $G^{\Gamma_{X,Y}}$ satisfy (*) with $n = n' + 1$. Moreover, if $n' > 1$, then $s(G^{\Gamma_{X,Y}}) = 3$. It is important that $G_{x,y}^{[r]} \neq 1$ for some $1 \leq r < 2n'$ implies $(G^{\Gamma_{X,Y}})_{x,y}^{[r]} \neq 1$ (see [20, II], Proposition 2.2), and $G_x^{[r]} \neq 1$ for some $1 \leq r \leq 2n'$ implies $(G^{\Gamma_{X,Y}})_x^{[r]} \neq 1$ (by analogous arguments). As an application, if $G_x^{[2]} \neq 1$ (and the case $s(G) = 3$ holds), then $(G^{\Gamma_{X,Y}})_x^{[2]} \neq 1$ for lines X, Y, and hence $p \leq 3$ by, for example, [27].

A similar construction is used when for a tree Γ and G the case $s(G) = 2$, $n = 3$ or the collineation case holds. Let X be a subspace of dimension $n' \geq 1$ of the projective space $\Gamma(x)$. Denote by Γ_X the regular subtree of Γ such that $x \in V(\Gamma_X)$, $\Gamma_X(x) = X$ and $\varphi_{z_0,z_1}(\Gamma_X(z_0)) = \Gamma_X(z_1)$ for each 1-arc (z_0, z_1) of Γ_X. Observe that the stabilizer in G of X is contained in $G_{\{V(\Gamma_X)\}}$ the stabilizer in G of the set $V(\Gamma_X)$. Let G^{Γ_X} denote the restriction of $G_{\{V(\Gamma_X)\}}$ on $V(\Gamma_X)$. Then $G^{\Gamma_X} \leq Aut(\Gamma_X)$, and Γ_X and G^{Γ_X} satisfy (*) with $n = n' + 1$. Moreover, if $n' > 1$, then $s(G^{\Gamma_X}) = 2$ and either $n' = 2$ or for Γ_X and G^{Γ_X} the collineation case holds. It is important that $G_x^{[r]} \neq 1$ for some $1 \leq r \leq n'$ implies $(G^{\Gamma_X})_x^{[r]} \neq 1$ (see [20, I], Proposition 4.1). In this connection note that if for Γ and G the collineation case with $n = 5$, $q = 2$ and (g) holds and $n' = 3$, then for Γ_X and G^{Γ_X} the collineation case with $n = 4$, $q = 2$ and (e) holds.

Besides the original proof of the equation $G_x^{[2]} = 1$ in the collineation case with $q = 2$, $n = 6$ (and hence, according to the previous paragraph, in the collineation case with $q = 2$, $n \geq 6$) by the track method, there is another one given in [21, IV]. In brief, arguments in [21, IV] are as follows. A careful analysis shows that in the collineation case with $n = 4$, $q = 2$ and (e) the stabilizer of a vertex has no subgroup isomorphic to $SL_4(2)$ (see Remark 7.1). Suppose now that for Γ and G the collineation case with $n = 6$, $q = 2$ and $G_x^{[2]} \neq 1$ holds. It can be shown (using triviality of certain cohomology groups of $SL_6(2)$) that the group $G_x/G_x^{[4]}$ has a subgroup isomorphic to $SL_4(2)$ which stabilizes a subspace X of dimension 3 of the projective space $\Gamma(x)$. By the above, for Γ_X and G^{Γ_X} the collineation case with $n = 4$, $q = 2$ and (e) holds. It follows the group $G_x^{[4]}$ acts trivially on $V(\Gamma_X)$. Hence $(G^{\Gamma_X})_x$ has a subgroup isomorphic to $SL_4(2)$, a contradiction.

References

1. M. Aschbacher, Some results on pushing up in finite groups, Math. Z. 177 (1981), 61-80.

2. B. Baumann, Über endliche Gruppen mit einer zu $L_2(2^n)$ isomorphen Faktorgruppe, Proc. Amer. Math. Soc. 74 (1979), 215-222.

3. B. Cooperstein, An enemies list for factorization theorems, Comm. Algebra 6 (1978), 1239-1288.

4. A. Delgado and B. Stellmacher, Weak (B, N)-pairs of rank 2, in *Groups and graphs: New results and methods*, Birkhäuser, Basel, 1985, pp. 58-244.

5. A. Gardiner, Arc transitivity in graphs, Quart. J. Math. Oxford (2) 24 (1973), 399-407.

6. A. Gardiner, Arc transitivity in graphs. II, Quart. J. Math. Oxford (2) 25 (1974), 163-167.

7. A. Gardiner, Doubly primitive vertex stabilizers in graphs, Math. Z. 135 (1974), 157-166.

8. D.M. Goldschmidt, Automorphisms of trivalent graphs, Ann. Math. 111 (1980), 377-406.

9. A.A. Ivanov, *Geometry of sporadic groups I. Petersen and Tilde geometries*, Cambridge Univ. Press, Cambridge, 1999.

10. A.A. Ivanov and S.V. Shpectorov, *Geometry of sporadic groups II. Representations and amalgams*, Cambridge Univ. Press, Cambridge, 2002.

11. W. Knapp, On the point stabilizer in a primitive permutation group, Math. Z. 133 (1973), 137-168.

12. U. Meierfrankenfeld, Eine Lösung des Pushing up Problems für eine Klasse endlicher Gruppen, Ph.D. Thesis, Univ. Bielefeld, 1986.

13. U. Meierfrankenfeld and B. Stellmacher, Pushing up BN-pairs of rank two, Comm. Alg. 21 (1993), 825-934.

14. R. Niles, Pushing-up in finite groups, J. Algebra 57 (1979), 26-63.

15. J.-P. Serre, *Arbres, amalgams,* SL_2, Astérisque 46, Soc. Math. de France, 1977.

16. F.G. Timmesfeld, Amalgams with rank 2 groups of Lie-type in characteristic 2, Preprint, Math. Inst., Univ. Giessen, 1984.

17. V.I. Trofimov, Stabilizers of the vertices of graphs with projective suborbits, Soviet Math. Dokl. 42 (1991), 825-828.

18. V.I. Trofimov, More on the vertex stabilizers of the symmetric graphs with projective subconstituents, in *Int. Conf. Algebraic Combin.*, Vladimir, 1991, pp. 36-37.

19. V.I. Trofimov, Graphs with projective suborbits, Math. USSR Izv. 39 (1992), 869-893.

20. V.I. Trofimov, Graphs with projective suborbits. Cases of small characteristics. I, Russian Acad. Sci. Izv. Math. 45 (1995), 353-398; II, Russian Acad. Sci. Izv. Math. 45 (1995), 559-576.

21. V.I. Trofimov, Graphs with projective suborbits. Exceptional cases of characteristic 2. I, Izv. Math. 62 (1998), 1221-1279; II, Izv. Math. 64 (2000), 173-192; III, Izv. Math. 65 (2001), 787-822; IV, Izv. Math. (to appear).

22. V.I. Trofimov and R.M. Weiss, Graphs with a locally linear group of automorphisms, Math. Proc. Cambridge Phil. Soc. 118 (2) (1995), 191-206.

23. W. Tutte, A family of cubical graphs, Proc. Cambridge Phil. Soc. 43 (1947), 459-474.

24. W. Tutte, On the symmetry of cubic graphs, Canad. J. Math. 11 (1959), 621-624.

25. R.M. Weiss, Über symmetrische Graphen und die projektiven Gruppen, Arch. Math. 28 (1977), 110-112.

26. R.M. Weiss, Symmetric graphs with projective subconstituents, Proc. Amer. Math. Soc. 72 (1978), 213-214.

27. R.M. Weiss, Groups with a (B, N)-pair and locally transitive graphs, Nagoya Math. J. 74 (1979), 1-21.

28. R.M. Weiss, s-transitive graphs, in *Algebraic methods in graph theory, II*, Colloq. Math. Soc. János Bolyai, 25, North Holland, Amsterdam, 1981, pp. 827-847.

29. R.M. Weiss, Graphs with subconstituents containing $L_3(p)$, Proc. Amer. Math. Soc. 85 (1982), 666-672.

30. R.M. Weiss, Graphs which are locally Grassmann, Math. Ann. 297 (1993), 325-334.

Computing in the Monster

Robert A. Wilson

Abstract

We give a survey of computational methods and results concerning the Monster sporadic simple group.

There are now three computer constructions of the Monster which are proving effective in answering real questions about this group. The first construction over the field of two elements is the fastest for calculations, and has been used to show the group is a Hurwitz group.

The second construction over the field of three elements, uses an involution centralizer as the heart of the construction, and has proved to be the most useful as far as calculations with subgroups is concerned. P. E. Holmes has used this construction to find explicitly four new conjugacy classes of maximal subgroups, as well as to eliminate various other possibilities for maximal subgroups.

The third construction over the field of seven elements uses the same generators as the first construction, which means that elements given as words in these generators can be investigated modulo 2 and modulo 7 simultaneously. This gives enough information in most cases to determine the conjugacy class of the element.

1 Introduction

The Monster is the largest of the 26 sporadic simple groups, and is of great interest for a variety of reasons, not least its still mysterious connections with modular forms, and quantum field theories. Until recently, its immense size has been a serious barrier to computation in the group. Thus it was too big for a computational existence proof, and its existence had to be proved 'by hand' [2] (see also [1]). Determination of maximal subgroups also had to proceed by theoretical arguments [15, 11, 12, 14].

The smallest matrix representations of the Monster have dimension 196882 in characteristics 2 and 3, and dimension 196883 in all other characteristics. Thus the smallest matrices which we could conceivably use to generate the Monster would require around 5GB of storage each, and on modern workstations with the best available algorithms it would take several weeks of processor time to multiply two such matrices.

Despite these obvious difficulties, I decided some years ago to attempt an explicit construction of these matrices, with no hope of ever being able to use them for any serious calculation. With the collaboration of Richard Parker, Peter Walsh and Steve Linton, this project was eventually successful [10]. The generating matrices were stored in a compact way, so that all the information and special programs needed would fit onto a single 1.44MB floppy disk. This construction was in characteristic 2 for speed, and therefore proceeded by gluing together 3-local subgroups.

However, it soon transpired that these 3-local subgroups were too small to contain many useful subgroups with which to attack the maximal subgroup problem, so I began an analogous construction in characteristic 3, using the much larger 2-local subgroups. This was eventually completed by Beth Holmes [6], who then used the construction to obtain a complete classification of subgroups of the Monster isomorphic to $L_2(23)$. This major achievement was then quickly followed by 10 more such classifications, each more astonishing than the last, including the discovery of four previously unknown maximal subgroups [7, 8, 3].

In the meantime, I had produced a third construction, in characteristic 7, again using the 3-local subgroups. The idea behind this is that the generators are the same as in the characteristic 2 construction, so that one can obtain information about elements 'modulo 14'. In particular, one can calculate character values modulo 14, in order to provide good conjugacy class invariants.

2 The 2-local construction

We present first the 2-local construction, as it is easier to describe than the earlier 3-local construction, and is closely related to the Griess construction [2]. We give only an overview, and refer the reader to [6] for details. The idea is to start with a subgroup $2^{1+24}{\cdot}Co_1$, which is one of the involution centralizers in the Monster. The 3-modular irreducible representation of degree 196882 for the Monster restricts to this subgroup as the direct sum of three irreducibles, of degrees 98304, 98280 and 298. The constituent of degree 298 is a representation of the quotient Co_1, obtained from the 24-dimensional Leech lattice representation of the double cover $2{\cdot}Co_1$ by taking a trivial representation off the top and bottom of the symmetric square. The constituent of degree 98280 is monomial, and can easily be constructed again from the Leech lattice. Finally, the constituent of degree 98304 is the tensor product of representations of the double cover, of degrees 24 (the Leech lattice again) and 4096 (obtained by a Clifford algebra construction from the Leech lattice).

Thus an element of this subgroup can be specified by three matrices (over $GF(3)$, or more generally, any field of characteristic not 2), of sizes 24, 4096, and 298, and a monomial permutation on 98280 points. In particular, the storage requirement for each element is around 3.6MB, rather than the 7.4GB required for a 196882×196882 matrix over $GF(3)$. Moreover, elements of this subgroup can be multiplied together relatively easily—the most time-consuming part of the calculation is multiplying together the 4096×4096 matrices, which takes around a minute, depending on hardware and software. Most importantly for the sequel, however, is that there is an easy algorithm for calculating the image of a vector of length 196882, under one of these elements. This takes less than a second.

Now to produce an element of the Monster not in this subgroup, we first centralize a second involution, and use a 'standard basis' method to conjugate the first involution to the second, normalizing the four-group they generate. It turns out that the conjugating matrix can be chosen to be one of two particular matrices, and it is easy to check that one of them does not extend our involution centralizer to the Monster, and therefore the other one does. This of course relies on the existence of the Monster, and of this particular representation. An independent existence proof using this explicit construction has not been attempted.

By careful change of basis in the representations described above, we can ensure that the final element is reasonably sparse. It turns out that we can write it as a monomial permutation on 147456 points, followed by 759 identical 64×64 matrices, and an 850×850 matrix. This takes up around 0.7MB, and the image of a vector under this element can be calculated in a fraction of a second.

It is important to realise that only the generators of the Monster can be stored in one of these two compact formats. For this purpose, we can regard every element of the involution centralizer $2^{1+24} \cdot Co_1$ as a generator. But every element of the Monster outside this subgroup has to be stored as a word in these generators and the final generator. Thus multiplying group elements together involves concatenating words, and can be problematical as unrestrained multiplication rapidly results in words which are too long to be useful. We shall see in Section 6 some techniques we have used to get around this major problem.

3 The 3-local constructions

The first computer construction of the Monster [10] was designed to produce the matrices over $GF(2)$, since calculation with such matrices is much faster than with matrices over any other field. The disadvantage, however, is that the maximal 2-local subgroups are no longer available as ingredients of the construction. Thus we decided to use maximal 3-local subgroups instead. Here again we give only a sketch of the construction, and refer to [10] for details.

We began with the normalizer of a cyclic group of order 3, generated by a $3B$-element. This group has the shape $3^{1+12} \cdot 2 \cdot Suz{:}2$. The restriction of the representation to this subgroup consists again of a tensor product part, a monomial part, and a small part. The small part is obtained by tensoring the complex Leech lattice with its dual, reducing modulo 2, and taking a trivial module off the top and bottom—the result is an irreducible representation of degree 142 over $GF(2)$.

The 'monomial' part is really only monomial for a subgroup of index 2, over the extension field $GF(4)$. Thus for the whole group it is induced from a 2-dimensional representation of a subgroup of index 32760.

The 'tensor product' part is again not exactly a tensor product: if we restrict to the subgroup of index 2, it is the direct sum of two (dual) tensor products over $GF(4)$, each tensor being the product of one 90-dimensional and one 729-dimensional representation. The latter is the natural irreducible representation of the split extension $3^{1+12}{:}2 \cdot Suz$, while the former is an indecomposable unitary module for $6 \cdot Suz$, with constituents of degrees 12, 66 and 12.

These technicalities greatly complicate the construction, as the underlying field is sometimes of order 2, and sometimes of order 4, and the field automorphism needs special treatment. Moreover, the construction of the 90-dimensional indecomposable module was quite difficult. Once these technical difficulties were overcome, however, we ended up with an efficient calculating tool for the Monster. The storage requirement for an element in our subgroup is around 270kB, as opposed to around 5GB for a full-size matrix.

As in the case of the 2-local construction, we can treat all elements of the subgroup $3^{1+12}\cdot 2\cdot Suz{:}2$ as generators, and we just need to find one other generator for the Monster. We chose another element of class $3B$, inside the normal 3-subgroup, and used a standard basis technique to find an involution swapping these two elements of order 3. By testing random products, we found the one possibility (out of 8) which extended our 3-normalizer to the Monster.

Again, by careful choice of basis we were able to write this extra element as a combination of a 'monomial' permutation on 87480 subspaces of dimension 2, two 324×324 matrices (repeated 11 and 55 times respectively), and a 538×538 matrix. The storage requirement is around 420kB for this generator.

A similar calculation can be done over any field of characteristic not 3, although it is easier if there is a cube root of unity in the field. For this reason, we repeated the calculations over the field of order 7, and obtained the same set of generators for the Monster in this different representation [17]. Over fields of characteristic bigger than 3, the dimension is 196883, as there is only one copy of the trivial module in the tensor product of the complex Leech lattice with its dual, modulo primes bigger than 3.

4 Basic calculations

In any of the above constructions, the basic operation that we can perform is to multiply a vector by a generator of the Monster. We can also work inside our chosen maximal subgroup to create new generators in this subgroup. An element of the Monster is stored as a word $x_1 t_1 x_2 t_2 \ldots$, where the x_i are in our maximal subgroup, and the t_i are equal to the extra generator (or its inverse, in the 2-local version).

An estimate of the order of the element represented by such a word is obtained by taking a 'random' vector, and applying the letters of the word in order, repeatedly until the original vector is returned. The number of times the word is scanned is then a divisor of the order of the element, and is extremely likely to be exactly that order.

To improve this estimate to an exact calculation, we pre-calculate two vectors, one of which is fixed by an element of order 71, while the other is fixed by an element of order 47 but not by an element of order 94. These were found by finding elements whose estimated orders were 71 and 94, so that their exact orders are also 71 and 94 (as the Monster has no elements of order more than 119), and adding up the 71 images of a random vector under the first, and the 47 images of a vector under the square of the second element. Now it is easy to show that no non-trivial element of the Monster fixes both of these vectors. Therefore the exact order of an element can be calculated by passing both of these vectors through the given word.

5 Random searches

The first serious calculations we attempted with the first (3-local) construction were to try to improve estimates for the symmetric genus of the Monster. By character calculations and using partial information on maximal subgroups, Thompson had shown that the Monster was a quotient of the triangle group $\Delta(2,3,29) = \langle x, y, z \mid x^2 = y^3 = z^{29} = xyz = 1\rangle$, but the challenge was to find the minimal value of n such that the Monster is a quotient of $\Delta(2,3,n)$.

It was easy enough to find elements of orders 2 and 3, in classes $2B$ and $3B$, inside our maximal subgroup. We then wrote down a long list of conjugates of these two elements, and checked the order of the product in each case. It did not take long to find products of order less than 29, and then to check many words in the given conjugates, to verify that they did in fact generate the Monster. By this method we quickly reduced the value of n to around 17.

The ultimate aim, however, was to reduce n to 7: from Norton's work on maximal subgroups [12] it seemed very likely that this was the minimal possible value. However, the probability that a random pair has product of order 7 is of the order of 10^{-8}, so we would need to look at around 100 million pairs to have a reasonable chance of success. This we did, using some 10 years of processor time. See [16] for more details.

6 Advanced calculations

The main difference between calculating in a group given by generators as matrices or permutations, and calculating in the Monster where elements are given as words in the generators, is that if you are not careful the elements you need are given by words whose length increases exponentially with time. Without a method of shortening words it is impossible to use standard methods for finding elements and subgroups with the required properties. This is the main reason, apart from sheer size, why we originally considered serious calculation in this group to be essentially impossible.

However, with experience, we found two methods of overcoming this obstacle. The first trick is a method of conjugating one involution to a commuting involution, by a short word. As we were initally tied to the given involution centralizer, this was called the 'post', so this method of conjugating one involution to another was dubbed 'changing post'.

The second trick, which is really the crucial ingredient which enables us to calculate in the Monster almost as easily as in a small matrix group, is a method of shortening words. Specifically, if we find a word in the generators, which commutes with the original $2B$-element, then it belongs to the original subgroup $2^{1+24}{\cdot}Co_1$. Therefore it can be written in the shorthand form as a combination of a 24×24 matrix, a 4096×4096 matrix, a monomial permutation on 98280 points, and a 298×298 matrix. It turns out that this shorthand form can be determined by calculating just 36 rows of the full 196882×196882 matrix for this element. Thus provided the word for this element is not too long, this standard form can be calculated fairly quickly.

By combining this trick with Ryba's method [5] for conjugating an involution in a group to an involution in a known subgroup, we can in principle shorten any word to one of length less than about 20. In practice, however, this is still too slow for arbitrary words, and its effective use is confined to the case described above. Details can be found in Holmes's Ph. D. thesis [3] (see also [7]).

The first trick is less elegant, but no less effective. It relies on the fact that all $2B$-elements in $2^{1+24}{\cdot}Co_1$ can be obtained from the central involution by a subset of the operations: (1) conjugate by t to make it a non-central involution of 2^{1+24}, (2) conjugate by an element of $2^{1+24}{\cdot}Co_1$, (3) conjugate by t again to move it outside 2^{1+24}, and (4) conjugate again by an element of $2^{1+24}{\cdot}Co_1$. The method of conjugating an arbitrary $2B$-element in this group to the central involution, basically consists of one application of the well-known dihedral group trick to conjugate our involution to a pre-calculated one in the quotient Co_1, and one random search in the Leech lattice, to find the correct conjugating elements to reverse the above operation. See [3] or [8] for details.

Another important principle for calculating in large groups is to do the required calculations in a proper subgroup if at all possible. In many places we need to work in particular subgroups to search for the particular elements we require. It is necessary therefore to find suitably small representations of these subgroups in which to perform such calculations. In some cases we created a permutation representation by permuting the images of a carefully chosen vector. In most cases, however, we chopped a suitable submodule out of the 196882-dimensional module using a type of condensation technique specifically adapted for the special form of the representation.

7 Maximal subgroups

The most effective method of classifying maximal subgroups of large simple groups in a computational setting is to choose an abstract amalgam generating the desired isomorphism type of subgroup, and to classify all embeddings of that amalgam in the large simple group. We then look at each embedding to decide whether it indeed generates the required subgroup.

For example, if we wish to classify subgroups isomorphic to $L_2(23)$, we use the fact that this group can be generated by the Borel subgroup 23:11 and the normalizer D_{22} of a torus, intersecting in the torus (of order 11). Thus we first find all types of 23:11 in the Monster (there is only one, up to conjugacy, and it can be found inside the involution centralizer). Next we find the normalizer of the cyclic group of order 11. This is in general not so easy, but in this case we find that, by choosing the element of order 23 carefully, it is generated by the part which is in the involution centralizer, and the extra generator t of the Monster. It is therefore possible to generate all necessary elements, that is the involutions inverting the element of order 11, by short words in the Monster generators. As a result we are able to investigate with relative ease the groups so generated, and find the unique conjugacy class of subgroups of the Monster which are isomorphic to $L_2(23)$. Moreover, we can use the normalizer of the cyclic group of order 23 to show that every $L_2(23)$ in the Monster centralizes a group S_3, and therefore its normalizer is inside the maximal subgroup $3^{\cdot}Fi_{24}$.

Other isomorphism types of simple subgroups of the Monster are not so easy to classify. The most successful calculation so far has been the classification of subgroups generated by two copies of A_5 intersecting in D_{10} (see [3, 8, 9]). This amalgam can generate $L_2(q)$, for any $q \equiv \pm 1 \pmod 5$, as well as $L_3(4)$, so if we can successfully classify such amalgams then we will have dealt with many of the remaining cases. Indeed, the cases $L_2(q)$ for $q = 9, 11, 19, 29, 31, 59, 71$ are all of this type, so eight cases can be dealt with in this manner. (In fact, the case $L_2(29)$ was treated earlier in a different way.)

This is easier said than done, however. It is hard to find representatives of the two classes of A_5 that need to be considered—eventually we found them by making a copy of $L_2(11) \times M_{12}$ in the Monster, itself no small undertaking, and taking suitable diagonal A_5s therein. Finding the normalizer in the Monster of the subgroup D_{10} was even more difficult, and involved working inside several different involution centralizers to find various parts of the required subgroup.

At the end of the calculation, after several months work, we found four new maximal subgroups by this method. In particular, we found explicitly maximal subgroups $L_2(59)$ and $L_2(71)$, thus answering a long-standing question as to whether these groups were subgroups of the Monster. This shows also that the maximal local subgroups 59:29 and 71:35 are not maximal subgroups of the Monster. In addition, we found new maximal subgroups $L_2(29)$:2 and $L_2(19)$:2.

To summarise the calculations to date, we have completely classified maximal subgroups of the Monster whose socle is isomorphic to one of the 11 simple groups $L_2(q)$, for $q = 9, 11, 19, 23, 29, 31, 59, 71$, $L_3(4)$, M_{11} or $U_4(2)$. This leaves just 11 cases to consider, namely $L_2(q)$, for $q = 7, 8, 13, 16, 17, 27$, $L_3(3)$, $U_3(3)$, $U_3(4)$, $U_3(8)$ and $Sz(8)$.

8 Traces and conjugacy classes

We tend to think of the trace of a matrix as being easy and quick to calculate, but that is only true if we actually have the matrix in front of us. To calculate the trace of a matrix which is only given as a word in some generators is a much more challenging problem. Indeed, the only reasonable method we could think of is essentially to calculate the matrix one row at a time, and extract the diagonal entries. This leads to a time of around 1 hour per letter of the word (depending of course on hardware and software) for the trace modulo 2.

On the other hand, the trace modulo 2 is not a very good conjugacy class invariant. It can only ever distinguish between different $2'$-parts of elements, since modulo 2 we have $Tr(x) = Tr(x^2)$. However, if we combine this invariant with the order, and the traces of powers of the elements, we can distinguish between most classes of odd-order elements in the Monster. The exceptions are irrational classes, where we cannot distinguish between elements which generate the same cyclic subgroup, and two other cases: we cannot distinguish between $3B$ and $3C$, or between $27A$ and $27B$.

To distinguish classes of even order, we need traces modulo an odd prime. This was the main reason why I decided to repeat the 3-local construction over the field of order 7, using exactly the same generators, so that the same words can be used in both representations simultaneously. Thus we can calculate the trace mod 2 and the trace mod 7 for the same element of the group, thus obtaining the character value modulo 14. This gives us a much better class invariant, which when combined with the order and the traces of powers, discriminates all classes of cyclic subgroups except $27A$ and $27B$. However, it is much more expensive to calculate traces modulo 7—days just to calculate one trace—so it can take weeks to identify the conjugacy class of an element by this method. With this apparatus my research student Richard Barraclough is in the process of producing a (partial) list of conjugacy class representatives. See also [4] for a description of a method and the results of some experiments designed to produce pseudo-random elements of the Monster for use in randomized algorithms.

9 Conclusion

At the time of writing, it is less than four years since the publication of our first paper [10] on constructing the Monster. During that time, we have effectively tamed the Monster, so that many computations are now feasible inside this huge group. This was beyond our wildest dreams in 1998, but now seems routine.

It is natural therefore to speculate on what further calculations might be possible. For example, could we provide an independent existence proof for the Monster? This seems hard at the present time, but may be possible. At least we could find elements satisfying a suitable presentation, and perhaps combine this with arguments concerning the 2-local geometry to produce an existence proof independent of [2].

Other problems worthy of attack include specific questions such as: Does the 196882-dimensional $GF(2)$-representation support an invariant quadratic form? If so, is it of $+$-type or $-$-type? There are also more speculative questions, for example concerned with classifying Norton's nets (see [13]), where computational assistance might be valuable. And, can we complete the determination of the maximal subgroups of the Monster? There are undoubtedly some hard cases still to crack, and they may take a huge amount of computer time, but it seems as though this aim is not completely unreasonable.

Many years ago, I used Moore's law (doubling of computer power every 18 months), plus a postulated doubling of software power every 18 months, and a doubling of our own brain power every 18 months (perhaps the least plausible assumption, but with hindsight the most important contribution), to estimate that we could determine the maximal subgroups of the Monster by the end of the second millenium AD. No-one seemed to take me seriously at the time, but maybe I was not so far off the mark.

References

[1] J. H. Conway, A simple construction for the Fischer–Griess monster group, *Invent. Math.* **79** (1985), 513–540.

[2] R. Griess, The friendly giant, *Invent. Math.* **69** (1982), 1–102.

[3] P. E. Holmes, Computing in the Monster, Ph. D. thesis, Birmingham, 2001.

[4] P. E. Holmes, S. A. Linton and S. H. Murray, Product replacement in the Monster, Preprint 2002/12, School of Mathematics and Statistics, The University of Birmingham.

[5] P. E. Holmes, S. A. Linton, A. J. E. Ryba and R. A. Wilson, A constructive membership test for black box groups, in preparation.

[6] P. E. Holmes and R. A. Wilson, A new computer construction of the Monster using 2-local subgroups, *J. London Math. Soc.*, to appear.

[7] P. E. Holmes and R. A. Wilson, A new maximal subgroup of the Monster, *J. Algebra* **251** (2002), 435–447.

[8] P. E. Holmes and R. A. Wilson, $L_2(59)$ is a subgroup of $\mathbb{M}$, in preparation.

[9] P. E. Holmes and R. A. Wilson, More new maximal subgroups of the Monster, in preparation.

[10] S. A. Linton, R. A. Parker, P. G. Walsh and R. A. Wilson, Computer construction of the Monster, *J. Group Theory* **1** (1998), 307–337.

[11] U. Meierfrankenfeld and S. V. Shpektorov, The maximal 2-local subgroups of the Monster and Baby Monster, in preparation.

[12] S. P. Norton, Anatomy of the Monster, I, in *The atlas of finite groups ten years on (ed. R. T. Curtis and R. A. Wilson)*, 198–214. Cambridge University Press, 1998.

[13] S. P. Norton, Netting the Monster, in *The Monster and Lie algebras (ed. J. Ferrar and K. Harada)*, 111–125. Ohio State Univ. Math. Res. Inst. Publ. **7**. de Gruyter, 1998.

[14] S. P. Norton and R. A. Wilson, Anatomy of the Monster, II, *Proc. London Math. Soc.* **84** (2002), 581–598.

[15] R. A. Wilson, The odd-local subgroups of the Monster, *J. Austral. Math. Soc. (A)* **44** (1988), 1–16.

[16] R. A. Wilson, The Monster is a Hurwitz group, *J. Group Theory* **4** (2001), 367–374.

[17] R. A. Wilson, Construction of the Monster over $GF(7)$, and an application. Preprint 2000/22, School of Mathematics and Statistics, The University of Birmingham.

